Commercial Satellite Communication

Commercial Satellite Communication

Stephan C Pascall BSc (Hons), PhD, C.Eng, MIEE
Head of Section, Spatial planning – DGXIII-C1, European Commission

David J Withers OBE, FIEE
Formerly Chief Engineer of the International Division of British Telecommunications

Foreword by Robert Verrue, *Director General, DG XIII, Telecommunications, International Market and Exploitation of Research, European Commission*

Focal Press
An imprint of Butterworth-Heinemann
Linacre House, Jordan Hill, Oxford OX2 8DP
A division of Reed Educational and Professional Publishing Ltd

 A member of the Reed Elsevier plc group

OXFORD BOSTON JOHANNESBURG
MELBOURNE NEW DELHI SINGAPORE

First edition 1997

British Library Cataloguing in Publication Data
A catalogue record for this book is available from the Brisish Library.

ISBN 0 7506 6123 55

Library of Congress Cataloguing in Publication Data
A catalogue record for this book is available from the Library of Congress.

Typeset by Laser Words, Madras, India
Printed and bound in Great Britain by
Hartnolls Limited, Bodmin, Cornwall

Contents

Foreword

Information, exchange of knowledge, and communications are of vital importance in economic activity. Policy makers are therefore concentrating on ways of acquiring, processing, storing and transmitting information. Telecommunications, as an information conveyor, are critical for influencing the 'nervous system' of modern society. Users' needs are continually evolving. Fortunately, systems which meet those evolving needs are also emerging. The established terrestrial telecommunications networks tend to be rather slow to respond to these new opportunities to serve society. Satellite communication as a transmission medium is particularly flexible and consequently it is more able to support such innovations.

National frontiers must not be allowed to hamper the development of a consistent communications system within the European Union and between the countries of the Union and its European neighbours. It has long been an aim of the Commission of the European Communities to eliminate such barriers. Excellent progress is being made in the lifting of restrictive national regulations within the European Union; this progress allows in particular the implementation of Community-wide satellite terminal networks, which are prominent in providing reliable, modern links to bridge gaps between the Union and neighbouring countries.

Flexibility and the ability to bridge gaps rapidly are some of the characteristics of satellite communications which have caused its dramatic development in recent years. Having long been essential to intercontinental communication, it is now becoming a major element for trans-European services, including the broader continental dimension which has developed from the revolutionary changes in Eastern Europe. For all of these reasons, I welcome this book, which will bring valuable insights in this technology to students and to engineers practising in the industry, and perhaps not least to major users of telecommunications services.

Robert Verrue
Director General, DG XIII,
The European Commission

Preface

Satellites provide the medium for a number of specialized markets in commercial telecommunications. In the course of more than 30 years they have come to supply a major share of long-distance links in the public telephone network and in this field the technology can be said to be mature. But for most other applications, satellite communication is still in a state of rapid development.

Over 20 years, ocean-going ships have come to depend for their communications on satellites. Airliners are beginning to follow their lead. The maritime market is developing to include many smaller craft that never voyage far from land, and the sea is a safer place as a result. Satellites have done much to expand television broadcasting, collecting material from outside events, distributing programmes to terrestrial broadcasting stations and cable networks and, not least, broadcasting directly to the public. Satellites have brought modern communication facilities to isolated communities that cannot be served economically in any other way. Very small earth stations are being used in large numbers to serve other niche markets, successfully competing with conventional transmission media even where the latter are most strongly entrenched. And a new revolution – satellite service to cars and hand-held radiotelephones – is about to begin.

This book outlines the history and reviews the technology of these telecommunication systems. But whereas that would suffice in more conventional media, satellite communication has other dimensions. Satellites must be established in orbit; they need launchers and elaborate control arrangements. Most satellites are costly and essentially international in coverage, and often in utilization; they need responsible organizations with adequate resources, technical and financial, to manage them. These areas are considered in some detail.

Finally, satellite systems make use of finite natural resources, radio spectrum and space, and above all the frequency bands that provide good radio propagation conditions between space and Earth and the geostationary orbit. Means have been developed to facilitate coordination of the use of spectrum and space so that unacceptable interference does not occur. Even so it has long been evident that the limits of these resources are being approached. Technical constraints have been agreed internationally, to ensure that the greatest use can be made of satellite communication by all countries. This regulatory dimension penetrates into many aspects of satellite system design, and this too is reviewed in this book.

Part 1 provides a brief historical introduction, recalling the state of long-distance communication as it was before the new transmission medium became available and outlining the early experimental years and the setting up of the first satellite systems.

Part 2 reviews the technology of the satellites themselves, treating them as platforms in space for the radio equipment that links together the users radio stations that are, with rare exceptions, on Earth. The three chapters of Part 2 are concerned with the design and operation of platforms, the orbits in which they revolve and the rockets that inject them into those orbits.

Part 3 is concerned with the principles of typical communication systems that are used with satellites. This subject falls naturally into six areas. The radio equipment, on Earth and on the platforms, is described in broad principle in Chapter 7.

Typical modulation systems, analogue and digital, are outlined in Chapters 8 and 9 respectively. The various radio propagation phenomena that degrade transmission between stations on Earth and stations in space, and the methods that are used to make allowance for such degradations, are considered in Chapters 5 and 10. And Chapter 6 outlines the constraints and procedures that have been agreed internationally for controlling interference, for maximizing the number of satellite systems that can use space and spectrum, and for ensuring an equitable sharing of those limited natural resources between the countries ready to use them.

Finally, Part 4 reviews the ways in which the satellite medium is applied in its three main areas of use. Chapters 11, 12 and 13 are devoted to services for fixed earth stations, large and small. Services to mobile stations on land, sea and air are reviewed in Chapter 14. And Chapter 15 deals with satellite broadcasting.

S C P
D J W

Acknowledgements

The authors express their grateful thanks to colleagues and friends in many satellite-related industries and organizations for their advice and information. Special acknowledgement is made to Wil Mittelbach of McDonnell Douglas Ltd, Bonn, for the provision of information on the Delta launchers, to Dr Eric J. Novotny of Hughes, Europe, to Christopher J. Faranetta of the American office of RSC Energia and to K. Bogus of the European Space Agency. Thanks are also due to the European Commission for facilitating the writing of this book.

Certain texts and drawings from ITU publications have been reproduced with the prior authorization of the Union as copyright holder. The sole responsibility for selecting extracts for reproduction lies with the authors alone and can in no way be attributed to the ITU. The complete volumes of the ITU material from which the texts reproduced are extracted can be obtained from

International Telecommunication Union
General Secretariat – Sales and Marketing Service
Place des nations
CH – 1211 GENEVA 20
Switzerland

Principal symbols used in the text

A (A_p)	absorption on a ray path (exceeded for percentage of the time)
A	aperture of an antenna
A_e	effective aperture of an antenna
A_e	area of exit plane of rocket motor
A_m (A_G)	absorption due to the medium (due to atmospheric gases)
a	length of semi-major axis of an elliptical orbit
B (B_{IF})	(IF) bandwidth
B	interference reduction factor
b	length of semi-minor axis of an elliptical orbit
b	channel bandwidth
C	carrier power level
C/N	ratio of carrier power to noise power
$(C/N)_u$	C/N due to uplink noise
$(C/N)_{im}$	C/N due to intermodulation noise
$(C/N)_d$	C/N due to downlink noise
$(C/N)_T$	C/N including noise from all sources
c	velocity of electromagnetic waves in free space $= 3 \times 10^8$ m s^{-1}
D	diameter of antenna
D, d	distance between earth station and satellite
$dB\ K^{-1}$	conventional designation of the value of the figure of merit of an antenna/receiver combination
E	the electric component of a wave
E_b	energy per bit of a digitally modulated carrier
e	the instantaneous voltage of a signal
F	reaction force operating on a rocket motor
F_p	peak frequency deviation
f, F	frequency (Hz)
f_c (f_m)	carrier (modulating) frequency
f_d (f_u)	downlink (uplink) frequency
f_l (f_u)	lower (upper) frequency limit of baseband
f_{pp} (f_r)	frequency deviation, peak-to-peak (r.m.s.)
f_{tt}	frequency of test tone
G	universal gravitational constant $\approx 6.67 \times 10^{-11}$ N m^2 kg^{-2}
G	gain
G_o (G_r) (G_t)	maximum (receiving) (transmitting) gain of an antenna
G_ϕ	gain of an antenna at $\phi°$ off-axis
g	peak-to-mean ratio of the power of an FDM aggregate of telephone channels
h	height of circular orbit above Earth
h_a (h_p)	height of apogee (perigee) above Earth
h_s (h_{ml})	height of earth station (melting layer) above mean sea-level
I	interference power level
I_{sp}	specific impulse of rocket system

i	angle of inclination of the orbital plane
k	Boltzmann's constant $= 1.38 \times 10^{-23}\,\mathrm{J\,K^{-1}}$
L	transmission loss
L_b (L_{bf})	(free-space) basic transmission loss
l	loading factor for an FDM aggregate of telephone channels
M	a number (e.g., of significant conditions of a carrier wave)
M (M_a)	(achievement) margin
M_E	mass of the Earth $\approx 5.977 \times 10^{24}$ kg
MW	molecular weight
m	peak phase deviation due to modulation
N, n	a number (e.g., of channels in a multiplex system)
N (N_w)	(weighted) noise power level
N_{bb} (N_{tc})	noise power in baseband (in top channel of FDM) after demodulation
N_i	interference expressed as an equivalent noise power
N_o, n_o	spectral noise power density
n	a noise power component
n_u (n_{im}) (n_d)	uplink (intermodulation) (downlink) noise power component
P	power level
P_e	probability of a bit error
P_e	advantage due to emphasis
P_w (P_s)	advantage due to (psophometric) noise weighting
p_i (p_o)	pressure on inner (outer) side of rocket motor wall
R	axial ratio of polarization of a wave
R (R_p)	rainfall rate (exceeded for percentage of the time)
R	bit rate
R_E	radius of the Earth (mean radius ≈ 6371 km)
R_{an}	rate of precession of ascending node
R_{ap}	rate of precession of argument of perigee
r	radius of a circular orbit
r_g	radius of the GSO
S (S_{tt})	power of signal (test tone) after demodulation
S	wave spreading factor
s	surface area of rocket motor walls
T (t)	time interval in hours (in seconds)
T	temperature
T	noise level expressed as a noise temperature (K)
T_B	brightness temperature (K)
T_c (T_m)	temperature of combustion chamber (propagation medium)
T_s	system noise temperature (K)
T_e (T_s)	noise temperature of earth station (satellite) receiver
t (t_a) (t_{at}) (t_{LNA}) (t_{sky}) (t_{sol})	noise temperature component (antenna) (atmosphere) (LNA) (sky) (Sun)
t_s (t_c)	symbol (chip) period in seconds
V_{ch}	characteristic velocity for an orbit
v	velocity of satellite in orbit
v_i (v_f) (v_e)	initial (final) (exit) velocity or relative velocity of products of rocket combustion
w (w_t) (w_r)	power (transmitted) (received)
γ	transmission gain
γ (γ_c)	specific attenuation (due to atmospheric gases)
δ	angle of elevation of satellite at a point on Earth
ΔV	velocity increment in launcher flight profile
ϵ	eccentricity of an elliptical orbit

η	efficiency factor
θ	angle measured relative to axis of satellite antenna
θ	exocentric angle
θ	latitude of earth station
θ	angle between electric vectors
θ_o	half-power beamwidth of antenna
λ	wavelength
τ	polarization tilt angle
Φ	power flux density
Φ_s	spectral power flux density
Φ_{ts}	single carrier PFD at satellite required to saturate transponder
ϕ	angle measured from axis of earth station antenna
ϕ_g (ϕ_t)	geocentric (topocentric) angle
$\omega(\omega_c)(\omega_m)$	$2\pi f$, where f = frequency $(f_c)(f_m)$

Principal abbreviations used in the text

ACTS	Advanced Communications Technology Satellite
ADPCM	adaptive differential PCM
ADS	automatic dependent surveillance
AES	aircraft earth station
AFC	automatic frequency control
AKM	apogee kick motor
AM	aeronautical mobile-satellite service
AMSS	amplitude modulation
ARQ	automatic repeat on request
ASVS	Air–Space Vehicle System
ATC	air traffic control
ATS	Applications Technology Satellite
Az-El	azimuth/elevation (antenna mount)
BER	bit error ratio
BPSK	binary PSK
BSS	broadcasting-satellite service
C	Celsius
CCIR	International Radio Consultative Committee (functions taken over by the Radiocommunication Assembly of ITU-R in 1993)
CCITT	International Telegraph and Telephone Consulative Committee (functions taken over by telecommunication standardization study groups of ITU-T in 1993)
CD	compact disc
CDMA	code division multiple access
CFDM	companded FDM
cos	cosine
CPSK	coherent PSK
CPFSK	continuous phase FSK
DA	digital audio broadcasting
DAB	demand assignment
dB	decibel
dBi	decibels relative to gain of isotropic antenna
dBm0	dB relative to 1 mW at a point of zero relative level
dBm0p	noise level, expressed in dBm0, psophometrically weighted
DBS	direct broadcasting by satellite
dBW	decibels relative to 1 W
$dB(W\,m^{-2})$	PFD in dBW per square metre
$dB(W\,m^{-2}(4\,kHz)^{-1})$	PFD in dBW per square metre per 4 KHz band
$dB(W\,m^{-1}(1\,MHz)^{-1})$	PFD in dBW per square metre per 1 MHz band
$dB(W(4\,KHz)^{-1})$	PFD in dBW per 4 KHz band
$dB(W(40\,KHz)^{-1})$	PFD in dBW per 40 KHz band

DCME	differentially coherent PSK
DCPSK	digital circuit multiplication equipment
DDI	direct digital interface
DNI	a TDM telephony system not using DSI
DPCM	differential PCM
DSC	digital selective calling
DSI	digital speech interpolation
DSSS	direct sequence spread spectrum (modulation)
DTH	direct to home (satellite broadcasting)
E	east
EBU	European Broadcasting Union
ECS	European Communication Satellite
EGC	Extended Group Call (an INMARSAT facility)
EHF	extra high frequency (the frequency range 30 to 300 GHz)
e.i.r.p.	effective isotropically radiated power
EPIRB	emergency position-indicating radio beacon
ESA	European Space Agency
ETSI	European Telecommunications Standards Institute
FANS	Future Air Navigation Systems (committee)
FDM	frequency division multiplex
FDMA	frequency division multiple access
FEC	forward error correction
FHSS	frequency hopping spread spectrum (modulation)
FM	frequency modulation
FS	(terrestrial) fixed service
FSK	frequency shift keying
FSS	fixed-satellite service
GaAs FET	gallium arsenide field effect transistor
GHz	Gigahertz (1 GHz = 1000 MHz)
GMDSS	Global Maritime Distress and Safety System
GPS	Global Positioning System
GSO	geostationary satellite orbit
G/T	figure of merit (of an antenna/receiver combination)
$(G/T)_e$	G/T of earth station receiver
$(G/T)_s$	G/T of satellite receiver
GTO	geostationary or geosynchronous transfer orbit
h	hour
HDTV	high definition TV
HF	high frequency (the frequency range 3 to 30 MHz)
HMET	high mobility electron transistor
HPA	high power amplifier
Hz	hertz (the basic unit of frequency)
ICAO	International Civil Aviation Organization
IDC	Intermediate Rate Digital Carrier (Eutelsat standard)
IDR	Intermediate Data Rate (Intelsat standard)
IEC	International Electrotechnical Commission
IF	intermediate frequency
IMO	International Maritime Organization
IMUX	input multiplexer
ISDN	integrated services digital network
ISL	inter-satellite link
ISO	International Standards Organization
ITU	International Telecommunication Union

ITU-R	Radiocommunication Sector of the ITU
ITU-T	Telecommunication Standardization Sector of the ITU
IUS	Inertial Upper Stage
K	degrees kelvin
kbit/s	kilobits per second
keV	kilo-electron-volt
kg	kilogram
kHz	kilohertz (1 kHz = 1000 Hz)
km	kilometre
kN	kilonewton
kW	kilowatt
LAN	local area network
LEO	low Earth orbit
LES	land earth station
LHCP	left hand circular polarization
LM	land mobile
LMES	land mobile earth station
LMSS	land mobile-satellite service
LNA	low noise amplifier
m	metre
MAC	Multiplexed Analogue Components (family of TV systems)
MAN	metropolitan area network
Mbit/s	megabits per second
MEO	medium altitude Earth orbit
MES	mobile earth station
MeV	mega-electron-volt
MHz	megahertz (1 MHz = 1000 kHz)
MIFR	Master International Frequency Register
mm	millimetre
MMSS	maritime mobile-satellite service
mN	millinewton
MPEG	Moving Picture Expert Group
ms	(terrestrial) mobile service
MS	millisecond
MSK	minimal shift keying
MSS	mobile-satellite service
mW	milliwatt
N	newton
N	north
NASA	National Aeronautics and Space Administration
NASDA	National Space Development Agency (Japan)
NCS	network coordination station
NSSK	North–South station-keeping
NTSC	National Television Standards Committee (name applied to a chrominance information transmission system)
OMJ	orthomode junction
OMT	orthomode transducer
OMUX	output multiplexer
OQPSK	offset QPSK
OTS	Orbital Test Satellite
PAL	Phase Alternate Line (a chrominance information transmission system)

PAM	Payload Assist Module
PCM	pulse code modulation
PFD	power flux density
PG	processing gain (of an FEC system)
PKM	perigee kick motor
PM	phase modulation
PSK	phase shift keying
PSTN	public switched telephone network
PTO	public telecommunications operator
pW0	picowatts at a point of zero relative level
pW0p	picowatts, psophometrically weighted, at a point of zero relative level
QPSK	quadriplex PSK
RARC	Regional Administrative Radio Conference (of the ITU)
RFA	radio-frequency amplifier
RHCP	right hand circular polarization
r.m.s.	root mean square
RR	Radio Regulations (ITU)
s	south
S	second
SAW	surface acoustic wave
SCPC	single channel per carrier
SECAM	Système Electronique Coupeur Avec Mémoire (a chrominance information transmission system)
SES	ship earth station
SHF	super high frequency (the frequency range 3 to 30 GHz)
sin	sine
SIS	Sound-in-sync
SMS	satellite multi-services (EUTELSAT)
SNG	satellite news gathering
SOLAS	(International Convention on) Safety of Life at Sea
SQPSK	staggered QPSK
SSPA	solid state power amplifier
SSR	secondary surveillance radar
SSTDMA	satellite switched TDMA
STS	Space Transportation System
tan	tangent
TDM	time division multiplex
TDMA	time division multiple access
TEC	total electron content (of a path through the ionosphere)
TTC&C	telemetry, tracking, command and communication
TV	television
TVRO	television receive-only (earth station)
TWT	travelling wave tube
TWTA	TWT amplifier
UDMH	ultra-high frequency (the frequency range 300 MHz to 3 GHz)
UHF	unsymmetrical dimethylhydrazine
UW	unique word
VAR	voltage axial ratio
VHF	very high frequency (the frequency range 30–300 MHz)
VSAT	very small aperture terminal
VSB	vestigial sideband (modulation)
W	watt

W	West
WARC	World Administrative Radio Conference (of the ITU)
WRC	World Radiocommunication Conference (of the ITU)
XPD	cross-polar discrimination
XPI	cross-polar isolation
μs	microsecond
μN	micronewton

Part 1
Introduction

1 Introduction

1.1 Telecommunications before satellite communication

In October 1945, Arthur C. Clarke [1] pointed out the possibilities of relaying telephone channels and broadcasting programmes from artificial satellites. He stressed in particular the special properties of a satellite orbit, equatorial and circular, about 42 000 km from the Earth's centre. The period of a satellite in this orbit is one sidereal day, and if the satellite moves eastwards, it keeps pace with the rotating Earth; to an observer on the Earth the satellite seems stationary in the sky. This is the geostationary satellite orbit (GSO). Clarke suggested that a single television transmitter on board a geostationary satellite could serve a whole nation, taking the place of the networks of VHF or UHF terrestrial broadcasting stations that were then being planned. Just three satellites could provide wideband relay facilities for the point-to-point links of a global telephone network.

It was widely thought at the time that Clarke's ideas were visionary and unlikely to be feasible for many years. Nevertheless, only 20 years later, on 28 July 1965, a geostationary satellite did indeed come into service to provide international links for the public switched telephone network (PSTN). A further 10 to 20 years were to elapse before there were satellites powerful enough to be used for the many other applications that are feasible today, including broadcasting. However, before considering these developments, it is of interest to recall the state of telecommunications and broadcasting in the period between 1945 and 1965, before satellites had arrived.

In the period 1945–65, broadcasting and most of the main telecommunication services were improving in technical quality in the developed countries. By 1965 at a national level, and especially in populated areas, services were broadly satisfactory and in many respects not greatly different from their present state. An exception was communication with road vehicles, which was still waiting for the revolution that was brought by cellular radiotelephones in the 1980s. However, none of these services operated satisfactorily over great distances. The need for improvement was felt most acutely in the PSTN.

Terrestrial communication between fixed points

The decade immediately following the Second World War was one of rapid advance in terrestrial transmission technology. Wideband underground coaxial cables and microwave radio relay systems were being developed for the telephone and similar services and were installed extensively, especially in the more developed countries. Submarine coaxial cables crossed narrow seas. Radio relay links predominated for the distribution of television signals to broadcasting stations. However, few international cable or radio relay systems existed except in Europe and North America, and the quality of transmission and switching systems set a limit of a few thousand kilometres on the distance that could be covered satisfactorily for telephony, except in particularly favourable circumstances.

Beyond the reach of cable and radio relay systems, high-frequency (HF) radio links were the available medium for point-to-point telephony. HF links, augmented to a minor degree by trans-oceanic telegraph cables with very low information capacity, were also used for telegraph services. When 'automatic repeat on request' (ARQ) telegraph equipment became available for HF links in the mid-1950s, use for telex became feasible. But HF radio, as a medium for long-distance commercial transmission, is seriously flawed.

The frequency range of the radio spectrum that supports regular long-distance ionospheric propagation is quite narrow, less than the bandwidth of a single satellite transponder, and the low-loss propagation mode which gives HF links a potentially world-wide reach also permits world-wide interference, limiting frequency reuse. Thus, the total quantity of telecommunication services that HF radio could supply was quite small. Secondly, long HF links are of very variable transmission quality, due to the vagaries of ionospheric propagation; quality is sometimes excellent, but it is often unacceptably bad. In addition, the cost per channel for equipment and for the skilled staff needed to operate it was high. A telephone call could not usually be connected on demand; it was necessary to make a reservation and wait until a circuit was free and fit for use. Moreover, charges for calls were high, although not high enough to cover the costs. As a result, potential users of the telephone service usually found that they could do without it; a message that was too urgent for airmail was sent by telegram.

The way forward to a global telephone network, generally agreed among the experts of the public telecommunications operators (PTOs) and in particular those of the British Post Office (BPO) and the American Telephone and Telegraph Corporation (ATT), lay in the development of long-distance wideband submarine cables. Papers by Buckley [2] when President of Bell Telephone Laboratories, Angwin [3] when Director General of BPO and Halsey [4] set out well the situation as it was then seen. By inserting repeater amplifiers at intervals along a coaxial cable to compensate for the transmission loss, it would be possible to extend the maximum range greatly and to increase the number of channels that a short cable could carry, thereby reducing the cost per circuit.

Progress in implementing this strategy was brisk. The first effective wideband submerged repeater was inserted into an existing cable between Anglesey and the Isle of Man in 1943. Extensive use of repeaters followed in the short cables across the shallow waters of the continental shelf between the UK and the rest of Europe. The more difficult task of designing and laying repeatered cables for deep water was achieved in 1949 in the Key West–Havana cable number 5; see Gilbert [5]. The big breakthrough came with the laying of Trans-Atlantic Telephony cable number 1 (TAT1) from Scotland via Newfoundland to Nova Scotia in 1956; a paper by Kelly and Radley [6] opened a symposium reviewing that project in 1957.

The TAT1 cable provided 35 telephone circuits of good quality between Europe and North America. There was an enthusiastic response from the public, and especially from business, to what was virtually a new service and the call rate rose rapidly. This was a valuable pointer to the public demand that could be expected if long-distance telecommunications services of acceptable quality were provided on other routes, by cable, or by any other transmission medium. A second cable system similar to TAT1 was laid in 1959 between France and North America and the Cantat cable between the UK and Canada followed in 1961.

In 1957 the governments of the British Commonwealth, collaborating through the Commonwealth Telecommunications Board (CTB), began to plan a chain of cable systems that would connect the countries of the commonwealth together.

The scheme that emerged was to extend the Cantat system across Canada to Vancouver, then by the Compac cable via Hawaii and Fiji to New Zealand and Australia. A second-stage would extend the system from Australia by the Seacom cable to Singapore and Hong Kong.

Before the planning could be completed, the launch of Sputnik 1, the first artificial Earth satellite, had shown that satellite communication should no longer be regarded as a remote possibility. However, the CTB decided that submarine cables and satellites were likely to be complementary, not mutually exclusive; it was agreed that the cable scheme should go ahead, as indeed it did, but the potentialities of satellites should be welcomed. These decisions have proved to be wise; to this day, submarine cables are the more economical transmission medium for the relatively small number of international routes that require telephone circuits or the equivalent in thousands or tens of thousands, whereas satellite systems cost less for smaller traffic loads.

1.2 The early satellite years

Demonstration and research

On 4 October 1957, the USSR amazed the world by launching Sputnik 1, the first artificial Earth satellite. This 84 kg spacecraft had been put into an elliptical, inclined orbit with initial heights above Earth at apogee and perigee of 942 and 231 km respectively. It was destroyed through atmospheric friction a few months later, but a long series of experimental satellites followed, nine of which were also of the Sputnik series.

There were strong and constructive reactions to the challenge of Sputnik 1, in particular in the USA, in many countries in Europe, in Japan and India. Increased effort was put into studies of the outer atmosphere using suborbital probes, the design of spacecraft and the systems, especially communication systems, that might make use of satellites. The launch by the USA on 31 January 1958 of a research satellite called Explorer 1 quickly demonstrated that the USSR did not have a monopoly of capabilities in spaceflight. A long series of US research satellites followed and one of their prime objectives was exploration of the environment in which artificial satellites have to operate.

It had long been known, from observations made at ground level and from high-altitude balloons, that the Earth is continuously bombarded by very energetic cosmic rays, by high-velocity solid particles (meteors) and by a broad spectrum of intense solar radiation. Charged subatomic particles emitted by the Sun, in particular during solar flares, were known to have a powerful effect on the Earth's ionosphere. Much of the power of the particles and the radiation was known to be absorbed by the atmosphere, but how much? How harsh would conditions be out there, in space? Would there be a high risk of mechanical damage to satellites from meteors? Could complex electrical equipment survive in space under continuous bombardment by charged and energetic particles? The delicate solar arrays on which artificial satellites would probably depend for their electrical power seemed particularly at risk. What effect would intense radiation have on the condition of the surface finish of the outer shell of a satellite, upon which the thermal balance of a radio station in space would to a considerable degree depend? What effect would extremes of temperature and the hard vacuum of space have on the bearing surfaces of mechanical devices?

Explorer 1 established the presence, not previously suspected, of the Van Allen radiation belts surrounding the Earth, made up of energetic protons and electrons, mostly of solar origin. In the lower Van Allen belt, the intensity of radiation was found to be strong enough to cause serious damage to unprotected solar cells. In most other respects, Explorer 1 and its successors sent back broadly reassuring reports on the space environment. The risk of damage from meteors, cosmic rays and solar particles outside the Van Allen belts seemed statistically acceptable and surface finishes capable of enduring intense radiation were identified.

The National Aeronautics and Space Administration

An Act of the US Congress, approved in July 1958 [7], established the National Aeronautics and Space Administration (NASA). One of NASA's objectives was 'the establishment of long-range studies of the benefits to be gained from, the opportunities for, and the problems involved in the utilization of aeronautical and space activities for peaceful and scientific purposes'. Another objective, no less significant for the development of satellite communication, was 'co-operation by the United States with other nations and groups of nations in work done pursuant to this Act and in the peaceful application of the results thereof'.

Satellite communications experiments

Among other fields of space activity at this time, the USA carried out a series of communication satellite experiments, involving NASA or the US Army, using passive reflectors in space, such as the Echo and West Ford projects, and active relaying satellites, such as Score and Courier. This programme, using active satellites, was continued by NASA with the six satellites of the Applications Technology Satellites (ATS) series with launchings running through the 1960s and was recently revived by the launch of the Advanced Communications Technology Satellite (ACTS) in 1993.

Syncom was another important early project, designed primarily to develop and refine techniques for launching satellites into the GSO. The Hughes Aircraft Company supplied three satellites. Syncom 1 was launched on 14 February 1963 into an orbit that was approximately geosynchronous, with an orbital inclination of 33.5°, but its communication subsystem failed in the final stages of orbit adjustment. Syncom 2 was launched on 26 July 1963 into a more accurate geosynchronous orbit, its inclination was 33.1°, and its communication system remained functional. Syncom 3 was highly successful, being launched on 19 August 1994 into an almost geostationary orbit, approximately circular, its orbital period almost exactly equal to one sidereal day and its inclination a mere 0.1°.

International collaboration in projects

The first important commercial application of satellite communication seemed bound to be for international telephone links, above all on the busiest routes in the world, namely those between North America and Europe. Public telecommunications services in Europe were operated by government monopolies, and it was perceived in the US that progress would require the involvement in experimental projects of European and Canadian PTOs. These foreign PTOs were eager for involvement. Around the beginning of 1961 NASA concluded memoranda of understanding with, among others, the PTOs of France and the UK. These documents were, in effect, agreements that the PTOs would build earth stations and participate, with US PTO earth stations, in tests of experimental satellites that were to be launched by NASA, each party meeting its own costs and all data obtained in the course of the tests being shared and published. The most important projects carried out under these agreements were Telstar and Relay.

The earth station in USA used for most of these tests, owned by ATT, was at Andover, Maine. It had a very large steerable horn antenna, 52 metres long, with an aperture 19 metres square, protected by a huge hemispherical radome 63 metres in diameter. A similar earth station was built for the French PTO at Pleumeur Bodou in Brittany. The BPO built an earth station at Goonhilly Downs in Cornwall having an antenna with a paraboloidal reflector 26 metres in diameter. Exceptionally low receiver noise figures, obtained by the use of maser amplifiers cooled to a temperature of a few degrees above absolute zero by immersion in boiling helium, enabled these earth stations to make use of the weak emissions radiated by these early satellites.

Two Telstar satellites were supplied by ATT, 77 kg in mass, equipped with a single transponder of 50 MHz bandwidth, receiving at 6390 MHz and transmitting about 2 W at 4170 MHz. In addition, NASA procured two Relay satellites, 78 kg in mass, the transponder having a bandwidth of 14 MHz, receiving at 1725 MHz and transmitting about 10 W at 4170 MHz. All these satellites were successfully launched between 10 July 1962 and 21 January 1964 into inclined, elliptical orbits with heights above Earth at apogee ranging from 5638 to 7422 km. All were successfully tested between US and UK and between US and France, using multi-channel telephony systems and colour television. These signals were extended into the inland networks from the earth stations at both ends for test telephone calls between officials and for nation-wide broadcasting of the television tests. Earth stations in Brazil and Japan also participated in the Relay tests.

The programme of tests was highly successful and its indirect results were very significant. In particular, it ensured that high-performance earth stations were built in countries that would play a key part in the establishment of an operational satellite system. It spread expertise. Furthermore, through the broadcasting of the trans-Atlantic television transmissions, not feasible with any terrestrial facilities then existing, it raised for satellite communication an approving awareness in the mind of a public on which international telephony had, at that time, little impact.

1.3 The emergence of operational systems

The US initiative

Satellite communication having been demonstrated as a feasible means of providing reliable links between fixed earth stations to interconnect terrestrial telephone networks, the next stage was to set up institutions with the financial and technical resources to establish operational systems. The US government gave a strong lead towards a single global system for this purpose. On 23 July 1961, President Kennedy made a statement on satellite policy in which he invited 'all nations to participate in a communications satellite system, in the interest of world peace and closer brotherhood among peoples throughout the world'. In the system foreseen, private ownership and operation of the US portion of the system was favoured, subject to safeguards, but there was tacit acceptance that this might not be found appropriate for other countries.

The Communications Satellite Corporation

In 1962 an Act of the US Congress [8] reaffirmed the position taken in the presidential declaration and authorized the establishment of the Communications Satellite Corporation (Comsat), a privately owned entity under a degree of governmental oversight. Comsat's main functions were to establish, in cooperation with the PTOs of other countries, a commercial communications satellite network, to set up the necessary earth stations in the US and, through them, to provide channels for the use of common carriers in the US.

The Intelsat interim agreements

Negotiations between interested governments followed in response to the US initiative, no doubt mindful of Resolution 1721 of the United Nations General Assembly which urged that 'communications by means of satellites should be available to the nations of the world as soon as practicable on a global and non-discriminatory basis'. On 20 August 1964 agreements (see, for example, [9]) were signed, setting up an entity that would provide the 'space segment' (that is, the satellites and the means for sustaining their usability and managing their use) for a global commercial communications satellite system. This entity was to be called the International Telecommunications Satellite Consortium (Intelsat). The representatives of 11 countries signed the agreements on the first day but all countries that were members of the International Telecommunication Union (ITU), that is, virtually every state in the world, were invited to accede to the agreements. The number of members now exceeds 130.

There were two agreements. One was between governments (the 'parties'). The other was between the national organizations (the 'signatories'), one per party, that were authorized to build earth stations (the 'earth segment') and to use the satellites of the system. For the USA, Comsat became the signatory. For most other countries the signatory was the PTO, often a department of government.

These were interim agreements, short, simple and subject to revision after a few years, as was appropriate for a venture with great potential but few certainties. Key provisions in the interim agreement between the parties were as follows:

(1) The ownership of the space segment was vested in the signatories in proportion to the contributions which they had agreed to make to its cost.
(2) Comsat was appointed to act as the manager of the space segment, under the direction of a committee of signatories' representatives.
(3) The establishment of a space segment with global coverage by late 1967 was set as an objective.
(4) The interim agreements were to be replaced by definitive agreements by the beginning of 1970.

The interim agreement between the signatories dealt mainly with financial, procurement and operating arrangements.

The start of operations

One of the first major decisions that the signatories had to address was the choice of orbit for the satellites; would the long transmission delay unavoidable with geostationary satellites be accepted by users for telephony or would the greater cost and the major technical problems posed by the use of a constellation of medium altitude satellites be inescapable? There had been subjective tests on laboratory simulations of telephone circuits via a geostationary satellite; see, for example, Riesz and Klemmer [10]. The results of these tests were fairly reassuring but some doubts remained. It happened that Comsat had signed a contract with the Hughes Aircraft Company on 16 April 1964 for the supply of a spacecraft similar to those of the Syncom experimental series. This spacecraft, called 'HS303' by the makers and 'Early Bird' by Comsat was designed to give two-way communication from the GSO for earth stations in Western Europe and

North America; see Figure 1.1. The Early Bird procurement was taken over by Intelsat and the satellite was launched on 6 April 1965 into a geostationary orbit. Eventually it was given the more prosaic name of 'Intelsat I'.

After preliminary tests the satellite links were put into regular use on 28 July 1965 for trans-Atlantic telephone service, occasionally interrupted for brief television transmissions. Users were asked to assess the quality of calls; their opinions, compared with customer assessments of calls made via submarine cable, were sufficiently reassuring for the geostationary orbit to be adopted by Intelsat for future satellites.

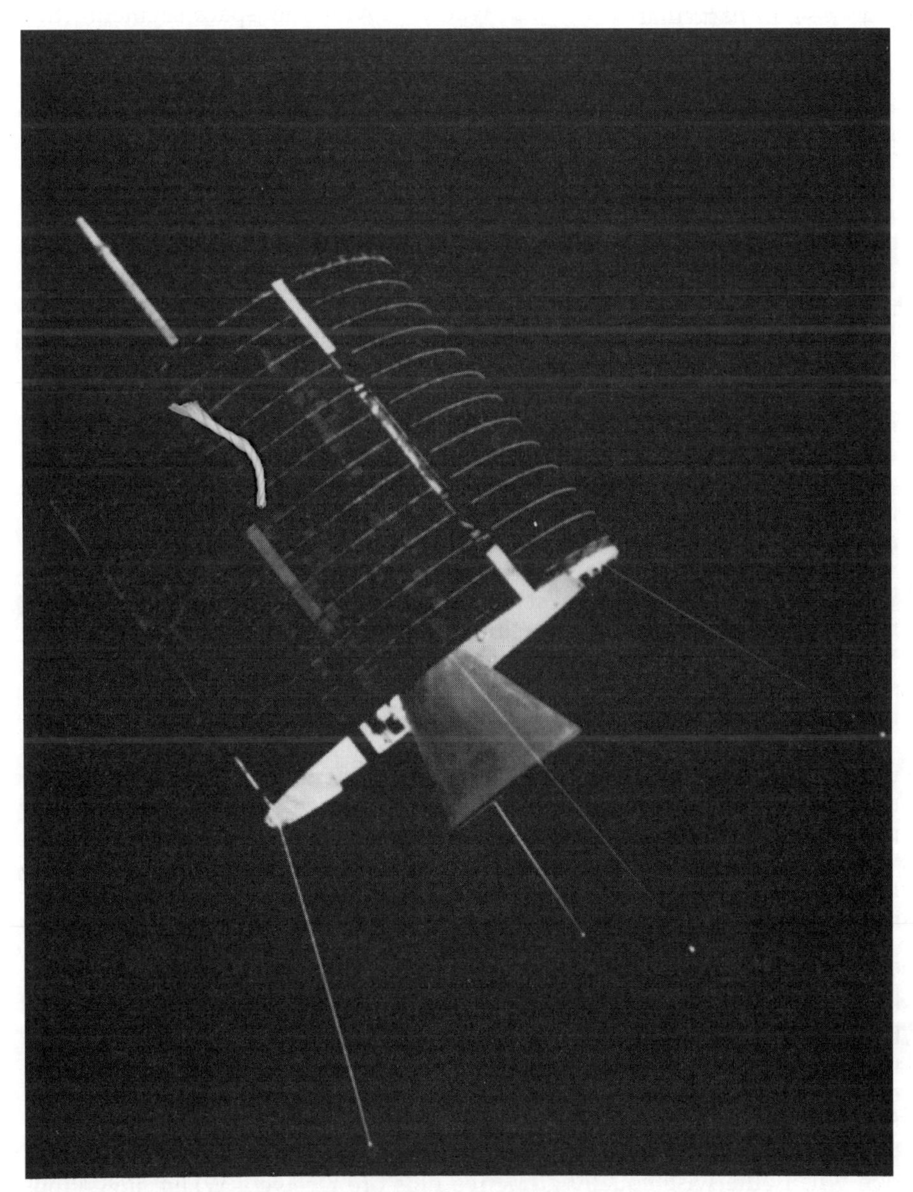

Figure 1.1 *Early Bird, later Intelsat I, the first commercial communication satellite. (Reproduced by permission of Intelsat.)*

The Intelsat definitive agreements

Negotiation of the terms of the Intelsat definitive agreements (for example, see [11]) began in 1969 and was concluded in May 1971. The new agreements are much more complex than the interim agreements of 1964, partly because by 1971 the scope for using satellites in communication was perceived to be much wider than had previously been thought and partly because the potential technological spin-off of space-oriented activities had also become evidently important. Some of the provisions of these agreements are of general interest, in view of the continuing importance of Intelsat, but also because they have formed a point of departure when the constitutions of other international satellite consortia have been drafted. In particular,

(1) Governments, through the two-yearly meetings of the Assembly of Parties, acquired an ongoing if limited role in the direction of the policy of the organization.
(2) The more immediate, and arguably the more important, decisions of the signatories in their commercial management of the system are taken at the frequent meetings of the Board of Governors, voting being weighted in accordance with the scale of a signatory's use of the system. However, many decisions on the longer-term issues of commercial policy were reserved for the annual Meeting of Signatories, where voting is unweighted, guaranteeing an influential role in the development of the system for the signatories of developing countries.
(3) Signatories lend capital to Intelsat as necessary, in proportion to quotas based mainly on the level of their use of the system. Interest is paid on these loans and the capital is repaid as space segment assets are amortized. Entities that are not signatories may use the space segment, subject to the approval of the government having jurisdiction over them. Utilization charges are paid by all users in proportion to their use of the system to meet running costs, the cost of interest on the loans and the repayment of capital.
(4) While the provision of a space segment for links between fixed earth stations remains the prime function of the Intelsat system, the agreements also permit the provision of virtually any other kind of non-military communication service by satellite, provided that the prime function of Intelsat is not harmed thereby.
(5) A special status is given to Intelsat relative to future satellite systems set up by other bodies. Parties and signatories undertake to advise Intelsat of any scheme they may have for setting up their own space segment for any purpose, or for acquiring such facilities from elsewhere. The potential disadvantage, technical or commercial, that such new facilities might cause to Intelsat would be assessed and Intelsat's response would be determined by the Assembly of Parties. (It may be noted that Intelsat's quasi-monopoly position has been eroded by precedent to a considerable degree since 1971.)
(6) There was tacit recognition of the pre-eminence that had been attained by US aerospace equipment manufacturers by 1971 and of the benefits that had flowed to one particular US corporation, namely Comsat, through its managerial position within Intelsat. The agreements contain various measures designed to ensure a wider spread of the expertise that would arise from future investment by Intelsat. Thus, when procuring goods and services, Intelsat contracts are to be awarded so as to stimulate world-wide competition when otherwise equally attractive bids are received. Where inventions are made by or on behalf of Intelsat, rights to those inventions become available to signatories and parties, subject to certain conditions and limitations.

> Finally Comsat would eventually cease to be the system manager. After a transition period, the signatories were to appoint a Director General who would be responsible to the Board of Governors for all managerial functions.

In 1973 the definitive agreements entered into effect and the formal title of Intelsat became the International Telecommunications Satellite Organization; the Intelsat name, however, was retained. Comsat continued to perform a number of management functions for Intelsat under contract until 1979, when the Intelsat Director General took over direct responsibility for all essential functions.

From its beginnings in 1965, the Intelsat space segment has grown greatly in both coverage and capacity. In 1967, when three second generation (Intelsat II) satellites were successfully launched, the capacity available in the Atlantic Ocean region was increased and service was inaugurated for the Pacific Ocean region. Coverage that was substantially global followed in 1969 when the Intelsat III satellites became available. Since then the space segment capacity, using successive generations of satellites in turn, has grown a hundredfold. There has been a corresponding growth of the number and variety of the earth stations dedicated to the system, and the diversity of the applications that have been found for the new medium.

The working methods of the organization as defined by the definitive agreements have changed little since 1979, although in 1995 consideration began to be given to ways in which the structure might desirably evolve, no doubt making the organization resemble more closely a conventional commercial corporation.

Intersputnik

Another agreement, setting up the Intersputnik organization as a second global satellite system, was also signed in 1971. As with Intelsat, the space segment is owned or leased by the organization and the earth stations are owned by the PTOs that use the space segment. The Intersputnik agreement addresses the direction, management, operation and funding of the system, although the financial principles on which funding is to be based are not precisely defined. The initial membership of Intersputnik consisted of Bulgaria, Cuba, Czechoslovakia, the German Democratic Republic, Hungary, Mongolia, Poland, Romania and the USSR. Several other countries joined Intersputnik subsequently, but there has not been a broad response, such as Intelsat received, to the open invitation that is contained in the Intersputnik agreement.

Other international space segment providers

Despite the existence of two global satellite networks primarily designed to serve the market for international links between national public telephone networks, several other systems have been set up to provide similar facilities for regional groupings of countries. Particular mention may be made of the systems set up by the Arab Corporation for Space Communication (Arabsat) and the European Telecommunications Satellite Organization (Eutelsat). Both organizations include many parties and signatories of Intelsat, but the Intelsat Assembly of Parties has agreed in both cases that these regional systems would not cause unacceptable economic disadvantage to Intelsat. Some private companies and consortia of companies, such as Panamsat and Orion, are also challenging the space segment cooperatives set up by PTOs for international links.

Arabsat

The agreement which created Arabsat was signed by about 20 Arab states in 1976. The primary purpose of the organization is to provide a space segment for conventional telecommunications and television links between earth stations at fixed locations. In addition, one or two television channels are provided, transmitted at higher power, for community reception. However, other functions, such as providing assistance to member states in the development of the earth segment, also fall within the organization's authority. Like the Intersputnik agreement, the Arabsat agreement deals mainly with the direction, management, operation and funding of the system.

Eutelsat

For Eutelsat, interim agreements were signed in 1977, followed by definitive agreements in 1982, the latter entering into force in 1985. As with Intelsat, the Eutelsat agreements are twofold, consisting of a convention agreed between states (the parties) and an operating agreement between PTOs, one per party, which are authorized to take responsibility for each state's support for and use of the system. More than 20 states of western and southern Europe joined Eutelsat initially and other European states have acceded since; there were 44 members by the end of 1995. In most respects the objectives, constitution and management principles of Eutelsat resemble those of Intelsat, although there have been recent changes of practice to accommodate the entry of competing PTOs, in place of government monopoly PTOs, in some member countries.

Domestic space segments

By no means all satellite systems are international in scope and control. The USSR set up the Orbita system using Molniya satellites for domestic services in 1965. Unlike all the other operational satellite systems mentioned above, which use geostationary satellites, the Molniya satellites were deployed in inclined, elliptical orbits, which provided better coverage of the territory of the USSR. Furthermore, it soon became evident that satellite links would be cost effective for domestic trunk networks in other very large countries and in smaller countries where the terrain was such as to make the deployment of terrestrial communication systems particularly costly. Canada set up Anik satellites for domestic services, operated by Telesat, in 1973. Several domestic systems were set up in the USA in 1974–75 and others since. The Indonesian system, Palapa, followed in 1976.

The widening market for satellite communications

Despite the emergence of optical fibre as a low-cost transmission medium for high-grade, high-volume terrestrial telecommunications, the break-even distance for satellite communication has continued to shorten. In the 1980s and 1990s, the number of countries that own satellites for their domestic communications has become too large to list here. In addition, many other countries have leased transponders from Intelsat and other space segment owners and use them for their domestic links.

Use for extending the PSTN is an important application for most of these national systems, but it is no longer the only use, and indeed this usage, in comparison with other applications, may be declining. Significant new markets for private telephone and data transmission networks have developed, some involving the use of earth stations located at the user's premises and using very small antennas. Satellite communication is already playing an important role in maritime communication, and its use by aircraft and road vehicles is developing. Finally, the distribution of television programme signals to terrestrial broadcasting outlets and directly to the public has become a major application for domestic satellite communication systems.

References

1. Clarke, A. C. (1945). Extraterrestrial relays. *Wireless World*, October.
2. Buckley, O. E. (1942). The future of transoceanic telephony. *JIEE*, **89**, Pt I, 454–461.
3. Angwin, Sir A. Stanley (1944). Inaugural address as President of the IEE. *JIEE*, **91**, Pt I, 15–20.
4. Halsey, R. J. (1944). Modern submarine cable telephony and the use of submerged repeaters. *JIEE*, **91**, Pt III, 218–236.
5. Gilbert, J. J. (1951). A submarine telephone cable with submerged repeaters. *BSTJ*, **30**.
6. Kelly, M. J. and Radley, Sir Gordon (1957). Foreword to a Symposium on the Trans-Atlantic Telephone Cable. *Proc. IEE*, **104**, Pt B, Suppl. 4.
7. *The National Aeronautics and Space Act* (1958). Public law 85-568, HR 12575, 29 July 1958.
8. *The Communications Satellite Act* (1962). Public Law 87-624, HR 11040, 31 August 1962.
9. *Satellite communications* (1964). HMSO Cmnd 2436, August 1964.
10. Reisz, R. R. and Klemmer, E. T. (1963). Subjective evaluation of delay and echo suppressors in telephone communication. *Bell Syst. Tech. J.*, **42**, 2919–2941.
11. *Satellite communications* (1971). HMSO Cmnd 4799, October 1971.

Part 2
Platforms in Space

2 Orbits and earth coverage

2.1 Basic orbital parameters

In a frame of reference which has its origin O at the Earth's centre of mass, and axes which are in directions fixed relative to the Earth's axis and the stars, and disregarding various perturbing forces, the orbit of an artificial Earth satellite is a closed ellipse. The major and minor axes of the ellipse can be made equal in length, making the orbit circular. The plane of the orbit contains the Earth's centre of mass, located at one of the foci of an elliptical orbit and at the centre of a circular one. In this frame of reference, and still disregarding the perturbations, the orbit will remain constant in shape, size and orientation indefinitely, the Earth rotating within it. Figure 2.1 shows a typical elliptical orbit, such as was used for the pioneering Telstar and Relay satellites.

The axes of the frame of reference are orthogonal. The OZ axis is perpendicular to the Earth's equatorial plane and coincident with its axis of rotation, north being reckoned positive. The OX axis lies in the equatorial plane and is directed towards a datum point in the constellation Pisces. The third axis, OY, also lies in the equatorial plane and is perpendicular to both OZ and OX. See Figure 2.2.

The same frame of reference is used by astronomers in defining the location of stars and other astronomical objects on the celestial sphere. The datum point in Pisces is aligned with the intersection of two planes, the Earth's equatorial plane and the plane in which the Earth revolves around the Sun, called the plane of the ecliptic. Right ascension, the equivalent for the celestial sphere of longitude on Earth, is measured eastwards from the OX axis. Astronomers use a scale of hours, minutes and seconds for right ascension, but it is more convenient to use a scale of 360 degrees in satellite communication. Declination, the equivalent of terrestrial latitude, is measured in degrees.

The Sun is seen from the Earth aligned with the OX axis on about 20 March each year, the time of the spring equinox in the Northern hemisphere, and accordingly the datum point is called 'the vernal equinox'. However, the orientation of the equatorial plane in space changes very slowly and this is causing the vernal equinox to move towards the constellation Aquarius at a rate of 1° in 71.8 years, a phenomenon called the precession of the equinoxes. This phenomenon was discovered by Hipparchus of Nicaea in the second century BC, at a time when the vernal equinox was in the constellation Aries, about 30° East of its present position. Another name for the datum, 'the first point of Aries', recalls its discovery.

Viewed from above the North Pole, the Earth rotates anticlockwise. Satellites that move in the same direction are said to have a direct orbit; otherwise the orbit is retrograde. The angle between the orbital plane and the Earth's equatorial plane is called the inclination of the orbit, and is denoted by i. If the angle of inclination is zero, the orbit is equatorial; if not, it is inclined. If the angle of inclination is 90°, the orbit is polar. These concepts are illustrated in Figures 2.1 and 2.3.

Perturbing forces have significant effects on the orbits of satellites. These forces include the gravitational attraction of the Moon and the Sun, the pressure exerted

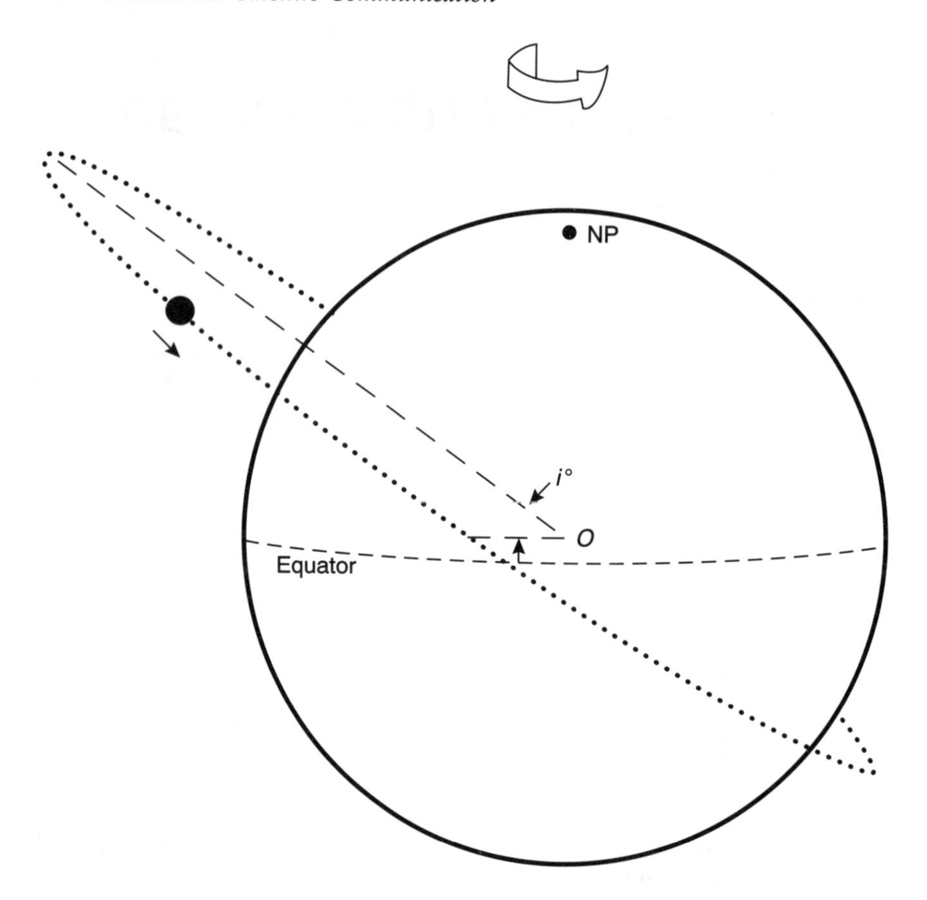

Figure 2.1 *A satellite in an elliptical orbit. The plane of the orbit is inclined at an angle i to the Earth's equatorial plane*

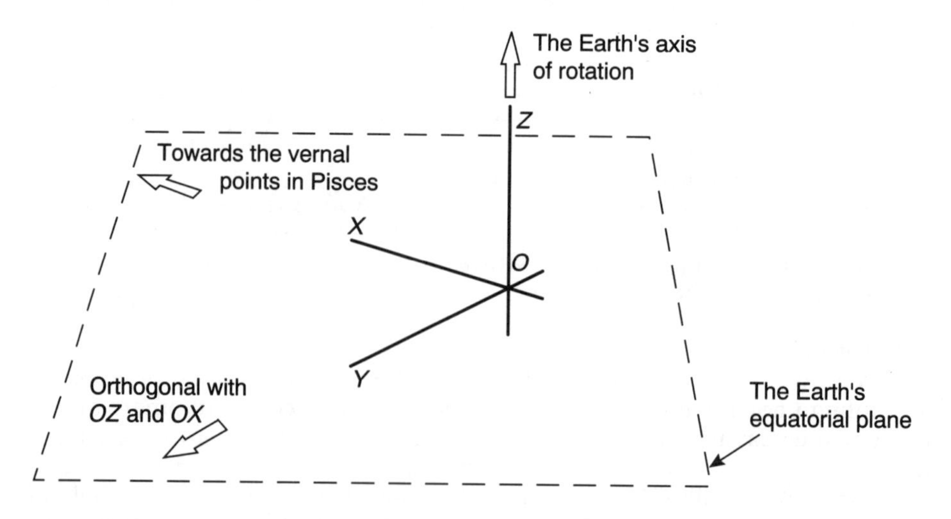

Figure 2.2 *The frame of reference used in defining satellite positions and orbits*

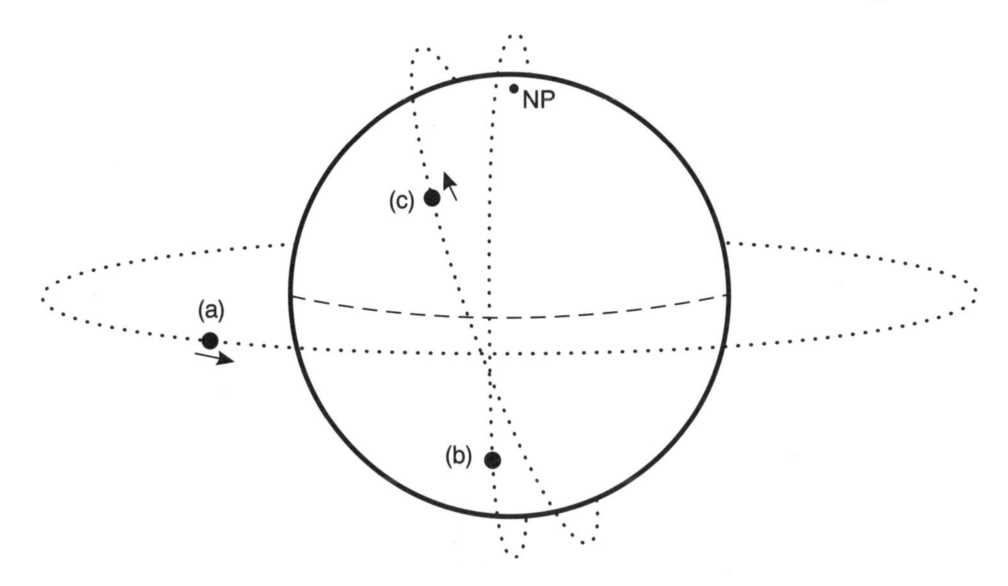

Figure 2.3 *Three kinds of circular orbit: (a) a direct, equatorial orbit, (b) a polar orbit, and (c) a retrograde, inclined orbit*

by the energy flux from the Sun, irregularities in the shape and density of the Earth and, at low altitude, the drag of the atmosphere. Furthermore, the rotation of the Earth itself has a dominant effect on the geographical area visible at any time from a satellite. However, these effects vary greatly with the type of orbit; they are reviewed for the three types of orbit that are currently of interest for communication systems in Sections 2.2, 2.3 and 2.4.

Period of revolution

The period of revolution, or orbital period, of a satellite is the time between successive passages of the satellite through an identifiable point on the orbit. For an inclined orbit, it is convenient to measure the period of revolution as the interval between successive passages of the satellite through the equatorial plane in a north-bound direction; this point on the orbit is called the ascending node. This datum is unsatisfactory for an equatorial orbit, for which it is customary to use successive passages of the satellite through a plane, perpendicular to the equatorial plane, containing the centre of the Earth, and fixed in orientation with respect to the stars.

Disregarding the effects of the perturbations, the period of revolution for a circular orbit is given by

$$t = 2\pi r^{3/2}(GM_{\mathrm{E}})^{-1/2} \quad \text{(seconds)} \tag{2.1}$$

Substituting known values (see below) for the constants and using more convenient units, we obtain

$$T \approx 1.4(r/R_{\mathrm{E}})^{3/2} \quad \text{(hours)} \tag{2.2}$$

whilst for an elliptical orbit, we have

$$T \approx 1.4(a/R_{\mathrm{E}})^{3/2} \quad \text{(hours)} \tag{2.3}$$

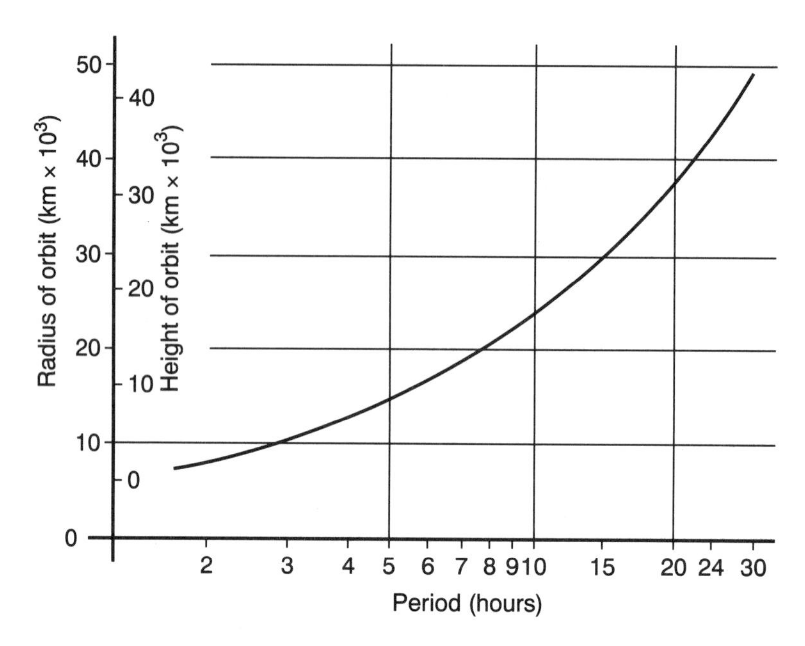

Figure 2.4 *The radius and height above ground of a circular satellite orbit as a function of the period of revolution*

where t and T are the period, expressed in seconds and hours respectively,
r and R_E are the radius of the orbit, if circular, and the radius of the Earth respectively, in metres,
a is half the length of the major axis of an elliptical orbit, in metres,
G is the universal gravitational constant ($\approx 6.67 \times 10^{-11}\,\mathrm{N\,m^2\,kg^{-2}}$)
M_E is the mass of the Earth ($\approx 5.977 \times 10^{24}$ kg)

The radius of a circular orbit, and its height h above the surface of the Earth, assuming that the Earth's radius is 6371 km, are shown as a function of the period in Figure 2.4 (6371 km is the mean radius of the Earth; the mean radius in the equatorial plane is 6378 km).

2.2 Geosynchronous and geostationary orbits

2.2.1 Basic orbital characteristics

The Earth's sidereal period of rotation, that is, the time taken for one complete rotation about its centre of mass relative to the stellar background, is one sidereal day, approximately 23 hours 56 minutes 4 seconds. If a satellite has a direct, circular orbit and its period of revolution, measured as above, is equal to one sidereal day, it will keep pace with the turning Earth; that is, it is a geosynchronous satellite. The radius of its orbit (r_g) will be 42 164 km (equation (2.1)) and its height above the Earth's surface will be about 35 786 km.

If this satellite's daily Earth track (that is, the locus of the points on the Earth's surface that are vertically below the satellite at any instant) is traced, it will show a figure-of-eight pattern as sketched in Figure 2.5(a). The maximum extent of the pattern in degrees of latitude, north and south of the equator, is equal to the angle of inclination of the orbit. Provided that the orbit is indeed circular,

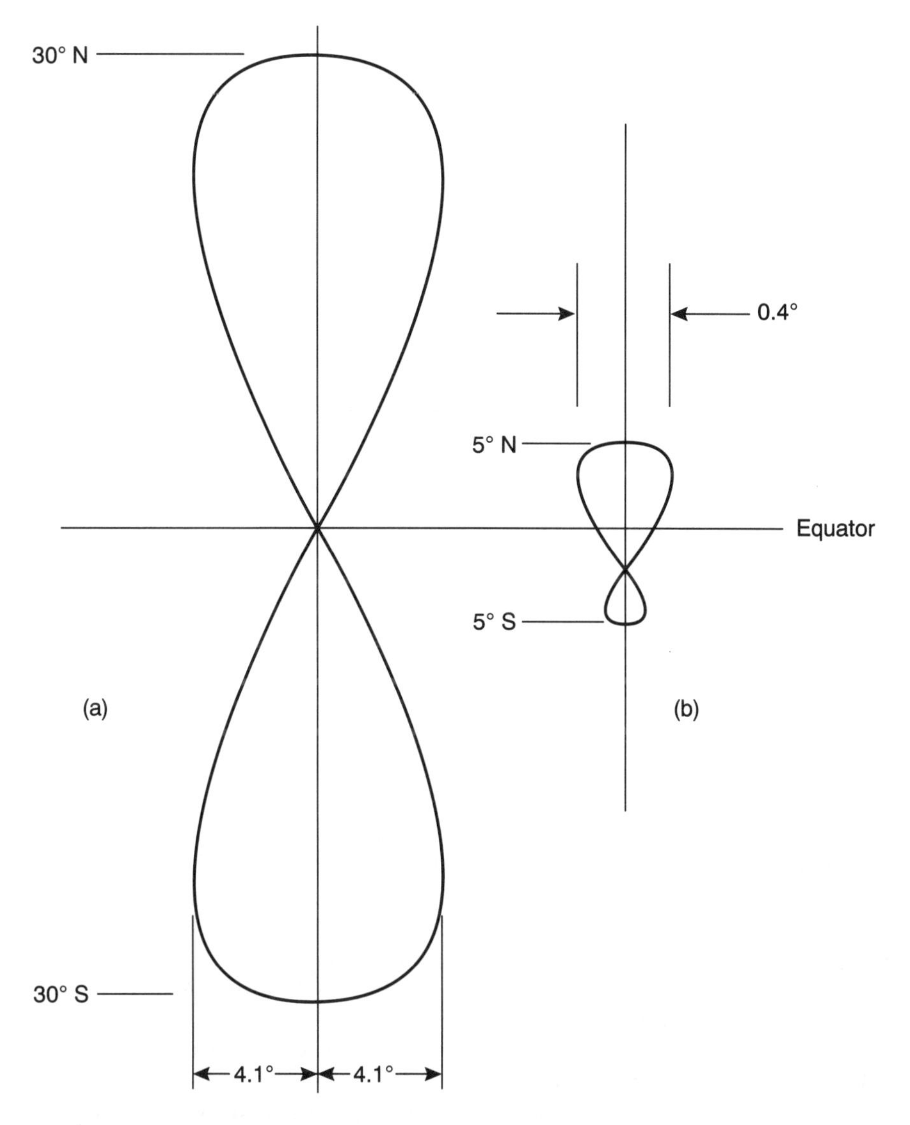

Figure 2.5 *A sketch of Earth tracks of geosynchronous satellites: (a) relates to a circular orbit having an inclination of 30°; (b) relates to a somewhat elliptical orbit having an inclination of 5°*

the north-going track crosses the equator at the same longitude as the south-going track and the pattern is symmetrical about that central line of longitude. However, if the orbit is elliptical, the cross-over point of the north-going and south-going tracks is no longer located in the equatorial plane and the pattern becomes asymmetrical; see, for example, Figure 2.5(b).

The maximum spread of the pattern, east and west of the central line of longitude, is given by

$$\text{maximum spread} = \pm \arcsin(\sin^2 \tfrac{1}{2}i / \cos^2 \tfrac{1}{2}i) \tag{2.4}$$

This function is plotted in Figure 2.6. The spread is substantial for large orbital inclinations. Thus, for the pattern shown in Figure 2.5(a), which relates to an

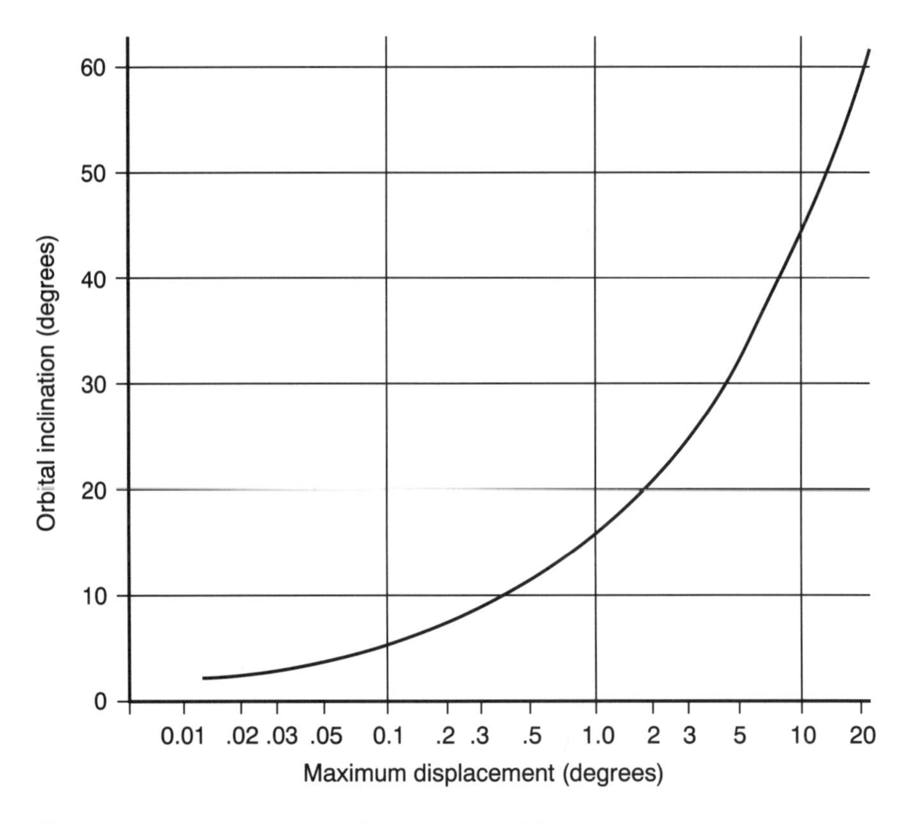

Figure 2.6 *The maximum displacement of the Earth track of a geosynchronous satellite from the central longitude, as a function of the orbital inclination*

orbital inclination of 30°, the maximum spread in longitude is ±4.1°. But for small angles of inclination, the spread becomes very small. When the inclination is zero, the Earth track pattern shrinks to a single point on the equator; in this case the satellite is stationary in space as seen from the surface of the Earth, and the orbit is said to be geostationary.

The geostationary satellite orbit (GSO), like other orbits, is unstable. There are orbital perturbations that are tending all the time to change its period, inclination and shape from the geostationary parameter set. These perturbations must be counteracted by frequent readjustment of the orbital parameters if the orbit is to remain even approximately geostationary; see Section 2.2.3. No actual satellite is ever perfectly geostationary for long, and purists prefer to apply the term geosynchronous to a satellite that conforms only approximately to geostationary orbital parameters. Nevertheless, geostationary satellites are so important in satellite communication that they have been given a privileged regulatory status by international agreement, subject to fairly close tolerances relative to the ideal parameters. Consequently, in practice, the term geostationary is usually applied to any commercial satellite that meets these tolerances.

Geosynchronous orbits of substantial inclination and eccentricity have Earth coverage characteristics which could be of interest in satellite communication; they are considered briefly in Section 2.3. However, the main interest lies in orbits that meet the regulatory geostationary tolerances and the remainder of Section 2.2 is concerned with orbits of this type.

Advantages and disadvantages of geostationary satellites

The GSO is better for most communication systems than any other orbit. The reasons for this are:

(1) Above all, one satellite can provide continuous links between earth stations. An inclined geosynchronous satellite can do this also, although the geographical area that can be served is more limited if the angle of inclination is large. The disadvantages of using satellites with an orbital period of less than one sidereal day for systems that are required to provide continuous connections are outlined in Section 2.4; they are severe.
(2) The gain and radiation pattern of satellite antennas can be optimized, so that the geographical area illuminated by the beam, called the footprint, can be matched accurately to the service area, yielding significant benefits.
(3) The geographical area visible from the satellite, and therefore potentially accessible for communication, is very large; see Figure 2.7. Applying equation (2.7), the diameter of the area within which the angle of elevation δ of a geostationary satellite is greater than 5° is about 16 960 km.
(4) If the orbit is accurately geostationary, earth station antennas of considerable gain can be used without automatic satellite tracking, reducing equipment cost and minimizing the operational attention required.
(5) The frequency assignments used in different geostationary satellite networks can be coordinated efficiently, the satellite footprints can be matched to the service areas, and earth station antennas usually have high gain. This enables a large number of satellites to share this single orbit, each at its assigned longitude, using the same frequency bands, without unacceptable mutual interference.

This fifth advantage is important, because the other four make the GSO so attractive that it has been chosen for almost all systems now in operation. In some arcs of the orbit, there are many satellites already in use, operating in the frequency bands that have good radio propagation characteristics, and many more have been proposed for launching, raising the possibility of saturation of the preferred frequency bands.

Nevertheless, the GSO has disadvantages and other orbits are to be preferred for some applications. The chief disadvantages of the GSO are:

(1) A satellite link, from earth station to earth station via a geostationary satellite, is very long. From equation (2.8), in the worst case, when the satellite has a low angle of elevation at both earth stations, the length d of the uplink might be about 41 100 km and similarly for the downlink. Consequently, both the transmission time and the free-space transmission loss are very large; see Chapter 5.
(2) As can be seen from Figure 2.7, the angle of elevation of the satellite as seen from earth stations in high latitudes is quite low, leading at times to degraded radio propagation and possible obstruction by hills, buildings, and so on. Beyond the Arctic and Antarctic Circles, coverage from the GSO is unsatisfactory and is further degraded at some times of day if the orbital plane has any inclination. Beyond latitude 75°, north or south, the use of geostationary satellites is scarcely feasible.

2.2.2 The geometry of geostationary satellite links

If a satellite is accurately geostationary, it is not essential to use the frame of reference defined in Section 2.1 to define its orbit for operational purposes. The

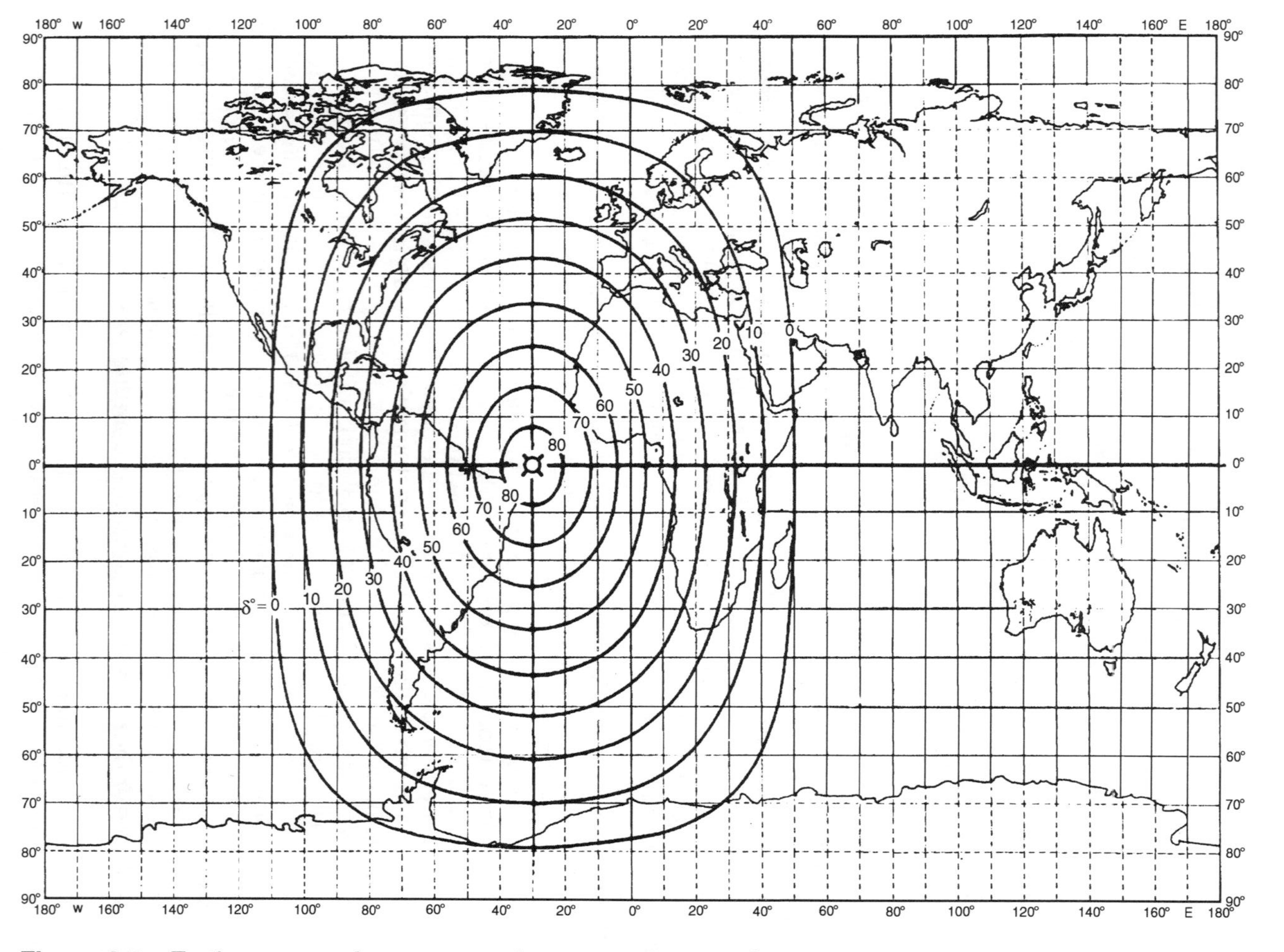

Figure 2.7 *Earth coverage from a geostationary satellite at 30° W longitude as a function of δ, the minimum angle of elevation of the satellite from an earth station. (From the ITU Handbook on Satellite Communications: Fixed-Satellite Service [1].)*

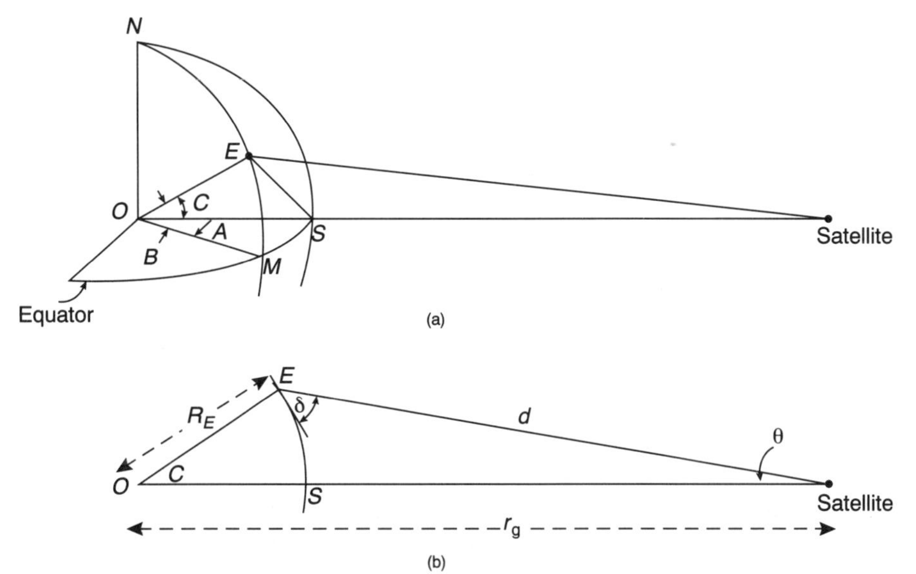

Figure 2.8 *The geometry of a link between a geostationary satellite and an earth station: (a) is a three-dimensional representation; (b) shows the plane which contains the centre of the Earth (O), the earth station (E) and the satellite (S)*

Earth's own frame of reference, namely longitude and, if necessary, the excursion in latitude due to orbital inclination, is sufficient and more convenient. The main elements are shown in Figure 2.8. Thus, disregarding trivial errors arising from irregularities in the Earth's shape and the effects of atmospheric refraction (which are significant only when the angle of elevation of the satellite is small) and assuming that the inclination of the orbital plane is zero, the key parameters can be calculated as follows:

(1) The geocentric angle (that is, the angle subtended at the Earth's centre) between the earth station and the subsatellite point (= the angle EOS, marked C in Figure 2.8) is given by

$$C = \arccos(\cos A \cos B) \quad \text{(degrees)} \tag{2.5}$$

(2) The azimuth of the satellite is one of the two parameters which define the direction in which the satellite is seen from an earth station, the other parameter being the angle of elevation. The azimuth, in degrees east of north, is most easily calculated by a two-stage process, an intermediate parameter a being calculated first:

$$a = \arcsin(\sin A / \sin C) \quad \text{(degrees)} \tag{2.6}$$

where a is the acute angle between the azimuth and the meridian through the earth station. However, there are ambiguities as to the quadrant. These can be resolved most readily by inspection, according to the relative geographical positions of the earth station and the subsatellite point; see Figure 2.9.

(3) The angle of elevation δ of the satellite as seen from the earth station is given by

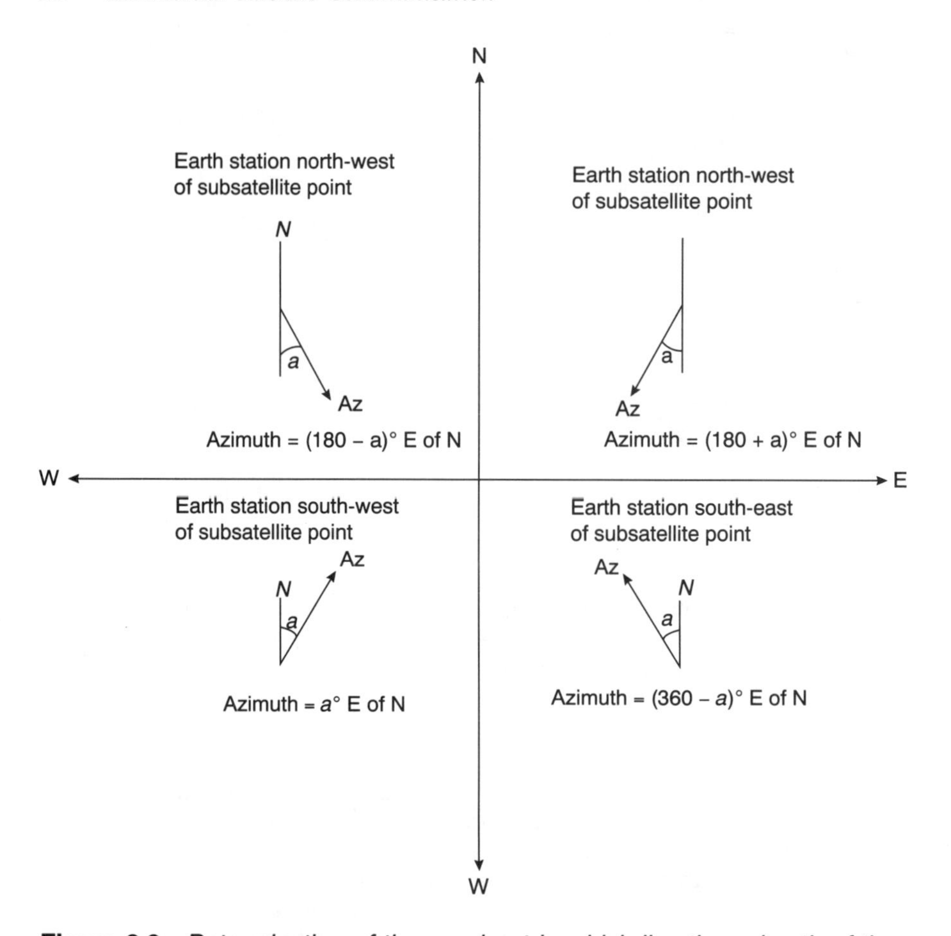

Figure 2.9 *Determination of the quadrant in which lies the azimuth of the satellite from an earth station*

$$\delta = \arctan[(\cos C - R_E/r_g)/\sin C] \quad \text{(degrees)} \tag{2.7a}$$

$$= \arctan[(\cos C - 0.15127)/\sin C] \quad \text{(degrees)} \tag{2.7b}$$

(4) The distance d to the satellite from the earth station is given by

$$d = (R_E^2 + r_g^2 - 2R_E r_g \cos C)^{1/2} \quad \text{(km)} \tag{2.8a}$$

$$= 42644(1 - 0.29577 \cos C)^{1/2} \quad \text{(km)} \tag{2.8b}$$

(5) The exocentric angle θ (that is, the angle subtended at the satellite) between the subsatellite point and the earth station is given by

$$\theta = 90 - (C + \delta) \quad \text{(degrees)} \tag{2.9}$$

(6) The topocentric angle ϕ_t (that is, the angle subtended at an earth station) between two geostationary satellites is illustrated in Figure 2.10; it is an important quantity when interference between networks using satellites in neighbouring locations in orbit has to be evaluated. The topocentric angle is rather larger than the geocentric angle (ϕ_g) between the two satellites and which, for geostationary satellites, is merely the difference between their longitudes. Using the symbols defined above and those shown in the figure, we obtain

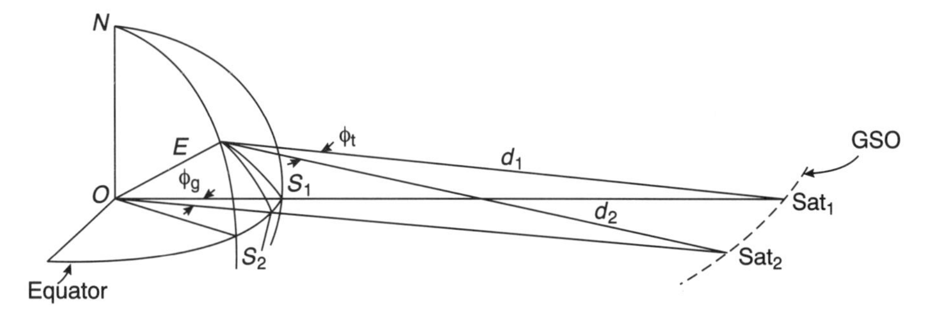

Figure 2.10 *The topocentric angle (ϕ_t) at an earth station (E) between two geostationary satellites separated by a geocentric angle of ϕ_g*

$$\phi_t = \arccos(d_1^2 + d_2^2 - 2r_g \sin^2 \tfrac{1}{2}\phi_g)/2d_1d_2 \quad \text{(degrees)} \tag{2.10}$$

2.2.3 Perturbations of geostationary orbits

The drag of the atmosphere on a satellite in a 24-hour circular orbit is negligible. However, there are three kinds of orbital perturbation which tend to move the parameters of the orbit of a geostationary satellite away from the nominal values. They are the gravitational effect of irregularities of the Earth's figure, the gravitational attraction of the Sun and the Moon and solar radiation pressure.

The effect of the Earth's triaxiality

The distribution of the mass of the Earth is not completely symmetrical. Basically the figure of the Earth is a sphere which is flattened at the poles. This oblateness causes no significant perturbation to the orbits of geostationary satellites, although it has very significant effects on satellites in inclined elliptical orbits; see Section 2.3. But the distribution of the Earth's gravitational field in the equatorial plane is also somewhat irregular, with maxima at longitudes 162° E and 12° W and minima at 77° E and 108° W; see Figure 2.11. Because of this three-dimensional departure from spherical symmetry, the Earth is said to be triaxial. The gravitational variation in the equatorial plane is very slight. Nevertheless, the period of revolution of a geostationary satellite is so precisely matched to the period of rotation of the Earth that this small lack of symmetry has a substantial impact on the orbital parameters.

The speed in orbit of an unperturbed geostationary satellite would be an unvarying 11 069 km h^{-1}. However, because of this equatorial irregularity, the gravitational force that a satellite experiences is not directed precisely towards the Earth's centre of mass. Consequently, the satellite is accelerated or decelerated, according to its orbital position, so that it tends to drift away from longitudes 12° W and 162° E and towards longitudes 77° E and 108° W. A satellite placed at 77° E or 108° W would not experience such forces; these are points of stable equilibrium. Conversely, 12° W and 162° E are points of unstable equilibrium.

These angular accelerations and decelerations are small, at most only about 2×10^{-3} degrees per day, and in most orbital locations the effect is less. However, the consequential change of velocity and the displacement of a satellite from its initial longitude are cumulative if not corrected and become significant within a few weeks at most longitudes.

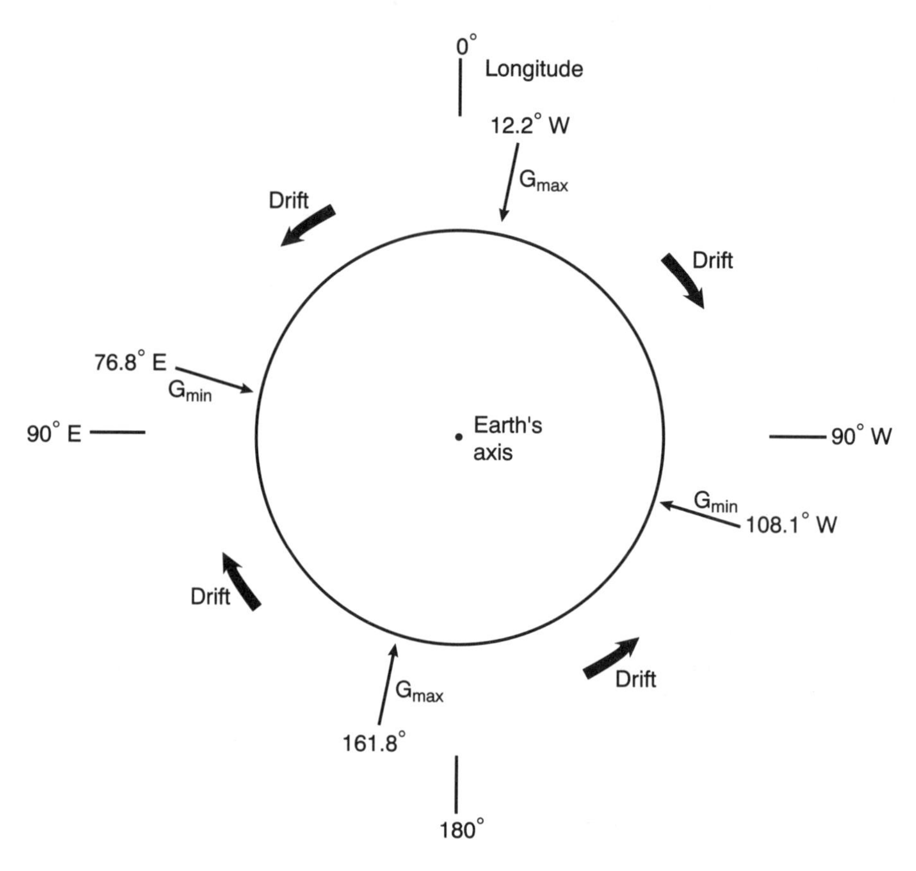

Figure 2.11 *An equatorial section through the Earth, seen from the north, illustrating the tendency of geostationary satellites to drift in longitude, due to non-uniformities in the Earth's gravitational field in the equatorial plane*

Lunar/solar gravitational effects

The Moon and the Sun exercise a gravitational attraction on artificial Earth satellites. The attraction is greater when the satellite is on the side of the Earth that is closer to the perturbing body and it is also a function of the positions of the perturbing bodies relative to the Earth's equatorial plane and relative to one another. The perturbation that this attraction causes varies in a complicated way over a period of 18.6 years, set by the periods of revolution of the Earth about the Sun and of the Moon about the Earth. The principal consequence for geostationary satellites is that the plane of the orbit tends to be tilted away from the equatorial plane, at a rate of about 0.8° to 0.9° per year in the present epoch. This trend converts a geostationary orbit into an inclined geosynchronous one. If uncorrected, the rate of increase of inclination will eventually decline, the inclination reaching a maximum of about 15°.

Solar radiation pressure

The effect of solar radiation pressure on the body of a satellite is to induce an annually varying eccentricity ϵ into an orbit that was originally circular, making it slightly elliptical. One consequence of this is that the longitude of the satellite oscillates with a daily cycle about a mean position. The magnitude of the eccentricity depends on the area of the cross-section that the satellite presents to the Sun. The eccentricity is generally small but ϵ might rise to 0.002 for current satellites with large solar panels, resulting in an apparent daily east–west movement about the mean longitude of up to about $\pm 0.25^\circ$.

Correction of perturbations

By international agreement (see Section 6.5, item 3), with few exceptions, a satellite operator must be able to maintain a geostationary satellite within $\pm 0.1^\circ$ of its nominal position in longitude, and the orbital inclination of some broadcasting satellites must not exceed 0.1°. The effects of these three perturbations on the station-keeping of satellites can be counteracted by the expulsion of matter from the satellite. Correcting for east–west satellite movement due to the Earth's triaxiality so as to meet these requirements demands small but frequent orbit adjustment, perhaps every few weeks. However, the mass of material that must be ejected to counteract the orbital inclination caused by the Moon and the Sun is quite considerable; see Section 4.5.1.

2.3 Inclined elliptical orbits

2.3.1 Basic orbital characteristics

The shape of an ellipse is characterized by its eccentricity ϵ, where

$$\epsilon = (1 - b^2/a^2)^{1/2} \tag{2.11}$$

and a and b are the semi-major and semi-minor axes of the ellipse. There are two foci, located on the major axis and separated from the origin of the ellipse by distance c, where

$$c = \epsilon a \tag{2.12}$$

For an Earth satellite with an elliptical orbit, one of the foci is located at O, the centre of mass of the Earth. The points on the orbit where the satellite is most and least distant from the Earth are called the apogee and the perigee respectively. The greatest and least distances from the surface of the Earth, the altitudes of apogee and perigee, h_a and h_p, are given by

$$h_a = a(1 + \epsilon) - R_E \tag{2.13}$$

and

$$h_p = a(1 - \epsilon) - R_E \tag{2.14}$$

These various terms are illustrated in Figure 2.12.

A satellite in a perfectly circular orbit has uniform speed round that orbit, but the speed of motion of a satellite in an elliptical orbit varies. As the satellite moves from apogee to perigee its potential energy falls and its kinetic energy, as revealed by its speed, rises. Correspondingly, the potential energy rises and the speed falls as the satellite moves from perigee to apogee. This variation of speed is conveniently expressed in the form of Kepler's second law of planetary motion. This states that each planet moves in such a way that a line joining it to the Sun would sweep out equal areas in equal periods of time.

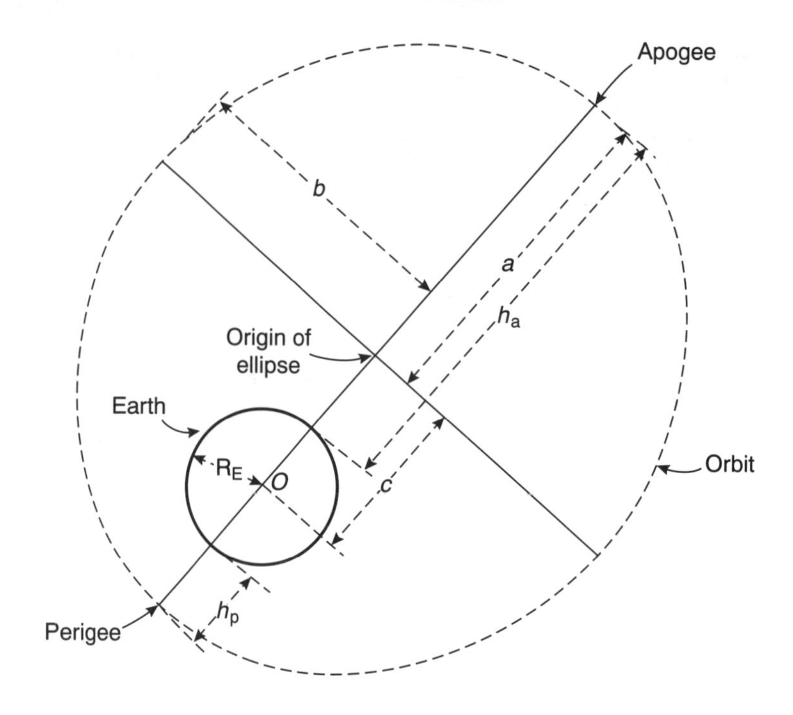

Figure 2.12 *The principal parameters of an elliptical Earth orbit*

Thus, in Figure 2.13, if the time taken by the satellite to move from N to M is the same as for the journey from K to J, then the sectors KOJ and NOM are equal in area and the ratio of the satellite speeds at the midpoints of the arcs NM and KJ (v_1, v_2) is related to the ratio of the distances of those midpoints to the centre of mass of the Earth (l_1, l_2) by

$$v_1/v_2 = l_2/l_1 \tag{2.15}$$

In describing the orbit of a non-geostationary satellite, unlike the geostationary case, it is not sufficient to use terrestrial longitude. Figure 2.14 shows an Earth satellite in a direct inclined elliptical orbit in the frame of reference $OXYZ$ that is shown in Figure 2.2. The point at which the satellite passes through the equatorial plane going northwards is called the ascending node. Likewise the satellite passes southwards through the plane at the descending node. The orientation of the plane of the orbit, relative to the stars, is defined by the angle, measured eastwards, between the OX axis and the line from the Earth's centre to the ascending node; this angle is called the longitude of the ascending node. The angle, measured eastwards, in the plane of the satellite's orbit, between a line from O to the ascending node and another from O to the perigee defines the attitude of the orbital ellipse relative to the equatorial plane; this angle is called the argument of perigee.

2.3.2 Perturbations of inclined elliptical orbits

The gravitational attraction of the Sun and the Moon and the pressure of solar radiation on the satellite body affect satellites with inclined elliptical orbits in much the same way as they affect geostationary satellites. However, these effects are small compared with the effect of the oblateness of the Earth on the argument

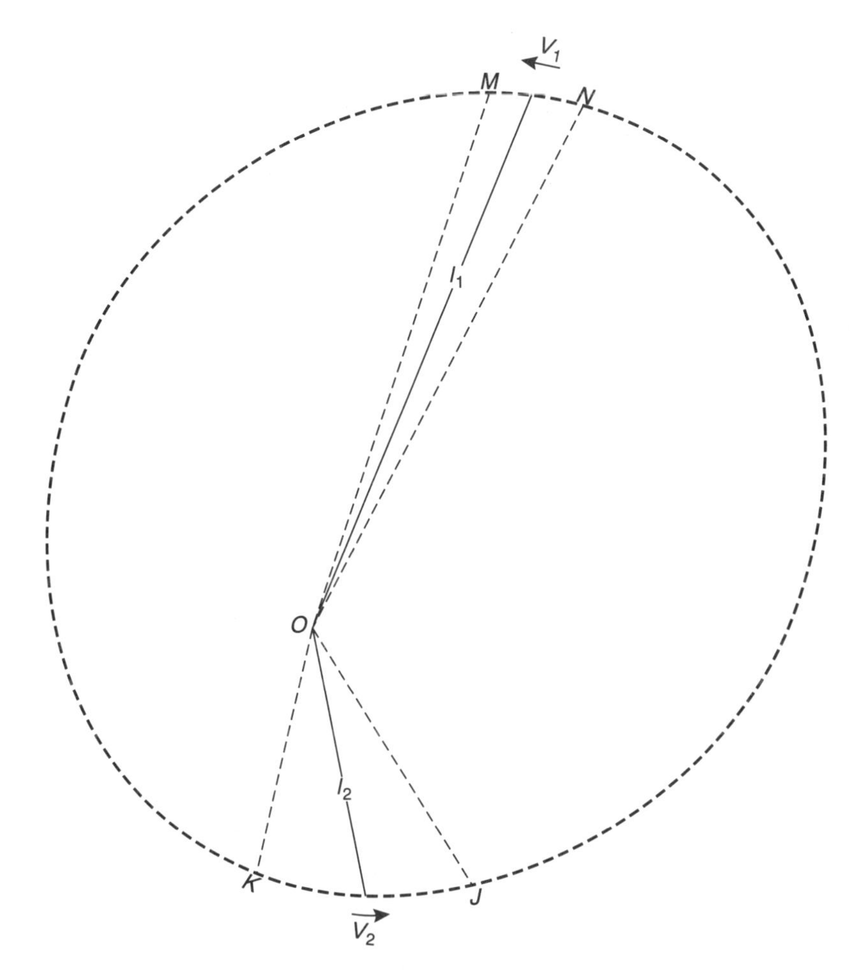

Figure 2.13 *The relationship between the speed of an Earth satellite round an elliptical orbit and its distance from the centre of mass of the Earth*

of perigee. Moreover, some elliptical orbits, having a quite small height at perigee, suffer considerable orbit modification through aerodynamic drag.

The effect of the oblateness of the Earth

The polar diameter of the Earth is 12714 km but the mean equatorial diameter is about 12756 km. This oblateness of the Earth's form, shown greatly exaggerated in Figure 2.15, is the source of the dominant perturbations for inclined elliptical orbits. These perturbations arise because, except when the satellite is exactly in the equatorial plane or in line with the Earth's axis, the force on it due to the Earth's gravity, shown as vector *SG* in the figure, is not directed exactly towards the centre of the Earth.

SG can be resolved into three orthogonal components. The largest component, *SO*, is indeed directed towards the centre of the Earth. Another, *SP*, is in the plane of the orbit and perpendicular to *SO*. The third component, *SV*, is perpendicular to the plane of the orbit.

Component *SP* causes the orbital ellipse to rotate in its own plane without changing the eccentricity of the orbit; this is called precession of the argument

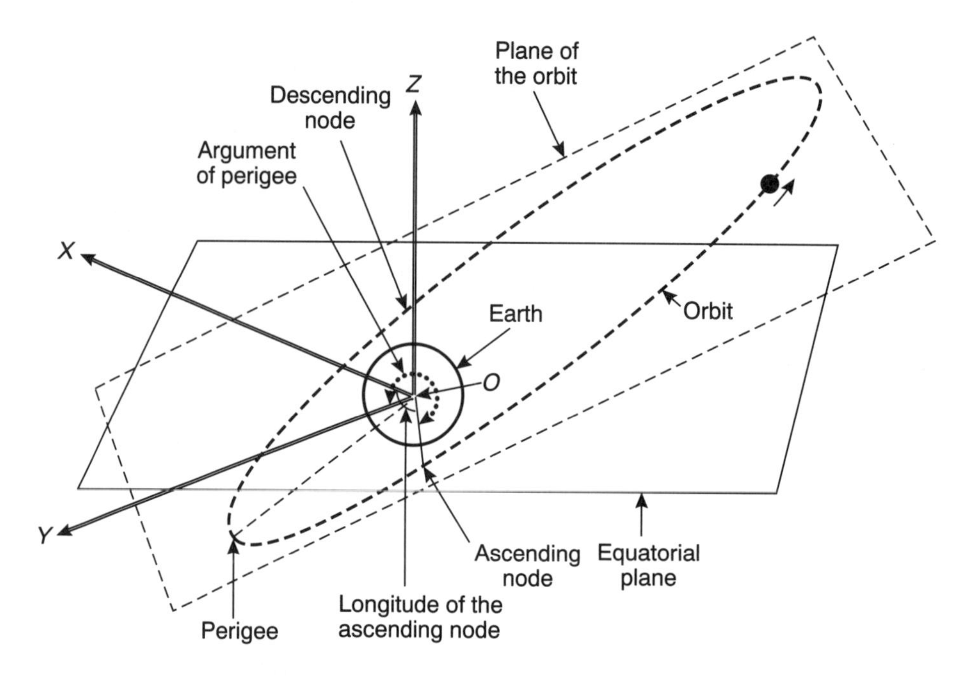

Figure 2.14 *An elliptical Earth orbit in its frame of reference*

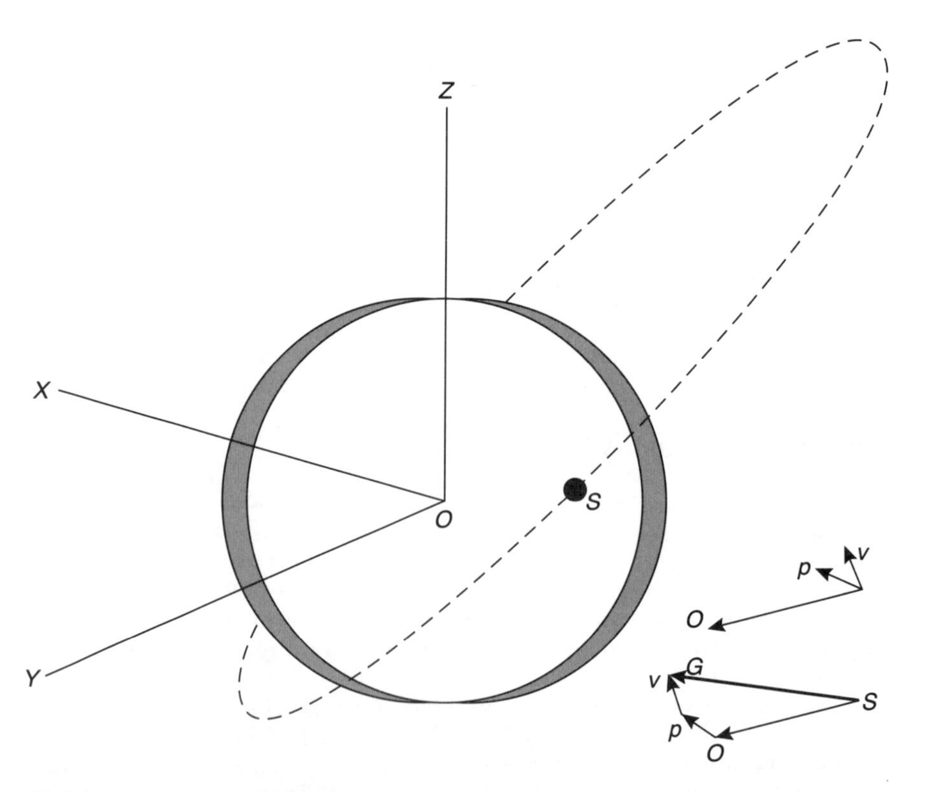

Figure 2.15 *Forces acting on a satellite due to the Earth's oblateness; the representation of the oblateness in this diagram is schematic and very greatly exaggerated*

of perigee. The rate of precession R_{ap} is given by

$$R_{ap} = 11.1T^{-7/2}(5\cos^2 i - 1)(1 - \epsilon^2) \quad \text{(deg/day)} \tag{2.16}$$

where T is the period in hours and i is the inclination of the orbit. If $i = 63.4°$, then $5\cos^2 i = 1$, and $R_{ap} = 0$. If i is greater than 63.4°, precession is in the same direction as the motion of the satellite round its orbit, and it is in the contrary direction if i is less than 63.4°.

Component SV of the forces acting on the satellite may cause the plane of the satellite's orbit to rotate, relative to the stars, about the Earth's axis; this is called precession of the ascending node. For a polar orbit, there is no precession. For a satellite with a direct orbit, the precession of the ascending node is westward and it is eastward for a retrograde orbit. The rate of precession R_{an} is quite fast for orbits with low inclination or a short period of revolution, but it is slow for most of the orbits considered in Section 2.3.3. The effect of this precession on a stable Earth track is to make the satellite rise above an earth station's horizon earlier (for a direct orbit) or later (for an retrograde orbit) each day.

The effect of aerodynamic drag

The effect of aerodynamic drag on a satellite orbit is significant at a height of 1000 km and does not become totally negligible until a height of 3000 km is reached. Drag reduces the energy of a satellite in an elliptical orbit with a low perigee. As a result, the length of the major axis is reduced, so the height at apogee and the orbital period become smaller. The height at perigee is little affected.

2.3.3 The Earth coverage of satellites in elliptical orbits

Satellites in orbits of substantial eccentricity spend most of each orbital period at a high altitude, close to the height of their apogee, from which they can cover a large footprint. In general they are of little use at low altitude, near to perigee. The systems that might find such orbits of value are national or regional in coverage rather than global. Thus it is necessary to choose an orbital period and to control precession of the argument of perigee to stabilize the Earth track, to ensure that the point on the Earth directly beneath the apogee should be consistently located at an appropriate point in the service area.

An orbital inclination of 63.4° enables this to be done. The orbits that are most likely to be useful for satellite communication fall into two groups, which can perhaps be called the high latitude coverage and the short orbital period groups.

High latitude coverage

A point on the surface of the Earth sweeps through right ascension at a constant rate of approximately $360°/24 = 15°$ per hour. A satellite in a direct elliptical orbit with a period of T (hours) sweeps through right ascension in the same direction as the Earth and at an average rate of $360°/T$ per hour, although the rate will be considerably less than the average near apogee and more than the average near perigee. For example, for a 12-hour orbit the average rate would be 30°. However, by careful choice of the parameters T and ϵ, it can be arranged that the satellite will be sweeping through about 15° per hour for a substantial fraction of the orbital period. During this time the satellite's Earth track will keep close to a chosen line of terrestrial longitude.

Furthermore, if it is arranged that the apogee occurs at the most northerly or most southerly limit of the orbit (that is, with an argument of perigee of 270° or 90° respectively), the Earth track will remain close to a line of latitude, numerically equal to the orbital inclination, for much of the time spent near that chosen line of longitude. Thus, an inclined elliptical orbit can provide a pseudo-geostationary satellite at a latitude around 60°, north or south, for some hours per revolution.

The USSR started to use an orbit of this kind in 1965 for its domestic Molniya satellites, because a truly geostationary satellite could not provide simultaneous service to all parts of that vast area at high latitudes. The same orbit is still used by Intersputnik to serve the CIS. The parameters of the orbit are approximately as follows:

$T \approx 12\,\text{hours}$	$i \approx 63.4°$
$\epsilon \approx 0.75$	argument of perigee $= 270°$
$h_a \approx 40\,000\,\text{km}$	$h_p \approx 500\ \text{km}$

The Earth track of the Molniya orbit, centred as an example on longitude 0°, is sketched in Figure 2.16. The satellite passes through apogee twice each day, at about the same location in the celestial frame of reference. However, the Earth rotates 180° during each satellite orbital period, causing these quasi-geostationary nodes to be separated by 180° of terrestrial longitude. At each apogee the satellite is seen from the Earth's surface to be within a few degrees of a central point around latitude 60° N and, for this example, at longitude 0° or 180° for a period of about eight hours. The geographical area that could be served from one of these pseudo-geostationary nodes, assuming an angle of elevation at earth stations not less than 5°, is also sketched in Figure 2.16. This area is very big, comparable with the coverage from a geostationary satellite but centred at a high latitude and extending right over one of the poles. For continuous service to these coverage areas, it is necessary to deploy three satellites, with appropriate timing and with ascending nodes at intervals of 120° in right ascension.

Smaller areas at high latitudes could be served with a high minimum elevation angle, an option which is not available using satellites in the equatorial plane. For this reason Molniya orbits have been considered recently for use with satellite broadcasting and land mobile systems for Northern Europe [2]. Other elliptical orbits with 63.4° inclination that have been studied have periods of about 24 hours [3, 4] such as the so-called Tundra and Sycomores orbits. The following parameters are typical of those considered:

Tundra	$h_a = 46\,000\,\text{km},$	$h_p = 25\,000\,\text{km},$	$\epsilon = 0.25$
Sycomores	$h_a = 50\,000\,\text{km},$	$h_p = 21\,000\,\text{km},$	$\epsilon = 0.35$

With arguments of perigee of 270° or 90°, these orbits provide a single quasi-geostationary satellite location lasting at least 12 hours per day around 60° north or south respectively. The angular movement seen from the Earth's surface around apogee is, however, rather larger than that for the Molniya orbit. Two satellites are needed to maintain continuous coverage of a single service area.

Short orbital period

Satellites in circular orbits with a height above the Earth of 8000 km have an orbital period of 4.7 hours; 12 satellites in phased orbits might be needed to provide continuous coverage of a service area that is continental in extent. A satellite with an elliptical orbit having a period of two hours might also have a height above the Earth's surface at apogee of 8000 km, depending on the eccentricity of its orbit. Provided that the orbital inclination was 63.4°, perhaps four satellites would be sufficient to provide continuous coverage of the service area.

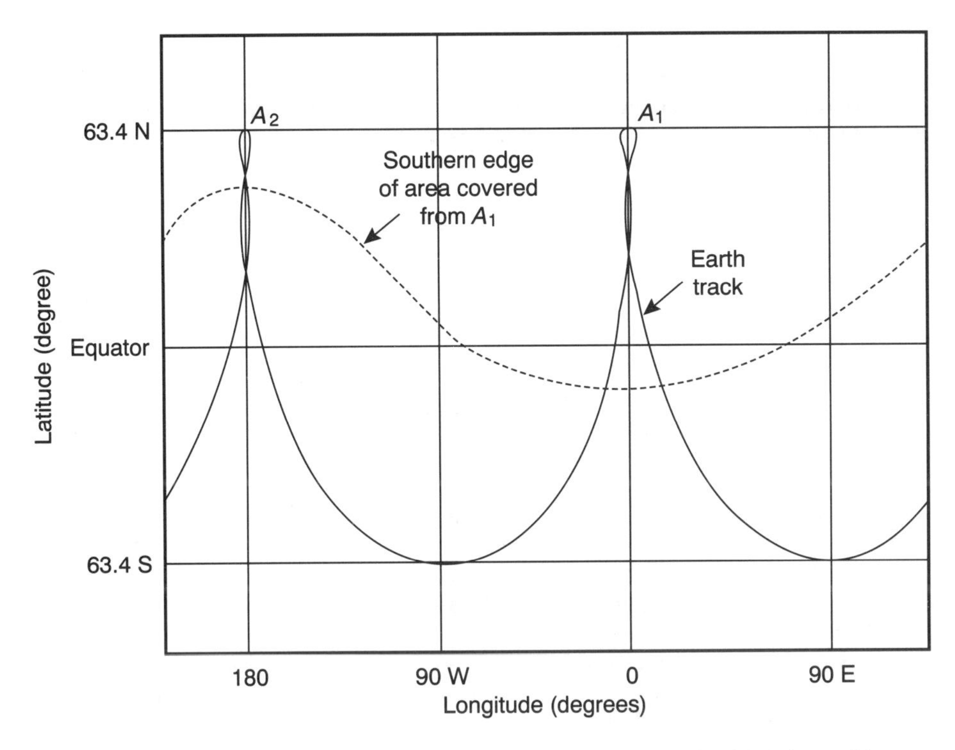

Figure 2.16 *A sketch of the Earth track of a satellite in the Molniya orbit, one apogee being assumed to be at longitude 0°. Also shown is the Southern edge of the coverage area from apogee node A_1 for a minimum elevation angle of 5°*

Now, a satellite in a 2-hour orbit must sweep through right ascension at a mean rate of 180° per hour. The rate of sweeping will be less near apogee, depending on the eccentricity, but it will not be nearly as small as 15° per hour, whatever value ϵ may have. Thus there is no pseudo-geostationary phase for an elliptical orbit with a short orbital period. Nevertheless, if there is a requirement to serve an area of continental extent using satellites at a height intermediate between those of the GSO and LEOs, such orbits might meet the need with the smallest number of satellites.

2.3.4 Advantages and disadvantages of inclined elliptical orbits

As indicated above, there are very large service areas for fixed links at high latitudes, such as the former USSR, for which the Molniya orbit is very suitable. Coverage of polar areas can be excellent. In addition, any of the inclined elliptical orbits with long orbital periods could be used to provide service to smaller areas at high latitudes at a high angle of elevation, enabling links to be provided for satellite broadcasting or to mobile stations on land, with minimal obstruction from buildings and foliage. Since the satellite remains close to the zenith throughout the active arc of each orbital period, it is feasible to use an antenna of limited gain at the road vehicle earth station without satellite tracking. Elliptical orbits compete for such applications with the GSO, providing better coverage but at the cost of additional satellites.

In addition, satellites in inclined elliptical orbits with short orbital periods compete with those in low- or medium-altitude circular orbits for applications for which the high transmission loss of the GSO is unacceptable, providing equivalent coverage with fewer satellites.

Nevertheless, the application of elliptical orbits seems likely to be limited to certain niche markets, for the following reasons:

(1) All of these elliptical orbits except those with 24-hour orbits, like Tundra and Sycomores, have a very low perigee. They are exposed to intense ionizing radiation twice in each orbital period when passing through the Van Allan belts. Solar cells and electronic components have to be protected from this radiation, increasing spacecraft cost and mass. In addition, atmospheric drag near perigee is significant, shortening lifetime in orbit.
(2) At the near-apogee operational phase of the 12-hour and 24-hour elliptical orbits, the distance between earth stations and the satellite is as great or greater than that for a geostationary satellite. Consequently the high space–Earth propagation loss and transmission delay pose just the same system problems.
(3) Doppler effects are severe, causing major problems for the transmission of signals with wide basebands, especially if the signals are digital and are interfaced directly with terrestrial networks; see Section 5.3.2.
(4) Where a geostationary satellite is a feasible alternative, the need to provide and launch two or three satellites into elliptical orbits to obtain continuous coverage adds to costs. However, this extra cost may be mitigated to some degree by lower per-satellite launching cost and it may be possible to use the same constellation of satellites to serve quite different areas during successive arcs of their orbits.
(5) In addition, where continuously available channels are to be provided, the difficult and costly problem of maintaining continuity during handover from one satellite to another several times a day has to be solved. The handover problem is much less difficult to solve for services to mobile stations, which usually occupy channels intermittently on demand.
(6) Techniques have not been developed yet for coordinating efficient simultaneous use of frequency bands by large numbers of different satellite systems using these orbits. Indeed, it may be doubted whether such systems will ever approach the relatively efficient use which geostationary satellite networks make of space and spectrum.

2.4 Medium-altitude and low circular orbits

Medium-altitude orbits

Geostationary satellites have great advantages for communications applications where polar coverage is not required. Satellites in 63.4°-inclined high-apogee elliptical orbits have some of the advantages of geostationary satellites and, in addition, they provide polar coverage. However, both of these types of orbit exhibit high space–Earth transmission loss and long transmission times.

In the early days of satellite communication, it was feared that one-way transmission times exceeding 250 ms might be an unacceptable impediment to telephone conversation. Moreover, the data communication systems then available, designed for terrestrial transmission media, could not operate with such long transmission times. As an alternative to geostationary satellites, consideration was given at that time to the use of satellites in a medium-altitude Earth orbit (MEO), circular and with orbital periods in the range 8–12 hours. The space–Earth transmission loss for these is perhaps 10 dB less than for

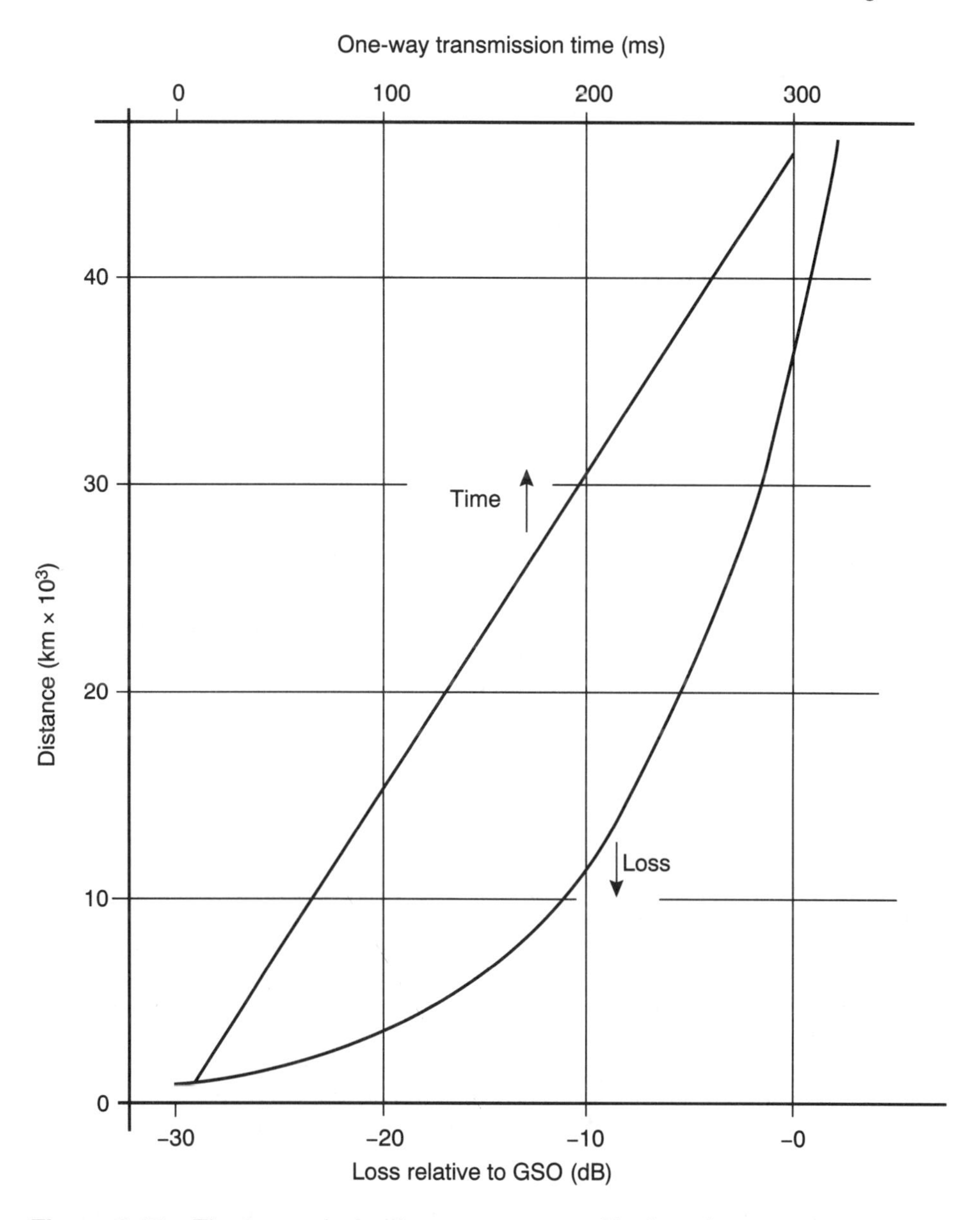

Figure 2.17 *The transmission loss on a space–Earth path, assuming free-space propagation conditions, as a function of distance between the earth station and the satellite, relative to the minimum loss for a geostationary satellite. The one-way transmission time from earth station to earth station is also shown*

geostationary satellites, and the transmission time is reduced to 100–150 ms; see Figures 2.4 and 2.17.

Experience has shown that, given good control of echo, a telephone connection which includes one hop in each direction via a geostationary satellite is acceptable to the public; see Section 5.3.1. Data transmission terminal units with ample memory for long transmission times soon became available, and lack of polar coverage has not raised significant problems in providing for the PSTN. Geostationary satellites seem likely to continue to dominate satellite communication with high-capacity links between fixed points. However, there has recently been a revival of interest in using medium-altitude orbits for serving mobile Earth stations, because, compared with the GSO, the transmission loss is lower.

It is fortunate that the GSO has been found acceptable for trunk telecommunications, because the use of lower orbits such as MEOs for this purpose would involve major additional problems and costs. These disadvantages are similar to several listed in Section 2.3.4 for elliptical orbits. Doppler effects would not be as severe as with high elliptical orbits but much worse than with geostationary orbits. A constellation of 8 to 12 satellites would probably have been needed to cover a single service area, with handing over from satellite to satellite every one or two hours. Efficient coordination of the use of space and spectrum seems likely to have been impossible. Added to this, the service areas that could have been covered by a constellation of medium-altitude satellites would be considerably smaller than those of geostationary satellites [5], leading to a considerably less satisfactory global telephone network.

Low orbits

A low Earth orbit (LEO) would provide a further reduction in space–Earth transmission loss relative to the GSO, and some low orbits provide polar coverage. Both features would be of great value for certain mobile communications applications. Figure 2.17 shows that the space–Earth transmission loss is up to 30 dB less for a satellite 1000 km above the ground (period $\approx$ 1.74 hours) when compared with a geostationary satellite.

The altitude of an LEO should be great enough to avoid substantial satellite deceleration due to atmospheric friction, but low enough to avoid the more intense levels of proton bombardment in the inner Van Allen belt. Altitudes between 780 and 1400 km are favoured, corresponding with orbital periods between 100 and 113 minutes.

Figure 2.18 shows the earth track of ten successive orbits of a satellite in a circular polar orbit with an altitude of 1000 km. Earth stations in shaded area A (4200 km in diameter) would see the satellite, in the absence of environmental screening, at an angle of elevation not less than 10°, while the satellite was passing through the equatorial plane. The coverage area has the same size and shape wherever the satellite is in the orbit, but its apparent size and shape would change with latitude, being distorted by the map projection used in the figure. Thus, at the South Pole, the coverage area at a single pass of the satellite is shown by shaded area B.

Figure 2.18 shows that a single LEO satellite in a polar orbit will have brief sightings of every part of the Earth's surface every day. There will be two or three of these glimpses per day near the equator, the number increasing as the poles are approached. The periods of visibility as seen from the earth station range from about ten minutes, the satellite passing overhead, down to a few seconds when the satellite appears briefly above the horizon. If the orbital plane of the satellite is given an angle of inclination differing from the 90° of the polar orbit, a similar pattern of Earth tracks is obtained but the geographical distribution of satellite visibility changes. Thus, one satellite with an orbital inclination of 50° would have better visibility between 60° N and 60° S latitude than a satellite in a polar orbit, but it would have no visibility at all of the polar regions.

Satellite visibility for earth stations could be improved by using more satellites. If more satellites, upwards of about 48, were included in the constellation

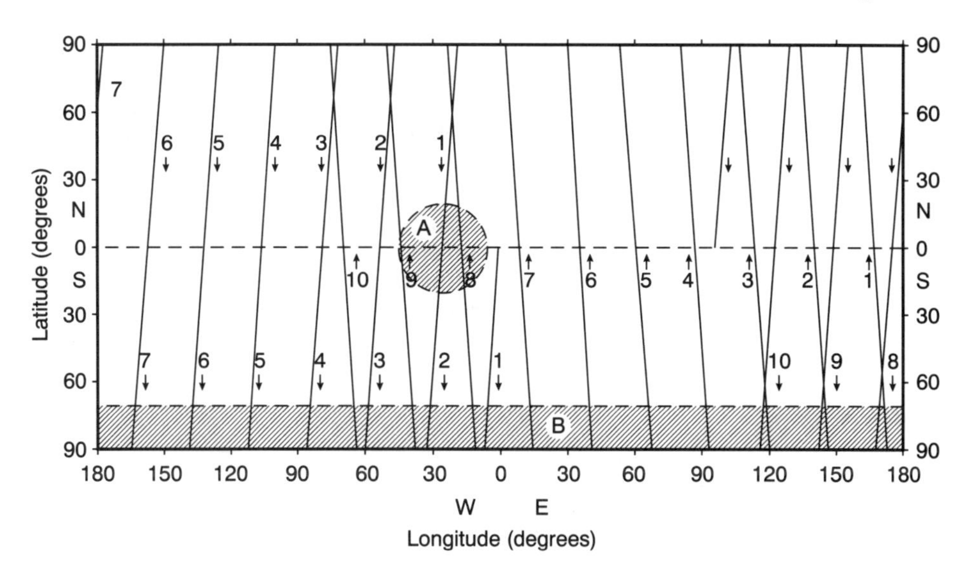

Figure 2.18 *The Earth track of ten revolutions of a satellite in a circular polar orbit having a height of 1000 km. Orbits are numbered from a descending node at 0° longitude*

in a carefully optimized pattern of orbits, each location on Earth could have continuous visibility of one or other of the satellites.

Orbital perturbations for MEOs and LEOs

Satellites in medium-altitude or low circular orbits are, of course, subject to orbital perturbations. For very low orbits, the aerodynamic drag is likely to be significant. However, some of the other perturbations, such as precession of the argument of perigee, resolve to zero if the orbit is circular or polar. In general, a perturbation is unlikely to have a serious effect on the operation of a multi-satellite constellation since it will usually affect all the satellites of the constellation in equal measure.

2.5 Eclipses of satellites

Communication satellites derive their electrical power from solar radiation. From time to time the sunlight illuminating a satellite is eclipsed by the Earth or the Moon. When this happens, housekeeping systems essential to satellite operation are sustained by means of batteries, which are recharged while the satellite is illuminated. Batteries usually provide power for the maintenance of normal operation during eclipses also.

In addition to the loss of the primary power supply, eclipses cause a profound thermal shock to the satellite and Earth sensors used in the control of satellite attitude may give false readings while the Sun is eclipsed, either of which may lead to incorrect operation of the satellite.

Eclipses of a geostationary satellite by the Earth occur at the vernal and autumnal equinoxes when the local time at the subsatellite point is about midnight; see also Figure 4.15. The duration of the eclipse varies from day to day, being brief on the 21st day before the equinox, growing longer every day until an eclipse duration of 69 minutes is reached at the equinox itself and then declining, reaching zero about the 21st day after the equinox. Most eclipses are total (that is, the Earth obstructs the whole of the Sun's disc as seen from the satellite) for most of their duration. Satellites in other orbits are eclipsed frequently, in many cases once per orbital period.

Eclipse of a satellite by the Moon is less regular and less frequent than eclipse by the Earth. On average there are two lunar eclipses of each geostationary satellite per year with a typical duration of 40 minutes. For other orbits the eclipses are briefer but there are more of them. Some lunar eclipses are partial, that is, the Sun's disc is not totally obstructed by the Moon at any time in the course of an eclipse.

References

1. *Handbook on Satellite Communications: Fixed-Satellite Service* (1985). Geneva: ITU.
2. Norbury, J. R., Smith, H., Renduchintala, V. S. M. and Gardiner, J. G. (1988). Land mobile satellite service provision from the Molniya orbit – channel characteristics. *Proc. Fourth International Conference on Satellite Systems for Mobile Communications and Navigation.* London: IEE.
3. Aghvami, H., Clarke, A., Evans, B. G., Farrell, *et al.* (1988). Land mobile satellites using the highly elliptical orbits – the UK T-sat mobile payload. *Ibid.*
4. Rouffet, D., Dulck, J. F., Larregola, R. and Mariet, G. (1988). Sycomores: a new concept for land mobile satellite communications. *Ibid.*
5. Dalgleish, D. I. and Jefferis, A. K. (1965). Some orbits for communication satellite systems affording multiple access. *Proc. IEE,* **112**, 21.

3 Launchers and launching

3.1 Introduction

A satellite may be launched into orbit by either a multi-stage expendable launch vehicle or a manned or unmanned reusable launcher. Additional rocket motors (perigee and apogee kick motors) may also be required. The process of launching a satellite as described in this chapter is based mainly on launching into equatorial circular orbits, and in particular the GSO, but broadly similar processes are used for other orbits. There are two techniques for launching a satellite into an orbit of the desired altitude, namely by direct ascent or by a Hohmann transfer ellipse.

In the direct ascent method, the thrust of the launch vehicle is used to place the satellite in a trajectory, the turning point of which is marginally above the altitude of the desired orbit. The initial phase of the ascent trajectory, the boost phase, is powered by the various stages of the launch vehicle. This is followed by a coasting phase along the ballistic trajectory, the spacecraft at this point consisting of the last launcher stage and the satellite. As the velocity required to sustain an orbit will not have been attained at this point, the spacecraft falls back from the highest point of the ballistic trajectory. When the satellite and final stage have fallen to the desired injection altitude, having in the meantime converted some of their potential energy into kinetic energy, the final stage of the launcher, called the apogee kick motor (AKM), is activated to provide the necessary velocity increase for injection into the desired circular orbit. See Figure 3.1.

The AKM is often incorporated into the satellite itself, where other thrusters are also installed for adjusting the orbit or the satellite's attitude throughout its operating lifetime in space. There is a description of the physical basis of the functioning of these various kinds of motor in Section 4.6.

The Hohmann transfer ellipse method enables a satellite to be placed in an orbit at the desired altitude using the trajectory that requires the least energy. First the launch vehicle propels the satellite into a low parking orbit by the direct ascent method. Then, the satellite is injected into an elliptical transfer orbit, the apogee of which is at the altitude of the desired circular orbit. Finally, at the apogee, additional thrust is applied by an AKM to provide the velocity increment necessary for the attainment of the required circular orbit. See Figure 3.2. This process is reviewed in more detail in Section 3.2.3.

In practice it is usual for the direct ascent method to be used to inject a satellite into a LEO and for the Hohmann transfer ellipse method to be used for higher orbits.

3.2 Expendable launch vehicles

3.2.1 Description and capabilities

Launch vehicles and their nose fairings impose mass and dimensional constraints on the satellites that can be launched. However, a number of different types of

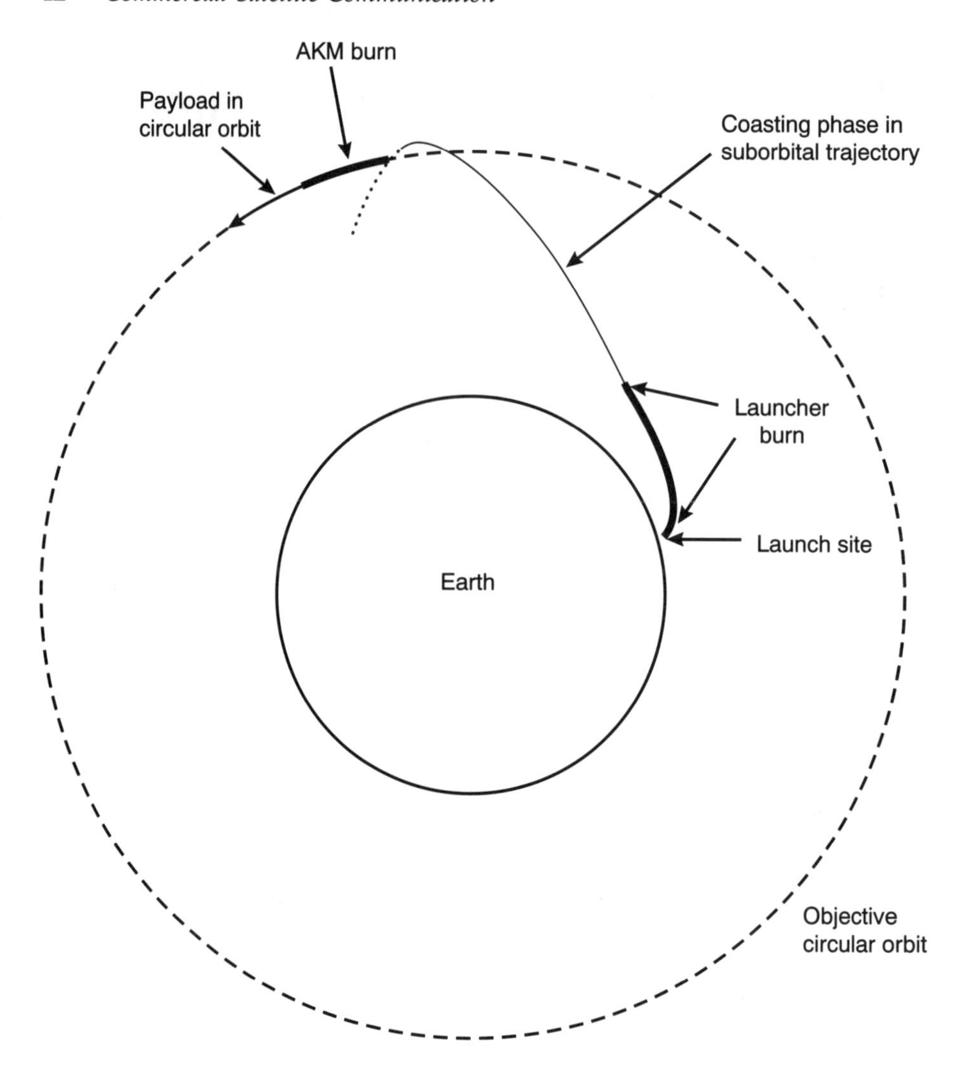

Figure 3.1 *Launching by the direct ascent method*

launcher are available for commercial use and the satellite designer ensures that the satellite will meet the constraints and capabilities of one of them, or preferably more than one.

A brief description of the major expendable vehicles currently used for launching commercial satellites follows in this section. It should be noted that a few of them have the capability of placing satellites directly into a high circular orbit; with the others, use is made of a Hohmann transfer elliptical orbit. When the objective is the GSO, the transfer orbit is called a geosynchronous or geostationary transfer orbit (GTO). All of these vehicles consist of several stages, mostly fuelled by bipropellant liquids, and some of them are assisted by solid rocket boosters strapped on to the first stage.

The dimensional constraint on the launcher payload, consisting of one or more satellites, is determined by the size and shape of the nose fairing which protects the payload while the launcher is within the atmosphere. Several different fairings are available for most launchers, accommodating satellites of different sizes and shapes after they have been prepared for launching by folding back such

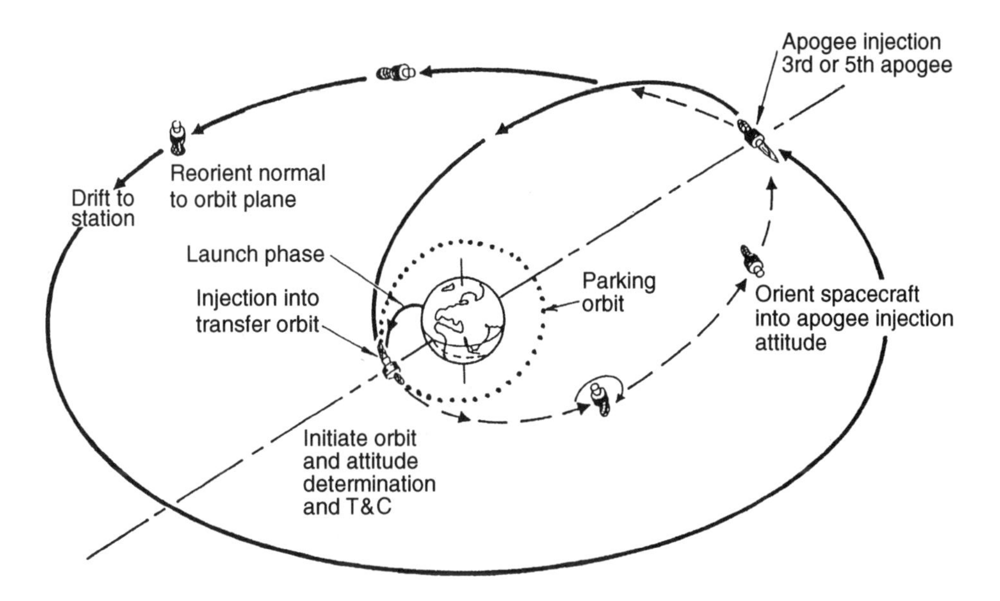

Figure 3.2 *Launching using the Hohmann transfer ellipse*

structures as solar arrays and large antennas. Examples of the dimensions of fairings that are available are given in Section 3.2.2.

Delta (USA)

The Delta series of launch vehicles has been extensively used and continuously upgraded for both civilian and military payloads since 1960. They are amongst the most reliable of intermediate capacity launchers. Delta II is the latest and most advanced version of the series to be in service, and the three-stage Delta II 7925 configuration is the most powerful. Delta II consists of five major elements: the first stage with nine strap-on solid rocket boosters, the interstage, the second stage, the third stage and the payload fairing; see Figure 3.3.

The first stage contains one RS-27 main engine and two LR 101-NA-11 vernier engines. The RS-27 is a single-start liquid bipropellant engine providing 921 kN thrust at sea-level. The solid fuel boosters strapped on to the first stage provide additional thrust during the boost phase. The interstage mates the first and second stages, jettisoning the first stage when it burns out. The second stage contains a restartable AJ10-118K engine which burns nitrogen tetroxide and Aerozine-50 (A-50) hydrazine propellant. Attitude and roll control during coasting and powered flight respectively is provided by a nitrogen cold gas system. Pitch and yaw control is achieved by hydraulically operated gimbals. The second stage houses the control and guidance equipment for the first two stages.

The third stage, containing a Star-48B solid rocket motor, is carried on a spin table connected to the top of the second stage. When the second stage has completed its burn, the third stage, not being steerable, is spun up on the spin table to provide gyroscopic directional stability during the rest of the launching process. When the spin-up is completed, the spin table releases the spinning third-stage/spacecraft combination, which moves away in the direction determined by the second-stage guidance and control system, and the third-stage burn begins.

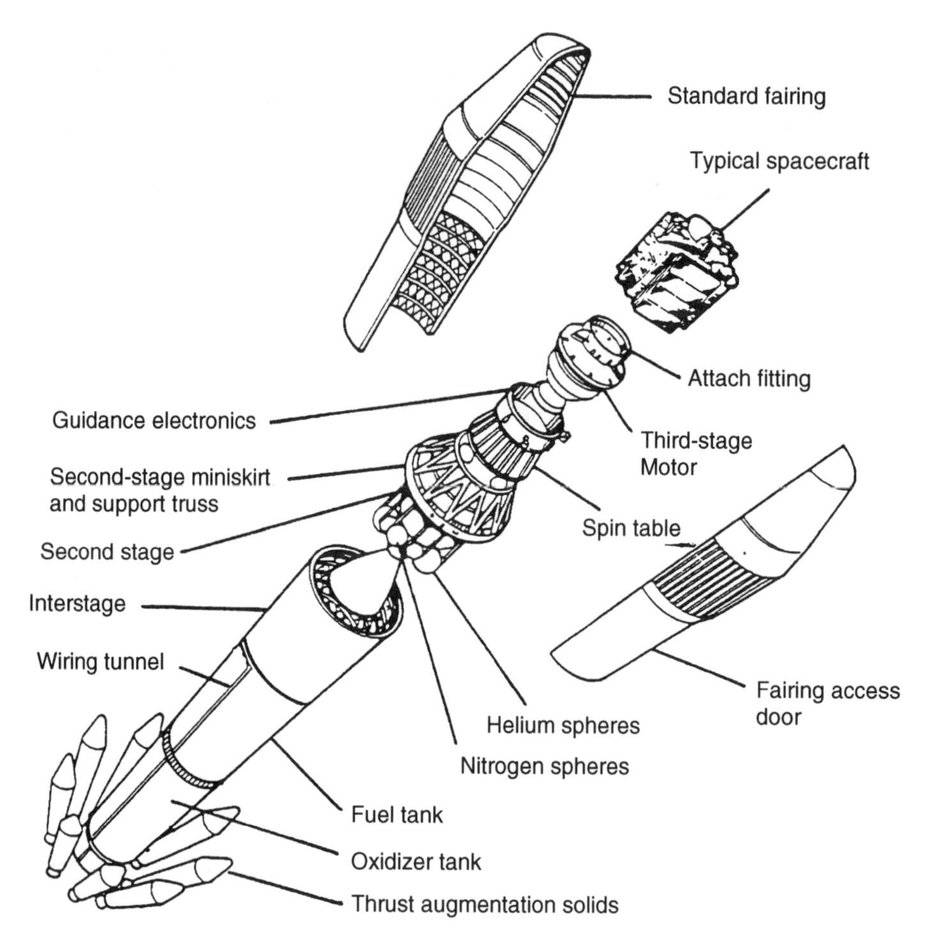

Figure 3.3 *A simplified view of the Delta II launcher illustrating its major components and subsystems. (Reproduced by permission of McDonnell Douglas Ltd.)*

The satellite, with its AKM, is attached to the top of the third-stage rocket. During flight in the atmosphere, they are protected by a fairing which is jettisoned at a minimum altitude of 125 km. When the third-stage burn is completed, the satellite and AKM, now in the Hohmann transfer orbit, are released to complete the final stage of the launch.

Delta II is 38.2 m high and has a mass of 230 tonnes at lift-off. Launched from Cape Canaveral, the Delta II 7925 can place 1819 kg into a GTO inclined at 28.7°, with heights at perigee and apogee of 185 km and 35 786 km respectively. For more details of Delta II, see [1].

A new generation of launchers, Delta III, is under development. It consists of two stages, both bipropellant systems using cryogenic liquid fuel, the first stage being assisted by nine strap-on solid fuel motors.

Atlas (USA)

The world's first communications satellite (Score) was launched by an Atlas vehicle in December 1958. The present Atlas family of launchers consists of four variants, namely Atlas I, II, IIA and IIAS. The Atlas is a reliable and versatile launch vehicle capable of delivering payloads to a wide variety of low and high circular orbits. The Atlas launcher, shown in Figure 3.4, consists of the Atlas booster, the Centaur upper stage and the payload fairing.

The booster consists of two stages, the lower part dropping away after burnout while the upper part, called the 'sustainer', continues to provide thrust. It is constructed of stainless steel and powered by Rocketdyne MA-5 engines for Atlas I or MA-5A engines for the Atlas II and its variants, utilizing RP-1 and liquid oxygen as propellants. The Atlas IIAS variant has its payload capability substantially enhanced by four strap-on Thiokol Castor IVA solid rocket boosters, fired two at a time in the first few minutes of a launch. For more details, see [2].

The Centaur upper stage is also constructed of stainless steel. It burns liquid hydrogen and liquid oxygen propellants. A stub adapter attached to the forward part of the Centaur tank supports the equipment module, spacecraft adapter and payload fairing. The Centaur avionics system provides guidance, flight control and sequencing functions for the Atlas booster. For more details on the Centaur upper stage, see [3].

The satellite to be launched is attached to the launcher with a spacecraft adapter and it is protected by a fibreglass fairing when at low altitude. The payloads deliverable into a GTO, inclined at 27°, with the Centaur upper-stage range from 2222 up to 3697 kg.

Titan III (USA)

The commercial Titan or Titan III is the 15th version of the Titan family of launch vehicles, used mainly for defence applications and manufactured by Martin Marietta (USA), which recently merged with Lockheed to form Lockheed Martin.

The Titan 1, first flown in 1959, used the same fuel and oxidizer as Atlas, namely RP-1 and liquid oxygen. In 1962 the Titan II version was developed. It employed Aerozine-50 unsymmetrical dimethylhydrazine (UDMH) as fuel and nitrogen tetroxide as an oxidizer (Aerozine-50 and nitrogen tetroxide ignite on contact and can be stored at normal temperatures; they are said to be 'storable fuels'). In 1964 a version designated Titan III-A was produced carrying a restartable upper stage (called the Transtage because of its limited size) for the delivery of defence payloads into the GSO. Another variant designated Titan III-B followed which featured longer fuel tanks and the restartable Agena upper stage.

After a long chain of further developments, mostly addressing defence requirements, the current Titan III commercial variants have emerged. The Titan-IIIT variant (see Figure 3.5), 30 m high and with an initial mass of 603 tonnes, can place 4767 kg into a GTO.

Proton (Russia)

The Proton launcher is one of the most capable and reliable heavy lift launch vehicles in operation today. It has an outstanding reliability record averaging about 96 per cent success rate over its latest 50 launches. It has flown more than 200 missions, including the launching of the Ekran, Raduga and Gorizont families of geostationary communications satellites. Proton launchers and launch services are offered by International Launch Services, a joint venture between Lockheed Martin of USA and Khrunichev Enterprise and RSC Energia of Russia. Khrunichev produces the three lower stages of the Proton, while RSC Energia produces the fourth stage.

Proton D-1 and D-1-E launcher variants have three and four stages respectively. Figure 3.6 shows both variants, together with the Proton D and KM. The first

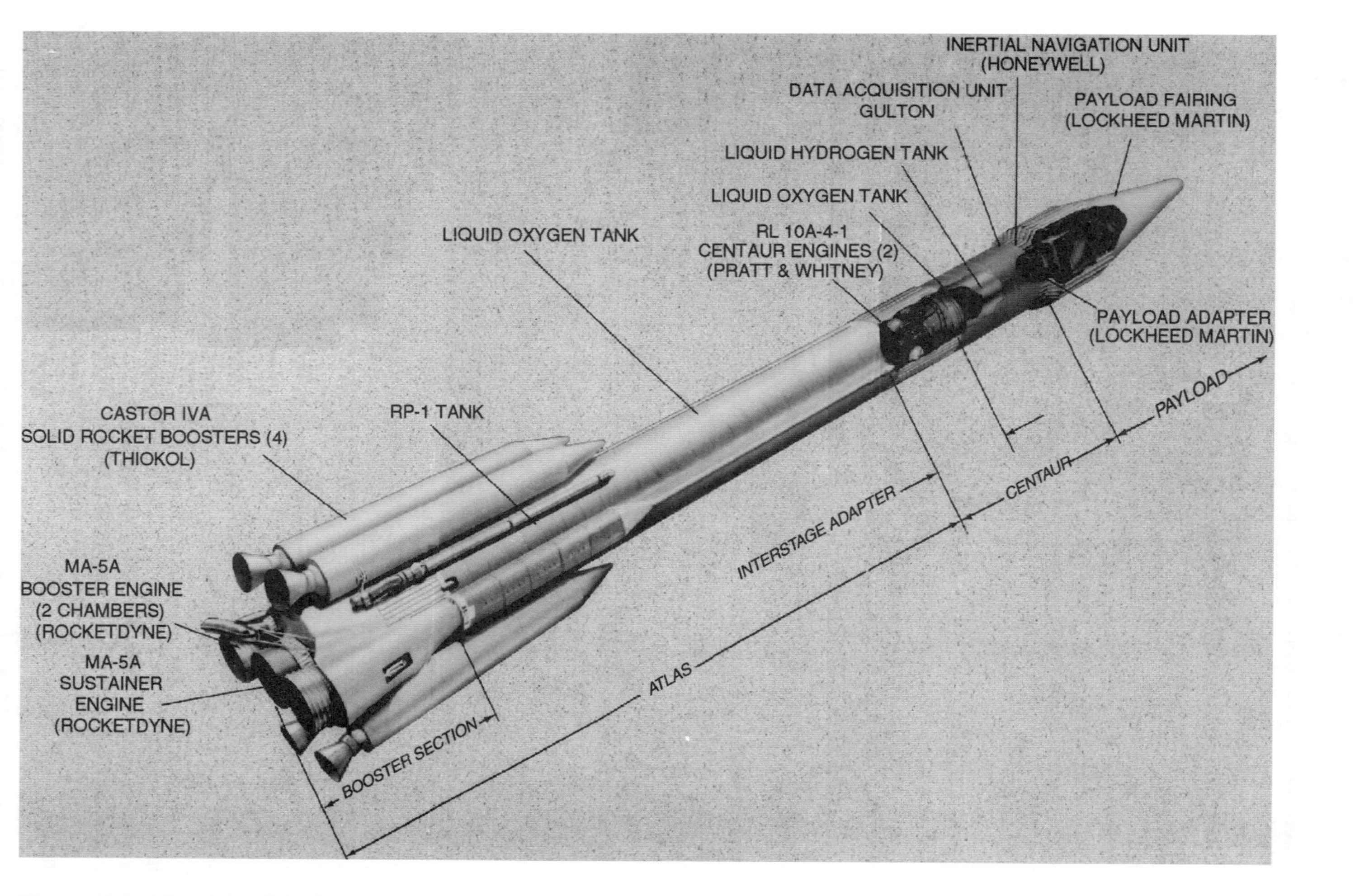

Figure 3.4 *The Atlas IIAS launcher with Centaur upper stage. (Reproduced by permission of ILS International Launch Services.)*

Figure 3.5 *A cutaway view of the commercial Titan launcher. (Reproduced by permission of Lockheed Martin Astronautics.)*

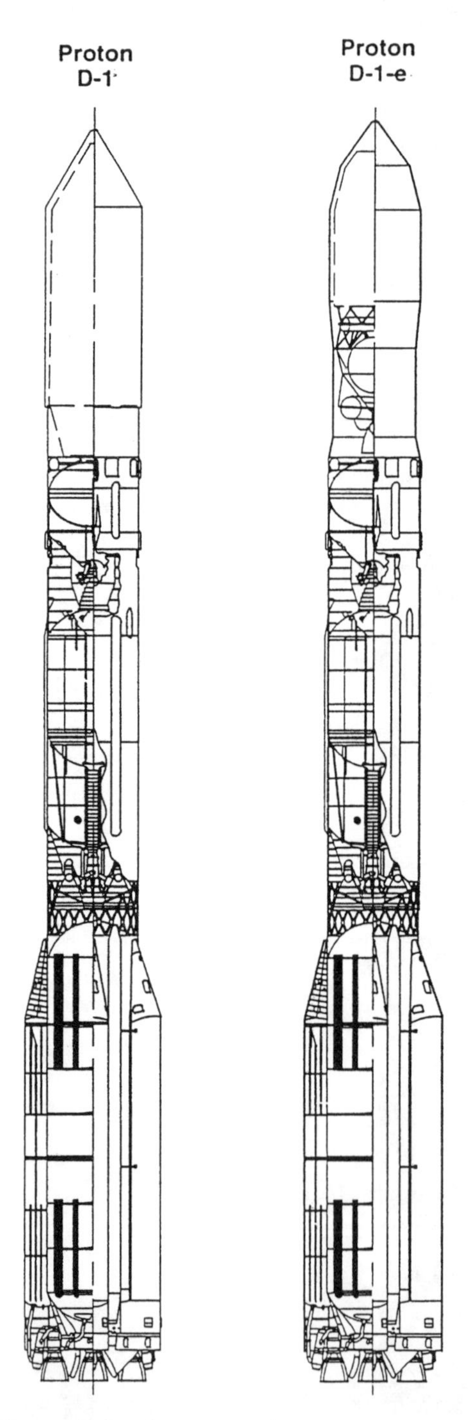

Figure 3.6 *A comparative cut-away view of the Proton D-1 and D-1-E launchers. (Reproduced by permission of ILS International Launch Services.)*

three stages are identical for both versions and use nitrogen tetroxide and UDMH as propellants. About 90 per cent of all Proton missions have used the four-stage variant. The fourth stage, code-named Block DM, is fuelled with liquid oxygen and synthetic kerosene. Two auxiliary control systems provide attitude control; each contains five engines, two with 49 N of vacuum thrust for pitch and roll control, one with 98 N thrust for yaw control and two with 25 N of thrust for axial loading to achieve propellant settling of the main engine.

For both of these variants the overall height is about 62 m. At lift-off the total weight of Proton is about 688 tonnes. The Proton vehicle has the capability of placing a maximum of 4500 and 2600 kg into a GTO and the GSO respectively. For more details on the Proton, see [4].

Zenit (Ukraine)

The Zenit launcher is available in two versions, the two-stage Zenit 2 and the three-stage Zenit 3. A general view of the two Zenit versions, together with a layout diagram of Zenit 2, are shown in Figure 3.7. The first stage of the Zenit is based on the engine used by the strap-on boosters of the Energia launcher; see Section 3.3.3. The third stage consists of the restartable fourth stage of Proton.

The first stage is powered by one four-chamber engine with common turbopumps. The second stage is powered by two engines, a sustainer single-chamber engine and a steering four-chamber engine with swivelling combustion chambers. The third stage is powered by one engine mounted on a gimbal. All stages

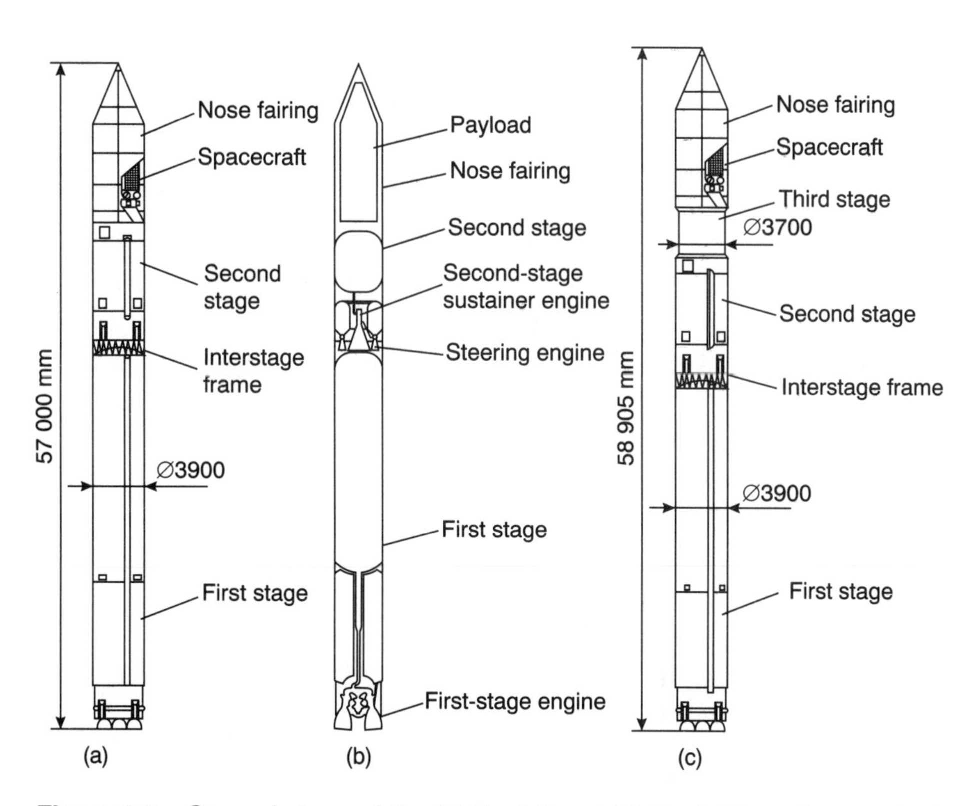

Figure 3.7 *General views of the (a) Zenit 2 and (c) Zenit 3 launchers and, at (b), a cutaway view of Zenit 2. (Reproduced by permission of RSC Energia, Kaliningrad, Russia.)*

use kerosene and liquid oxygen propellants. The principal characteristics of these launchers are shown in Table 3.1.

The Zenit 2 can be used to place spacecraft weighing up to 15.7 tonnes into a 200 km orbit. The Zenit 3 can place spacecraft of up to 2 tonnes directly into the GSO or 4.5 tonnes into the GTO. A fairing fitted on the upper end of the second stage of Zenit 2 and the third stage of Zenit 3 provide 20 and 67 cubic metres of space respectively for payload. The dimensions of the fairings are shown in Figure 3.8.

A version of the Zenit 3 launcher, designated Zenit 3SL is part of the Sea Launch System described in Section 3.4.2. The Zenit 3SL can place a payload of 5.25 tonnes into the GTO in a standard mission profile, launched from the equator.

Ariane (Europe)

The Ariane family of launch vehicles has been developed by the European Space Agency (ESA). The Ariane launchers are commercially marketed, being produced

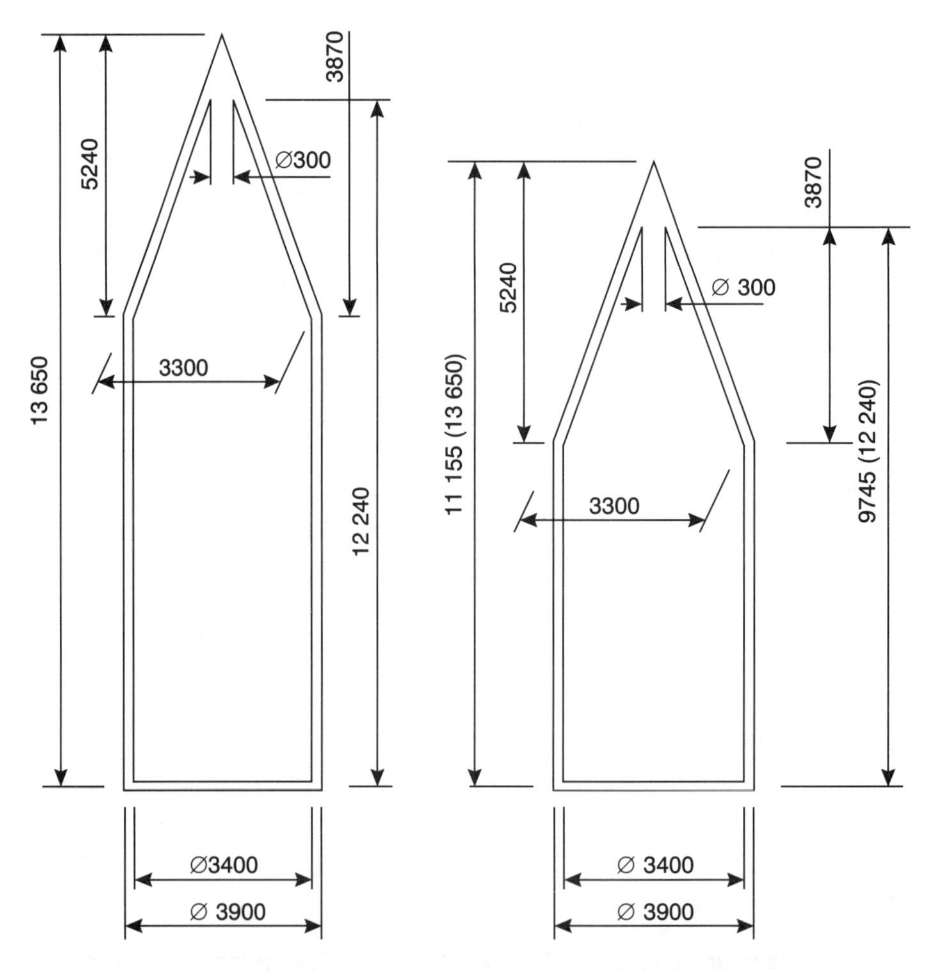

Figure 3.8 *The dimensions of the Zenit 2 and Zenit 3 payload fairings. (Reproduced by permission of RSC Energia, Kaliningrad, Russia.)*

Table 3.1 *Principal characteristics of the Zenit launch vehicles*

	Zenit 2	*Zenit 3*
Mass at launch (tonnes),		
including spacecraft of 13.8 tonnes:	459	
including spacecraft of 2 tonnes:		466
Maximum spacecraft mass (tonnes):	15.7	6
Overall length (metres):	57	58.9 (61.4)
Overall diameter (metres),		
first and second stages and fairing:	3.9	3.9
third stage:	–	3.7
Thrust of engines (kN),		
first stage, on Earth/in vacuum:	7260/7900	7260/7900
second stage, sustainer/steering:	834/78	834/78
third stage:		8.3
Maximum duration of burn (seconds),		
first stage:	140–150	140–150
second stage, sustainer:	200–315	300
second stage, steering:	300–1100	375
third stage:	–	660

and launched by the Arianespace Company which is based in France. Arianespace is the first commercial space transportation company in the world.

The Ariane family of launchers consists of four evolutionary models designated 1, 2, 3 and 4. Ariane 2 and Ariane 3 are modified variants of Ariane 1, while Ariane 4 incorporates a number of major technological improvements, including an upgraded propulsion system. These launchers consist of three stages and a vehicle equipment bay; see Figure 3.9. The fairing which accommodates one or more satellites is mounted on the equipment bay. Each stage includes propellant tanks containing fuel and oxidizer, one or more engines and various electromechanical equipment. The stages are interconnected by skirts which incorporate pyrotechnic devices for stage separation.

The Ariane 4 launcher is offered in six versions differing by the number and type of strap-on boosters and the size of the fairings. It provides for single- or dual-launch configurations for the injection into the GTO of payloads ranging from 1900 to 4700 kg. The Ariane dual-launch external carrier structure (Spelda) is used for putting two entirely independent payloads into orbit. For more details on the Ariane launchers, see [5, 6]. An Ariane 4 launcher on the launch pad at the Guiana Space Centre, about to launch Intelsat VI flight spacecraft 601, is shown in Figure 3.10.

From 1996 onwards Ariane 5 will gradually replace Ariane 4. It is an entirely new design. The two-stage lower part consists of a cryogenic core rocket and two large strap-on solid propellant boosters. The core rocket has a Vulcain main engine, fuelled with 131 tonnes of liquid oxygen and 28 tonnes of liquid hydrogen; it develops an average thrust of 1130 kN during a 10-minute burn. Each booster can be loaded with over 200 tonnes of propellant; they deliver a thrust of more than 11 000 kN at lift-off. For details of Ariane 5 booster facilities, see [7]. The upper part of the launcher consists of a storable-fuel upper stage, the vehicle equipment bay, payload adaptors and a fairing 4.57 metres in diameter. See Figures 3.11 and 3.14.

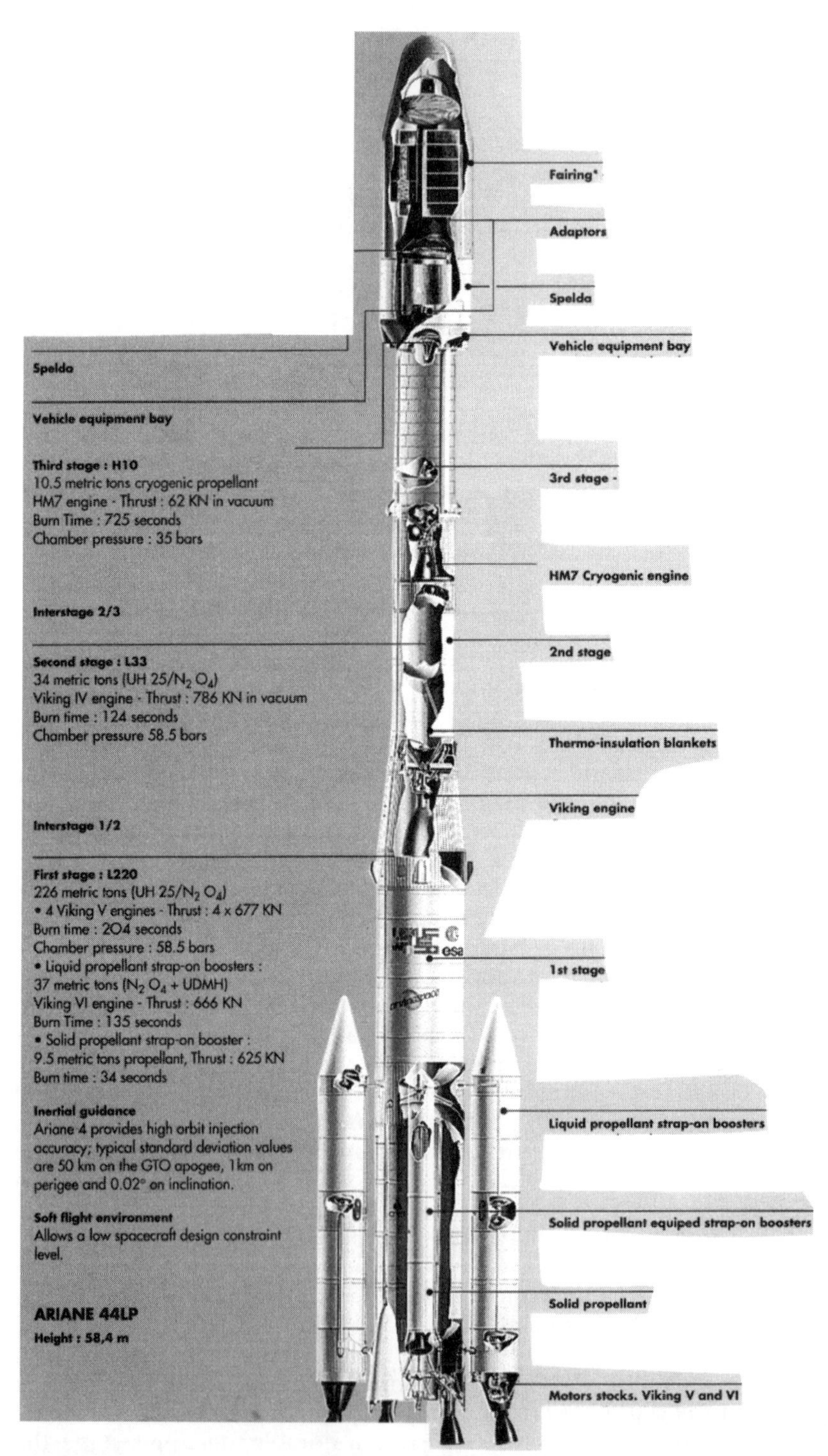

Figure 3.9 *The Ariane 4 launch vehicle. (Arianespace copyright; reproduced by permission.)*

Figure 3.10 *An Ariane 4 launcher at the Guiana Space Centre on 29 October 1991, about to launch Intelsat VI flight model 601. (Reproduced by permission of Intelsat.)*

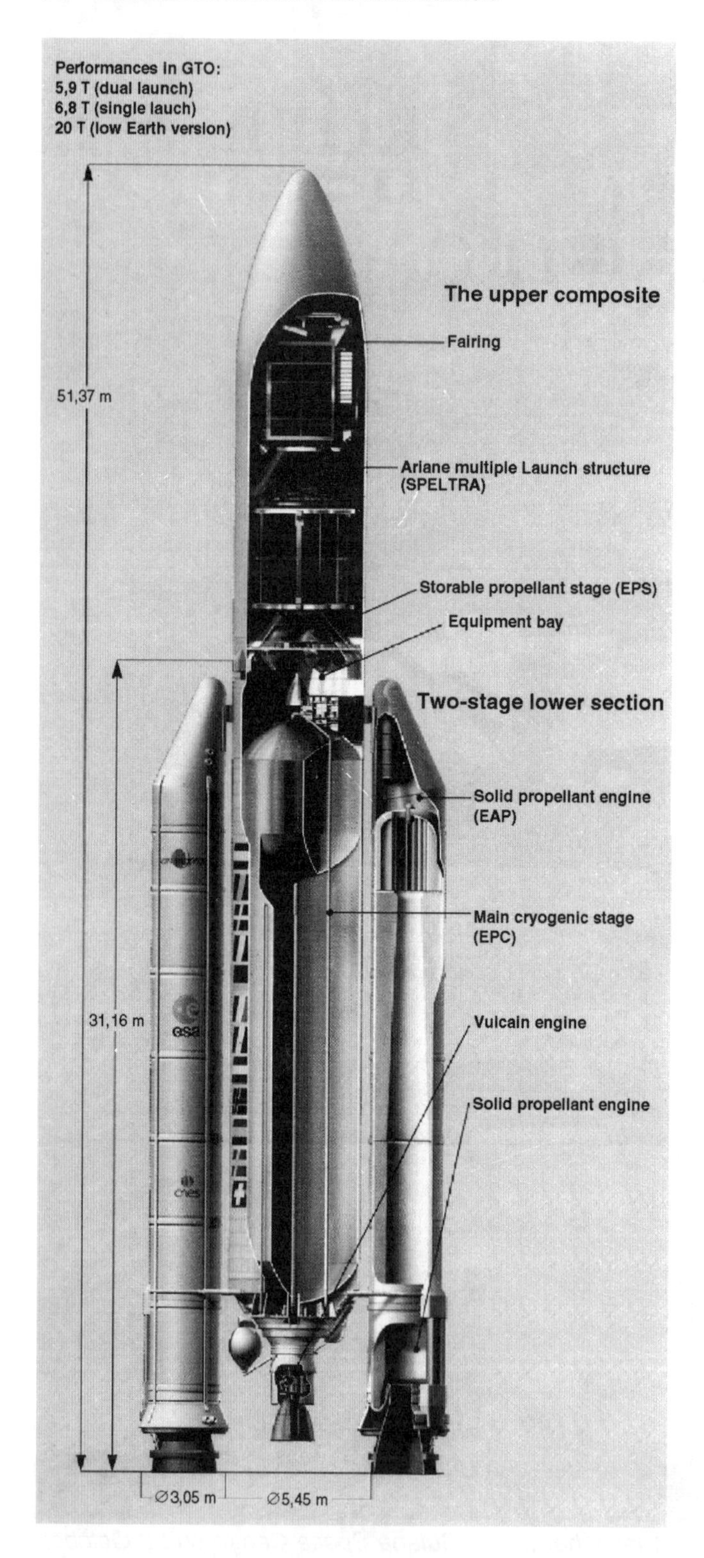

Figure 3.11 *The Ariane 5 launch vehicle. (Arianespace copyright; reproduced by permission.)*

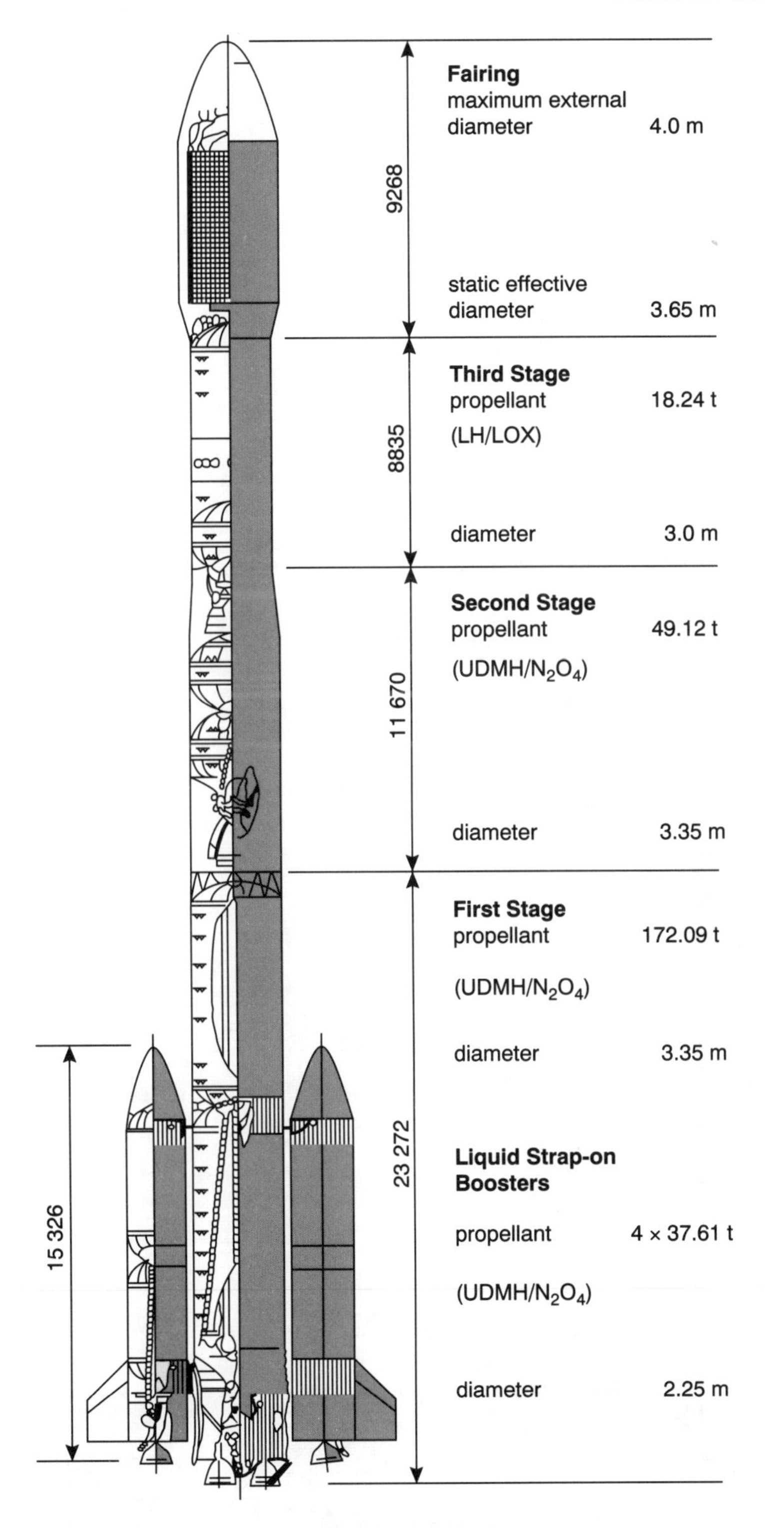

Figure 3.12 *The LM-3B launcher with details of its main technical characteristics. (Reproduced by permission of the China Great Wall Industry Corporation.)*

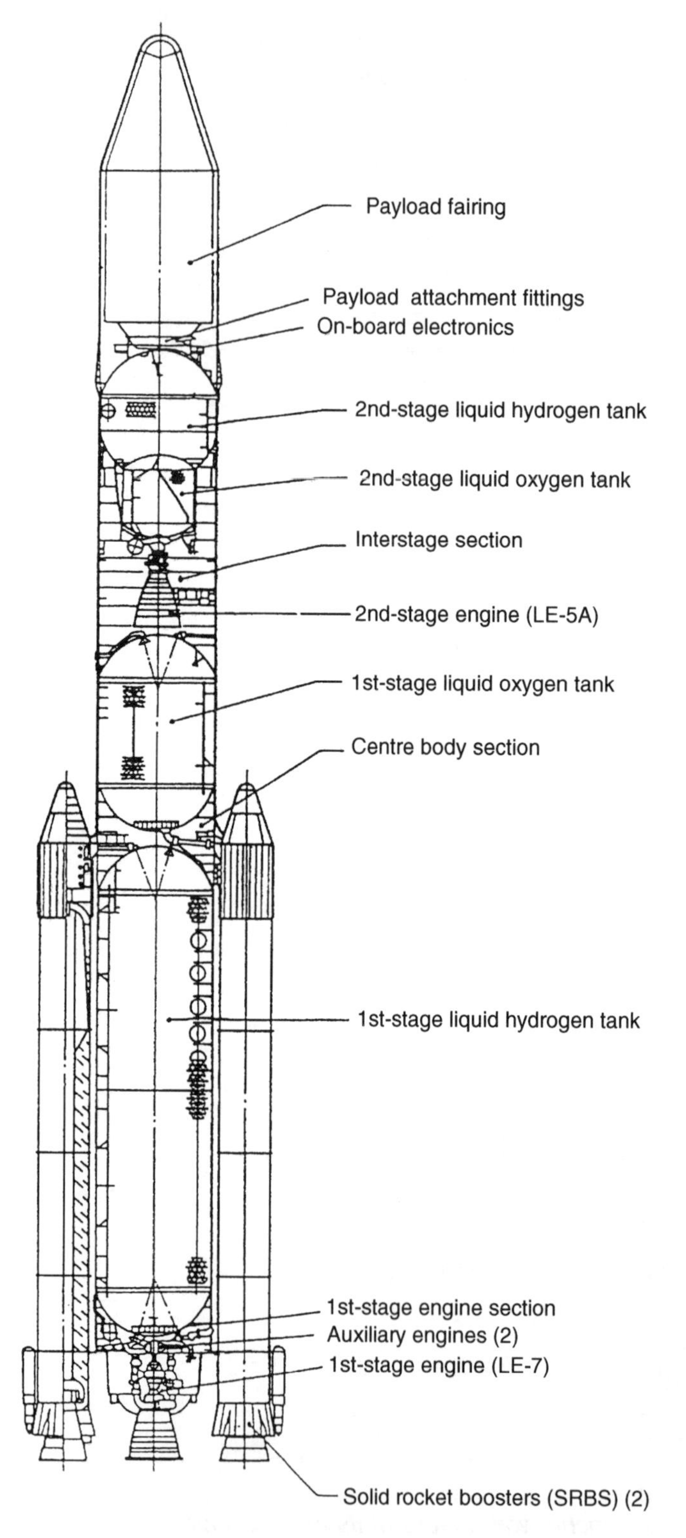

Figure 3.13 *The main elements of the H-II launch vehicle. (Reproduced by permission of the National Space Development Agency of Japan.)*

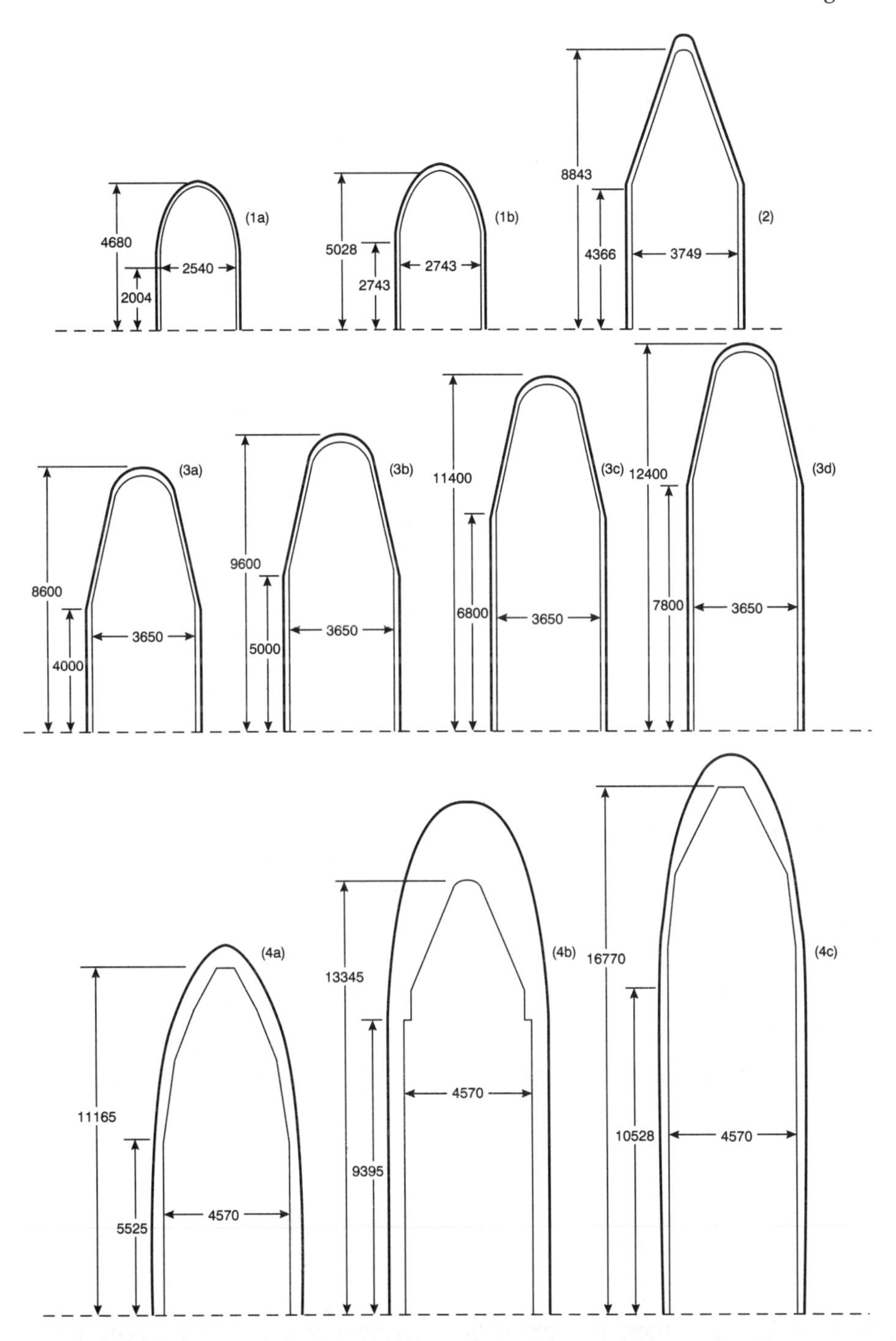

Figure 3.14 *Examples of the space available for payload under the fairings of expendable launch vehicles. Examples 1a and 1b are for Delta II. Example 2 is for Delta III. Examples 3a and 3b are for Ariane 4; examples 3c and 3d are also for Ariane 4 with the addition of a Spelda to permit a double payload to be launched. Examples 4a, 4b and 4c are for Ariane 5, with the short fairing, the long fairing and the short fairing and a Speltra. All dimensions on the figure are in millimetres*

Ariane 5 stands about 52 metres high, depending on the choice of fairing. It can place a single-satellite payload of 6800 kg into a GTO (or 2-satellite and 3-satellite payloads of 5900 and 5500 kg respectively) or a 20 tonne payload into a low Earth orbit. With such a large payload mass capability and such large fairings, it is likely that most Ariane 5 missions will be for multiple launches, and various payload carrier devices, called Speltra, Silma and Mini-Silma, have been developed for this purpose.

Long March (China)

The development of the Chinese space capability commenced in the late 1950s, leading to the firing of the first launcher in 1964 and the orbiting of the first Chinese satellite in 1970. There are now seven variants in the Long March launch vehicle family.

The Long March 1 (LM-1) is a three-stage launch vehicle suitable for placing small satellites into orbit. It was used to place the first Chinese satellite into orbit in 1970. Its commercial version designated LM-1D is specially designed for the current generation of small satellites. The Long March 2 (LM-2) is a two-stage launcher. An improved version designated LM-2C had achieved 100 per cent launch success with 14 launches by 1993. It is now offered commercially for orbiting small low Earth orbit recoverable payloads.

The Long March 3 (LM-3) is a three-stage launcher designed primarily to place satellites into the GTO. The first two stages are based on the first and second stage of the LM-2C. The third stage is a relatively new development utilizing liquid oxygen and liquid hydrogen as propellants with a high-altitude restart capability. The first successful flight of LM-3 was in 1984. An LM-3 launched the Apstar-1 satellite into a GTO on 21 July 1994.

In 1990 the more powerful two stage commercial variant LM-2E was successfully launched. The LM-2E incorporates four liquid-fuelled strap-on boosters and uses the subsystems of the LM-2C and the first and second stages of LM-3. The LM-2E can place 9200 kg into low Earth orbit and about 3140 kg into the GTO using a PAM upper stage (designated PAM D4). An LM-2E successfully launched the Optus B1 satellite on 14 August 1992.

Strap-on booster technology has also been applied in the development of the LM-3B variant. It consists of an LM-3A core stage on which four boosters are strapped; see Figure 3.12. The boosters are identical to the ones used by the LM-2E. The GTO injection capability of the LM-3B is 4800 kg. Two recently developed variants, the LM-3A and LM-2E/HO, can place 2300 kg and 4500 kg into the GTO respectively.

H-II (Japan)

The Japanese space programme was inaugurated with the launching in 1970 of its first rocket, code-named Lambda-4S. The National Space Development Agency (NASDA) went on to develop the N-I and N-II launch vehicles, with a substantial input of US expertise, and these were used in the period 1975–86 to launch a number of light satellites. This phase has been followed by independent development in Japan of more powerful launchers, H-I and H-II.

H-I is a three-stage vehicle capable of launching a 550 kg payload into the GSO. Development of H-I started in 1981, based in part on technological know-how acquired from the development of the N series of launchers. Both the second and third stages together with the inertial guidance system have been developed and manufactured in Japan. The second stage is equipped with a high-performance liquid oxygen and liquid hydrogen propulsion system. This second stage has a high-altitude restart capability, which permits the launch of a satellite to a medium-altitude orbit by a two-stage launcher. In the period between 1986 and 1992, H-I vehicles launched nine payloads, placing 13 satellites (including shared payloads) into orbit.

The first stage of the basic H-II vehicle uses the Mitsubishi LE-7 engine with liquid oxygen and liquid hydrogen as propellants, providing 848 kN of thrust at sea-level. It uses the higher performance-for-weight characteristics of a closed-cycle two-stage combustion design, cycling some precombustion gases through the engine to increase pressures and engine rotation. Four newly developed solid rocket boosters, each providing 1560 kN of thrust, are attached to the first stage; see Figure 3.13. The guidance and control of the first stage is performed by the hydraulically steerable nozzles of the LE-7 engine and of the solid rocket boosters, controlled by an inertial guidance computer. Two auxiliary engines are also provided to control attitude.

The second stage uses a liquid oxygen and liquid hydrogen engine designated LE-5A, providing 120 kN of thrust in vacuum. The guidance and control of the second stage is performed by the hydraulically steerable nozzle of the LE-5A engine. The LE-5A has a restart capability and can be tilted to control pitch and yaw during powered flight. The height of the H-II launcher is 50 m, its mass at lift-off is 260 tonnes, and it is capable of putting a 2000 kg satellite into the GSO. For more details on the H-II launcher, see [8].

3.2.2 Payload fairings

Several fairings are available for most launchers, providing for a variety of payload dimensions. Some examples, from the options available for the Ariane 4 and 5 and Delta II and III launchers, are shown in Figure 3.14. The bold outline indicates the external dimensions of the fairing and the inner line indicates the space normally available for payload. See also Figures 3.7 and 3.12, which illustrate fairings available for the Zenit 2 and 3 and LM-3B launchers.

These fairings have all been designed to accommodate a payload consisting of a single satellite, of mass within the capabilities of the launcher. The larger fairings can, alternatively, accommodate a framework which holds two or more smaller satellites, to be released individually with their final rocket stages, typically at the apogee of their transfer orbits.

3.2.3 Flight profiles

Every launch mission is characterized by a fictitious velocity, termed the characteristic velocity V_{ch}, which is the sum of all the velocity increments needed to place the payload into the desired orbit. Therefore

$$V_{ch} = V_{leo} + \sum \Delta V \tag{3.1}$$

where V_{leo} = circular velocity in a given parking orbit;
$\Sigma \Delta V$ = the sum of all other velocity increments needed to achieve the final orbit.

In Table 3.2 the circular orbit velocity at selected orbital altitudes and the characteristic velocity needed to achieve these orbits are listed, together with the perigee and apogee velocities for a transfer orbit between a 161 km parking orbit and apogee at geostationary altitude. The perigee and apogee velocities provide an indirect indication of the velocity increment (ΔV) required for injection into transfer orbit (at perigee) and into circular orbit (at apogee) respectively.

Table 3.2 *Velocity parameters and orbital periods*

Orbital altitude (km)	*Circular orbit velocity* ($km\,s^{-1}$)	*Characteristic velocity required* ($km\,s^{-1}$)	*Orbital period* (hours: minutes)	*Velocity of GTO* ($km\,s^{-1}$)
0	7.91	7.91	1:24.48	
161	7.81	8.00		Perigee = 10.25
200	7.77	8.01		
500	7.60	8.17	1:34.62	
1000	7.34	8.41	1:45.12	
2000	6.88	8.78		
5000	5.92			
10000	4.93			
35786	3.07	10.80	23:56.07	Apogee = 1.60

Before a satellite can be launched, the time slot during which launching will be feasible, referred to as the launch window, must be defined. The launch window is normally constrained by the requirements imposed by the power and thermal control systems of the satellite.

A generalized flight profile

A typical flight profile of a three-stage launcher, used to inject a payload into the GSO, is shown in Figure 3.2. The satellite launcher is fired vertically. After a short period of vertical flight, the trajectory starts slowly to curve towards the horizontal direction, although the vehicle does not reach level flight until a height is reached that is virtually beyond the atmosphere; in this way the effects of aerodynamic drag are minimized. During this, the 'boost phase', the lower stages of the launcher inject the third stage and the payload into a circular parking orbit, usually located about 160 km above the Earth. As they burn-out, the various stages are jettisoned. The remaining stage of the launcher, the AKM and the satellite, then coast in the parking orbit. At a point where the parking orbit cuts the equatorial plane (the ascending or descending node), the final stage of the launcher is ignited to thrust the AKM and the satellite into a GTO, an elliptical orbit that has its apogee at the same height as the GSO.

In general, the boost trajectory, the parking orbit and the GTO are inclined to the equatorial plane. The inclination of the orbit depends on the latitude of the launch site and the launch azimuth. For example, a launch azimuth of 90° (that is, due east) results in a transfer orbit inclination that is equal to the latitude of the launch site. Using any other launch azimuth increases the inclination of the orbit. Exceptionally, a launch site on the equator could launch directly into a GTO of zero inclination. The payload is injected into the GTO while it is in horizontal flight and crossing the equatorial plane. That ensures that the perigee of the GTO will be in the equatorial plane, and in consequence, the apogee of the GTO will also be positioned over the equator. This facilitates the next phase, namely injection of the satellite into the GSO.

Injection into the GSO is not usually performed at the first apogee of the GTO. Time is required, for example, for gathering precise data on the orbital parameters by ground telemetry, tracking and command (TT&C) stations. However, the satellite attitude is adjusted in a series of manoeuvres at successive apogee passes using the satellite's own attitude control thrusters, placing the satellite/AKM assembly in the orientation that will be required for the AKM burn. Then the satellite and its AKM are spun up to ensure that this attitude will be maintained while the AKM burn is taking place.

In order to inject the satellite correctly into the GSO, the thrust of the AKM burn must reduce the orbital inclination to zero and provide just enough impulse to increase the satellite's velocity to that required for the GSO. From Table 3.2 it can be seen that the GTO velocity at apogee is 1.60 km s^{-1} and the geostationary orbital velocity is 3.07 km s^{-1}. Therefore the velocity increment required just to circularize the orbit at apogee is 1.47 km s^{-1}. However, an additional velocity increment is required to remove the inclination of the GTO if the launch site is not equatorial. For example, an inclination of 28.3° resulting from a due east launch from Cape Canaveral calls for a total velocity increment of 1.83 km s^{-1} to both circularize the orbit and rotate the orbit into the equatorial plane, as shown in Figure 3.15. The latitude of the launch site is therefore very important for a geostationary launch; the higher the latitude, the larger is the incremental velocity required at the apogee for injection into the geostationary orbit.

On successful completion of the AKM firing, the satellite is in a near-geostationary orbit, slowly drifting in longitude. The drift orbit is normally somewhat elliptical with an inclination more than zero. Three main sequences take place during the final phase of the launch mission, namely attitude acquisition, orbital station acquisition and preparation for operational use.

During attitude acquisition the satellite attitude stabilization mode is changed from the spin-stabilized mode of the transfer orbit to the final operational mode. The manoeuvre carried out during attitude acquisition depends on whether the spacecraft is to be spin-stabilized or three-axis stabilized in its operational phase. The transition to a three-axis stabilized mode usually consists of eliminating the spin that was given to the satellite in the GTO, ensuring that the satellite's Sun and Earth sensors lock on to their targets, deploying the solar arrays and spinning up the momentum wheel.

When the satellite attains its operational attitude, the station acquisition manoeuvre can commence. The purpose of the station acquisition manoeuvre is to drift the satellite to its final operational longitude and to adjust the

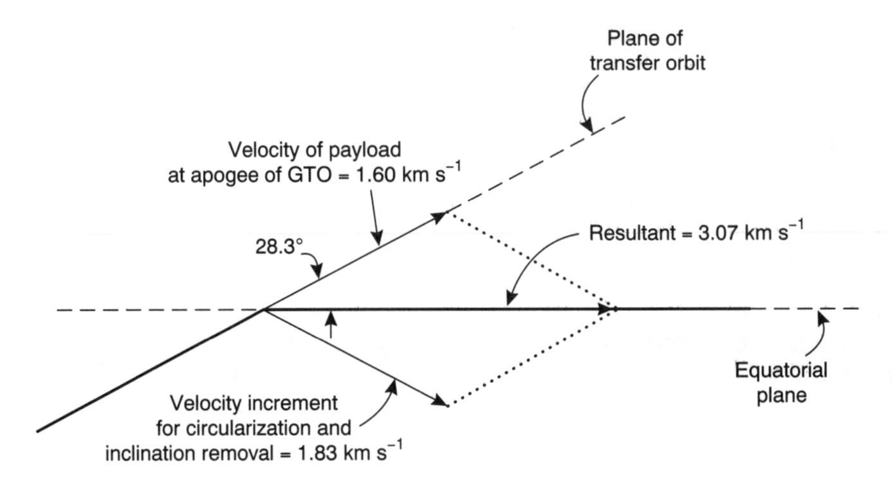

Figure 3.15 *The velocity increments required at the apogee of a GTO to circularize the orbit and remove 28.3° of orbital inclination*

orbit to geostationary parameters by circularizing it and removing any remnant of inclination. These manoeuvres are usually carried out by the low-thrust propulsion devices of the attitude and orbit control systems of the satellite itself although many recent satellites, equipped with a liquid-fuelled AKM, can use the same thruster for correction of orbital inclination. After the satellite is placed in its final operational longitude, the initialization operations can begin.

Specific flight profiles

If the launch process is examined in greater detail, it is found to differ with every launcher and every launching site and it varies to some degree with the nature of the payload. However, the following example, giving some details of launching by a Delta II 7925 launcher from Cape Canaveral, may be seen as typical.

The Delta II vehicle can place payloads into a variety of orbits, including low Earth orbits and the GTO, from dedicated dual-launch pads at Launch Complex 17 at Cape Canaveral Air Force Station, Florida. For both types of mission, the first-stage engine and six of the nine solid rocket motors are ignited for lift-off, the three remaining solid motors being ignited after the first six have burned out. The main engine continues to burn until it cuts off (MECO) 265 seconds after ignition. Then, after a short coasting phase, the first stage is jettisoned, followed by second-stage ignition.

For a two-stage LEO mission, the second stage burns for 350 to 380 seconds, at the end of which time the second-stage engine cuts off for the first time (SECO 1). Coasting to the desired LEO altitude then takes place, following a Hohmann transfer trajectory. About 2900 seconds after SECO 1, the second-stage engine is reignited and burns till the necessary orbital velocity is reached. At this point,

Table 3.3 *Event times for a Delta II flight into the GTO*

Event	*Time* (s)
Main engine ignition	$T+0$
Solid motor ignition (6 solids)	$T+0$
Solid motor burnout (6 solids)	$T+63$
Solid motor ignition (3 solids)	$T+66$
Solid motor separation (3/3 solids)	$T+67/68$
Solid motor burnout (3 solids)	$T+130$
Solid motor separation (3 solids)	$T+133$
Main engine cutout (MECO)	$T+265=M$
Blow stage I/II separation bolts	$M+8$
Stage II engine ignition	$M+13$
Fairing separation	$M+37$
Stage II engine cutout (SECO1)	$M+360=S_1$
Stage II engine restart	S_1+616
Stage II engine cutout (SECO2)	$S_1+686=S_2$
Fire spin rockets, start stage III sequencer	S_2+50
Separate stage III	S_2+53
Stage III ignition	S_2+90
Stage III burnout	S_2+177
Spacecraft separation	S_2+290

the second stage cuts off again (SECO 2). About 250 seconds later, spacecraft separation begins.

For a typical three-stage mission to the GTO, the payload is first placed into a parking orbit with an altitude of 185 km, inclined at 28.7°, using the first burn of the second stage. A coasting period then follows until the equatorial plane is being crossed, when the second stage is reignited and continues to burn till the second cutoff point (SECO 2) is reached. After this, the third stage is spun up, followed by separation from the second stage and the ignition of the third-stage engine, in order to achieve the GTO. A nominal flight from lift-off to spacecraft separation in the GTO lasts 26.7 minutes. A typical Delta II 7925 flight profile into the GTO indicating the sequence of events and the event times is shown in Table 3.3. For more details of the Delta II launch mission, see [1].

3.3 Reusable launch vehicles

3.3.1 Reusable space vehicle projects

Burnt-out upper stages of expendable launchers and many small pieces of debris remain in decaying transfer orbits for many years. Some burnt-out second stages and much small debris will also remain in low orbits for a long time. The rest crashes back to Earth, where it falls into the sea or litters deserts. What remains in space adds to the growing space junk hazard for operational satellites and all future space operations (see also Section 6.5); and satellite launching might well cost less if the major launcher components were not discarded after only one flight.

There are projects in USA, Russia, Europe and Japan which have as their aim the development of vehicles that could journey into space and return, all or much of their structure being reusable. Perhaps none of these projects is being pursued primarily to provide a means of launching communication satellites. Certainly none at present offers launching services for commercial satellites and some vehicles now planned lack the capacity to lift such payloads into space. Nevertheless, this activity gives rise to the expectation that, at some time in the not very distant future, commercial launch services will be available from reusable vehicles at less cost than those of expendable launchers, and without their environmental disadvantages. Perhaps it will also be possible to service failed satellites while in orbit or bring them back to Earth for major repairs and renovation. An outline of these various projects follows.

The Space Transportation System (USA)

The US government has a fleet of manned and mainly reusable space vehicles which are capable of lifting a payload of up to 29.5 tonnes into a parking orbit, inclined at 28.5°, with an altitude of up to 431 km. This is called the Space Transportation System (STS). Expendable perigee kick motors (PKMs) and AKMs may be used to raise satellites, typically via Hohmann transfer orbits, to orbits at a working height. There is a description of the STS and associated upper stages in Section 3.3.2.

The STS came into service in 1981 and a number of commercial satellites were launched by STS vehicles in the early 1980s. However, following the disastrous mishap involving the Challenger vehicle in 1986, it was decided that the STS should be, in general, reserved for government and scientific activities.

In 1994 NASA began exploring options for a replacement for the STS by 2010. By using the latest materials and techniques it is expected that a design providing launch into a low orbit by a single-stage vehicle would be feasible in that timeframe, using a vertical take-off and landing either vertically or, like an aeroplane, horizontally. Tests of a number of systems including a one-third scale model of a system called the Delta Clipper have already begun. See [9].

The Air-Space Vehicle System (Russia)

Energia is the most powerful operational launch vehicle in the world today, capable of launching about 100 tonnes of payload into space. In particular it can launch the Buran spaceplane, enabling the latter to acquire a low Earth orbit with the aid of its own rocket engines. The four first stage booster units of Energia are recoverable for reuse and Buran can land like an aeroplane at the end of a mission under the control of its crew, or automatically. The main purposes for which this heavy lift vehicle was developed were to ferry personnel and supplies to Mir, the Russian space station and to retrieve or repair satellites already in orbit. However, Energia can also carry into space a side-mounted cannister containing an upper stage and a payload compartment suitable, for example, for large heavy spacecraft or groups of communication satellites to be placed in orbit. Energia flew for the first time on 15 May 1987 carrying a spacecraft mock-up, and for the second time on 18 November 1988 carrying an unmanned version of Buran. For more details of Energia and Buran, see Section 3.3.3.

Consideration is being given in Russia to the development of a more compact winged spaceplane designed to ferry personnel and their luggage into space. This compact shuttle craft could be placed in orbit on top of a Proton launcher.

Hermes (Europe)

ESA is developing a small manned reusable space shuttle called Hermes. It is designed to be launched on top of the Ariane 5 rocket. If the programme is fully funded, the first launch would probably be around the year 2000. Hermes is a delta-wing space plane able to carry two pilots and up to four other crew members for maintaining and servicing space stations. It would be 18 m long, with a wingspan of 10 m, designed to carry 4500 kg of payload.

HOPE (Japan)

The Japanese government is funding the development of a manned reusable spaceplane to be launched on top of the H-II rocket: it is called HOPE, an acronym for 'H-II Orbiting Plane'. HOPE will be 12.2 m long, with a 10 m wingspan, and will weigh 11 tonnes. It would be able to carry up to 4000 kg of men and luggage into low Earth orbit.

HOPE is the first of a series of planned Japanese spaceplane developments. It will be followed by a large spaceplane, leading eventually to the development of a craft which could operate either as a space shuttle or as a supersonic airliner.

HOTOL (United Kingdom)

In the mid-1980s the British government co-funded British Aerospace for the initial research and development of an unmanned reusable spaceplane called HOTOL, an acronym for 'horizontal take-off and landing'. It is designed to take-off and land horizontally like a conventional aircraft and to be able to place satellites into orbit and to repair those that are already in service. Use of the design as the basis for a supersonic commercial airliner is also foreseen. Government funding ceased in 1988, but it is intended that HOTOL should be developed further by an international consortium led by British Aerospace and Rolls-Royce.

3.3.2 The Space Transportation System

The STS, considered as a satellite launcher, consists of the Space Shuttle and one or more upper stages. The Space Shuttle vehicle itself consists of the three main elements shown in Figure 3.16, namely the orbiter, the external tank and two solid rocket boosters.

The Space Shuttle

The orbiter is basically a delta-wing aeroplane with a large cargo bay in which a payload, consisting, for example, of several satellites and the necessary upper

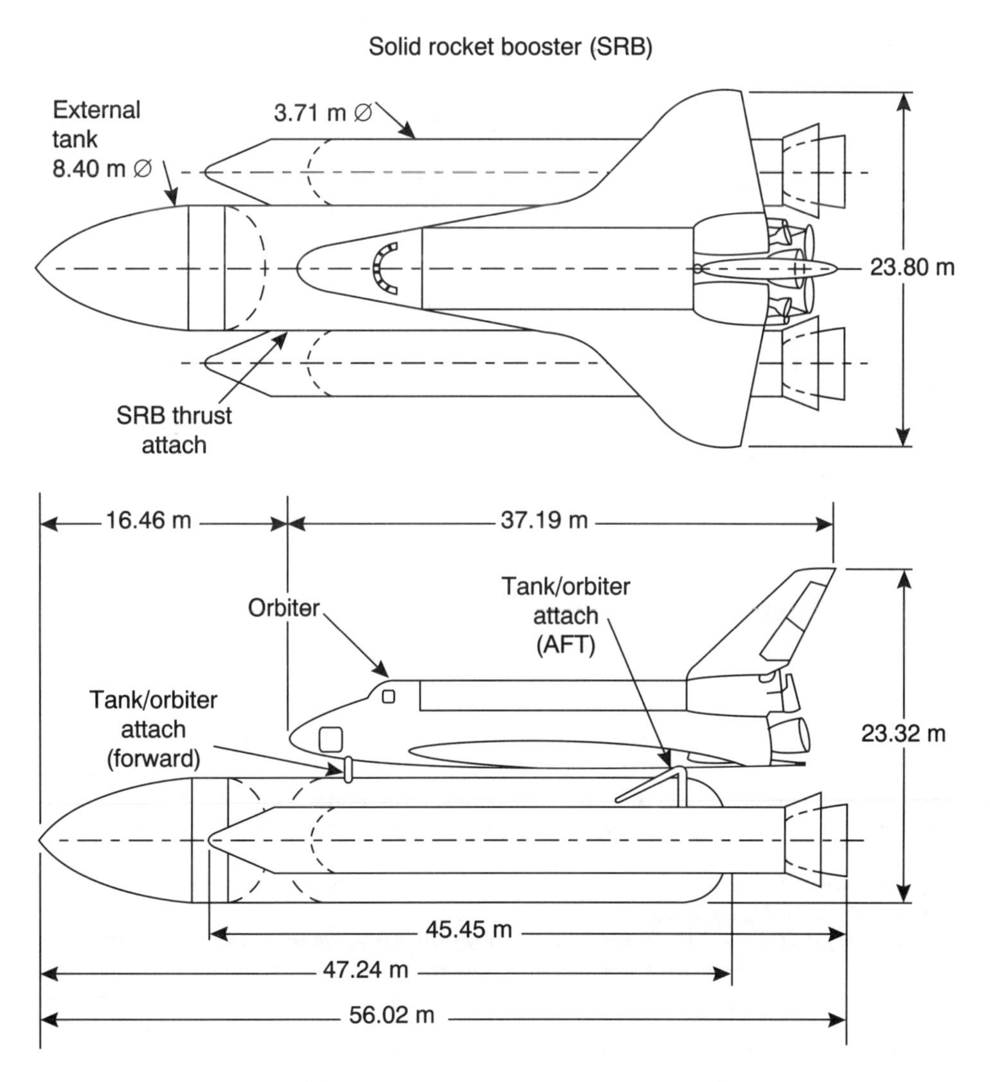

Figure 3.16 *The US Space Shuttle, configured for launching. (A NASA drawing.)*

stages, can be carried. It is launched into a trajectory by means of its three rocket motors, fuelled by liquid oxygen and liquid hydrogen from the external tank, assisted by the solid rocket boosters. The total maximum thrust is 28 500 kN. The boosters burn-out in about two minutes, at a height of about 43 km and the empty cases are released; they burn up on re-entering the Earth's atmosphere. The external tank is empty at a height of 116 km, eight minutes after the engines are ignited, and then it too is released; parachutes are used to check its descent as it falls into the sea, to be recovered for refurbishing and reuse.

The orbiter continues in powered flight, using the two orbital manoeuvring system thrusters (thrust equals 26.7 kN) and storable bipropellant fuel that it carries on board, until it reaches the parking orbit. After the payload has been removed from the cargo bay and despatched to higher orbits, the orbiter returns to Earth, using its orbit manoeuvring thrusters to decelerate it into a descending trajectory. The resistance of the atmosphere decelerates the orbiter further, enabling it to land unpowered, like a glider.

Upper stages

To achieve geostationary orbit from the orbiter parking orbit, two velocity increments are required. An increment of about 2.4 $\mathrm{km\,s^{-1}}$ is necessary to inject the payload into the GTO. Then an increment of 1.8 $\mathrm{km\,s^{-1}}$ is necessary at apogee to remove the orbit inclination and place the satellite into the GSO. These increments can be provided by propulsion units built into the satellite, by special independent propulsive stages (upper stages) or a combination of the two. The process of raising a satellite from the parking orbit to the GSO is basically similar to that described for launching by an expendable launcher in Section 3.2.3.

There are a number of types of upper stage available in the USA, such as the Payload Assist Module–Delta Class (PAM–D) or its uprated version PAM-DII, both manufactured by the McDonnell Douglas Corporation. For more details on the PAM upper stages see [10, 11]. The Inertial Upper Stage (IUS) is a solid propellant, three-axis stabilized upper stage intended to boost single or multiple spacecraft into higher orbits or escape trajectories. With most satellites currently being launched by expendable launchers, the only upper stage currently appearing in the Shuttle manifest is the IUS.

3.3.3 The Energia/Buran System

The Energia launcher

Energia is a two-stage launch vehicle in which the second stage provides the structural frame to which the first stage booster units and the payload are attached. In the standard configuration the first stage consists of four side-boosters which use liquid oxygen and hydrocarbon propellants. Each booster is about 40 metres long, 4 metres in diameter, and is powered by an RD-170 four-chamber rocket engine. Each engine develops thrusts of 7250 and 7900 kN at ground level and in a vacuum respectively.

The second stage is 60 metres long and about 8 metres in diameter. It employs four single-chamber liquid rocket engines that use liquid oxygen and liquid hydrogen propellants. Each engine develops 1450 and 1860 kN of thrust at ground level and in a vacuum respectively. For a more detailed description of the Energia launcher, see [12, 13].

This two-stage vehicle is used to inject into space either the Buran spaceplane or a side-mounted cannister with cargo space for a payload beneath a protective fairing; see Figures 3.17. and 3.18. Buran is equipped with rocket engines capable of thrusting it into low Earth orbits; see below. Cannisters, which also have upper-stage engines, are available with various combinations of fuel capacity and payload capability; Figure 3.19. shows Energia carrying a cannister designed to inject a spacecraft into an interplanetary trajectory.

The first- and second-stage engines are ignited together just before lift-off, simultaneous ignition of the two stages increasing orbit injection reliability by avoiding engine ignition in flight. The total initial thrust is thus 35 300 kN. When the first stage boosters are depleted, they separate from the second stage and land by parachute on designated sites 400 km down-range for recovery and reuse. The

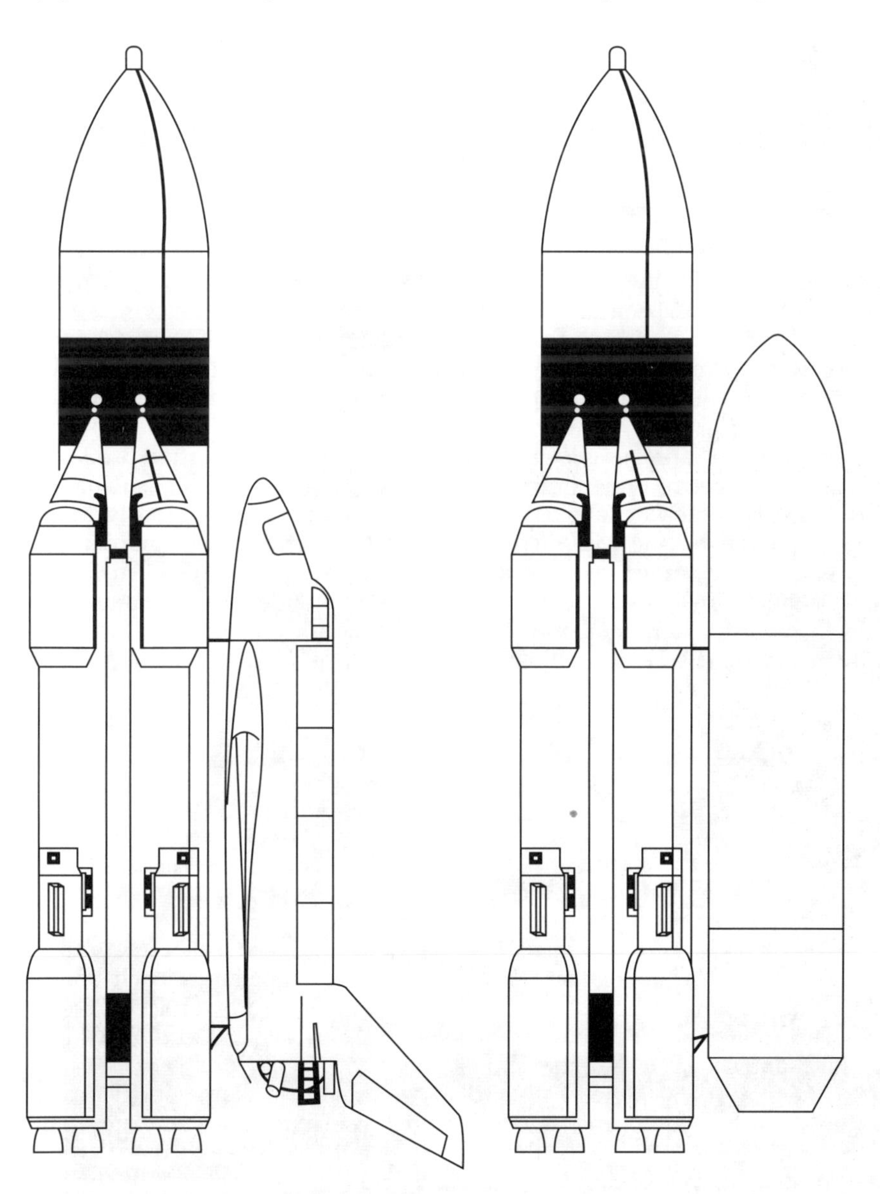

Figure 3.17 *The Energia launcher carrying (a) a Buran spaceplane and (b) a cargo cannister. (Reproduced by permission of RSC Energia, Kaliningrad, Russia.)*

Figure 3.18 *The Energia launcher carrying a Buran spaceplane on the launch pad. (Photo courtesy RSC Energia, Kaliningrad, Russia.)*

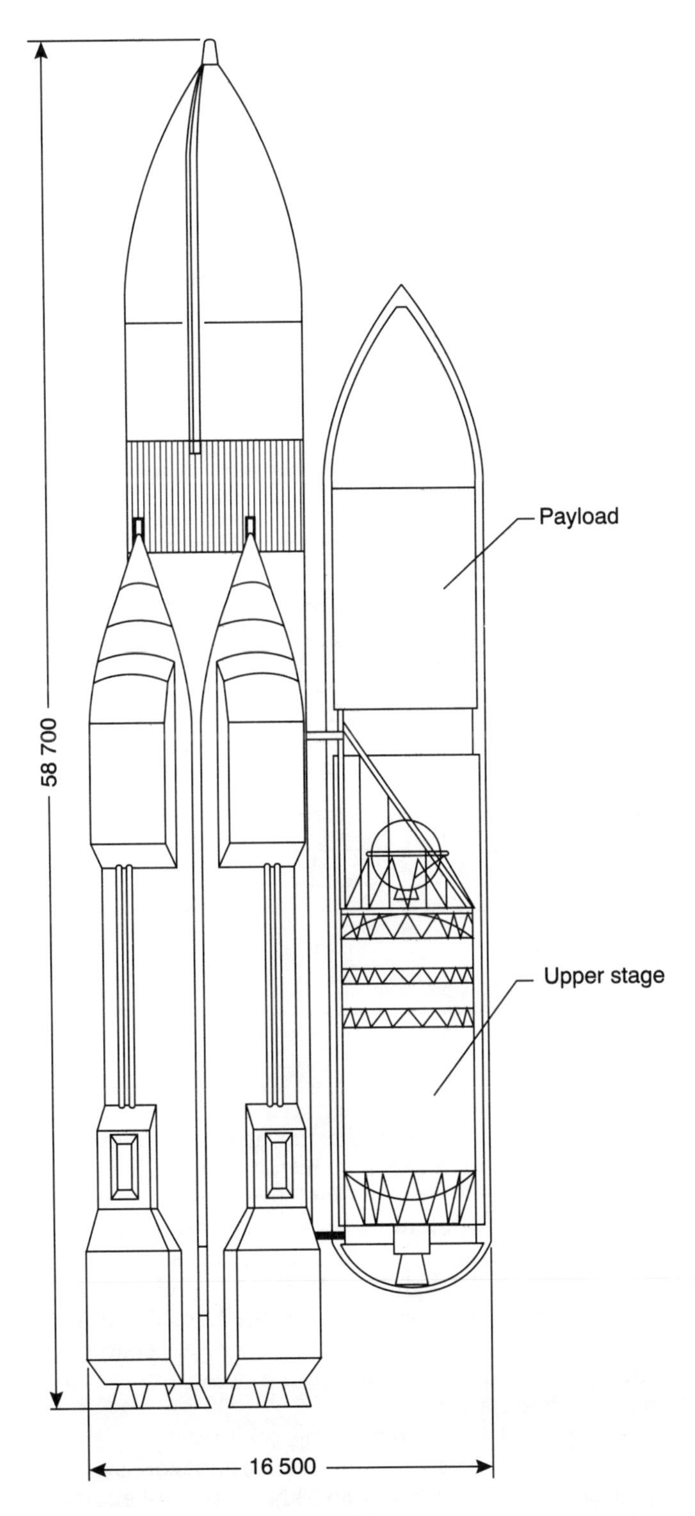

Figure 3.19 *The Energia launcher carrying a cargo cannister, the latter being cut away to show the upper stage. (Reproduced by permission of RSC Energia, Kaliningrad, Russia.)*

second stage accelerates the payload to a suborbital velocity; it is then jettisoned. The propulsion system of the Buran or the upper stage injects the payload into the desired orbit. The upper stage is fuelled with up to 70 tonnes of liquid oxygen and liquid hydrogen, its main engine develops a thrust of 100 kN in vacuum and its maximum payload is 17 tonnes.

A smaller version of the standard Energia, designated Energia M, has been developed and validated. It can place spacecraft into various orbits, including the GSO and into interplanetary trajectories. Energia M consists of a second-stage core unit and two of the first-stage boosters of the standard Energia. A liquid oxygen/hydrogen upper stage or an increased capacity version of the launcher provides the aceleration for the final injection. The basic characteristics of both the standard version and the M version of Energia are listed in Table 3.4.

Table 3.4 *The main characteristics of the Energia launcher*

	Standard version	*Version M*
Mass at lift-off (tonnes)	2400	1050
Height (metres)	60	50.8
Maximum cross-section dimension (metres)	18	18
Maximum payload mass into LEO (tonnes) (circular, inclination 50.7°, height 200 km)	102	34
Payload dimensions (metres)		
diameter	4.35–6.7	5.1
length	32.5–42	12–22
Maximum payload mass into GSO (tonnes)		
oxygen/hydrocarbon fuelled upper stage	18	7
oxygen/hydrogen fuelled upper stage	22	–

The Buran spaceplane

The reusable Buran spaceplane is 36.3 metres long, the fusilage is 5.6 metres in diameter, the wingspan is 24 metres and its mass is about 100 tonnes. It can be flown under the control of a pilot to place payloads in low Earth orbits, or retrieve them. Up to ten people, crew and passengers, can be accommodated. It can also be flown in an automatic configuration, under the control of computers on the ground. The cargo compartment measures 4.7 metres by 18.3 metres and can carry up to 30 tonnes into an orbit of 200 km altitude and 51.6° inclination. The Buran can achieve a higher orbit, up to 450 km using its standard fuel reserves and up to 1000 km with extra fuel tanks. Its nominal mission duration is seven days but this can be extended to 30 days with extra fuel.

Buran is powered by a unified propulsion system using ecologically friendly propellants of modified kerosene and liquid oxygen. The propulsion system consists of two manoeuvring engines with a thrust of 86.3 kN, 38 control engines with thrusts of 3.8 kN, and eight attitude control engines with thrusts of 196 N. The spaceplane can provide 30 kW of electrical power to the payload and a datalink to Earth of 6 Mbits/s.

3.4 Principal space launch centres

3.4.1 Land-based launching

Most satellite launchings have taken place from one or other of the following launch facilities:

(1) The USA launches satellites from two main locations, Cape Canaveral in Florida and the Vandenberg Air Force Base in California.

The Cape Canaveral facility consists of the NASA John F. Kennedy Space Center and the Cape Canaveral Air Force Station. The relatively low latitude of Cape Canaveral (28.5° N) permits satellites launched eastwards to take some advantage of the Earth's rotation, reducing the velocity increment necessary for direct equatorial orbits. The outlook eastwards is over the Atlantic Ocean, providing a relatively safe area for jettisoned launcher units and facilitating recovery where reuse is feasible. Cape Canaveral is used for all USA geostationary launchings, lunar and planetary missions, and scientific and applications satellites.

The Vandenberg Air Force Base launching site is a large coastal military complex located about 200 km north-west of Los Angeles. Vandenberg is ideally located for polar orbit missions.

(2) The main European launch site is the Guiana Space Centre in French Guiana. The Centre's position at 5.2° N latitude facilitates launches to orbits of all inclinations and enables almost maximum advantage to be taken of the Earth's rotation for direct equatorial orbits. The consequential reduction of the energy needed for orbit plane change manoeuvres at apogee in transferring from the GTO to the GSO can be translated into a geostationary transfer orbit mass gain of about 15 per cent compared with launches from Cape Canaveral.

(3) For the Russian Republic, satellites are launched from two main launch centres, the Baikonur and the Northern cosmodromes.

The Baikonur Cosmodrome lies north of Tyuratam, east of the Aral Sea, in the Republic of Kazakhstan. It is a huge complex of launch pads, towers, assembly buildings and control centres linked to nearby dormitory areas and supporting industrial complexes. The Cosmodrome measures 161 km from east to west and 90 km from north to south. The Proton launch complex consists of four launch pads, together with launch vehicle, spacecraft and integration facilities. There are also facilities for launching the Energia heavy launcher.

The Northern Cosmodrome is located near the town of Plesetsk, south of Archangel, within the compounds of a substantial military rocket base. It is used for launching mostly military satellites into high-inclination orbits for telecommunications, navigation, reconnaissance and other purposes. The Northern Cosmodrome is the world's busiest launch site.

(4) The principal launch sites of the People's Republic of China are the Jiuquan, Taiyuan and Xi Chang launch centres, the Xi Chang centre being the one most used for launches into the GSO.

The Xi Chang Launch Centre is situated 64 km north of Xi Chang city, at a latitude of 28.4° N in a high altitude (1826 m above sea-level), isolated mountainous corner of Sichuan province in Southern China. It is the main launching site for China's entry into the international commercial satellite launching market. There are three operational launch pads handling up to nine Long March geostationary missions per year.

(5) Japan's Tanegashima Space Centre is situated on the Southern part of Tanegashima Island near Uchinoura in the prefecture of Kagashima. The facilities

include the Takesaki Range for small rockets and the Osaki Range. The Osaki launch complex of the Osaki Range was used for the launch of H-I vehicles until the programme was terminated in 1992; after renovation it will be used as the launching facility for the future J-I Japanese launcher. To satisfy the requirements of the new H-II launcher, the Yoshinobu launch complex has been constructed next to the Osaki launch complex. The Yoshinobu firing test facilities are used for performing static firing tests of the LE-7 engine which is the central element of the H-II launcher.

3.4.2 The Sea Launch System

Many existing launch sites have been located in geographical positions which have been chosen because they are particularly suitable for launches into either polar or equatorial orbits. Consequently both the US and Russia operate two launching sites, one as near to the equator as possible for equatorial orbit launches and one as high in latitude as possible for polar orbit launches. To overcome this costly duplication, the Boeing Commercial Space Company of the USA, in collaboration with a Norwegian shipbuilding company and Russian and Ukrainian aerospace companies, are building a sea-based alternative called the Sea Launch System. A further advantage of launching at sea lies in greatly reduced terrain overflight constraints, providing much wider choice of launching azimuth.

The system consists of the Zenit 3SL launcher, a floating launch platform and an assembly, command and control ship. (For more details of the Zenit launcher, see Section 3.2.1.) Harbour facilities for the ships and facilities for rocket assembly will be built on the west coast of the United States. The system is designed to place spacecraft into a variety of orbits and is capable of puting a 5250 km payload into a GTO. The first two stages of the launch vehicle, an adapted Zenit launcher, will be built by NPO-Yuzhnoye of Ukraine. The third stage is the Block DM unit built by RSC-Energia of Russia. The fairing will be constructed by Boeing.

Launch vehicle components will be assembled in specially designed hangars in the harbour complex. The launchers and the satellite will be assembled horizontally in the command and control ship before sailing to the designated launch site. A launcher with its payload will then be transferred in the horizontal position to the floating launch pad and raised to a vertical position for fuelling and launching. During launch, the crew of the launch pad will transfer to the ship, which will initiate and control the launch from a position about 5 km away from the floating launch pad. For more details, refer to [14].

References

1. McDonnell Douglas Commercial Delta Inc. *Commercial Delta II User Manual.* McDonnell Douglas, 5301 Bolsa Avenue, Huntington Beach, CA 92647, USA.
2. International Launch Service. *Atlas Mission Planner's Guide.* Published by General Dynamics Commercial Launch Services Inc, 5001 Kearny Villa Road, San Diego, CA 92123, USA.
3. General Dynamics Convair Division. *Centaur Mission Planner's Guide.* General Dynamics, PO Box 80847, San Diego, CA 92138, USA.
4. International Launch Services. *Proton User's Guide.* Prepared by Lockheed-Khrunichev-Energia International Inc, 2099 Gateway Place, San Jose, CA 95110, USA.
5. Arianespace. *Ariane 2 and Ariane 3 User's Manual.* Issued by Arianespace, BP 177, 91006, Evry, France.
6. Arianespace. *Ariane 4 User's Manual.* Issued by Arianespace, BP 177, 91006, Evry, France.
7. Sartini, P. and de Dalman, J. (1993). The Ariane 5 Booster Facilities. *ESA Bulletin,* No. 75.

8. NASDA, H-II Launch Vehicle No. 3, Feb 95.
9. Dornheim, M. A. (1995). DC-X holds promise; big questions remain. *Aviation Week and Space Technology*, August.
10. McDonnell Douglas Astronautics Company (1985). *PAM-D User's Requirement Document.*
11. McDonnell Douglas Astronautics Company (1985). *PAM-DII User's Requirements Document.*
12. Gubahov, B. T. *The Energia Versatile Rocket–Space Transportation System: The Space Vehicle for Today and Tomorrow.* Kaliningrad, Russia: RSC Energia.
13. Semenov, Y. P. (1990). The Energia blast-off. Mockba 'Mawnhochtpoehne'.
14. Boeing Commercial Space Company (1996). *The Sea Launch User's Guide.*

4 Satellite technology

4.1 Introduction

A satellite consists of two distinct hardware elements, namely the payload and the platform. The payload, in the case of a communications satellite, consists of the antennas and the electronic devices that are used for relaying signals from one earth station to another. The platform consists of all the systems which enable the payload to operate in the required manner. The technological aspects of the satellite platform are described in this chapter, together with the space environment in which it operates. For a more detailed account, see [1–3]. The payload is treated in Chapter 7.

4.2 The space environment

4.2.1 The nature of the environment

The physical conditions under which satellites operate place constraints on the types of material that can be used for their construction and the design of the devices that are incorporated in them. The purpose of this section is to describe this environment and in particular those characteristics which affect satellite design.

The cosmic environment

Interstellar space is occupied by gas consisting mainly of hydrogen atoms; the density is extremely low, about three atoms per cubic centimetre. However, there are subatomic particles called cosmic rays moving through space at enormous velocities, having their origin outside the solar system; they are mostly hydrogen and helium nuclei and electrons. Small solid particles of the order of 1 mm in diameter (meteoroids) move through the solar system with velocities of 20 to 70 $km\,s^{-1}$, and dust concentrations create a micrometeorite environment. The concentrations of cosmic rays and meteoroids are low, but both kinds of particle are very energetic and their potential for damage to satellites is great.

In the frame of reference defined in Section 2.1, an artificial Earth satellite moves through space on a course determined largely by the velocity that had been given to it when it was injected into its orbit and by the Earth's gravitational field, modified by the gravitational fields of the Sun and the Moon. However, an observer located on the satellite would perceive almost no gravity, although a mild centrifugal force would be continuously present if the satellite had been set spinning, for example to provide attitude stability. In addition, mild inertial and centrifugal forces are present in any satellite during the course of attitude or orbital corrections.

The sky, seen from a satellite, is black and very cold, about 3 K in temperature, in virtually all directions except those of the Sun and the Earth. The Sun's disc has a diameter of about 0.5° at an Earth satellite. It normally exhibits a temperature of about 5800 K, but see below. The Earth occupies a much larger part of a satellite's sky, the disc having a diameter of 17.6° as seen from the GSO, rising to perhaps 120° from a low orbit, and the Earth exhibits a mean temperature of about 255 K. From time to time the radiation from the Sun is cut off, wholly or partially, by an eclipse by the Earth or the Moon; see Section 2.5.

Solar emissions

Emissions from the Sun include a wide spectrum of radiation and several kinds of high-velocity charged particles.

With a normal surface (photosphere) temperature of about 5800 K and a diameter of 1.392×10^6 km, the Sun emits 3.83×10^{26} W of power under steady-state conditions. The radiated spectrum (see Figure 4.1) extends from gamma rays to radio waves, conforming closely to black-body characteristics over most of its range but with a higher level at radio frequencies. When the Sun is disturbed, this excess radiation in the radio spectrum is greatly increased, and at times, for the brief duration of a solar flare, the level of radiation, in particular gamma rays and X-rays, is considerably increased. There is a correlation between the incidence of flares and the sunspot cycle.

A plasma, consisting chiefly of protons, streams out continuously from the Sun; it is caused by the thermal energy of the ionized gas which forms the corona

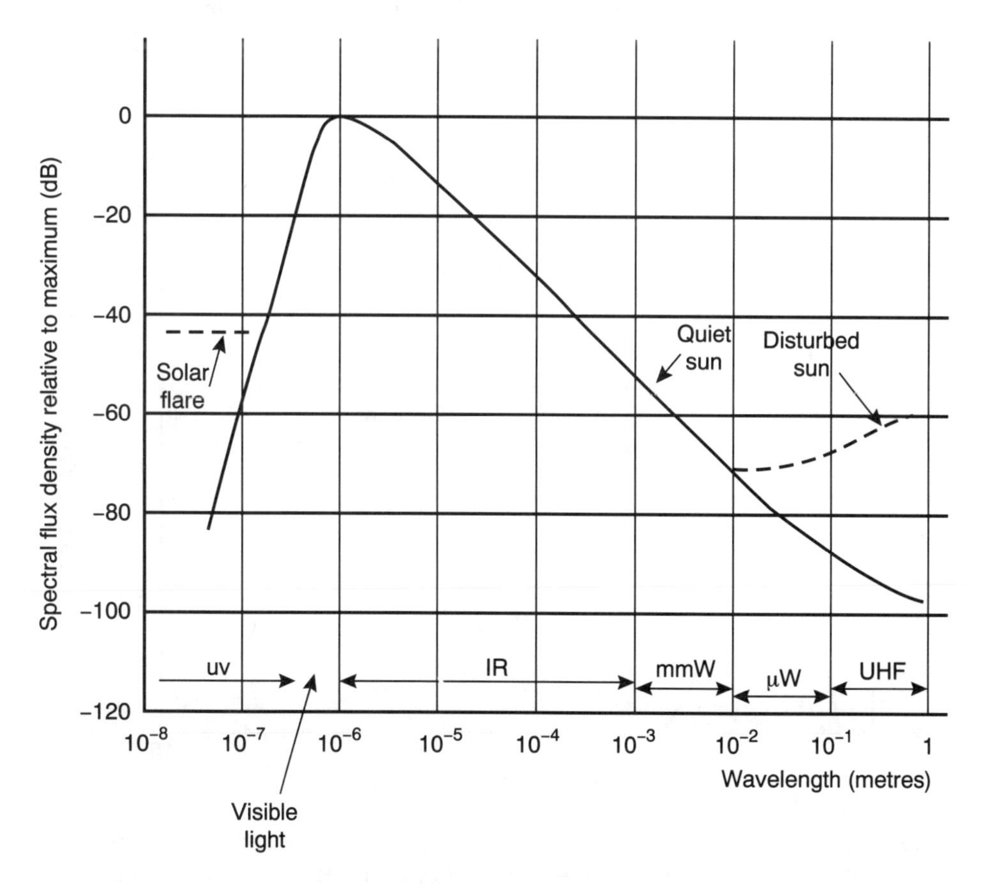

Figure 4.1 *The spectral characteristics of solar radiation.*

surrounding the Sun and which cannot be contained by the Sun's gravitational field. It is called the solar wind. The speed of the solar wind increases as it moves away from the Sun, reaching speeds between 200 and 900 km s^{-1} near the orbit of the Earth. The density of the plasma decreases with distance, measuring about eight particles per cubic centimetre near the Earth's orbit. The density is considerably increased during a solar flare, but whereas the increase in radiation due to the flare is perceived on earth within eight minutes, about three days elapse before the enhancement of the solar wind reaches Earth. For a more detailed description of solar conditions, see [4].

Effects due to the Earth

As mentioned above, the mean temperature of the Earth, measured from space, is about 255 K; this radiation includes a significant component at radio frequencies.

At a height of 1000 km, the atmospheric pressure is very small indeed compared with the pressure at sea-level, although it is not so low that the aerodynamic drag that it causes is totally negligible. At greater altitudes, in say the GSO, the condition is what would be described as extremely hard vacuum by the standards of sea-level conditions.

The region of space surrounding the Earth and dominated by the Earth's magnetic field is called the magnetosphere. All the orbits of interest for satellite communication lie within the magnetosphere. The closed magnetic lines of the magnetosphere trap high-energy particles. In 1958 an experimental package designed by James A. Van Allen and installed on the Explorer I spacecraft revealed that the Earth's magnetic field traps high-energy electrons and protons

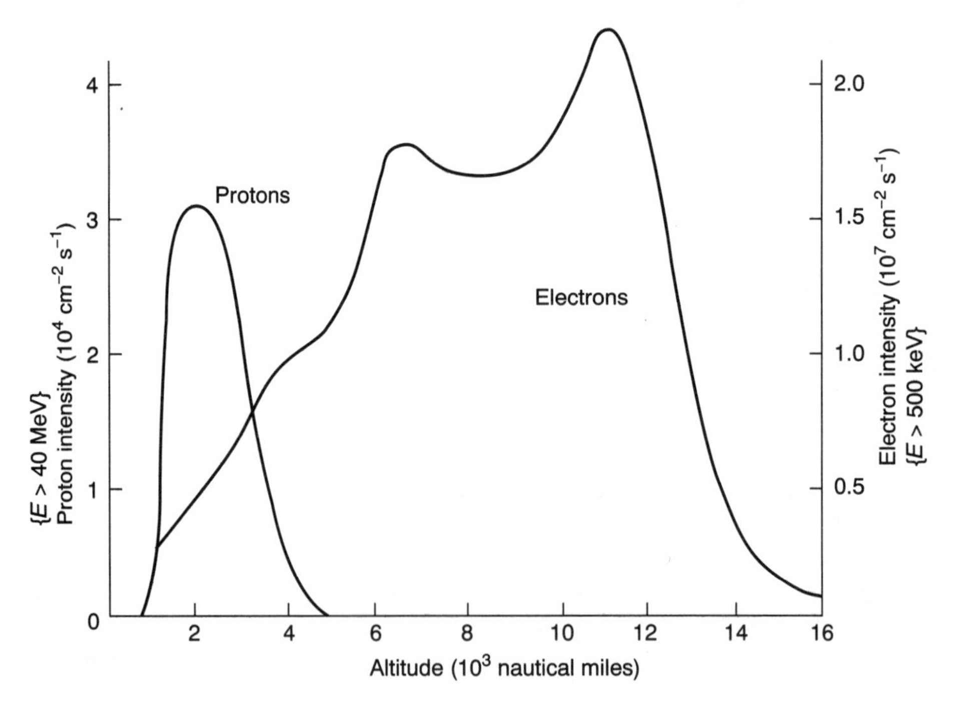

Figure 4.2 *The intensity of subatomic particles in the Van Allen belts. (From* Spacecraft Systems Engineering, *edited by Peter Fortescue and John Stark. Reprinted by permission of John Wiley and Sons Ltd.)*

in two belts centred about the Earth's magnetic axis. These belts, known as the 'Van Allen belts', have the particle density distribution with height shown in Figure 4.2. More recently the Solar, Anomalous and Magnetospheric Particle Explorer (SAM-PEX), a US/German satellite launched in 1992, revealed an additional radiation belt within the inner Van Allen belt at an altitude of several hundred kilometres.

The inner Van Allen belt is believed to consist of high-energy protons with energies in excess of 40 MeV. The outer belt contains electrons with lower energies, in the range 10–50 keV. Consequently the outer belt is considered as a 'soft' belt, while the inner one is a 'hard' one. These two belts consist of particles originating from the Sun. In contrast, the new belt discovered by SAM-PEX collects particles from interstellar space that originate from stellar winds, supernova explosions and the Big Bang. These latter are high-energy particles of oxygen, nitrogen and neon which were electrically charged by the ultraviolet radiations from the Sun as they approached the Solar System.

Charged particles reaching the neighbourhood of the Earth, not captured by one of the radiation belts but nevertheless under the influence of the Earth's magnetic field, may penetrate more deeply into the atmosphere. Here they add to the level of ionization in the ionosphere at a height of about 100 km, in particular in polar regions and in the fringes of the polar regions where they cause auroral phenomena.

Finally, the radiation from the Sun which reaches the Earth ionizes the atmospheric gases, forming the ionosphere.

4.2.2 The effects of the environment on satellites

Radiation of energy at radio frequencies from the Sun and the Earth adds to the noise entering satellite radio links via earth station and satellite receiving antennas; see Section 5.2. The Earth's ionosphere, produced by radiation and particulate bombardment of the atmosphere by the Sun, has effects on radio propagation between satellites and earth stations which are negligible above 10 GHz but quite significant below 3 GHz; see Chapter 5. However, this present section is concerned with the effects of the space environment on the satellites themselves.

Ultraviolet radiation may change the molecular bond structure of materials, causing transparent materials to become opaque. The emissivity and absorptivity of surface coatings may be affected, thereby degrading the operation of thermal systems, and the mechanical properties of materials may be changed, leading to the failure of mechanical structures.

Electromagnetic radiation, in particular X-rays and gamma rays, penetrate deeply into materials, causing ionization or changing the crystalline structure, thus effecting their electrical conduction properties. The efficiency of operation of electronic components may be degraded, perhaps causing total failure. Ionization particularly effects organic compounds such as insulators, adhesives and lubricants, which may lose their specific properties. The effects of X-rays and gamma rays are particularly severe during enhanced solar activity.

High-energy charged particles from the Sun and from deep space, and in particular the particles trapped in the Van Allen radiation belts, cause the formation of recombination centres in semiconductors, leading to the failure of such devices. Bombardment by these high-energy particles has a severe effect on the performance of solar cells in the absence of protective shields. Accordingly, satellites in circular orbits operate at heights above, below or between the belts, so as to avoid the effects of high-energy proton bombardment.

The micrometeorite environment of space can inflict serious mechanical damage or performance degradation on solar cells or thermal radiative surfaces, the properties of which may change due to the sand-blasting effect produced by impacts.

Extreme vacuum conditions encourage solid substances to sublimate, leading to the physical removal of atoms from the surfaces of materials used in the spacecraft. These atoms may condense on adjacent cooler surfaces, degrading their optical properties or causing electrical short-circuits. Lead, magnesium, zinc and cadmium sublime at lower temperatures than other metals and they cannot be used for spacecraft surfaces.

An extreme vacuum also changes the mechanical properties of materials. For example, glass becomes more resistant to fracture, whereas steel and molybdenum develop a greater tendency to fracture and creep. Soft metals such as aluminium and magnesium become weaker due to the absence of surface oxidation, although the strength of some metals is enhanced by the absence of corrosion.

The surfaces of a satellite that are illuminated by the Sun are liable to gain a lot of unwanted heat; reflective surface finishes must be used to minimize this heat gain. On the other hand, quite large amounts of waste heat are generated in devices in the satellite and this must be removed from the hot spots and, ultimately, rejected from the satellite altogether. Within the spacecraft, both radiation and conduction heat-transfer mechanisms are available. However, the absence of a surrounding atmosphere prevents the transfer of heat between the spacecraft and its environment by conduction or convection, leaving radiation as the only available mechanism. Thus, the waste-heat radiators must be located on surfaces of the satellite, typically the north and south faces, which always look out on to a cold sky.

Fluid and powder lubricants cannot be used in space because constituents vaporize, rendering them ineffective. Conventional metal bearings cannot operate in space as the thin layer of air between their moving surfaces, present on Earth, disappears in space and the surfaces bind together in a cold welding process. Soft metal or alloys are used instead in joints which have only limited use or a restricted range of operation and bearings made from special compounds such as nickel-bonded titanium are employed where the usage is heavier.

On the positive side, the space environment has properties that permit certain technologies to be applied that would not be feasible on Earth. An extreme vacuum has good electrical insulating characteristics, hence electrical conductors can be placed closer in space than on Earth or can carry higher voltages without electrical arcing appearing. In space there is no atmospheric corrosion, no wind, no vibrations external to the spacecraft, so large lightweight weak structures can be deployed and operate for a long time, floating in the gravity-free stillness of space.

4.3 Satellite structures

4.3.1 Satellite configurations

The satellite structure is the skeleton designed to support all the other systems in the spacecraft. It protects vulnerable components from the space environment, isolates appendages and provides alignment for attitude control. It must also meet certain other requirements such as stiffness, inertial ratios and direction of principal inertial axis. It must withstand the stresses that arise, in particular during the launch phase, and be compatible with the launcher in terms of mass, size and interface requirements. In general, the design of a satellite structure must meet the mission requirements of the satellite with the minimum weight and maximum reliability.

The shape and layout of a satellite is the result of a compromise between conflicting constraints and requirements. The overriding requirement of the first generation of satellites was simplicity of design and operation. These satellites were symmetrical. Their attitude was stabilized by spinning the body about the axis of symmetry. They had few moving parts or none. Their solar cells were mounted on the outer surface of the body, covering most of it. The communications antennas were simple and had little directivity.

A typical example of a first-generation satellite design is the Intelsat I (Early Bird), shown in Figure 1.1, which was launched on 6 April 1965. The cylindrical body is about 75 cm high (excluding the antenna), the diameter is about 60 cm and the mass is 38.6 kg. The solar cells provided 33 W of DC power for the operation of its two receivers and its 6 W travelling wave tube (TWT). The antenna consists of six dipoles phased to illuminate the Northern hemisphere from the GSO. Early Bird provided a full-time operational service for 3.5 years. Other first-generation types of commercial satellite are Intelsat II and Intelsat III.

Increased communication capacity requirements and the availability of launch vehicles with higher payload capacity led to the development of the bigger and more complex second-generation spacecraft. These were either large spin-stabilized spacecraft with despun antenna farms providing steerable beams such as Intelsat IV (see Figure 4.3) or box-like three-axis stabilized structures with large deployable solar panels and antenna farms providing steerable shaped beams such as the Intelsat V (see Figure 4.4).

Requirements for still more communication capacity, more power and smaller steerable spot beams are stimulating the development of third-generation satellites which typically consist of a polyhedral framework supporting compact payload packages and a superstructure to which very large deployable solar panels and antennas are attached. An early example of such designs is the TDRSS satellite shown in Figure 4.5.

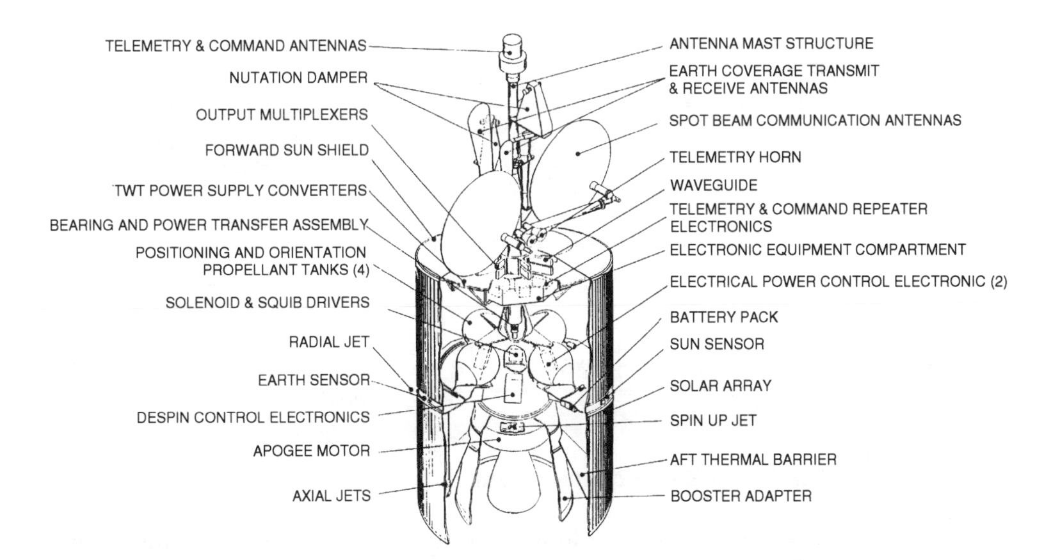

Figure 4.3 *Intelsat IV, an early example of a second-generation spin-stabilized satellite. (Reproduced by permission of Intelsat.)*

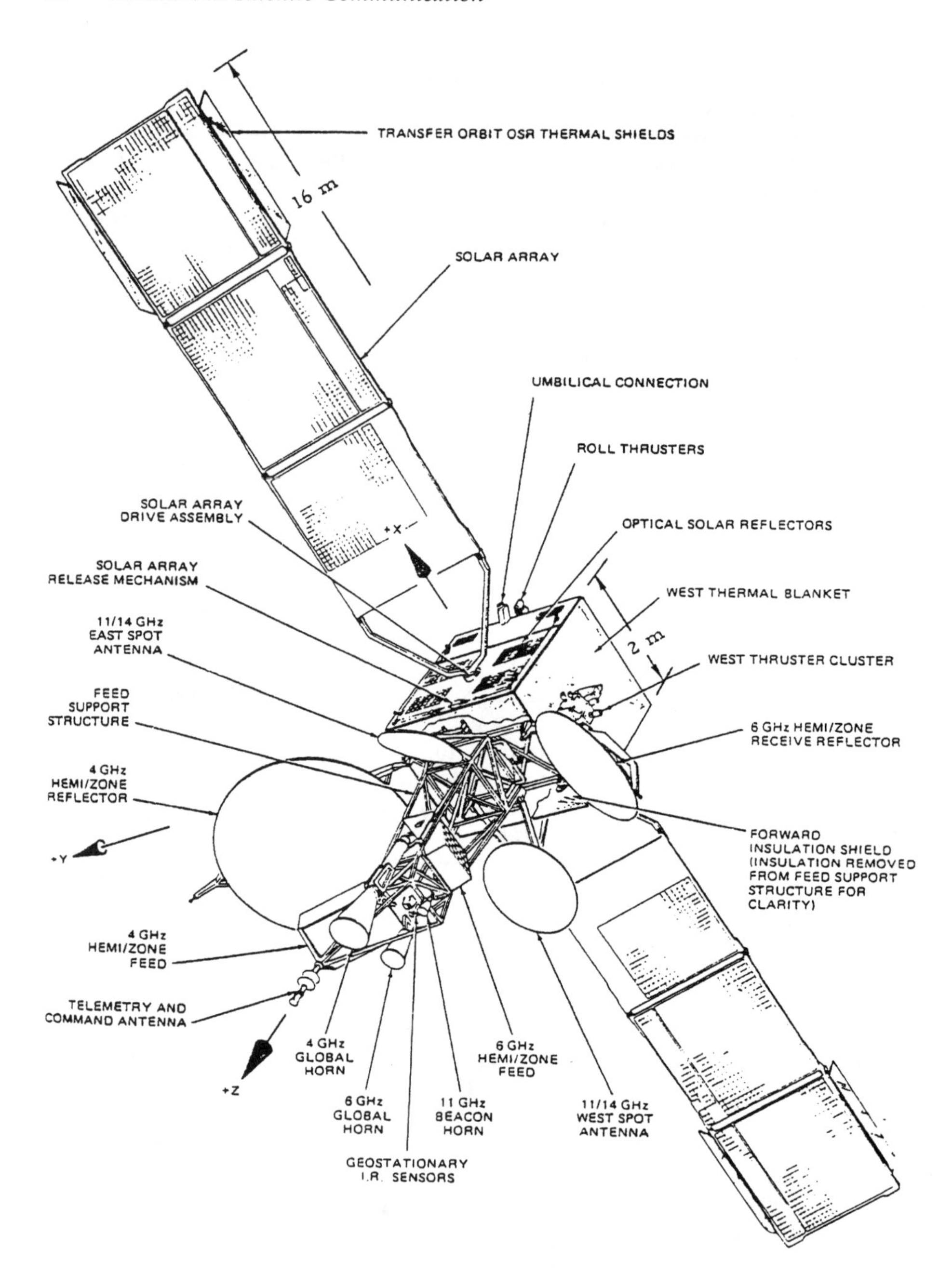

Figure 4.4 *Intelsat V, an early example of a second-generation three-axis stabilized satellite. (Reproduced by permission of Intelsat.)*

4.3.2 Structural design requirements

The satellite structure is designed to meet certain requirements of strength, stiffness and tolerance of vibration. The satellite must be physically strong enough to bear the ground qualification and acceptance loads, as well as launch and in-orbit loads, vibration tests, and shock. Parts of the structure must be stiff to ensure that

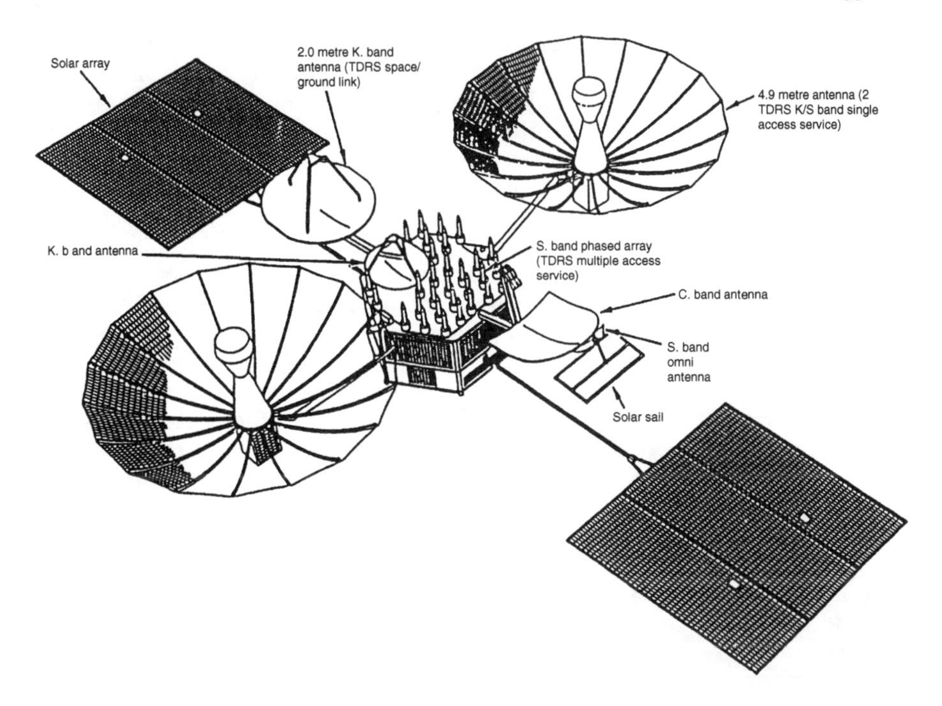

Figure 4.5 *The TDRSS satellite, an early example of a third-generation satellite design. (Reproduced by permission of Columbia Communications Corporation.)*

under all loading conditions the resulting deformations and displacements are within acceptable levels and exhibit no elastic instability. The effects of vibration have to be considered to ensure that fundamental resonant frequencies do not appear below certain frequencies (typically below 30 and 10 Hz in the longitudinal and lateral axes respectively). In general, the basic principles applied in structural design are as follows:

(1) Materials with high specific stiffness and strength are utilized.
(2) There are joints in the structure to enhance damping.
(3) Areas of stress concentration are minimized.
(4) Surface irregularities such as sharp corners and crevices are avoided.
(5) Eccentric load paths are avoided.
(6) Residual tensile stresses are avoided.
(7) Contact between different metals is avoided.

4.3.3 Typical structural designs

The structure of a spacecraft is normally divided into two main components, namely primary and secondary. The primary structure is the basic satellite skeleton containing the main load-bearing elements to which the satellite equipment is attached. The secondary structure includes all other structural elements which provide cover or additional support, such as antennas, appendages and solar arrays.

The satellite structure is constructed from non-magnetic materials which combine minimal mass with high strength and stiffness, low outgassing characteristics, and which involve no major fabrication problems. Metals such as aluminium, high-strength steels, titanium, magnesium, beryllium, and fibre composites are widely used in spacecraft construction. Beryllium and carbon-fibre-reinforced plastics are particularly attractive, offering particularly low mass characteristics coupled with high stiffness and low thermal distortion. However, their high cost and the toxicity of beryllium dust limit their use.

Aluminium honeycomb sandwich panels are extensively employed in the construction of satellite structural components. The panels are constructed from an aluminium honeycomb core, on each side of which an aluminium sheet typically 0.5 to 1 mm thick is bonded. This construction offers particularly high strength and stiffness. Where mass saving is particularly important, the aluminium sheets of the honeycomb panels can be replaced by graphite epoxy or other composite materials.

Figures 4.6 and 4.7 show exploded views of three-axis and spin-stabilized satellites, illustrating the main structural elements and other subsystems. Both types of satellite incorporate an AKM, housed within the central tube of the primary satellite structure, to transfer the spacecraft from the GTO to the GSO

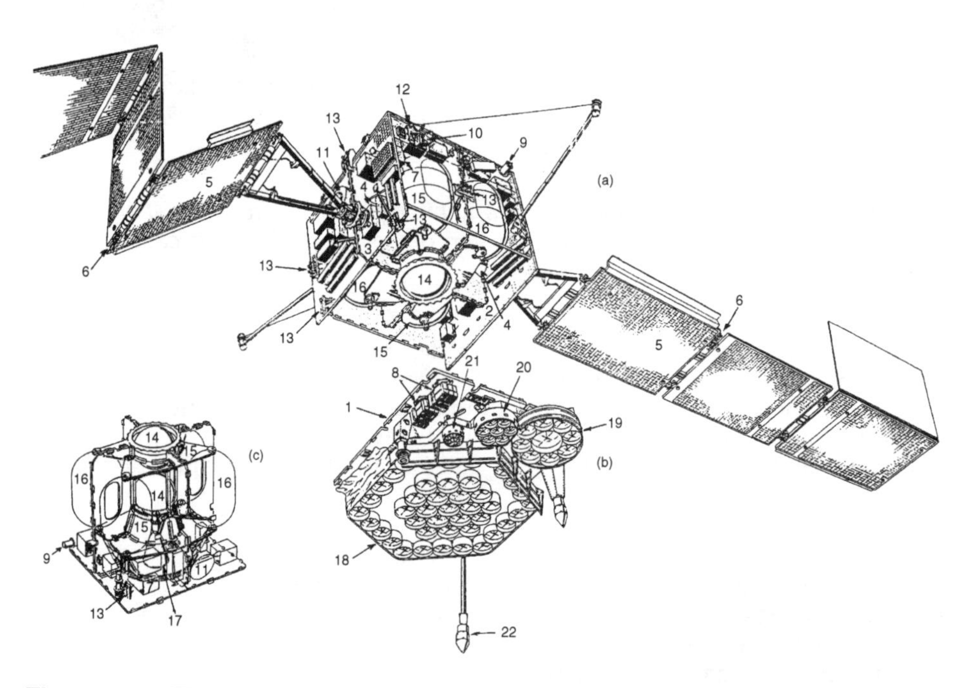

Figure 4.6 *The Inmarsat 2 satellite, a second-generation three-axis stabilized satellite, showing (a) a general view, (b) the communications floor, exploded away from the main body, and (c) the central structure, withdrawn from the main body and inverted to reveal areas not visible in (a). (Reproduced by permission of Inmarsat.) Key: 1, Earth-facing wall structure. 2, North-facing wall structure. 3, South-facing wall structure. (Most payload and TTC&C subsystem components are mounted on these three walls.) 4, Solar array drive mechanisms. 5, Solar panels. 6, Solar array sun sensors. 7, Batteries. 8, Infrared two-axes Earth sensors. 9, Sun acquisition sensor. 10, Earth sun sensor. 11, Fixed momentum wheels. 12, Gyro. 13, Thruster modules. 14, Pressurant (helium) tanks. 15, Fuel (monomethyl hydrazine) tanks. 16, Oxidant (nitrogen tetroxide) tanks. 17, Apogee kick motor. 18, L band transmit antenna. 19, L band receive antenna. 20, C band transmit antenna. 21, C band receive antenna, 22, TTC&C omnidirectional antenna*

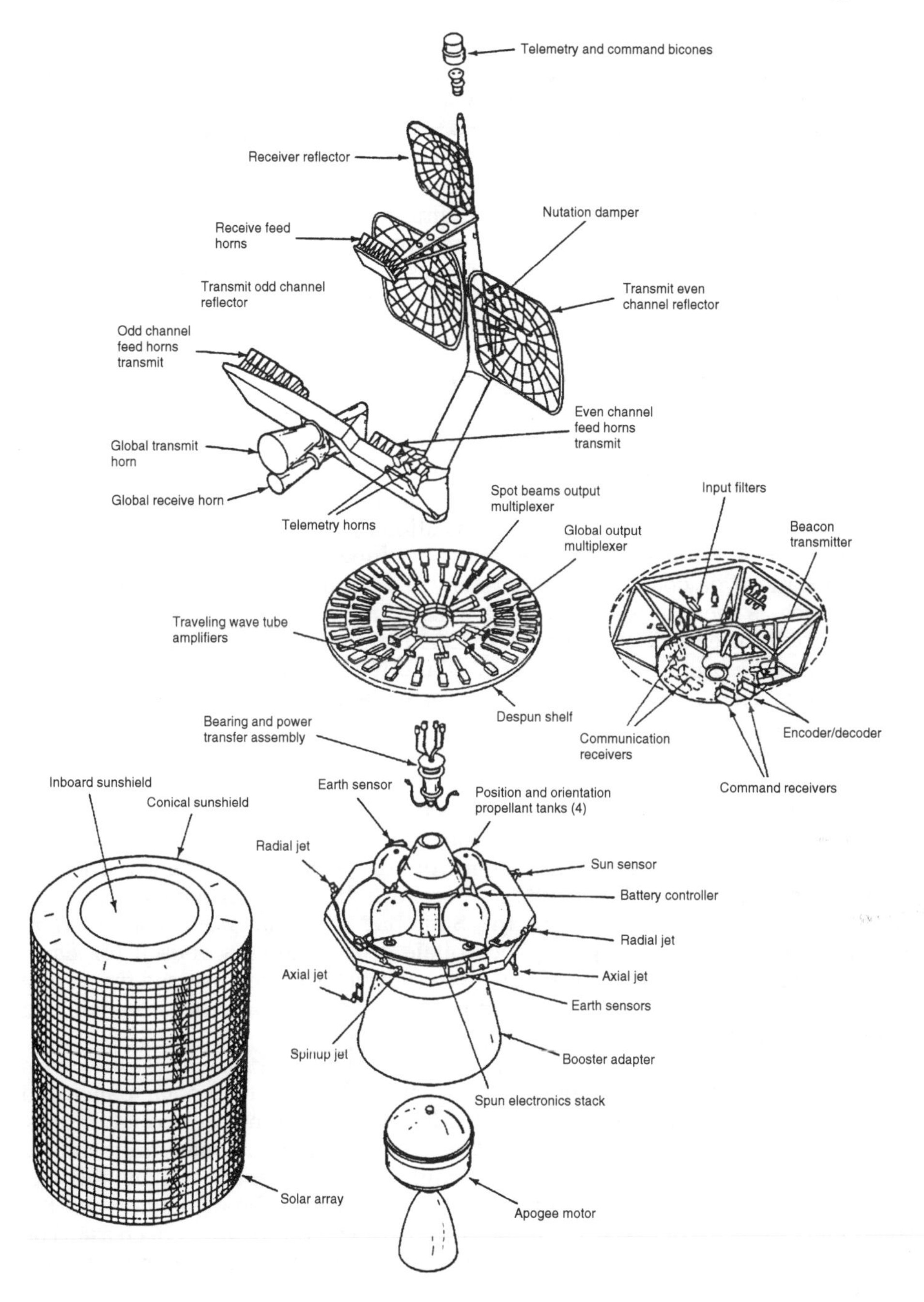

Figure 4.7 *An exploded view of the structure of Intelsat IVA, a spin stabilized satellite. (Reproduced by permission of Intelsat.)*

during the launching process. This primary structure is normally constructed from aluminium or magnesium sheet metal and stiffened with aluminium alloy rings which interface with the launch vehicle adapter, the antenna and the AKM. In cases where mass saving is of vital importance, the central tube is constructed from lightweight fibre-reinforced plastic. For a three-axis stabilized spacecraft, two honeycomb sandwich bulkheads attached to the central cylinder form the north and south sides of the spacecraft respectively; they support the solar array drives. Honeycomb sandwich panels also form the east and west bulkheads, which provide support for the reaction control equipment. Where the honeycomb sandwich is particularly heavily loaded, both the honeycomb core and its covering sheets may be locally reinforced. Equipment units are attached to the honeycomb panels with cylindrical aluminium inserts, bonded in the honeycomb core. The solar array structures and other large appendages are made of carbon-fibre-reinforced plastics. Pressure vessels are usually made of titanium and stainless steel.

4.4 The electrical power supply

4.4.1 Energy sources and power systems

The power requirement of a geostationary satellite may be as high as 5 kW. Most of the power is required for the communications payload; for example, approximately 76 per cent of the 4.7 kW requirement of Intelsat VIIA is designated for the communications payload [5]. Other systems also utilize considerable amounts of power, such as the attitude control system.

The supply of electric power in a spacecraft involves several basic elements. There must be a primary source of energy and means for converting that energy into electrical energy. A device is needed for storing the electrical energy to meet peak demands and to provide power during eclipses. Finally a system is required for conditioning, regulating and distributing the electrical energy at the required levels.

Only two primary sources are suitable for missions with lifetimes in space measured in years, namely radiation from the Sun and the spontaneous decay of radioactive material. The radioactivity option is preferred where the power requirement is high, perhaps 10 kW or more, or when an interplanetary mission takes the spacecraft too far from the Sun. But solar radiation, converted to electrical power by an array of photovoltaic cells, is always chosen for Earth satellites used for telecommunications. Rechargeable batteries are used for storing electrical power during the sunlit portion of the orbit, to provide power during eclipses and at peak demand periods.

A simplified diagram of a typical satellite power subsystem is shown in Figure 4.8. Primary electrical power is generated by the main solar panel and it is supplied to the power utilization loads through the main distribution bus. Bus voltage limiters ensure that bus voltages do not exceed the maximum voltage rating of the consuming subsystems. Limiter resistors energized by the bus load limiters provide variable shunt loads to stabilize the bus voltage when the solar panel output voltage rises or the power utilization loadings decrease. The spacecraft batteries charge/discharge control functions are provided by a battery controller device.

When the distribution bus voltage is reduced to a predetermined level due to decreased solar panel output, the battery discharge regulator is activated to provide a minimum regulated bus voltage from the battery. The battery voltage needs to be greater than the bus voltage for two main reasons. The discharge regulator has a certain impedance, causing a voltage drop across its terminals. And the maximum battery voltage required for full recharge is considerably higher than the battery discharge output voltage. Usually a small area of solar panel connected in series with the main panel provides the required recharging voltage boost and limits the maximum recharge current. This simplifies the

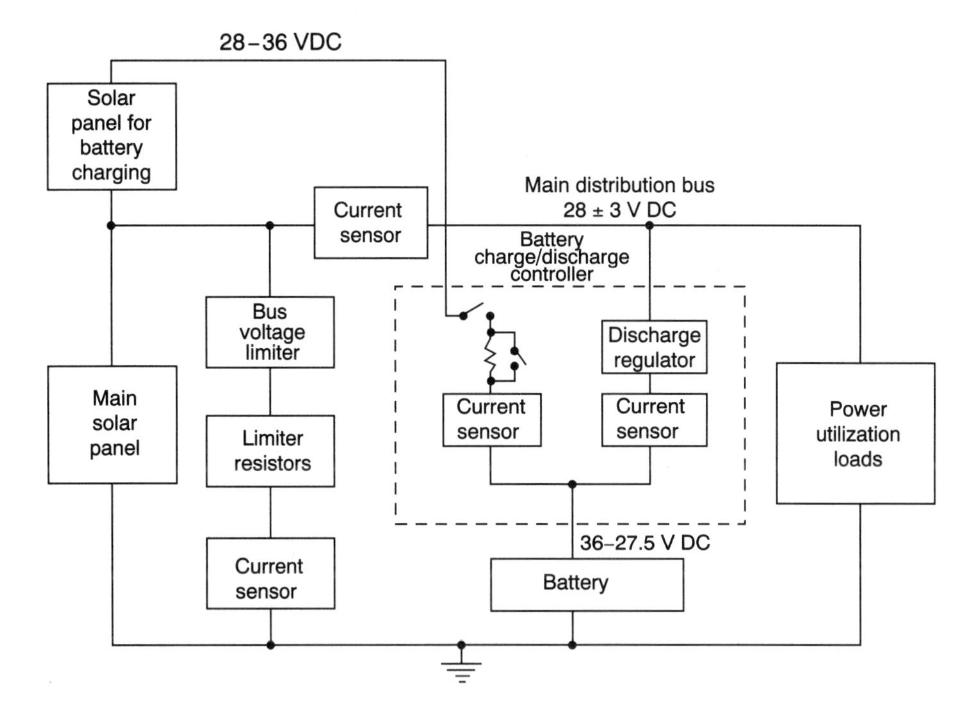

Figure 4.8 *The block diagram of a conventional electrical power subsystem*

charge control functions of the controller to the simple on–off and rate-change switching shown in Figure 4.8.

4.4.2 Solar cells

A solar or photovoltaic cell converts incident solar radiation into electric energy by the photovoltaic effect as illustrated in Figure 4.9. The current versus voltage characteristics of a solar cell, when it is illuminated and when it is dark, is shown in Figure 4.10. The curve of the non-illuminated solar cell indicates that it behaves as a power sink like any other semiconductor diode; its resistance

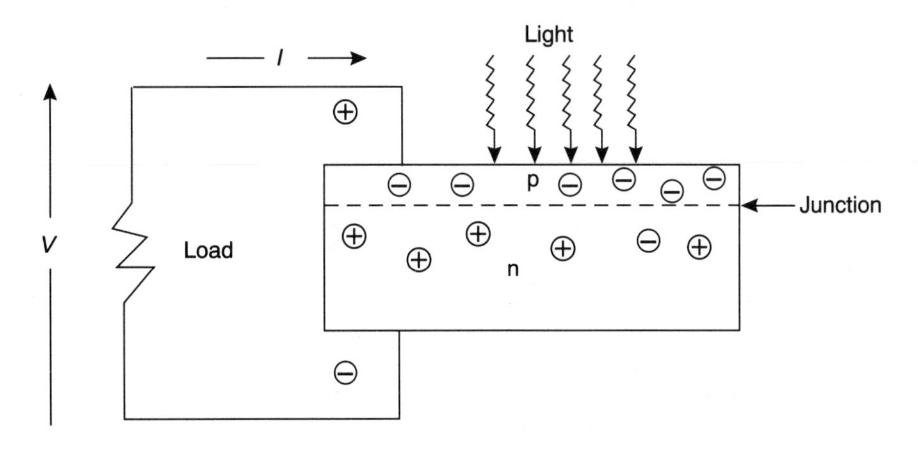

Figure 4.9 *A simplified diagram illustrating solar cell operation*

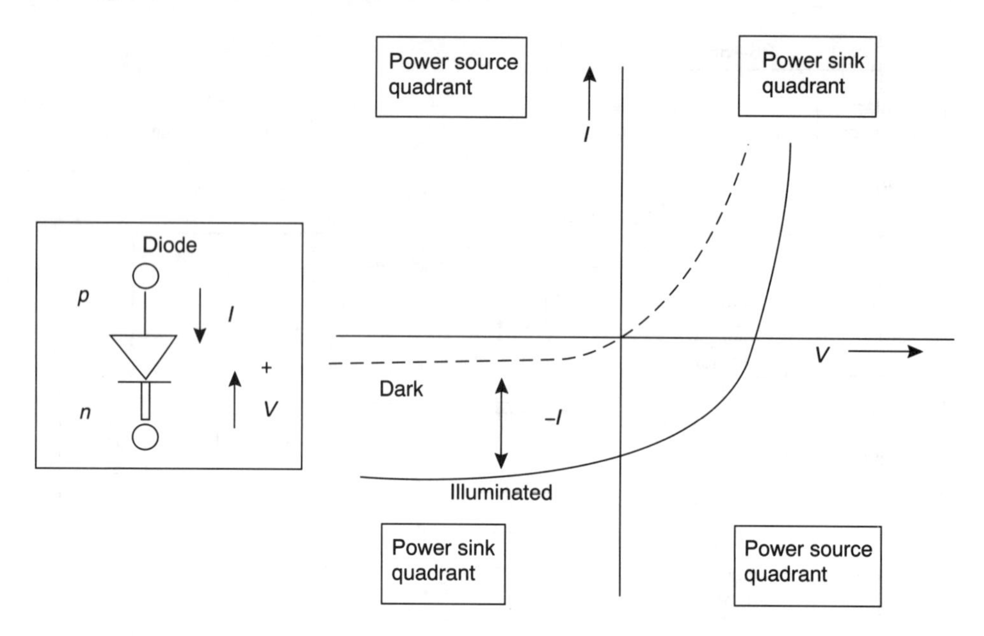

Figure 4.10 *Solar cell current/voltage characteristics in the dark and illuminated states*

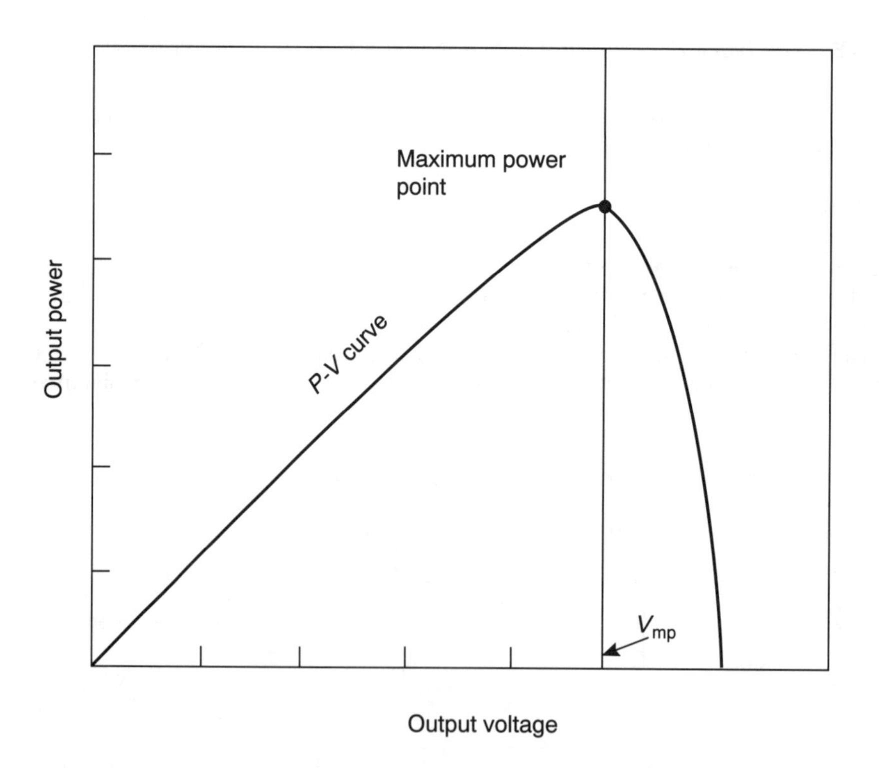

Figure 4.11 *The power/voltage characteristic of typical solar cells (reproduced from* Solar Cell Array Design Handbook *by H. R. Rauschenbach, © Van Nostrand Reinhold 1980, all rights reserved)*

is low when the current flows in the forward direction and high when it flows in the reverse direction. When the solar cell is illuminated, incident photons are absorbed, ionizing the atoms of the semiconductor material and producing electron/hole pairs. Electric current is generated when the electrons and holes reach the junction before recombining.

llumination results in the voltage/current curve being shifted in the minus-current direction in proportion to the strength of the incident illumination. The characteristic I/V output graph shown in Figure 4.11 is obtained by inverting the power source quadrant of Figure 4.10. The output voltage of a silicon cell under typical load conditions is about 0.5 V.

Solar cells can be produced from a number of semiconductor materials, although only silicon and gallium arsenide are widely used at present. A typical silicon cell consists of a thin wafer of ultra-pure single-crystal silicon (p-type) in which the junction is produced by the diffusion of phosphorus (n-type) through one of its faces.

The solar cells of a geostationary satellite may be cycled in temperature between 110 and 340 K. The efficiency of solar cells reduces with increasing temperature, so their temperature has to be kept as low as possible in order to obtain the maximum power output. The variation of efficiency with temperature for gallium arsenide, germanium and silicon solar cells is shown in Figure 4.12.

The efficiency of solar cells in space also decreases with the passage of time. This is due to the damage caused to the cell semiconductor material, due mainly to solar flare protons and the trapped electrons of the Van Allen radiation belts. The cells are also bombarded in space by low energy particles which produce defects near the $p-n$ junction, increasing the current at the no-illumination state and decrease the open circuit voltage. Typically, the power output is reduced by

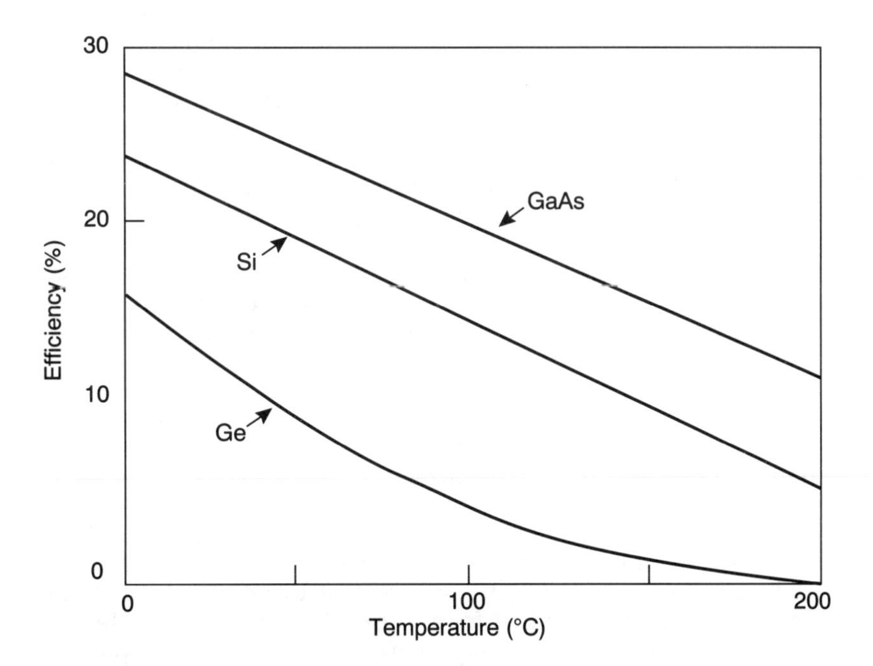

Figure 4.12 *The theoretical efficiency of germanium, silicon and gallium arsenide solar cells as a function of temperature. (Reproduced from* Solar Cell Array Design Handbook *by H. R. Rauschenbach, © Van Nostrand Reinhold 1980, all rights reserved.)*

10 per cent in the first two years of exposure, and a further 10 per cent in the following five years. Radiation damage caused by the lower-energy particles is considerably reduced by placing a glass cover over each solar cell. The amount of radiation falling on the cell is reduced as the thickness of the glass employed is increased. However, the cover glass also acts as a filter which reduces the total energy absorbed by the cell, thus reducing its operating temperature and diminishing somewhat the loss of power output caused by the presence of the cover. The cover glass is usually made of quartz, sapphire, or silica doped with cerium.

The conversion efficiencies of modern silicon and gallium aluminium arsenide solar cells is about 16 and 18 per cent respectively at 273 K, compared with their theoretical maxima of 24 and 27 per cent respectively. For more details, see [6–10].

4.4.3 Solar arrays

The power output of a single solar cell is very low and the individual cells must be connected in a series–parallel matrix as shown in Figure 4.13 to satisfy the voltage, current and power requirements of the spacecraft. The solar cell matrices are mounted on lightweight honeycomb or fibreglass substrates to form solar arrays.

Satellites utilize either dual-spin or three-axis attitude stabilization. In the dual-spin configuration the spinning drum is covered with solar cells. In the three-axis stabilized configuration cells are mounted on deployable flat panel appendages which are continuously oriented to obtain maximum exposure to solar radiation. In spin-stabilized satellites the individual cells are usually arranged in a series–parallel configuration of which the series string can typically vary from 66 to 71 cells. Modules are constructed with three strings in parallel connected together in a series–parallel network as shown in Figure 4.13(ii). This arrangement minimizes the power loss resulting from the failure of a single cell.

The power output from a solar cell decreases by a cosine function as the angle of incidence of solar radiation departs from the perpendicular. In a dual-spin design only the solar cell strings located in the centre of the illuminated half-cylinder receive radiation at perpendicular incidence. On either side of the centre string, the output of each successive string decreases by the cosine function of the angular deviation of the Sun line from the normal. Thus, at best, the total panel output is equal to that of an equivalent projected flat area of solar cells, the dimensions of which are the length and the diameter of the complete cylindrical solar array.

However, in the GSO the Sun line varies seasonally by $\pm 23.5°$ relative to the normal to the spin axis obtained at the solstices. Thus, the equivalent projected flat area decreases further as a cosine function of the incident sun angle. As the maximum length and diameter of spin-stabilized satellites are constrained by the dimensions of the launch vehicle fairing, the maximum equivalent projected flat area exposed to the sun and thus the maximum power generating capability of such satellites is also limited. A further disadvantage of spinning solar arrays is the increased cost that arises from the requirement to provide π times the number of operating solar cells required for a three-axis stabilized satellite. One important advantage of spinning solar arrays, however, is that thermal control is greatly simplified as their temperature varies only within a range of about 270 to 290 K over the normal range of Sun angle variation; with deployable solar panels the temperature range is much greater.

Three-axis stabilization consumes additional power and requires a complex attitude control system, as discussed in Section 4.5, but the associated deployable

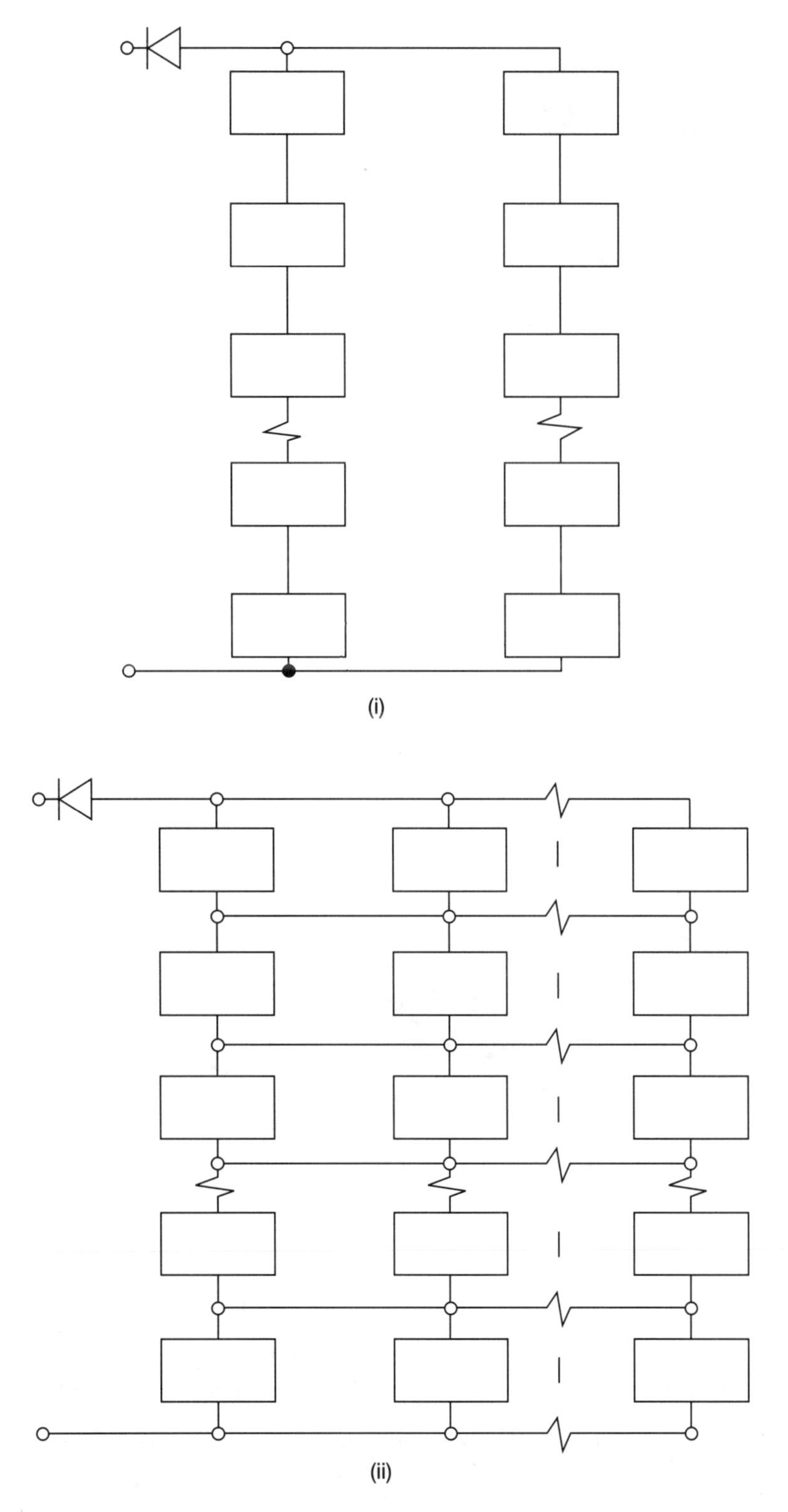

Figure 4.13 *Solar array cell connection configurations (i) a series–parallel matrix and (ii) a parallel–series matrix*

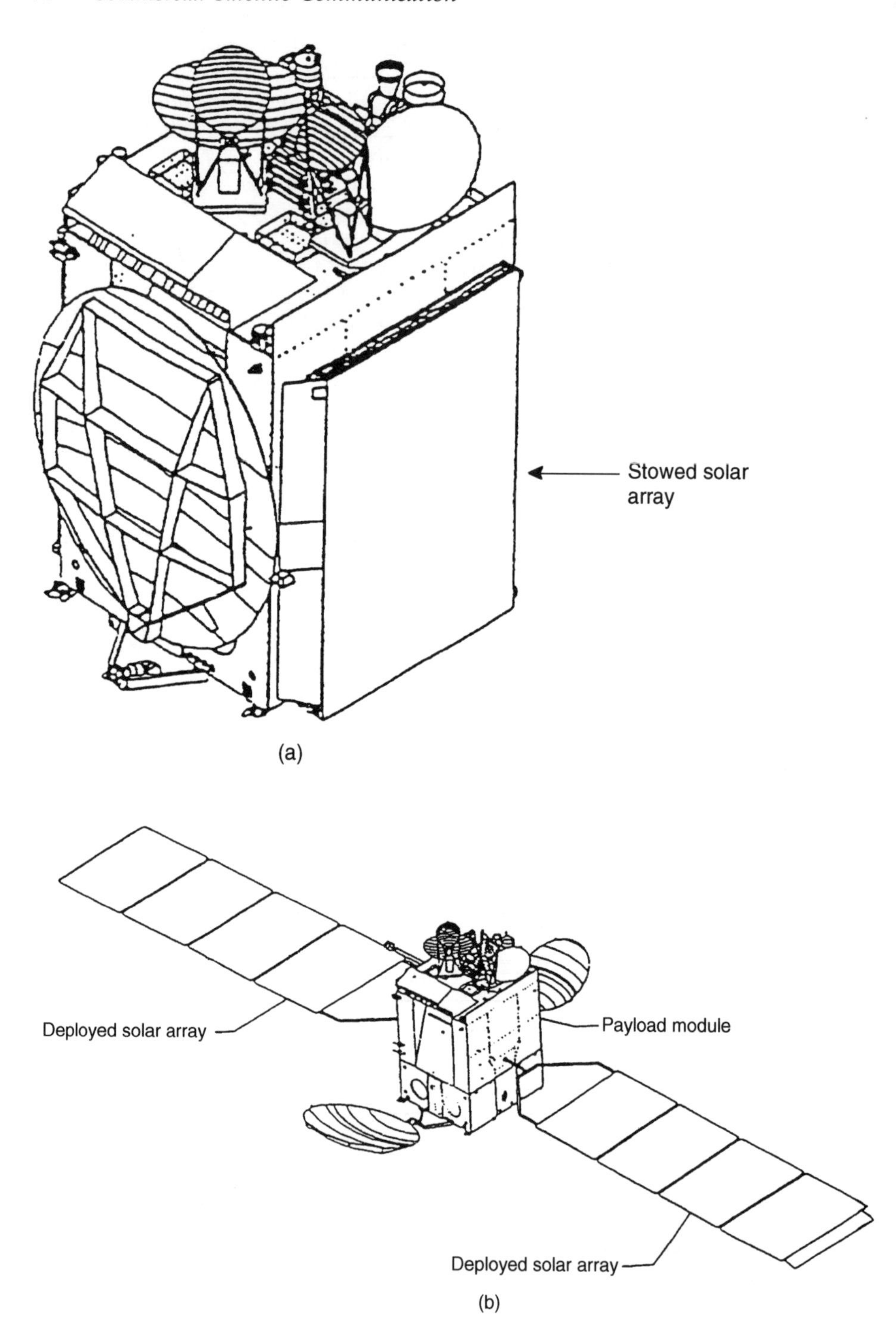

Figure 4.14 *Solar arrays (and also C band antenna reflectors) of the Intelsat VIII satellite (a) in the launch state and (b) the in-orbit state. (Reproducted by permission of Intelsat.)*

solar panels provide more efficient solar power conversion. During launch and while the satellite is spinning in the transfer orbit, the arrays are stowed and only the outer sections (one per array) are exposed to the Sun to provide power for the satellite. Once the satellite is in the GSO, a pulley-and-cable actuator system or an extendible boom is used to deploy the array. A solar array drive mechanism is employed, first to acquire an orbit and then to track the Sun, maintaining the face of the array approximately normal to the solar radiation as the satellite moves around its orbit. The drive mechanism also includes means for transmitting power from the rotating arrays (one revolution per orbit) to the equipment in the spacecraft body. The deployable solar arrays of the Intelsat VIII satellite, in the launch and the deployed configurations, are shown in Figure 4.14.

4.4.4 Energy storage

Storage batteries are used to provide power during eclipses and to satisfy peak power demands, which may be much greater than the average available solar power. In the GSO there are two eclipse seasons per annum, centred on the vernal and autumnal equinoxes respectively; each eclipse period lasts for 45 days with a maximum shadow time of 1.2 hours as shown in Figure 4.15. For satellites in lower orbits, the number of eclipses is greater.

There are many types of storage battery commercially available but only four, namely silver–zinc, silver–cadmium, nickel–cadmium and nickel–hydrogen, have been qualified for space applications to date. For communication satellites, nickel–cadmium batteries have been extensively used in the past but they

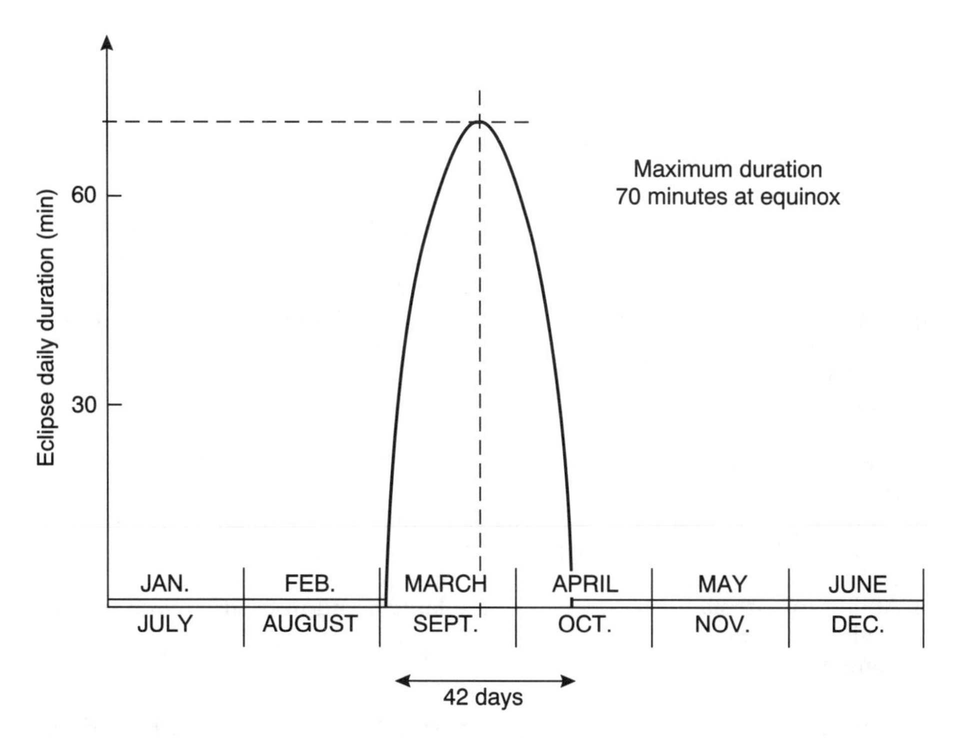

Figure 4.15 *Dates and durations of eclipses of the Sun by the Earth for geostationary satellites. (From* Satellite Communications Systems *by G. Marel and M. Bousquet, © 1993. Reprinted by permission of John Wiley and Sons Ltd.)*

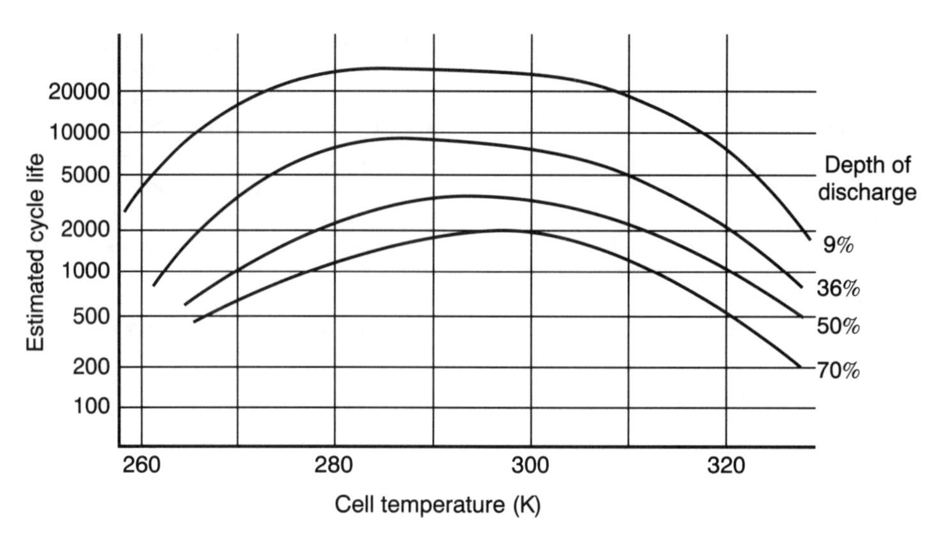

Figure 4.16 *Depth of discharge versus cycle life for 12 ampere-hours nickel–cadmium cells at various temperatures. (Reproduced by permission of Hughes Electronics.)*

are being replaced by the more efficient nickel–hydrogen batteries. Lithium metal sulphide and sodium/beta-alumina/sulphur batteries offering high energy densities are attracting great interest as advanced energy storage devices for future space applications. The primary characteristics of these various storage devices are listed in Table 4.1. The life of a battery is highly dependent upon both the working temperature and the depth of discharge as shown in Figure 4.16.

In general, the main requirements of the storage batteries for space applications are:

(1) To receive and deliver power at high rates. This enables the spacecraft to utilize the extra power delivered by the solar cells when the satellite comes out of eclipse and to provide the extra power needed at peak periods.
(2) To survive a large number of charge–discharge cycles under various conditions. There are 90 charge–discharge cycles per year in the GSO where the battery is being charged during the orbital day and discharged during the orbital night.
(3) High recharge efficiency.
(4) Leakage-free seals, able to withstand a large number of extreme thermal and pressure changes.
(5) Capability of operating in a gravity-free environment in all positions.
(6) High usable energy capacity for low per-unit weight and volume.
(7) Stability under long-term overcharge conditions.
(8) Ability to withstand the stresses of the launch and in-orbit environments.
(9) High reliability.
(10) Long life.

As can be seen from Table 4.1, the most attractive characteristics of the silver–zinc battery are high energy density and good high-temperature discharge performance. For an analysis of the mass characteristics of various space power systems, see [11]. However, this battery has a relatively low cycle life capability which limits its utilization to short-duration missions. Furthermore, control of

Table 4.1 *Primary characteristics of various storage batteries for space applications*

Type	*Energy density* (W-h kg^{-1})	*Cycle life at different depths of discharge*			*Temperature* (K)	*Thermal characteristics*	*Charging/ discharging characteristics*	*Space qualified*	*Life expectancy* (years)
		25%	*50%*	*75%*					
Silver/ zinc	110−133	2000	400	75	283−313	Good high temperature discharge performance. Reduced performance at low temperature.	Current/voltage limit. Control of overcharge is critical.	Yes	2.5 to 4.5
Silver/hydrogen	80−100	>718000			283−313			No	
Nickel/cadmium	22−30	21000	3000	800	263−313	Good discharge performance at low/ moderate temperatures. Reduced discharge performance at high temperatures.	Current limit. Overcharge at moderate rates not critical but needs to be minimized for long life.	Yes	3 to 7
Nickel/ hydrogen	50−80	>15000	>10000	>4000	263−313		Relatively unlimited for overcharge.	Yes	
Silver/cadmium	60−70	3500	750	100	273−313			No	10 (goal)
Lithium/metal sulphide	160−200	....	300−500	....	723			No	10 (goal)
Sodium/ beta-alumina/ sulphur	160−200	....	500−1000	....	623			No	10 (goal)

overcharge is critical with this type of battery as oxygen produced during overcharge does not readily recombine.

Nickel–cadmium batteries have much longer life and increased tolerance to overcharge exposure at moderate charge rates due to their better oxygen recombination abilities. The charge control functions needed with these batteries are much simpler than those needed for silver–zinc. The main disadvantages are lower energy density and reduced high-temperature performance, which impose a weight penalty and thermal design constraints on the satellite.

Nickel–hydrogen batteries have improved in-orbit lifetimes and reduced mass compared with nickel–cadmium. This makes them the battery of choice for use in the large long-life geostationary satellites such as the latest Intelsat, Astra and Optus series, despite their relatively higher cost.

4.4.5 Power conditioning and control

The power conditioning and control system (PCCS) ensures that the output of the solar array and the storage battery satisfy the power requirements of the various subsystems. The PCCS also controls the battery charge–discharge function, which is essential in order to maximize the battery lifetime. In essence, the PCCS is an assembly of various types of regulators, DC-to-DC or DC-to-AC convertors and control circuits.

Table 4.2 *The advantages and disadvantages of power systems with regulated and unregulated buses*

Type	*Advantages*	*Disadvantages*
Regulated	Some loads can be connected directly to the bus. Low bus impedance. Regulator/converter units with lighter load characteristics. The operating point of the solar array is fixed.	Solar array output is partly wasted. Limited buffering of devices from bus. A number of types of regulator are required.
Unregulated	More tolerant of single point failures. Devices are buffered from bus noise. Simple interface.	Operation of solar array at maximum power point may be prevented by unit switch-on surge current. Complex load regulator/ converter devices. If devices are required to operate over a wide bus voltage range, input filters are required which entail considerable weight penalty.

PCCSs can be classified into two main categories: dissipative and non-dissipative. In dissipative systems the power requirement of the satellite is less than the output of the solar arrays, the unused power being dissipated by the system. In non-dissipative systems almost all the power provided by the solar arrays is used to satisfy the operational needs of the spacecraft.

Dissipative systems can be categorized according to whether the bus voltage is regulated or unregulated. In general, the different voltage levels and regulation profiles which are required at different points of the spacecraft subsystems differ from the ones provided by the bus. The bus voltage therefore needs to be further regulated, increased, decreased or inverted by utilizing regulators and DC-to-DC and DC-to-AC convertors. The main advantages and disadvantages of regulated and unregulated bus systems are listed in Table 4.2.

4.5 Orbit and attitude control

4.5.1 Orbit control

For a communications satellite to accomplish its mission, it must first acquire and then maintain its specified orbit within close limits. The initial acquisition of a geostationary orbit is described in Section 3.2.2; other kinds of orbit are acquired in a similar way. The orbital perturbations which make subsequent corrections of the parameters of the orbit necessary are described in Section 2.2.3. The final stages of the launching process and all of the in-service orbital corrections are carried out by firing thrusters on board the satellite in the appropriate directions to obtain the desired incremental velocity vectors. While the satellite is on-station and operating, it must also be correctly oriented, as described in Section 4.5.2, so that its antennas and its solar arrays can function as intended; this orientation of the satellite attitude in space also facilitates the adjustment of the orbital parameters.

In order to maintain the satellite orbit inclination at zero, the gravitational forces due to the Sun and the Moon should be counteracted by the north–south station-keeping (NSSK) propulsion system, which provides thrust to the north or the south at the appropriate phase of the orbit. The inclusion of the NSSK system and its fuel on board a satellite carries a mass penalty, which can be as high as 15 per cent of the spacecraft mass when conventional hydrazine technology is used; the penalty may be even larger if a very long lifetime in orbit is foreseen. Utilization of an electric propulsion system can reduce the mass penalty to about 7 per cent for a 7-year geostationary orbit mission. Various thruster options are reviewed in Section 4.6.

The forces arising due to the triaxiality of the Earth and solar radiation pressure act along the plane of the orbit, resulting in a relative east–west satellite motion. A correction can be provided by operating thrusters in an easterly or a westerly direction. The propellant mass required for east–west station-keeping is normally in the range of 3 to 10 per cent of the mass of the satellite, depending on the satellite configuration and the correction strategy employed.

4.5.2 Attitude control

The attitude of a satellite is normally described in terms of the same set of coordinates as is used for ships and aeroplanes, namely roll, pitch and yaw. As shown in Figure 4.17, with the satellite in the attitude that its designer intended for normal operation, the yaw axis points to the centre of the Earth, the roll axis points in the direction of the orbital velocity vector and the pitch axis is perpendicular to the orbital plane; for a geostationary satellite the pitch axis points by convention towards the south. Once a satellite has been established in

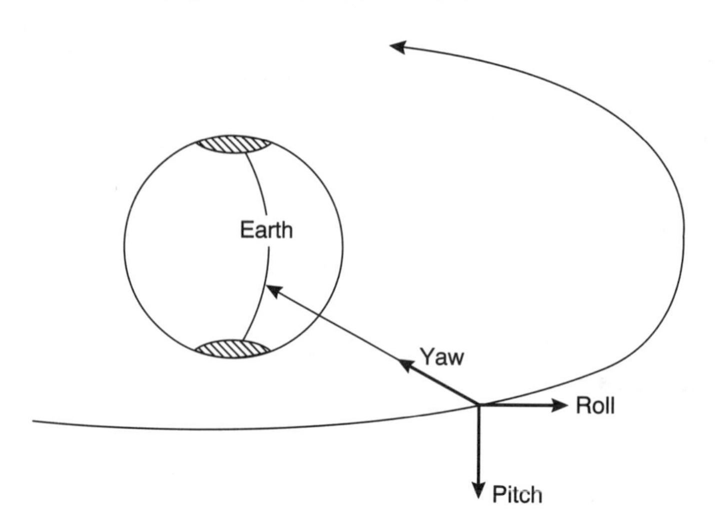

Figure 4.17 *Satellite attitude coordinates, roll, pitch and yaw*

its intended attitude in orbit, the maintenance of that attitude can be facilitated by spin stabilization or three-axis stabilization.

Spin-stabilized satellites

In a spin-stabilized configuration the satellite is spun around a rotationally symmetrical axis which tends to be fixed in vertical space by the momentum provided by the spin. According to Newton's laws the angular momentum of a body remains constant unless external torques are applied on it.

Although internal torques cannot change the overall momentum of a body, they can change the amount of kinetic energy that it possesses. If these internal torques dissipate energy, say in the form of heat, the body will then tend towards the minimum energy state. For spin-stabilized spacecraft this minimum energy state is the required mode of spin only if the inertia about this spin axis is greater than the inertia about any other axis. This condition is only fulfilled if the body is disc-shaped (oblate). The placement of large antennas above the spinning body make a typical satellite body pencil-shaped (prolate), tending to spin stably about a transverse axis rather than the axis of symmetry.

A solution to this problem was patented by the Hughes Aircraft Company of the USA under the name 'Gyrostat' but is generally known as the dual-spin configuration. In this configuration the antennas are despun mechanically against the rotation of the outer body (rotor); see Figure 4.3. It is convenient to include the communications payload on the antenna platform; thus, Figure 4.18 shows the despun platform of an Intelsat IV satellite, the antennas being above and many components of the 12 transponders being below. A damped pendulum (or nutation damper) is placed in this despun part, stabilizing the complete satellite body about the spin axis of the spun part.

The despun platform of a spin-stabilized satellite maintains an almost constant orientation relative to the Earth when the spin axis of the rotating body is normal to the orbital plane (that is, the spin axis is parallel with the pitch axis). Consequently, the roll and yaw orientation of the spacecraft being fixed in relation to the Earth, only pitch angle measurement and control is needed to ensure that the correct attitude is maintained. An infrared Earth horizon sensor with a relatively narrow field of view placed on the spinning body can provide the

Figure 4.18 *The despun platform of Intelsat IV, with antennas above and many of the components of the 12 transponders visible below. (Reproduced by permission of Intelsat.)*

necessary pitch measurements. From these, the antenna pitch orientation relative to the Earth can be obtained by comparing the spin phase at which the Earth is detected with the phase of an index relating the despun part containing the antennas to the spun part. By applying a suitable amount of torque to the motor connecting the spun and the despun parts of the satellite, the pitch orientation is controlled.

External forces cause a gradual change of the angular vector orientation which has to be measured and then corrected. This can be achieved by employing two narrow beam infrared pitch control earth sensors, inclined a few degrees above and below the spin plane. In this position, when the spin axis is normal to the

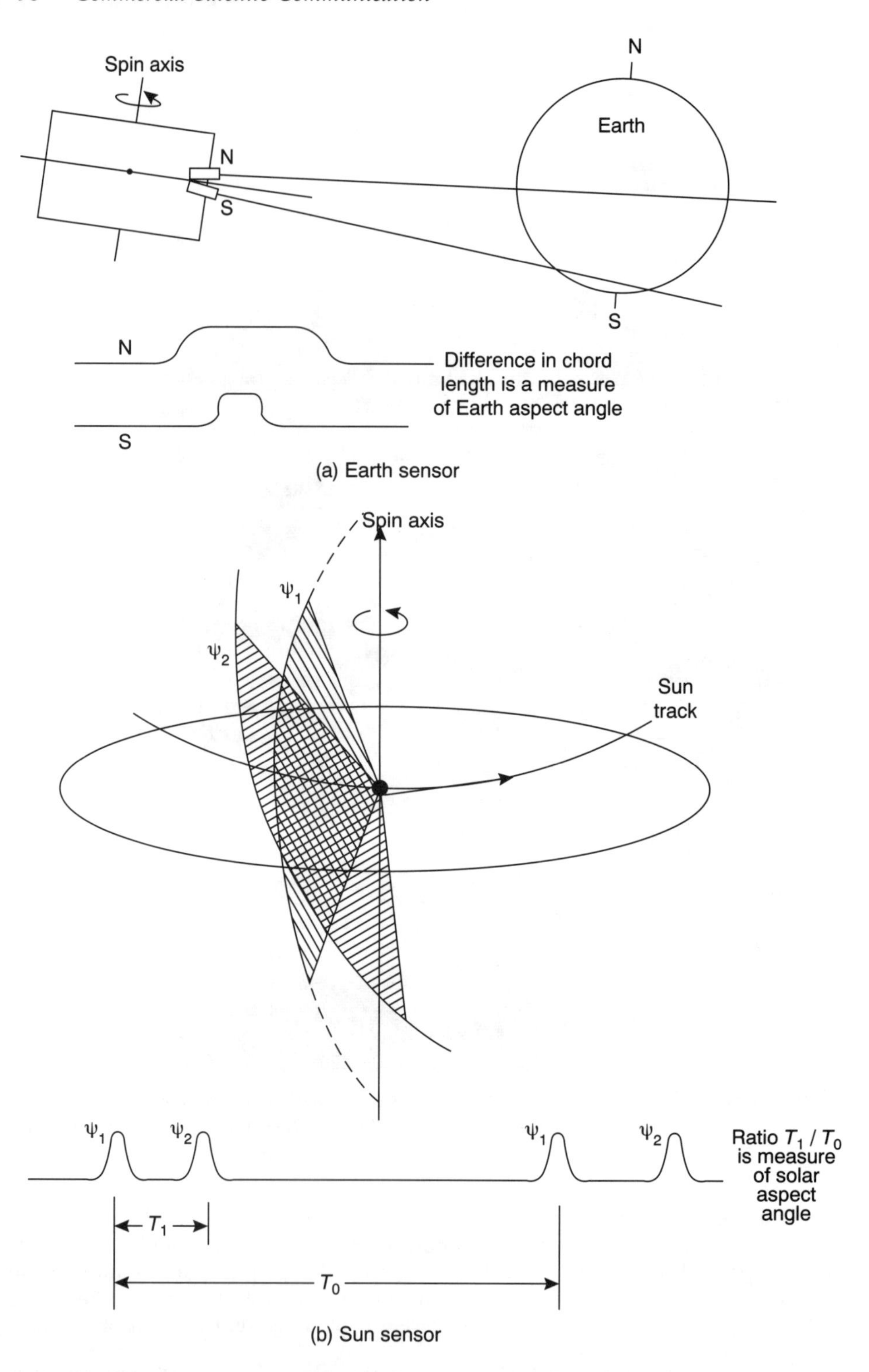

Figure 4.19 *Attitude measurements of a dual-spin satellite (a) using an Earth sensor and (b) using a Sun sensor. (Reproduced by permission of Hughes Electronics.)*

orbital plane, the sensor scans a chord-like section of the Earth's disc between the pole and the equator as shown in Figure 4.19(a). The width of the sensed chord is compared with the nominal width and from this the Earth aspect angle is determined, enabling the roll attitude of the satellite to be computed.

A Sun sensor on the spinning part of the satellite enables a second, independent measurement of roll error to be made during most of the day. Additional attitude information can be provided by measuring the angle between the Earth and the Sun. The solar aspect angle is measured by two-slit Sun sensors arranged in a Vee position. The azimuth angle between the Earth and the Sun is calculated from the time between the two respective pulses, as indicated in Figure 4.19(b). An accurate determination of the spin axis attitude is achieved from these measurement.

For a spin-stabilized satellite, all station-keeping and attitude control manoeuvres can be carried out with a combination of one radial and one axial thruster jet, as shown in Figure 4.20. These jets are positioned in such a way that advantage is taken of the spin motion of the spacecraft to control the direction in which the thrust is made. A brief pulse of thrust from the spinning radial jet enables a velocity change to be made in the orbital plane by appropriately phasing the pulse in relation to the position of the Earth or the Sun. A pulse from the axial jet produces a change of attitude of the roll or the pitch axis, depending on the phase of the body spin at which it occurs. A long thrust from the axial jet, having a duration of several or many rotations of the spinning body, made at the appropriate time of day, produces a change of orbital inclination. To provide long-life reliability through a measure of redundancy, four or six jets are provided in total, but this is to be compared with the 12 to 20 jets normally required on a three-axis stabilized satellite. For a full description of Intelsat IV, a typical spin-stabilized satellite, see [12].

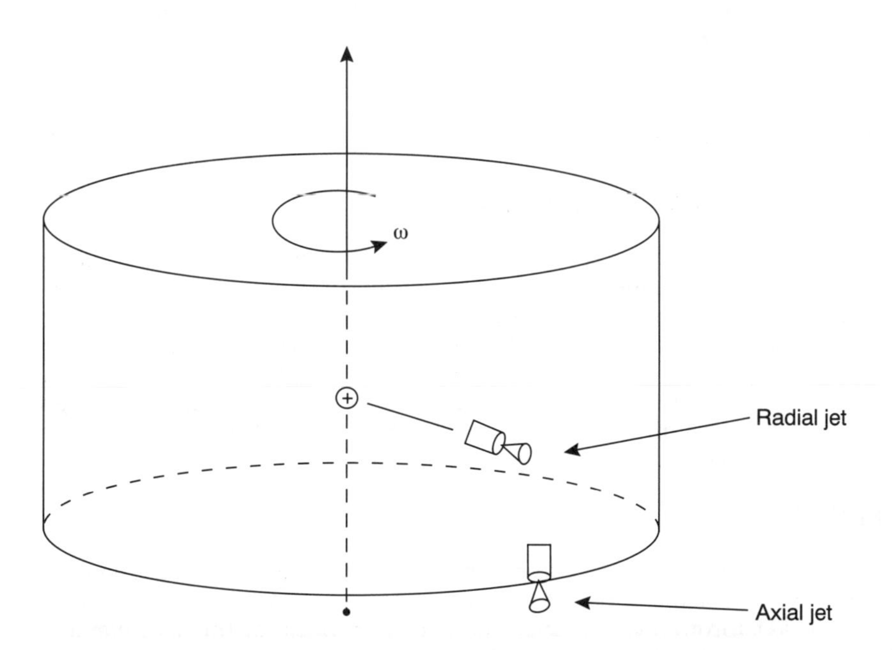

Figure 4.20 *The locations of the two thruster jets required for the execution of all the thrusting manoeuvres required for a spin-stabilized satellite*

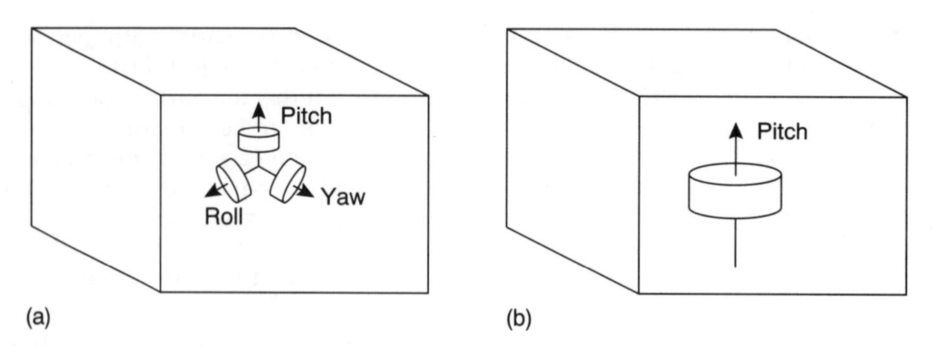

Figure 4.21 *Types of three-axis stabilized satellite: (a) zero momentum and (b) momentum bias*

Three-axis stabilized satellites

As the need for higher satellite power grew, the relative inefficiency of the cylindrical solar arrays led most spacecraft manufacturers to develop three-axis or body stabilized configurations. The solar arrays of the body stabilized satellites are pointed continuously at the Sun and the antennas point continuously at the Earth. In this type of spacecraft the solar array drive is normally aligned with the pitch axis.

Body stabilized satellites can be divided into two categories according to their momentum states, namely zero momentum and momentum bias. The zero-momentum configuration has independent control loops for each of the three axes. Consequently each control loop needs an attitude error sensor and the necessary control forces are produced by their respective dedicated reaction wheels as shown in Figure 4.21(a). Excess momentum accumulated in each of the three wheels is dumped using thrusters or magnetic torquers. In the momentum-bias configuration, a pitch-wheel running at high speed is employed to provide gyroscopic stiffness; see Figure 4.21(b).

4.6 The propulsion system

4.6.1 Rocket propulsion principles

Rocket motors, whether part of a massive booster or a low thrust device used for satellite in-orbit attitude adjustment, operate on the principle that mass is accelerated and expelled, thus creating a reaction force in accordance with Newton's third law. The rocket motor is a vessel containing matter and energy. The matter, which is originally at rest relative to the vessel, is transformed into a gaseous form by the release of kinetic energy. This gas escapes through the nozzle at high velocity while the remaining non-gaseous part of the rocket undergoes a time rate of change of momentum, resulting in a reaction force or thrust.

Rocket thrust

An equation for the rocket thrust can be derived by considering a rocket system consisting of two elements: a rocket of mass m and an exhaust gas of mass δm leaving the rocket through the nozzle. The net momentum of this two element combination can be expressed by the following equation:

$$N = mv + \delta m(v - v_e) \qquad (4.1)$$

where N = net momentum,
v = rocket velocity,
v_e = relative velocity of exhaust gas leaving the nozzle.

The dynamic decrease in rocket mass corresponding to a proportional increase in rocket exhaust gas mass can be expressed as

$$d(\delta m)/dt = -dm/dt \tag{4.2}$$

Assuming that the exhaust velocity v_e is constant, the acceleration of the gases through the nozzle is zero; hence

$$dv_e/dt = 0 \tag{4.3}$$

The net momentum can be differentiated with respect to time and the result equated to zero. Furthermore, by considering that δm is very small and in the limit approaches zero, and by applying equations (4.2) and (4.3), the following equation can be derived:

$$m(dv/dt) = -(dm/dt)v_e \tag{4.4}$$

As the first part of equation (4.4) can be considered as a reaction force F acting on the rocket, then the equation can be rewritten

$$F = -(dm/dt)v_e \tag{4.5}$$

The negative sign in equation (4.5) reflects the fact that the nozzle mass flow is taken as a positive number and the exhaust velocity and the reaction force act in opposite directions.

The pressure forces acting on the internal and external walls of a rocket motor, releasing kinetic energy by transforming matter into gaseous exhaust, are shown in Figure 4.22. The sum of these forces produces the thrust which propels the rocket. Consequently, this thrust can be expressed by the following equation:

$$F = \int_s p \, ds \tag{4.6}$$

where p = pressure acting on the rocket walls,
ds = vector element of the rocket surface area,
s = total wall surface area of the rocket motor,

or

$$\int_s p \, ds = \int_{s_i} p_i \, ds_i + \int_{s_o} p_o \, ds_o \tag{4.7}$$

where s_i = inside wall surface area of the rocket motor,
s_o = outside wall surface area of the rocket motor,
p_i = pressure acting on the inside wall surface area of the rocket motor,
p_o = pressure acting on the outside wall surface area of the rocket motor,
ds_o = vector element of the outside wall surface area,
ds_i = vector element of the inside wall surface area.

By applying the momentum theorem to the combustion gas contained within the internal walls of the rocket, the integral of equation (4.7) can be evaluated. In

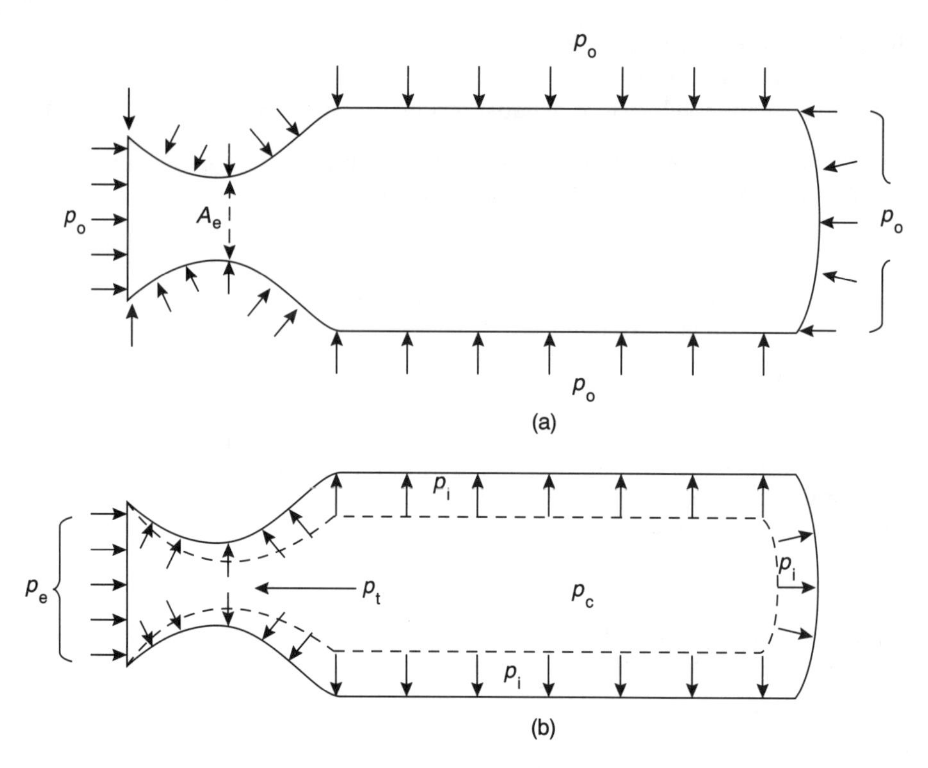

Figure 4.22 *An illustration of the production of thrust by the action of pressure forces on the internal and external walls of a rocket motor*

particular, by equating the axial time rate of change to the total force acting on the gas contained within the internal walls of the rocket and by assuming axially symmetrical pressure forces, the following scalar equation can be derived:

$$\int_{p_i} p_i \, ds_i = A_e p_e - \dot{m} v_{e_a} \tag{4.8}$$

where A_e = the area of the nozzle exit plane,
$\dot{m}$ = exhaust gas mass flow rate,
v_{e_a} = the average axial component of the exhaust velocity relative to the rocket.

Thus $\dot{m} v_{e_a}$ is the axial flux of momentum through the exit plane. The exhaust velocity v_e normally consists of two orthogonal components, the velocity component along the rocket axis, v_{e_a}, and the velocity component perpendicular to the rocket axis, v_{e_p}. For simplicity, it is assumed that $v_{e_p} = 0$ and therefore $v_e = v_{e_a}$. Hence equation (4.8) can be written as

$$\int_{p_i} p_i \, ds_i = A_e p_e - \dot{m} v_e \tag{4.9}$$

The integral of the pressure forces acting on the outside wall area of the rocket in equation (4.7) can be evaluated by considering the resultant force due to the atmospheric pressure p_o (see Figure 4.22(a)) and the fact that the sum of these forces is zero over the outer surface of the closed container at rest.

Hence

$$\int_{p_o} p_o \, ds_o = -p_o A_e \tag{4.10}$$

It is assumed that the pressure forces in equation (4.7) are axially symmetrical. By substituting equations (4.8) and (4.10) into equations (4.6) and (4.7), the following equation for computing the rocket thrust is obtained:

$$F = \dot{m} v_e + A_e (p_e - p_o) \tag{4.11}$$

It is assumed that v_e and p_e are averaged over the nozzle exit area. For a more detailed derivation of the rocket thrust equation, see [13, 14].

The first term of the right-hand side of equation (4.11) represents the net momentum increase in the gases accelerated and ejected through the nozzle and it is called the momentum thrust. The second term is the resultant of the pressures acting over the nozzle exit area A_e and it is called the pressure thrust. In this context equation (4.11) can be represented as follows:

$$\text{total thrust} = \text{momentum thrust} + \text{pressure thrust} \tag{4.12}$$

Exhaust velocity

It can be shown that the thrust F is maximum when the pressure thrust is zero, or $p_e = p_o$ in equation (4.11); see [13, 14]. The nozzle geometry determines the value of $p_e - p_o$ and consequently the value of the pressure thrust. Rocket motors use nozzles which initially decrease and then increase their cross-sectional area (De Laval type nozzles). A nozzle is termed 'correctly expanded' when its geometry is such that $p_e = p_o$. The exhaust velocity at the rocket nozzle exit plane, v_e, can be calculated approximately using the following relationship:

$$v_e \approx k f(A_e/A_t)(T_c/\mathrm{MW})^{1/2} \tag{4.13}$$

where k = a constant,
A_e = exit plane area,
A_t = thrust plane area,
T_c = combustion chamber temperature,
MW = molecular weight of exhaust gases,
$f(\cdot)$ = a function of $(\cdot)$

The principal arithmetic parameters of rocket propulsion, including the ones contained in equation (4.13), are shown in Figure 4.23.

The performance of a rocket is highly dependent on the exhaust velocity v_e; see equation (4.11). To obtain high rocket performance, high exhaust (exit) velocity must be achieved. The exhaust velocity is one of the major indicators of propellant performance. The relatively high performance of the liquid oxygen/liquid hydrogen bipropellant system arises from the high exhaust velocity that this combination produces, due to the low molecular weight of the exhaust gases.

For all currently operational rocket motors, the momentum thrust is much greater than the pressure thrust, hence the thrust equation (4.11) can be reduced to

$$F = \dot{m} v_e \tag{4.14}$$

and therefore

$$v_e = F/\dot{m} \tag{4.15}$$

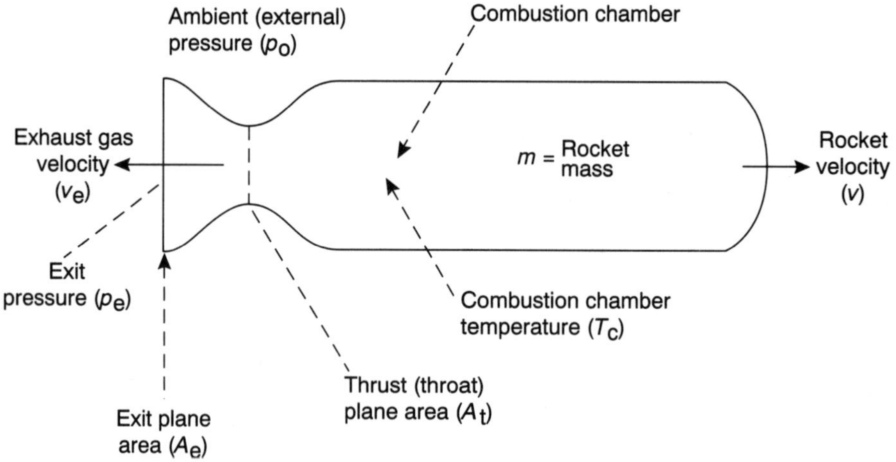

Figure 4.23 *The principal arithmetical parameters of rocket propulsion*

Specific impulse

The performance of a rocket motor can also be expressed by a quantity called the 'specific impulse' I_{sp}. It measures the amount of thrust produced per unit mass flow rate of propellant. Hence, by definition,

$$I_{sp} = F/\dot{m}g_o \tag{4.16}$$

where g_o = the reference acceleration due to gravity.

Alternatively I_{sp} shows the amount of time during which a propellant can deliver a thrust equal to the initial weight of the propellant. Hence I_{sp} can be expressed as follows:

$$I_{sp} = \left(\int F\,\mathrm{d}t\right) \Big/ \left(g_o \int \dot{m}\,\mathrm{d}t\right) \tag{4.17}$$

The dimension of I_{sp} is time, measured in seconds.

I_{sp} can be expressed in terms of the exhaust velocity by substituting equation (4.15) into equation (4.16) to obtain

$$v_e = g_o I_{sp} \tag{4.18}$$

Rocket equation

If we ignore the effects of gravity during a rocket burn, the rocket thrust is used only to accelerate the rocket, and hence

$$F = v_e(\mathrm{d}m/\mathrm{d}t) = -m(\mathrm{d}v/\mathrm{d}t) \tag{4.19}$$

where $\mathrm{d}m/\mathrm{d}t = \dot{m}$. The negative sign shows that the expulsion of propellant, which leads to a decrease in rocket mass, produces (a positive) acceleration. By rearranging equation (4.19),

$$\int_{v_i}^{v_f} \mathrm{d}v = -v_e \int_{m_i}^{m_f} \mathrm{d}m/m \tag{4.20}$$

where v_i = initial velocity of rocket,
v_f = final velocity of rocket,
m_i = initial rocket mass,
m_f = final rocket mass.

By integrating the terms of equation (4.20) from their initial values to their final values, the so-called 'rocket equation' is obtained, as follows:

$$m_i/m_f = e^{\Delta V/v_e} \tag{4.21}$$

where ΔV = the vectorial velocity increment. The term m_i/m_f is called the mass ratio. By substituting equation (4.18) into equation (4.21), the rocket equation can be expressed in terms of I_{sp} rather than v_e, as follows:

$$m_i/m_f = e^{\Delta V/g_o I_{sp}} \tag{4.22}$$

The rocket equation can also be expressed in terms of the velocity increment that the rocket burn produces. From equation (4.22), the velocity increment can be expressed as follows:

$$\Delta V = g_o I_{sp} \ln(m_i/m_f) \tag{4.23}$$

The rocket equation provides the ratio of the initial to final mass of the rocket or spacecraft as a function of the velocity increment ΔV resulting from the rocket burn as a function of either the exhaust velocity v_e or the specific impulse I_{sp}. It also shows that, by increasing the I_{sp}, the rocket mass ratio decreases for a fixed ΔV, thus less propellant is consumed.

Specific impulse/satellite mass trade-offs

The impact of the differences of specific impulse on effective payload can be seen most clearly when considering the choice of thruster system for maintenance of orbit and attitude during a satellite's working lifetime in orbit. In general, a total ΔV budget is fixed for each GSO satellite mission in orbit. For example, the ΔV budget for the Intelsat IV and IVA satellites was 432 m s^{-1}. Most of this, namely 358.4 m s^{-1}, was required for north–south station-keeping. Consequently the only parameter that could be increased in order to reduce the mass ratio was the I_{sp}; see equation (4.23). The effect on mass ratio that could be obtained by using thruster systems having various values of I_{sp} for a payload of this magnitude is shown in Table 4.3. For more details, see [15].

Table 4.3 *The impact of varying I_{sp} for Intelsat IV and IVA satellites on mass ratio*

m_i/m_f	I_{sp} (secs)	*Type of motor*
1.02	3000	Electric propulsion
1.14	300	Hydrazine, electrically augmented
1.18	220	Hydrazine
1.33	150	Hydrogen peroxide

Rocket propulsion devices

The rocket motor contains its own propellants, consisting of a fuel and an oxidizer. Rocket motors can be classified into three categories according to the type of energy contained in their propellants and used to produce gases of high pressure and temperature. Motors utilizing the chemical energy of the propellants are called chemical rocket motors. The ones that use electricity to raise the temperature of a propulsive gas or to generate magnetic fields that accelerate electrically charged particles are called electric rocket motors or electric engines. Finally, those utilizing the energy released during a nuclear reaction are called nuclear rocket engines.

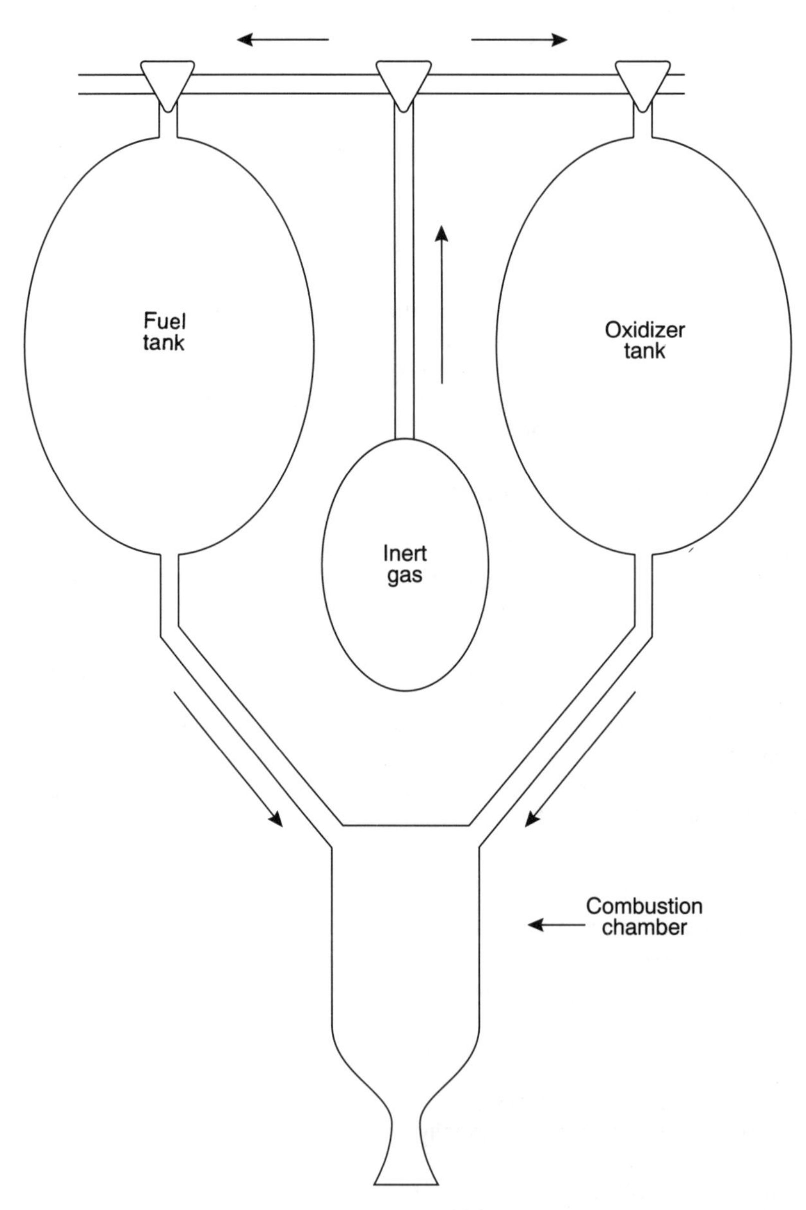

Figure 4.24 *A simplified drawing of a liquid bipropellant rocket motor*

Chemical rocket motors are further subdivided into three categories, namely solid, liquid and hybrid. The solid motor, as the name implies, employs a solid propellant which is a mixture of fuel and oxidizer materials. The surface of the solid propellant burns to produce the propelling gas at high temperature and pressure. Liquid motors use a liquid fuel and a liquid oxidizer which are fed to the combustion chamber from different storage tanks as shown in Figure 4.24. Hybrid motors are usually a combination of a solid fuel and a liquid oxidizer component. The principal characteristics of the main rocket motors are compared in Table 4.4.

Table 4.4 *Principal characteristics of the main rocket motors*

Motor type	*Thrust (N)*	*Operating time (approx.)*	*Main applications*	*Status*
Chemical solid	up to 10^7	up to minutes	Launch vehicles, boosters, AKMs, PKMs.	Extensively used.
Chemical liquid	up to 10^7	up to minutes	Launch vehicles, stabilization and control, satellite unified propulsion.	Well developed and in growing use.
Chemical hybrid	up to 10^5	up to minutes	Launch vehicle upper stages.	Experimental flights.
Electric engines	10 to 10^{-6}	up to years	Satellite station-keeping, satellite stabilization and control, orbit transfer.	Increasingly used for satellite station-keeping.
Nuclear	up to 10^2	up to hours	Interplanetary.	Under development.

4.6.2 The Chemical rocket motor

In a chemical rocket the fuel and the oxidizer react in the combustion chamber, releasing chemical energy, producing combustion products in the form of gases at temperatures between 2000 and 3000 K. These gases thermally expand within the limited space of the combustion chamber, raising the pressure, which forces the gases to accelerate and escape through the nozzle, generating a reaction force. Any chemicals that react to produce high-temperature gases can be used as rocket propellants. However, in practice, propellants with highly reactive combustion products, such as hydrogen fluoride, are not used.

Liquid monopropellant motors

In monopropellant motors, hydrogen peroxide or hydrazine are decomposed, producing high-temperature gases. The monopropellant rocket motor has the

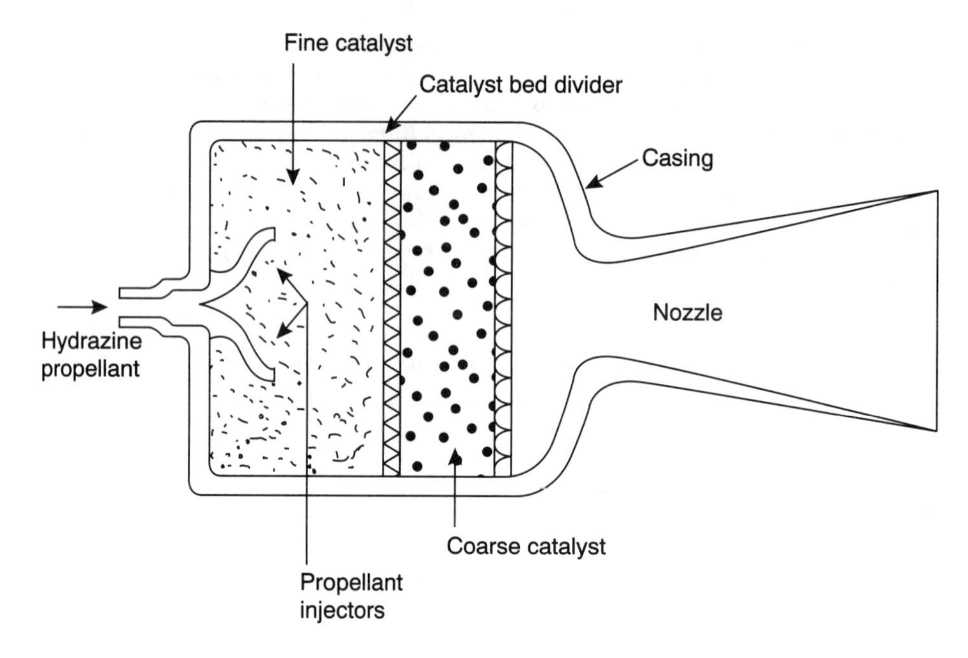

Figure 4.25 *A simplified drawing of a hydrazine thruster*

advantage of simplicity, but at the expense of a relatively low specific impulse, about 250 seconds.

In a monopropellant hydrazine catalytic thruster, liquid hydrazine is injected into a catalyst bed of iridium, decomposing the hydrazine into a hot gas of about 1200 K consisting of nitrogen, hydrogen and ammonia, which is expelled through a conical nozzle. The space forming the catalytic bed is usually filled with small balls which offer the maximum propellant contact area within the confined area of the combustion chamber. Figure 4.25 shows a hydrazine thruster; the catalyst consists of two sizes of porous balls consisting of iridium on aluminium oxide, offering a surface area of about 1 m^2 for catalytic decomposition.

Hydrazine thrusters can offer specific impulses of around 220 seconds, depending on the temperatures of the catalyst and the propellant, the starting conditions (hot or cold) and the mode of operation (continuous or pulsed). Thrusts of 0.5 to 20 N can be obtained, the maximum value being set by the rate at which hydrazine can be decomposed.

In electrothermal monopropellant catalytic hydrazine thrusters the gases produced by catalytic decomposition are superheated to a temperature of about 2300 K before ejection through the nozzle. This increases the velocity of ejection and consequently the specific impulse rises from about 220 to about 300 seconds. Heating is provided by electric elements which require several hundred watts of electric power. The thrust obtainable is limited to less than 1 N. For a more detailed description of hydrazine thrusters, see [16–18]

Liquid monopropellant propulsion systems

A representative monopropellant propulsion system consists of four spherical fuel tanks connected to the thrusters through filters and latch valves. The filters remove all particles larger than a certain size, typically 10–20 μm, to prevent valve contamination. The tanks are placed in the four corners of the satellite structure and are connected in pairs to ensure the minimum spacecraft imbalance as fuel is consumed. Each tank is divided into two parts by a diaphragm, separating a pressurizing gas such as nitrogen or helium from the fuel, normally hydrazine; the pressurizing gas serves to force the propellant to the thrusters (the 'blowdown mode') when the latch valves permit it. Latch valves are placed between the tanks and the thrusters to provide isolation in case of thruster failure. Cross connection of all feeding lines between the two latch valves minimizes the effect of a valve failure. Solenoid valves permit control of individual selected thrusters. Heaters are used to prevent the freezing of space-exposed fuel pipes and to maintain the desired operating temperature of thrusters. Pressure transducers placed between each tank and the thruster clusters provide a pressure reading which, together with temperature measurements given by temperature sensors, permit the amount of fuel remaining in each tank pair to be calculated.

All the components of the system, with the possible exception of the thrusters, are constructed from titanium or stainless steel; the former offers lightweight and chemical compatibility with hydrazine, while the latter has better thermal conductivity and lower cost.

Liquid bipropellant motors

Bipropellant fuels are classified as those that must be stored at cryogenic temperatures, such as liquid oxygen and liquid hydrogen, and those that are stable at ambient temperatures on Earth; the latter are said to be 'storable'. Some propellants ignite spontaneously on contact (hypergolic) but others need an ignition device. Propellant combinations are chosen on the basis of their particular characteristics which include the combustion temperature, the molecular weight of the combustion product, the propellant combination density, handling and storage characteristics, and price.

In large liquid rocket motors such as the ones used in launch vehicles, part of the energy contained in the combustion gases is used to drive turbine pumps which force the propellants into the combustion chamber. In the more simple liquid propellant rocket motors used in the satellite propulsion system, pressurization of the propellant tanks provides enough power for this purpose. The temperatures attained in the combustion chamber (2000 to 3000 K) are above the melting point of the majority of metals used in rocket motor construction. Special materials such as graphite or ceramics are employed for the engine parts that reach particularly high temperatures. Various techniques are used to keep the temperatures of metal components sufficiently low, such as cooling, insulation and heat sinks.

Liquid bipropellant propulsion systems

The bipropellant propulsion systems generate hot gases in the combustion chamber for ejection through the nozzle, usually by spontaneous ignition on contact of the oxidant and fuel elements of the hypergolic propellants. The principal cryogenic propellants are liquid oxygen and liquid hydrogen. Typical storable propellants consist of monomethylhydrazine as the fuel and nitrogen tetroxide as the oxidizer. The latter reaction produces a gas mixture consisting of hydrogen, nitrogen, carbon monoxide, carbon dioxide and water. Bipropellant propulsion systems yield specific impulses of around 290 to 320 seconds, with thrusts ranging from a few newtons for orbit control motors, to thousands of newtons for orbit injection motors.

For use on board a satellite for final orbit acquisition (AKM) or orbit control in service there is a trade-off between the extra performance provided by a bipropellant system due to its relatively high specific impulse value compared to a monopropellant system, versus the greater dry mass that results from the duplicate tanks and hydraulics that are needed. In general, bipropellant systems become cost effective for satellites above 1000 kg in mass in final orbit and where a single set of propellant tanks is used for injection into orbit and attitude and orbit control (unified propulsion).

Solid propellant motors

In a solid propellant rocket the propellant is wholly contained within the combustion chamber. The motor is basically a pressure vessel, partly filled with blocks of solid propellant, termed the grain, with a nozzle at one end, and an ignition device at the other end; see Figure 4.26. When ignited the propellant burns at its surface and the shape of the burning surface determines the thrust characteristics of the motor. The solid propellant can be either of double base such as nitrocellulose and nitroglycerine, or a composite such as ammonium perchlorate, polyurethane or polybutadiene. The former category combines fuel and oxidizer in one molecule, while the latter is a mixture of fuel and oxidizer. Solid propellants provide lower specific impulses, in the range 200 to 260 seconds, compared with liquid propellants. They are, however, easier to handle, store and operate.

Solid propellant rockets are classified in accordance with their grain geometry, as the change in their burning surface geometry with time determines their thrust–time profile. In this context, grains can be neutral, progressive or regressive. In a neutral grain, the burning surface area remains constant and

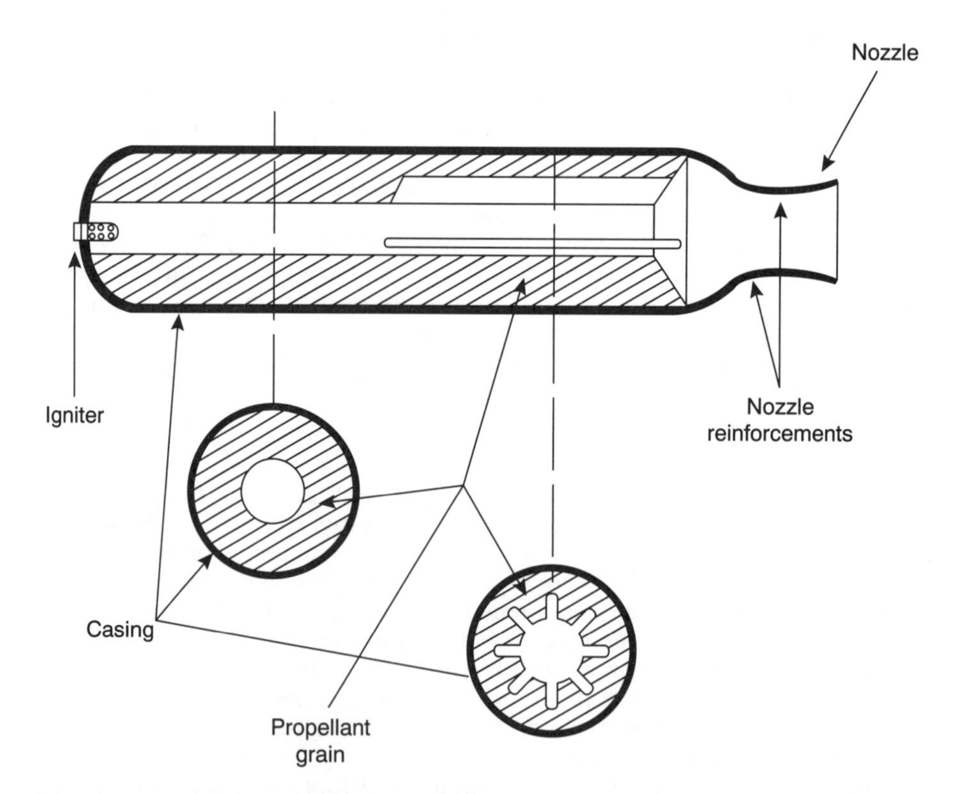

Figure 4.26 *A simplified diagram of a typical slotted tube grain solid rocket motor*

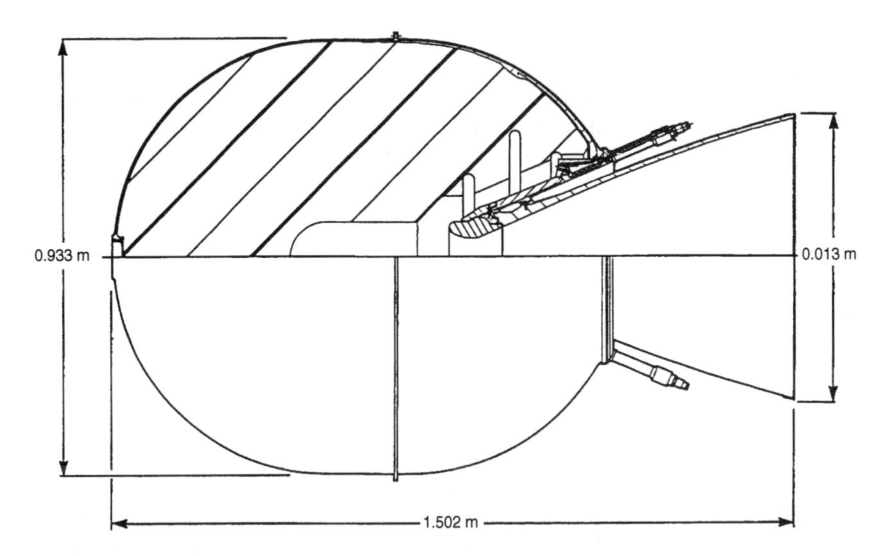

Figure 4.27 *A schematic diagram of the Thiokol Star 37XFP solid propellant rocket motor. (Reproduced by permission of the Thiokol Corporation.)*

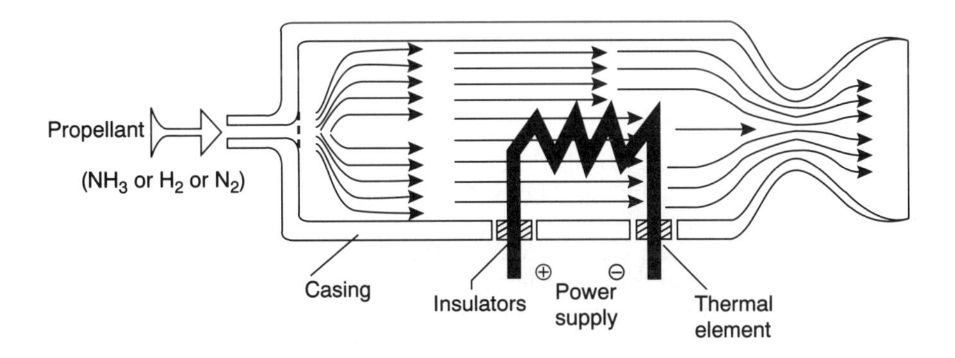

Figure 4.28 *A simplified diagram of a resistojet thruster*

consequently the thrust is almost constant. In a progressive grain, the thrust increases with time, while in a regressive grain, it decreases with time.

Solid propellant apogee motors which are integrated in the satellite structure have been used extensively to provide the large velocity increment needed to inject the satellite from the GTO into the GSO. An AKM such as the Thiokol Star 37F (the current model, 37 XFP is shown in Figure 4.27) was used to provide the incremental velocity needed for satellites such as Intelsat V; see [19]. The propellant usually consists of polybutadiene carboxide or hydroxide, to which aluminium powder is added to facilitate combustion. It is contained in a case of titanium and a wound shell of composite material such as epoxy-impregnated Kevlar. The nozzle is constructed of low-density materials that can withstand temperatures of around 3800 K, such as a skeleton made of carbon fibres and a carbon matrix which binds the fibres. The inner surface of the casing is coated with insulation materials which protect the casing from the high temperatures generated during the burn. The propellant is ignited by an electrically controlled device located either near the throat of the nozzle or in the front end, as shown in Figure 4.27.

Hybrid propellant motors

The hybrid rocket engine usually contains its fuel in solid form within its combustion chamber, while the oxidizer is fed from pressurized tanks, sustaining an energy releasing reaction at the fuel surface.

4.6.3 Electric rocket propulsion

In an electric rocket propulsion system propulsive thrust is generated by the expulsion of a propellant which has been accelerated by electrical heating or by the application of electrical and magnetic forces. Compared with chemical propulsion, electric propulsion is a state of the art technology which can provide high specific impulses of 1000 to 10 000 seconds, providing thrusts which are low, less than 0.1 N, but which can be sustained over long periods. Electric propulsion systems are now accepted as the systems of choice for satellites where very precise pointing and positioning is required and mass is at an especial premium. They offer reliability together with large mass savings compared with conventional satellite propulsion systems although they require large amounts of electrical power. In this respect, a quantity termed the specific power is also specified along with the specific impulse. Specific power is the ratio between the electric power input and the thrust produced and it varies between 25 and 50 W mN^{-1} depending on the thruster type. For a more detailed analysis of electric propulsion systems, see [20, 21].

Resistojet thrusters

Resistojet thrusters are gas expulsion devices where the expellant is gas which has been heated electrically; see Figure 4.28. Thermal storage resistojets use a heating element with high thermal capacity, supplied continuously with electric power at a low level, keeping the temperature of the heating element almost constant; propellant is fed through the heating element in short pulses to provide the required thrust. Fast heat-up resistojets contain a fast response, low thermal capacity, heating element which is provided with high-level pulsed electric power, synchronized with the pulses of propellant gas. Common propellants used are ammonia, hydrogen and, to a lesser extent, nitrogen. Specific impulses around 550 seconds can be achieved at temperatures of 2500 K with a power consumption of about 10 W.

Arcjet thrusters

The arcjet thruster uses the thermal energy of an electric arc discharge to heat a propellant gas which is then accelerated and expelled by gas-dynamic expansion in a nozzle. A typical low-power thruster comprises an anode, which forms the chamber, and a nozzle and cathode in the shape of a rod with a conical tip; see Figure 4.29. The anode is constructed of a high melting point material such as tungsten or tungsten–rhenium alloy while the cathode is typically constructed from thoriated tungsten. An arc discharge is generated in the chamber where a gas such as catalytically decomposed hydrazine, argon or ammonia is heated to about 20 000 K, yielding theoretical exhaust velocities between 25 000 and 30 000 m s^{-1}. However, the efficiency of the arcjet thruster is low because of heat transfer to the electrodes and the energy losses associated with ionization and dissociation. Excessive erosion of the electrodes is another problem encountered with this type of thruster.

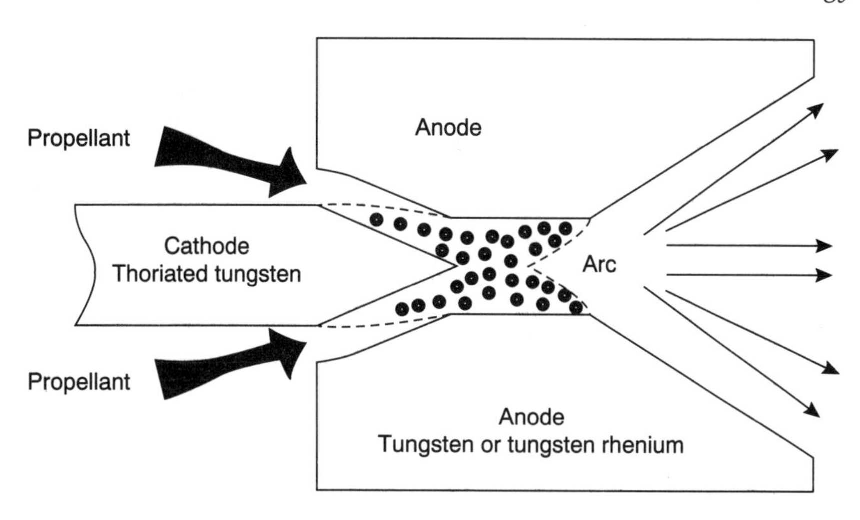

Figure 4.29 *An outline drawing of an arcjet thruster*

Arcjet thrusters have been flown on US-built satellites such as Telstar-4 and four are to be used on each Intelsat VIII satellite for north–south station-keeping. Each high-performance arcjet provides a mission average specific impulse of 517 seconds. Additional information on arcjet thrusters is provided by Smith *et al.* [22] and Ghislanzoni *et al.* [23].

Ion thrusters

Ionic propulsion is achieved by the acceleration of positively charged particles (ions) in an electrostatic field. A neutralizer in the form of an electron gun is employed to eject the same amount of electric charge of the opposite sign, to ensure that the voltage potential of the satellite relative to its environment is not excessively increased. Heavy metals that maintain a liquid form at storage temperatures, such as mercury and caesium, and also xenon, are the propellants used; see [24].

A 'field emission' ion thruster consists of two shaped metal plates placed together as shown in Figure 4.30. The liquid metal propellant is fed by capillary action into the space between the plates, reaching their sharp edge. The high local electric field generated by the large potential difference between the anode and the emitter plates, reinforced by the point effect at the sharp edge of the emitter, extracts electrons from the propellant atoms, thus generating ions. The ions are accelerated into an ion beam which provides thrust. Specific impulses between 8000 and 10 000 seconds can be obtained at the expense of electrical consumption of several kilowatts. Parallel operations of field emission ion thrusters can yield thrusts of about 10 mN.

'Prior ionization' ion thrusters first remove the electrons from the propellant atoms in an ionization chamber and then accelerate the resulting positive ions by a grid of high negative potential as shown in Figure 4.31. The ions are generated either by bombarding a cloud of propellant atoms with an electron gun (Kaufman engine) or by inducing excitation using a radio-frequency field at about 1 MHz. This type of ion thruster yields specific impulses of 2000 to 3000 seconds. With an electric power consumption of 60 to 600 W, thrusts of 2 to 20 mN can be obtained with this type of thruster. See [25].

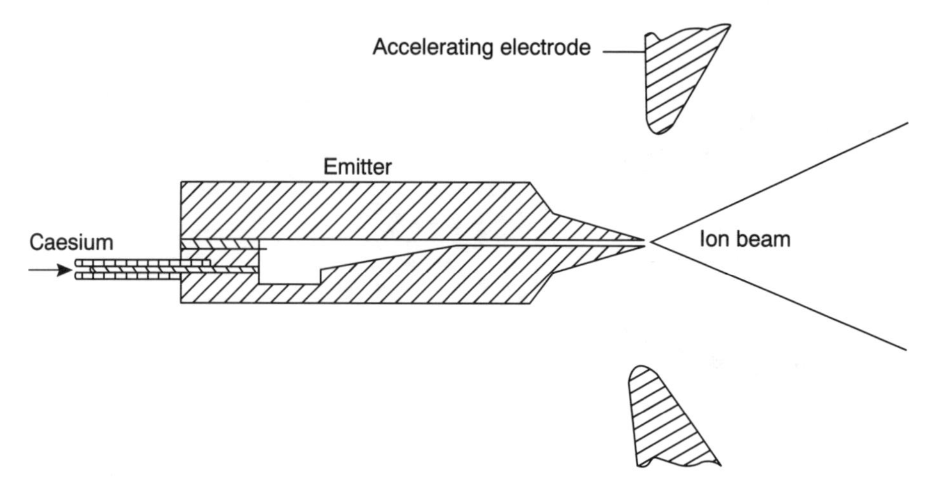

Figure 4.30 *An ion thruster using the field emission technique. (From* Satellite Communications Systems *by G. Maral and M. Bousquet, © 1993. Reprinted by permission of John Wiley and Sons Ltd.)*

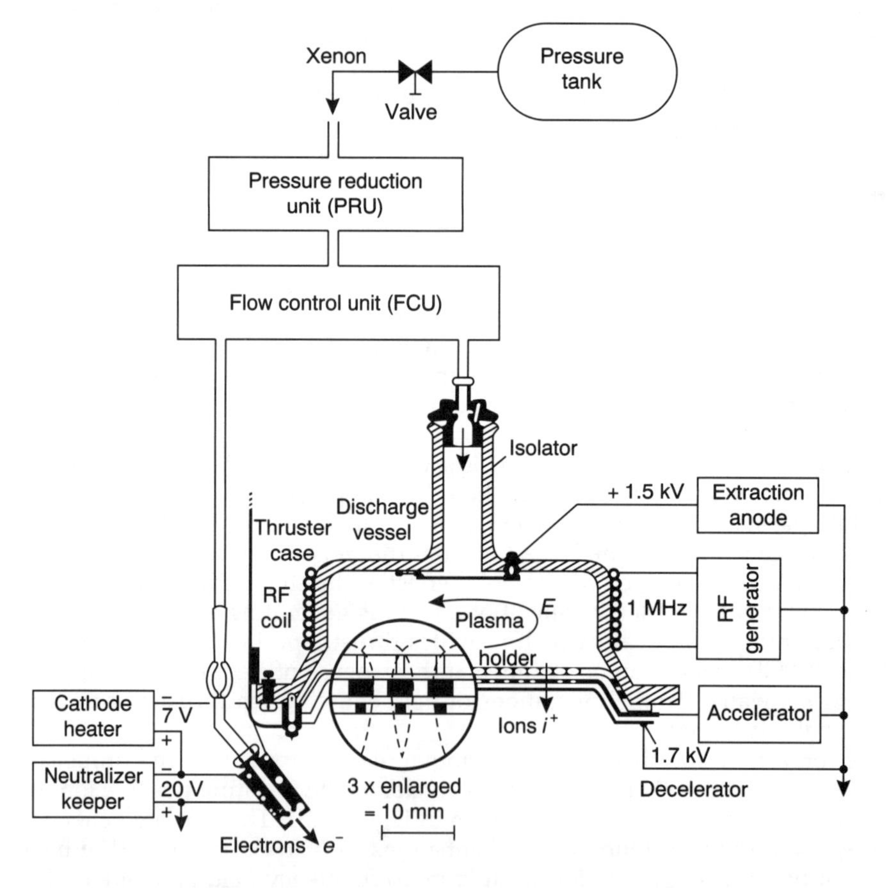

Figure 4.31 *A schematic diagram of an ion thruster using the prior ionization technique. (From* Satellite Communications Systems *by G. Maral and M. Bousquet, © 1993. Reprinted by permission of John Wiley and Sons Ltd.)*

Significant mass savings can be achieved by using ion propulsion instead of conventional hydrazine for north–south station-keeping for geostationary satellites. In this context Intelsat is developing an ion propulsion system based on xenon, and xenon thrusters will be used in the Hughes HS702 satellite series.

Colloid thrusters

Colloid thrusters operate on principles similar to those of ion thrusters but with the difference that the charged particles are solid particles in a powder form or particles formed from vapour condensation. Colloidal particles are much larger than ions. Although a relatively high accelerating voltage is required, a good thrust/weight ratio is provided.

Plasma thrusters

A plasma thruster is basically a capacitor with a rod, usually made of Teflon, as the dielectric material. An electric source charges the capacitor, increasing the voltage potential across the plates until the dielectric insulation of the Teflon rod breaks down, causing a spark to pass through it. This results in the ionization of a layer of the dielectric material and the acceleration of plasma by the autogenerated electromagnetic field. In service the capacitor is repeatedly charged and discharged, providing a pulsed propulsive thrust. As the propelled plasma is electrically neutral, an electron gun is not necessary. Plasma thrusters are simple devices providing specific impulses between 1000 and 5000 seconds. However, they are associated with problems of electromagnetic compatibility and pollution on the spacecraft surface.

4.6.4 Satellite propulsion system configurations

The propulsion requirements of the earliest satellites were met by cold gas systems as they required the minimum electrical power from the satellite, and fulfilled the total impulse requirements of the limited operational lifetime of the spacecraft. As the mission requirements became progressively more demanding, the low specific impulse provided by the nitrogen propellant, about 60 seconds, involved an unacceptable mass penalty. Hydrogen peroxide provided a specific impulse of about 130 seconds with less tankage mass and size. However, hydrogen peroxide is prone to spontaneous chemical breakdown causing operational problems as the lifetime of satellites was extended. Hydrazine used in conjunction with the Shell 405 catalyst offered specific impulses of 150 to 225 seconds and replaced cold gas systems for long-duration missions. Bipropellant system development followed with operational systems offering specific impulses in excess of 300 seconds.

The development of electric propulsion systems with specific impulses above 3000 seconds, at the expense of electrical power consumption, brought new possibilities. The use of lightweight efficient solar arrays make possible the operational application of electric propulsions in satellites.

Thus, a number of system options are available now for attitude and orbit control of satellites and solid and bipropellant liquid motors are available for AKMs. Typical choices from these options are as follows.

Solid apogee motor and hydrazine propulsion

The combination of a solid AKM and monopropellant hydrazine catalytic propulsion on-station was extensively used for communication satellites up to 1990. This propulsion system combination offers simplicity although at the expense of relatively low specific impulse and lack of precise control during injection into the GSO. This combination is currently cost effective for satellites with mass not exceeding 1000 kg in orbit.

Solid apogee motor and bipropellant propulsion

The replacement of hydrazine by a bipropellant system for on-station control increases the specific impulse of the system and hence provides mass gains which become cost effective for satellites over 1000 kg. However, the complexity of the bipropellant system, coupled with the disadvantages associated with the solid apogee motor, restricts the cost effectiveness of such combination in comparison with the unified bipopellant propulsion system.

Unified bipropellant propulsion

In a unified bipropellant propulsion system the propellants required for orbit injection and attitude and orbit control are stored in a common set of tanks. A high-thrust apogee motor of about 400 N is used for orbit injection, while motors of lower thrust are used for attitude and orbit control. Both usually use monomethyl hydrazine fuel with a nitrogen pentoxide oxidizer. This system provides a greater overall specific impulse at the expense of a greater system dry mass. The system becomes cost effective for satellites above 1100 kg. For a satellite of around 1100 kg, the unified system causes a 5 per cent increase in net satellite mass. A unified propulsion system was used in the Intelsat VII and VIIA spacecraft, first launched in 1993 and 1994 respectively.

A typical system is shown in Figure 4.32. It uses monomethyl hydrazine fuel and nitrogen tetroxide oxidizer contained in titanium tanks. The helium pressurant is contained in a separate tank. Fine mesh screens and quadrant capillary arms are used in the propellant management apparatus, ensuring a bubble-free thruster operation. During launch, the propellant tanks are isolated from the thrusters by pyrotechnically operated valves, normally closed. After the spacecraft is separated from the upper stage of the launch vehicle, the pyrotechnic valves are fired open and the 22 N thrusters can be operated. Orbit injection is carried out by the 490 N apogee engines after other normally closed pyrotechnic valves are fired open to provide regulated pressure. In orbit, station-keeping is carried out by the 22 N thrusters.

Liquid apogee motor and monopropellant on-station propulsion

The Intelsat VIII satellite, first launched in 1995, utilizes two bipropellant liquid AKMs and monopropellant thrusters for on-orbit attitude control and station-keeping. The AKMs use hydrazine fuel and nitrogen pentoxide oxidizer, operating in a pressure-regulated mode. After injection into the GSO, the oxidizer and pressurant tanks are isolated and hydrazine is used alone to feed the monopropellant propulsion system in the blowdown mode. Four high-performance arcjets providing a mission average specific impulse of 517 seconds are used for north–south station-keeping.

The Intelsat K satellite, launched in 1992, also used two bipropellant liquid motors for orbit injection. The monopropellant on-station propulsion system

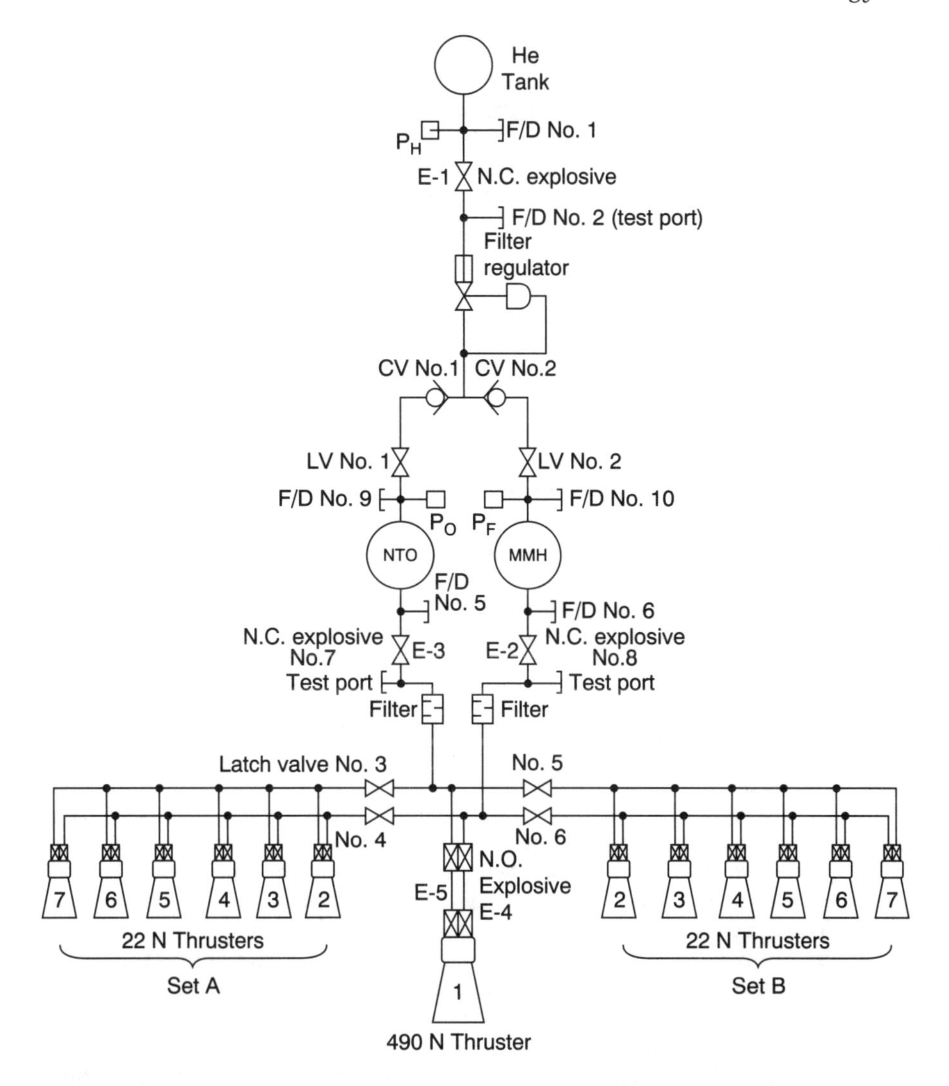

Figure 4.32 *A schematic diagram of a unified propulsion system. (From* Satellite Communications Systems *by G. Maral and M. Bousquet, © 1993. Reprinted by permission of John Wiley and Sons Ltd.) Key: LV, latching valve. CV, check valve. F/D fill/drain valve. MMH, monomethyl hydrazine. NTO, nitrogen tetroxide. He, helium. N.O., valve, normally open. N.C., valve, normally closed*

incorporated improved performance electrothermal hydrazine thrusters, which offer better performance than the commonly used bipropellant engines.

4.7 The thermal control system

4.7.1 Principles of temperature control

Spacecraft are constructed from components and materials which function well in the Earth's hospitable environment. In the extreme conditions of space, the thermal control system of a satellite must provide another environment in which those components can also function well. This involves providing a balance, at a suitable temperature, between heat generated within the satellite plus the heat gain from solar radiation and heat lost from the satellite by radiation into space.

In the early satellites, thermal control was achieved mainly by the utilization of a combination of surface coatings and finishes which provided the desired radiation equilibrium. As satellites grew both in size and complexity and their missions became longer, the thermal control requirements became more demanding. To satisfy these increasing demands, a number of thermal control concepts have been developed which deploy both passive and active hardware devices. For an analysis of the design trends in communication satellite thermal subsystems, see [26].

4.7.2 Passive thermal control hardware

Passive devices control temperatures by the use of conductive and radiative heat paths. This is implemented mainly by the application of thermal insulation, the use of surface coatings, heat sinks and phase change materials. However, passive temperature control devices carry relatively large mass and surface area penalties, their control response is rather limited and their performance degrades with time.

Thermal insulation

Thermal insulation is used in a satellite to reduce the heat flow rate per unit area between two boundary surfaces at particular temperatures. The insulation medium is either a single homogeneous material such as low-conductivity foam or a multilayer system of insulation where each layer behaves as a low-emittance radiation shield separated by spacers of low conductance.

These shields are usually made of polyester or polyamide films typically 25 μm thick, the surface(s) of which are metallized with gold or aluminium to reduce emittance. The spacers are either discrete, made of materials such as foam, plastic or paper, or crinkled 25 μm polyester film, metallized on one side, sandwiched between the layers. Multilayer insulations are employed to protect particular spacecraft components by minimizing heat flows, reducing temperature fluctuations caused by external time-varying radiative heat fluxes and minimizing temperature gradients caused by received radiative heat from varying directions. For example, multilayer insulation is used to reduce the heat flow to cryogenic propellant tanks in order to minimize the evaporation of the stored propellant.

Surface finishes and coatings

The outer surfaces of a satellite are radiatively coupled to space, the ultimate heat sink available in orbit. However, these surfaces also receive heat energy from external sources, mainly the Sun and the Earth. Consequently their radiative properties must be chosen to balance at a desired temperature the heat radiated into space, the internal dissipations and heat received from external sources. The two most important properties of the surface are the emittance and the solar absorptance. Emittance describes the ability of the surface to radiate thermal energy in the infrared region of the spectrum. Solar absorptance describes the amount of solar energy the surface absorbs. On the outer surface of a satellite a combination of coatings and finishes can be used in a variety of geometric patterns to obtain the desired average values for emittance and solar absorptance. Values of the emittance and absorptance of various materials and finishes are shown in Figure 4.33.

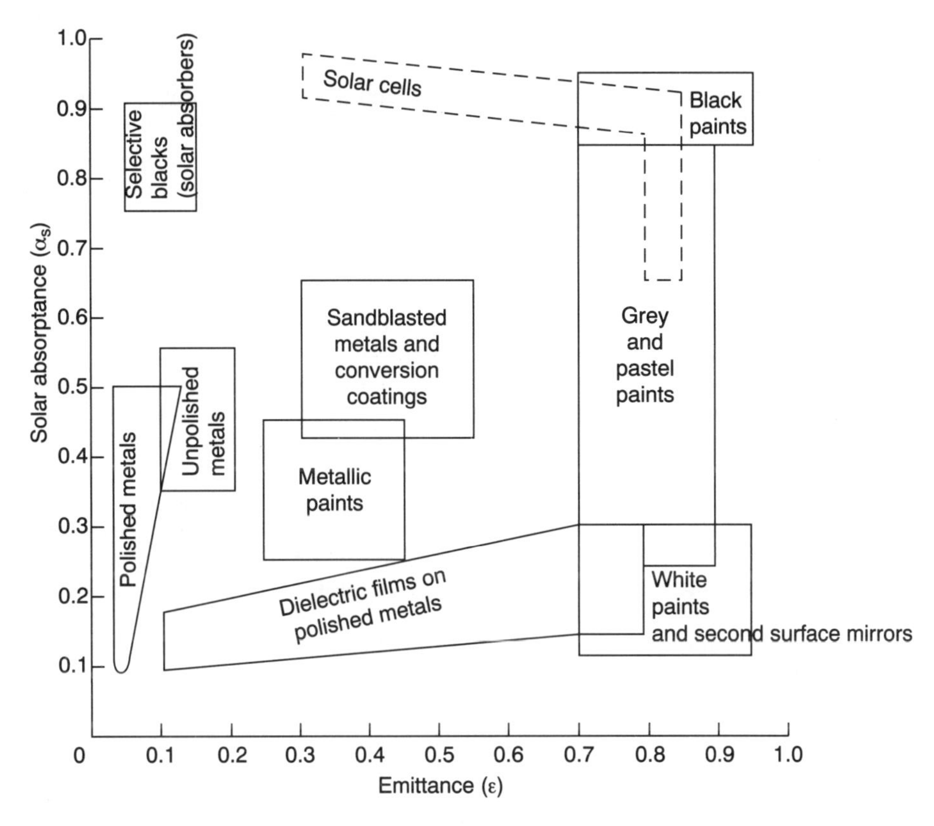

Figure 4.33 *Values for solar absorptance and emittance for various materials and finishes*

Heat sinks

Heat sinks are passive elements with relatively high thermal mass and capacity, thermally coupled to components or equipment which need to be temperature controlled. They usually limit local temperature rises by transferring heat to suitable adjacent locations by conduction or radiation and reduce temperature variations of components or equipment subject to cyclical variations of power generation or dissipation. The use of heat sinks in spacecraft design is limited due to the associated relatively high-mass penalty.

Phase change materials

Materials that are able to transform from one physically distinct and mechanically separable state to another one are called phase change materials. When a component has to be kept below a particular maximum temperature, a phase change material with a melting point close to this temperature may be thermally coupled to it. When the temperature of the component is increasing through the melting point of the protective phase change material, the latent heat needed for the phase change provides certain thermal inertia to any further temperature increase. However, when all the protective material has melted the temperature cannot be further prevented from rising. Phase change materials are particularly suited to keeping cyclically operating electronic components at a constant temperature and to store energy isothermally for later release.

4.7.3 Active thermal control hardware

Active temperature devices are used to overcome the limitations of passive devices. Active devices monitor the temperature of the spacecraft components and are activated when predetermined temperature limits are reached. Heat pipes, louvres and heaters are mainly used for active thermal control.

Heat pipes

Heat pipes are used to transfer heat from internal hot spots to remote radiator surfaces, thus reducing overall temperature differentials and keeping radiative temperatures to manageable levels. The heat pipe in its simplest form is a self-contained passive temperature control device which requires no power or moving parts for its operation. It has high thermal conductance, having regard to its mass, and it transfers heat from a source to a sink with little leakage at intermediate locations.

A heat pipe consists of an elongated container, closed at both ends, the inner walls of which are lined with a capillary wick saturated with a working fluid, usually ammonia or methanol; see Figure 4.34. Functionally, a heat pipe can be divided along its length into three parts, the evaporator, the adiabatic part and the condenser part. The heat generated by equipment adjacent to the evaporator part of the pipe evaporates fluid contained in the wick, leading to an increase in vapour pressure there. This pressure increase propels vapour through the adiabatic part to the condenser part. The latter is coupled to a relatively cool external surface of the satellite, which radiates heat into space; vapour condenses, releasing the latent heat of evaporation of the working fluid. The excess fluid which therefore accumulates at the condenser part of the heat pipe is carried back to the evaporator part by the capillary action of the wick to complete the cycle. For a more comprehensive description of heat pipes, see [27–31].

Thermal louvres

A simple and reliable method of active temperature control is provided by louvres which vary the effective emittance of a satellite radiator. A louvre consists

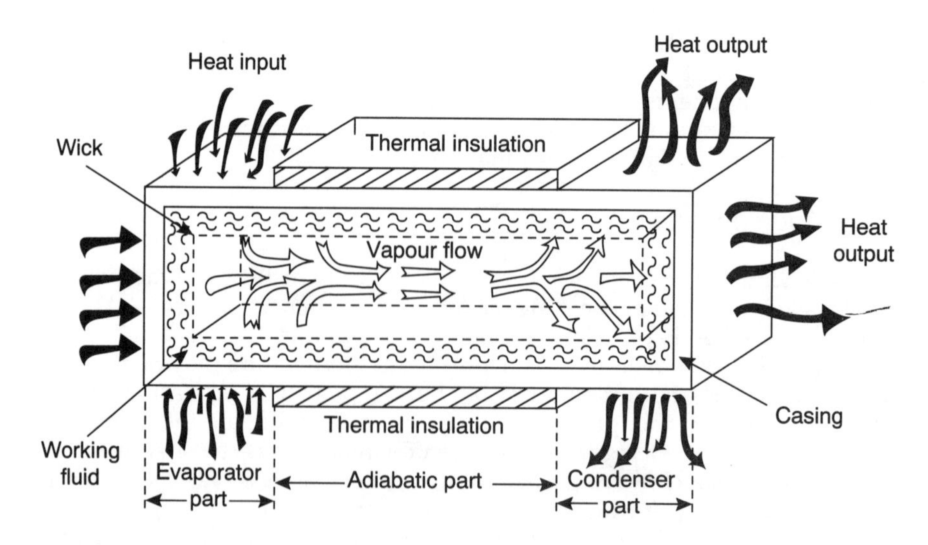

Figure 4.34 *A schematic diagram of a basic heat pipe*

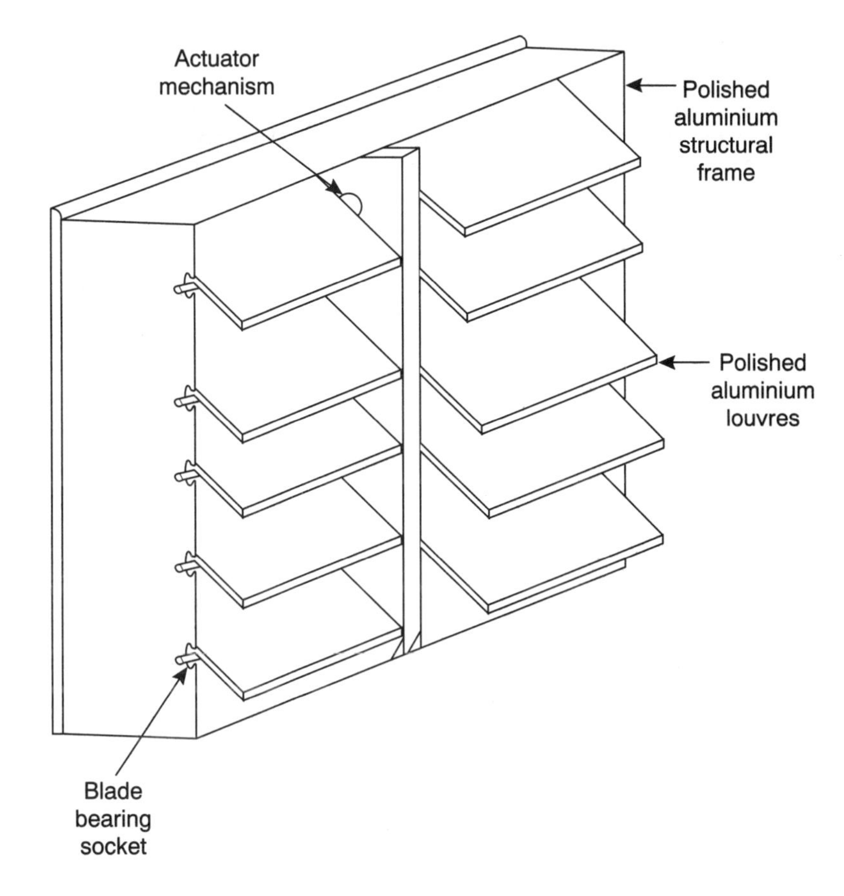

Figure 4.35 *A simplified diagram of a thermal louvre*

of a number of polished aluminium blades arranged like a venetian blind over a radiator of high emittance, as shown in Figure 4.35. The blades close at low temperatures thus reducing the heat radiated into space. As the temperature increases, mechanical sensors cause the blades to open, thus increasing the heat radiated into space. In this way a certain degree of temperature control is achieved.

Electrical heaters

Electric heaters are used to maintain minimum component, equipment or structure temperatures during cold conditions. The heaters are normally thermostatically controlled resistance elements or etched foil strips contained between two thin Kapton films.

4.7.4 Intelsat IV thermal control

To illustrate the practical application of the principles of thermal control, it is convenient to use as an example a satellite model which is not too complex. Intelsat IV is a convenient example.

As a satellite moves around the Earth in its orbit, it is subjected to the thermal conditions imposed by deep space and the Sun. The satellite also passes through the shadow of the Earth and remains in darkness for up to 70 minutes daily during two seasons of the year.

In addition, the satellite has to dissipate heat produced by its electronic components and, for a limited period, by the firing of the AKM. Two sources of heat arise from the AKM firing: the exhaust plume and the heated motor casing. The former is a few hundred metres long and originates a few metres away from the location of sensitive electronic components. Heat from the AKM casing is conducted through the satellite structure, reaching the electronic equipment approximately 30 minutes after the AKM burn, a process called 'soak back'.

The objective of the Intelsat IV thermal control system is to maintain the satellite electronic components near terrestrial room temperatures. This is achieved mainly by:

(1) minimizing the cross-satellite thermal gradients by using the rotating solar drum of the satellite as a steady temperature boundary;
(2) designing the equipment located outside the solar drum to tolerate a temperature range from 125 to 430 K;
(3) using heaters occasionally to raise the temperature of components reaching their lower design limit;
(4) radiating heat from the upper (forward) solar panel.

The antenna components which are particularly susceptible to thermal distortion are insulated to prevent heat escaping. The spot beam antennas are not insulated because metallic insulating material would interfere with their radio characteristics. The sun-shield radiator on the end of the satellite which is to the north incorporates second surface mirrors made of quartz which provide a low solar absorptance and high infrared emittance, thus lowering the spacecraft temperatures during the summer. However, it may be noted that ultraviolet radiation increases the solar absorptance of these materials, causing the satellite to run hotter at the end of its life. A thermal barrier at the opposite end of the satellite consisting of stainless steel and nickel foil isolates the apogee motor from other parts of the satellite and keeps the apogee motor casing relatively cool.

Thermal design considerations greatly influenced the satellite configuration. For example, thermal constraints dictated that the heavy electronic equipment is placed near the sun-shield radiator at the Northern end, although this increases the mass of the satellite structure.

4.8 Telemetry, tracking, command and communication

4.8.1 The TTC&C functions

The telemetry, tracking, command and communications (TTC&C) system provides means for monitoring and controlling the satellite in orbit. The telemetry subsystem supplies measurements of various parameters to an earth station which is responsible for satellite management. The tracking subsystem provides facilities by which the satellite orbit can be determined. The command subsystem provides the means by which the satellite is controlled. And the communications subsystem provides links between the satellite and the earth station for the transmission of the various signals necessary for these purposes. These functions are usually grouped together and referred to as the telemetry, tracking, command and communication (TTC&C) system. In communication satellites the TTC&C system is normally independent of the payload. An omnidirectional antenna provides the necessary coverage for telemetry and command information to be exchanged between the ground and the satellite irrespective of the attitude of the latter.

4.8.2 Telemetry

Satellites employ a number of devices, sensors or transducers, which measure physical parameters in the spacecraft, such as pressures and temperatures, and convert them into electrical signals for transmission via the communications subsystem to Earth. In a typical communications satellite measurements of a few hundred parameters need to be kept under observation. As most of these measurements produce signals that vary only slowly, requiring only a few hertz of bandwidth, the signals can be multiplexed into a single carrier before transmission. Two methods of multiplexing are used, either frequency division multiplexing (FDM) or time division multiplexing (TDM).

In FDM, the signal voltage produced by a sensor usually varies the frequency output of a voltage-controlled subcarrier oscillator so that the oscillator frequency deviation represents the amplitude of the output voltage of the sensor. For each sensor output signal a different subcarrier frequency is used. Subsequently the outputs of all the subcarrier oscillators are combined and fed into a single amplifier.

In TDM, the sensor outputs are fed into a commutator which samples in sequence the output of each sensor several times per second, producing a series of pulses which represent the information content. Time division multiplexing is only used with particularly low-frequency signals due to its inherent limited frequency response.

4.8.3 Tracking

The orbit parameters of a communication satellite need to be determined and monitored during the launch and the on-station phases of its mission. This is usually achieved by radio tracking and in the case of geostationary satellites by ranging also.

Satellites operating in LEO need to have their orbit parameters determined so that their passage over individual earth stations can be accurately predicted. This can be achieved simply by using a radio transmission from the satellite as a beacon in order to measure, at a number of orbital points, the ground antenna elevation and azimuth angles. From these successive measurements, carried out by a number of earth stations, a suitably programmed computer on the ground calculates the orbital parameters of the satellite. In general, the telemetry downlink carrier acts as the beacon for these measurements.

For geostationary satellites, angle measurements alone are not enough to provide the required accuracy for transfer orbit and station-keeping operations. In this case the required precision is obtained by using range and range-rate measurements. An uplink RF carrier is modulated by one or more ranging sidetones. The time interval between the transmission and the reception of these sidetones through the satellite radio link is measured by one or more earth stations. As this interval is proportional to the range between the measuring earth station and the satellite, the required ranging information can be computed.

4.8.4 Telecommand

Two types of commands are used in the control of a satellite, namely pulse commands and value commands. Pulse commands are discrete ON/OFF relay commands represented by a pulse of predetermined width, typically 50–100 ms, produced by the command decoder on receipt of an appropriate command instruction. Value commands are digital words usually 16 bits long which are fed to a particular register on receipt of an execute message.

4.8.5 Communications

The FDM or TDM signals produced in the telemetry subsystem are transmitted to the monitoring station on Earth, using carrier frequencies set aside for the purpose, by means of radio links which are, in all significant respects, similar in principle to the links that earth stations use for communication through the payload of the satellite; and likewise, in the reverse direction for the telecommand signals.

References

1. Fortescue, P. W. and Stark, J. P. W. (eds.) (1995). *Spacecraft Systems Engineering*, 2nd ed. Wiley.
2. Chetty, P. R. K. (1991). *Satellite Technology and its Applications*, 2nd edn. TAB Professional and Reference Books.
3. Maral, G. and Bousquet, M. (1993) *Satellite Communication Systems*, 2nd ed. Wiley.
4. Phillips, K. J. H. (1995). *Guide to the Sun*. George J. Flyyn. Science Boom and Films.
5. Neyret, P., Betaharon, L., Dest, L. *et al.* (1992). The Intelsat VIIA spacecraft. *Proc. AIAA 14th International Communications Satellite Systems Conference*, Washington, DC, USA.
6. Ralph, E. L. (1989). Photovoltaic space power history and perspective. *Proc. IAF International Conference on Space Power*, Cleveland, Ohio, USA.
7. Landis, G. A. Bailey, S. G. and Flood, D. J. (1989). Advances in thin film solar cells for lightweight space photovoltaic power. *Proc. IAF International Conference on Space Power*, Cleveland, Ohio, USA.
8. Vèrié, C., Beaumont. B., Freundlich, A. *et al.* (1989). GaAs solar power cells for space applications. LPSES, Centre National de la Recherche Scientifique, Parc du Sophia Antipolis, rue Bernard Gregory, 06560, Valbonne, France.
9. Flores, C. and Palletta, F. (1989). Alternative solar cell assessment. CISE Report No. 5160, 13-10-1989, CISE Technolgie Innovative. IFIAR, Segregate (Milano), via Reggio Emilia, Milano 20134, Italy.
10. Hill, R. and Pearsall, N. M. (1989). Review of indium phosphide solar cells. Newcastle Photovoltaics Applications Centre, Newcastle Upon Tyne Polytechnic, Newcastle upon Tyne NE1 8ST, UK.
11. Kenny B., Cull R. C. and Kan Kan M. D. (1990). An Analysis of space power system masses. *Proc. 25th Intersociety Energy Conversion Engineering Conference*, Reno, Nevada, Vol. 1, pp. 484–489
12. Jilg, E. T., McCaskell, A. M., Neill, D. V. and Sattelee, A. A. (1972). The Intelsat IV spacecraft etc. *COMSAT Technical Review*, **2**(10).
13. Seifert, H. S. (1959). *Space Technology*. Wiley.
14. Cornelisse, J. W., Schöyer, H. F. R. and Wakker, K. F. (1979). *Rocket Propulsion and Spaceflight Dynamics*. Pitman.
15. Owens, J. R. (1976). Intelsat satellite onboard propulsion systems past and future. *Proc. AIAA/CASI 6th Communication Satellite Systems* Conference, Montreal.
16. Zafran, S., Murch, C. K. and Grabbi, R. (1977). Flight applications of high performance electrothermal thrusters. *Proc. AIAA/SAE 13th Joint Propulsion Conference*, Orlando, Florida.
17. Grabbi, R. and Murch, C. K. (1976). High performance hydrazine thruster (HiPEHT) development. *Proc. AIAA/SAE 12th Propulsion Conference*, Palo Alto, California.
18. Jackson, F. A., Stansel, J. C., Jortner, D. and Hagelberg, C. F. (1966). An operational electrothermal propulsion system for spacecraft reaction control. *Proc. AIAA 5th Electric Propulsion Conference*, San Diego, California.
19. *Star Rocket Motors Handbook*. Thiokol, Elkton Division.

20. Perrotta, G., Cirri, G. and Matticari, G. (1992) Electric propulsion for lightsats: a review of applications and advantages. *Tacsats for Surveillance, Verification and C3I*, AGARD Conference Proceedings 522, Brussels, Belgium.
21. Saccoccia, G., Deininger, W., Paganncci F. and Scortecci, F. (1994). Development of satellite electrical propulsion systems. Preparing for the future. *ESA's Technology Programme Quarterly*, **4**(2).
22. Smith, C. R., Roberts, K., Davies and Vaz, J. (1990). Development and demonstration of a 1.8 kW hydrazine Arcjet thruster. *Proc. AIAA 21st International Electric Propulsion Conference*, Orlando, Florida.
23. Ghislanzoni, L., Petagna, C. and Trippi, A. (1990). The application of Arcjet propulsion systems for geostationary satellites. *Proc. AIAA 21st International Electric Propulsion Conference*, Orlando, Florida.
24. Feam (1977). A review of the UKTS electron bombardment mercury ion thruster. *Proc. ESTEC Conference on Altitude and Orbit Control Systems*, Noordwijk.
25. Bober, A. S. and Maslennikov, N. A. (1995). Stationary plasma thrusters. *Space Bulletin*, **2**(1).
26. Chalmers, D. R., Burke, P. H. and Case G. W. (1988). Design trends in communication satellite thermal subsystems. *Proc. 3rd Space Thermal Control and Life Support Systems.*
27. Dunn, P. D. and Reay D. A. (1982). *Heat Pipes*, 3rd edn.
28. Brost, O., Groll, M. and Mûnzel W. D. (1978). Technical applications of heat pipes. *Proc. 3rd International Heat Pipe Conference.*
29. Bhati, R. S. (1986). Heat pipe. SABCA Report 8307.
30. Dobrau, F. (1987). Heat pipe research and development in America. *Proc. 6th International Heat Pipe Conference.*
31. Delil, A. A. M. (1977). Theory and design of conventional heat pipes for space applications. NLR-I-77001, 1977.

Part 3
Satellite Communication Technology

5 Earth–space radio propagation

5.1 Signal attenuation and distortion

5.1.1 Terminology

If an emission having a power of w_t watts is delivered to a transmitting antenna and the emission is then received some distance away, the receiving antenna delivering w_r watts to the receiver input stage, then, assuming that the antennas are correctly matched to their respective transmitter and receiver and disregarding any losses in the antennas,

$$10 \log w_t / w_r = L \quad \text{(dB)} \tag{5.1}$$

where L is called the transmission loss.

If G_t and G_r are the gains in decibels relative to an isotropic antenna (dBi) of the transmitting and receiving antennas respectively in the relevant directions, then

$$L_b = L - G_t - G_r \quad \text{(dB)} \tag{5.2}$$

where L_b is called the basic transmission loss. Thus, $L_b = L$ for a link between isotropic antennas. If the transmission path is entirely in free-space, L_b equals the basic free-space transmission loss L_{bf}:

$$L_{bf} = 20 \log 4\pi d / \lambda \quad \text{(dB)} \tag{5.3a}$$

$$= 20 \log 4\pi d f / c \quad \text{(dB)} \tag{5.3b}$$

$$\approx 32.5 + 20 \log F + 20 \log D \quad \text{(dB)} \tag{5.3c}$$

where d and D = the distance separating the antennas, in m and km respectively,
f and F = the frequency in Hz and MHz respectively,
c = the velocity of light in free-space $\approx 3 \times 10^8$ m s^{-1},
λ = the wavelength in m.

If the medium through which the emission is propagated is not free-space, L may be greater than L_{bf}; this loss relative to free-space, due to absorption in the medium, has the symbol A_m.

$$A_m = L - L_{bf} \quad \text{(dB)} \tag{5.4}$$

It is sometimes convenient to express the attenuation of the emission due to the spreading of the wavefront with distance from the source in the form of the power flux density (PFD), represented by Φ, in the vicinity of a distant point of interest, typically a receiving antenna. Thus,

$$\Phi = 10\log w_t + G_t - 10\log 4\pi d^2 \quad (\text{dB(W m}^{-2}\text{)}) \tag{5.5}$$

where the symbols are as previously defined. The term $4\pi d^2$ is called the wave spreading factor. The term $10\log w_t + G_t$ is called the equivalent isotropically radiated power (e.i.r.p.) of the signal; its dimensions are decibels relative to 1 watt (dBW).

For much of its length, the ray path between a satellite and an earth station is indeed in what is virtually free-space. However, below the tropopause (the level about 12 km above sea-level where the troposphere ends and the stratosphere begins) the medium may have substantial impact on the propagation of microwave and millimetre wave radio signals, and in the ionosphere (from about 60 km to about 800 km above the ground) the free ions present there interact with radio waves, especially at the lower end of the range of frequencies of interest for satellite communication. These tropospheric and ionospheric effects include absorption and scattering and cause loss of signal strength 'additional to free-space' and other unwanted phenomena. The subject is a complex one; a comprehensive treatment of it is to be found, for example, in [1].

5.1.2 Absorption by rain

Below 7 GHz, absorption over a space–Earth path due to rain is small but heavy rain causes significant absorption above 10 GHz and this absorption becomes a major factor in system design above 15 GHz. A system designer aims to provide enough downlink carrier power to ensure that the end-to-end channel performance falls short of agreed criteria for no more than an acceptably small percentage of the time. Strategies for dealing with rain absorption are considered in Section 10.3.1 but the provision of a rain margin, that is, an assignment of satellite power in excess of the minimum required to secure the specified link performance under ideal conditions, is the most common approach.

Nevertheless, it is economically important to keep rain margins small; if satellite power is spent unnecessarily, the amount of information that the satellite could carry is materially reduced. So that excessive rain margins are not required, Intelsat for example requires data on rain absorption to be declared for proposed new earth stations for operation above 10 GHz, to ensure that particularly high absorption predictions are matched by higher specified earth station performance.

It is therefore necessary to estimate what margin of power will protect a channel from the effects of rain. But how? Rainfall rates vary from region to region and from year to year and topographical features in the vicinity of an earth station can have a substantial effect on the local rainfall.

The best predictions of future absorption are based on measurements made in the past of the absorption suffered by satellite signals at the earth station in question over a period of several years, at the appropriate angle of elevation and desirably at the appropriate frequencies. Such information may be available, for example, when a new earth station is to be built at a location where other earth stations have been in use for a long period. If absorption records are available but at the wrong frequency, ITU-R Recommendation PN.618 [2] provides an equation for scaling the available measurements at frequency f_1 to the wanted frequency f_2, as follows:

$$A_2 = A_1(\phi_2/\phi_1)^{\alpha} \quad (\text{dB}) \tag{5.6}$$

where A_1 and A_2 are the absorptions at f_1 and f_2 respectively (dB),

$\alpha = 1 - H(\phi_1\phi_2A_1)$,

$\phi(f) = f^2/(1 + 10^{-4}f^2)$ where f is in GHz,

$H(\phi_1\phi_2A_1) = 1.12 \times 10^{-3}(\phi_2/\phi_1)^{0.5}(\phi_1A_1)^{0.55}$.

In the absence of a long series of measurements of absorption, a long programme of rainfall rate measurements at or near the earth station using a rapid response rain gauge with an integration time of the order of one minute could provide a firm basis for estimating future rain absorption. More usually, however, it is necessary to predict the absorption from the available meteorological records. This introduces substantial uncertainties.

Prediction of short-term rain absorption statistics at an earth station from rainfall rate statistics

ITU-R Recommendation PN.618 [2] recommends a procedure for predicting the absorption due to rain at an earth station for 1 per cent of an average year and for smaller percentages of the time, for use where no adequate series of local measurements of absorption is available. The factors involved are a matter of broad interest and a simplified account of the procedure is as follows.

Stage 1. Calculate the length of the slant path that is exposed to rain

Heavy rain falls from massive cloud formations. Such clouds extend upwards through the 0°C isotherm level. Above that level, the water present is in the form of ice, usually hail. Hail scatters some of the energy of radio waves, causing minor loss of power from the wave, but there is no absorption. However, as the hailstones fall through the 0°C isotherm level, they melt to form raindrops. For this reason the 0°C isotherm level is also called the melting layer, its height above sea-level having the symbol h_{ml}. Water in the form of raindrops absorbs energy from the wave.

Thus it is necessary to determine d_s, the length of the part of the ray path that lies between the earth station antenna, at height h_s above sea-level, and h_{ml}. If d_s, h_s and h_{ml} are all expressed in kilometres,

$$d_s = (h_{ml} - h_s)/\sin\delta \quad \text{(km)} \tag{5.7}$$

where δ is the angle of elevation of the satellite as seen from the earth station; see Figure 5.1.

There have been extensive studies of the way in which the value of h_{ml} varies, seasonally and with geographical location; see for example [3] and the COST 210 Report [4]. As yet this parameter has not been satisfactorily characterized. ITU-R Recommendation PN.618 [2] recommends that the following values be used:

$$h_{ml} = 3.0 + 0.028\theta \text{ km} \qquad \text{for } \theta < 36^\circ \tag{5.8a}$$

$$h_{ml} = 4.0 - 0.075(\theta - 36) \text{ km} \qquad \text{for } \theta > 36^\circ \tag{5.8b}$$

where θ is the latitude of the earth station.

Stage 2. Determine the rainfall rate $R_{0.01}$ that is exceeded for 0.01 per cent of an average year

If adequate local rainfall rate records are available, they should be used to provide a value for $R_{0.01}$. Otherwise it is necessary to rely on the available data which is derived, ultimately, from meteorological records.

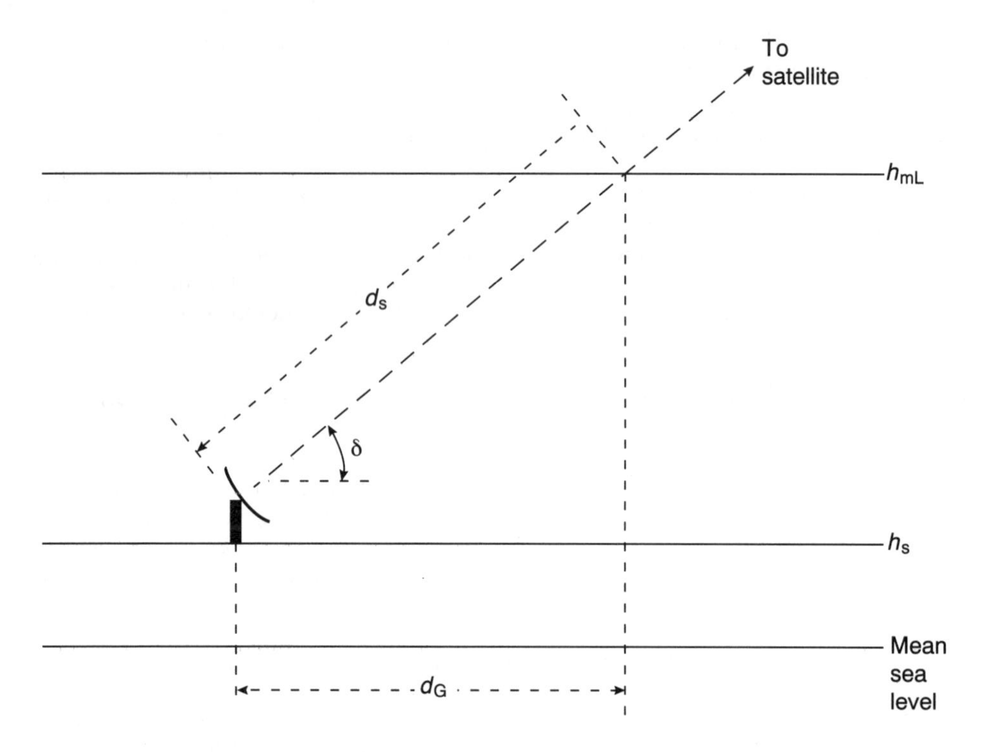

Figure 5.1 *A model for an Earth–space ray path in rain*

Maps in ITU-R Recommendation PN.837 [5] divide the world into fifteen rain climatic zones, A to Q. Zone A corresponds to the driest deserts and polar areas, where heavy rain scarcely ever occurs. Zones N, P and Q correspond to the tropical areas, where rainfall is heaviest. For example, Table 5.1 shows R_p, the rainfall intensity exceeded in an average year for various percentages p of the time, in Zone E (a typical temperate climate) and Zone P (a typical tropical rain

Table 5.1 *Sample rainfall rate probabilities (ITU-R Recommendation PN.837 [5])*

Percentage of time the rate is exceeded in average year (%)	*Zone E (temperate climate)* ($mm\,h^{-1}$)	*Zone P (tropical rain forest)* ($mm\,h^{-1}$)
1.0	0.6	12
0.3	2.4	34
0.1	6	65
0.03	12	105
0.01	22	145
0.003	41	200
0.001	70	250

forest climate). These data have been taken from Recommendation PN.837. Thus, for example, for Zones E and P, $R_{0.01}$ is 22 and 145 mm h^{-1} respectively.

Stage 3. Make allowance for the non-uniformity of heavy rain

The value of d_s calculated in Stage 1 tends to exaggerate absorption for high rainfall rates, expecially when the angle of elevation of the satellite as seen from the earth station is low. High rainfall rates are usually associated with thunderstorms, where the heaviest rain is often found in rain cells, typically only a few kilometres in diameter, moving with the wind through a wide area of much less heavy rain. See Figure 5.2.

Thus, if the probability is 0.01 per cent that the rainfall at any given point will exceed R mm h^{-1}, then a rain cell containing rainfall of that intensity may be present somewhere in the propagation path for, perhaps, 0.05 per cent of the time, but the loss over the whole path for that period may be little more than that due to d_c, the section of the path which actually lies within the rain cell. If the angle of elevation of the satellite is low, this section may be quite short compared with d_s. Consequently, in calculating the absorption for a transmission path suffering a very high rainfall rate, it should not be assumed that that rate applies to the whole length of the path below the melting layer.

ITU-R Recommendation PN.618 offers an empirical method for reducing this overestimate. A reduction factor $r_{0.01}$ is calculated, where

$$r_{0.01} = 1/(1 + d_G/L_0) \tag{5.9}$$

where $d_G = d_s \cos \delta$ km (see Figure 5.1),
$L_0 = 35e^{\beta}$, where $\beta = (-0.015R_{0.01})$ if $R_{0.01} < 100$ mm h^{-1}.

In calculating $r_{0.01}$, the value 100 mm h^{-1} should be used instead of $R_{0.01}$ when the estimated value of $R_{0.01}$ exceeds 100 mm h^{-1}.

The cellular distribution of heavy rain in a thunderstorm gives rise to the possibility of using earth stations in site diversity as a means of maintaining communication despite severe rain absorption. There is a brief discussion of the use of this technique in Section 10.3.1.

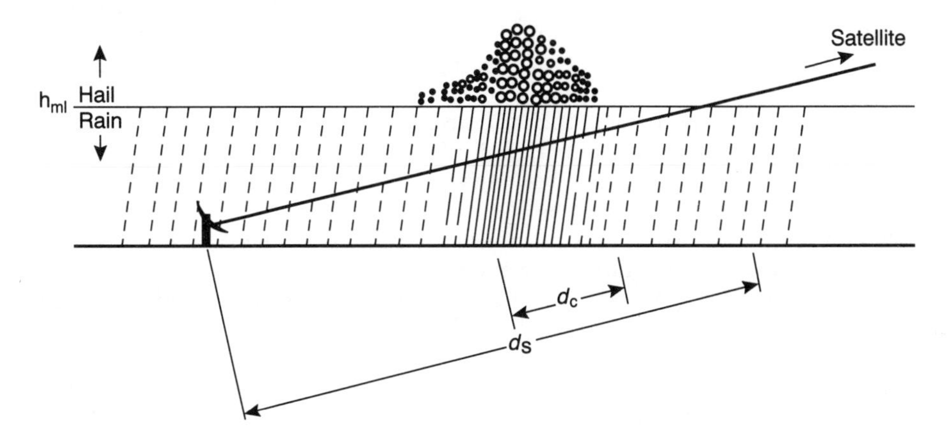

Figure 5.2 *A ray path through rain, including a rain cell*

Stage 4. Predict $A_{0.01}$, the absorption for 0.01 per cent of an average year

$$A_{0.01} = \gamma_R d_s r_{0.01} \quad \text{(dB)} \tag{5.10}$$

where γ_R is the specific attenuation corresponding to a rain rate of $R_{0.01}$ mm h^{-1}.

$$\gamma_R = k(R_{0.01})^{\alpha} \quad (\text{dB km}^{-1}) \tag{5.11}$$

where k and α are coefficients which vary with frequency and the polarization state of the wave. Values for k and α are to be found in ITU-R Recommendation PN.838 [6].

Stage 5. Predict absorption for values of p other than 0.01 per cent

Having calculated $A_{0.01}$, the absorption arising for p% of an average year, where p lies between 1.0 and 0.01%, can be predicted from

$$A_p / A_{0.01} = 0.12 p^{-b} \tag{5.12}$$

where $b = (0.546 + 0.043 \log p)$. Figure 5.3 shows estimates of the rain absorption as a function of frequency for 0.1 per cent of an average year for two earth stations, one in rain climate Zone E at latitude 45° and the other in Zone P at latitude 20°. It has been assumed that the earth stations are at sea level, with the angle of elevation of the satellite equal to either 30° or 10° and that polarization is either linear with a tilt angle of 45° or circular.

Worst-month rain absorption estimates

Climate is seasonal. If the rainfall rate for p% of an average year is R_p, the rainfall will be at this level for ap% of the rainiest month, where a is greater than unity. A standard of service that would be acceptable though significantly degraded for p% of the year if bad conditions were spread evenly over the twelve months may be found unacceptable if most of the degraded hours and minutes were concentrated into just a few months. For this reason most of the performance objectives adopted for international telecommunication channels relate to a stated percentage of the worst month of an average year, not to the year as a whole.

For earth stations for which long-term measurements of absorption on space–Earth paths or local rainfall rate records of adequate quality are available, statistics for the worst month can be extracted from the data and used for worst-month predictions of absorption using the method outlined above. But comprehensive meteorological rainfall rate data are not available on a monthly basis. Thus, where local records are not available, it is necessary to estimate worst-month absorption statistics from average year predictions, that is, using A_{ap}. The coefficient a is not a constant and its values are not, as yet, well defined; it may vary with climate and it is known to vary with p.

The determination of values for a is under study and an account is given in CCIR Report 723 [7] of the state of the work. See also ITU-R Recommendations PN.618 [2] and PN.581 [8]. Until better information becomes available and in the absence of data specific to the area in which the earth station is located, the report suggests that a should be assumed to equal 5.5 when p is 0.01%. Similarly, when p is 0.1%, a may be assumed to equal 4, and when (p) is 1.0 %, (a) may be assumed to equal 2.5.

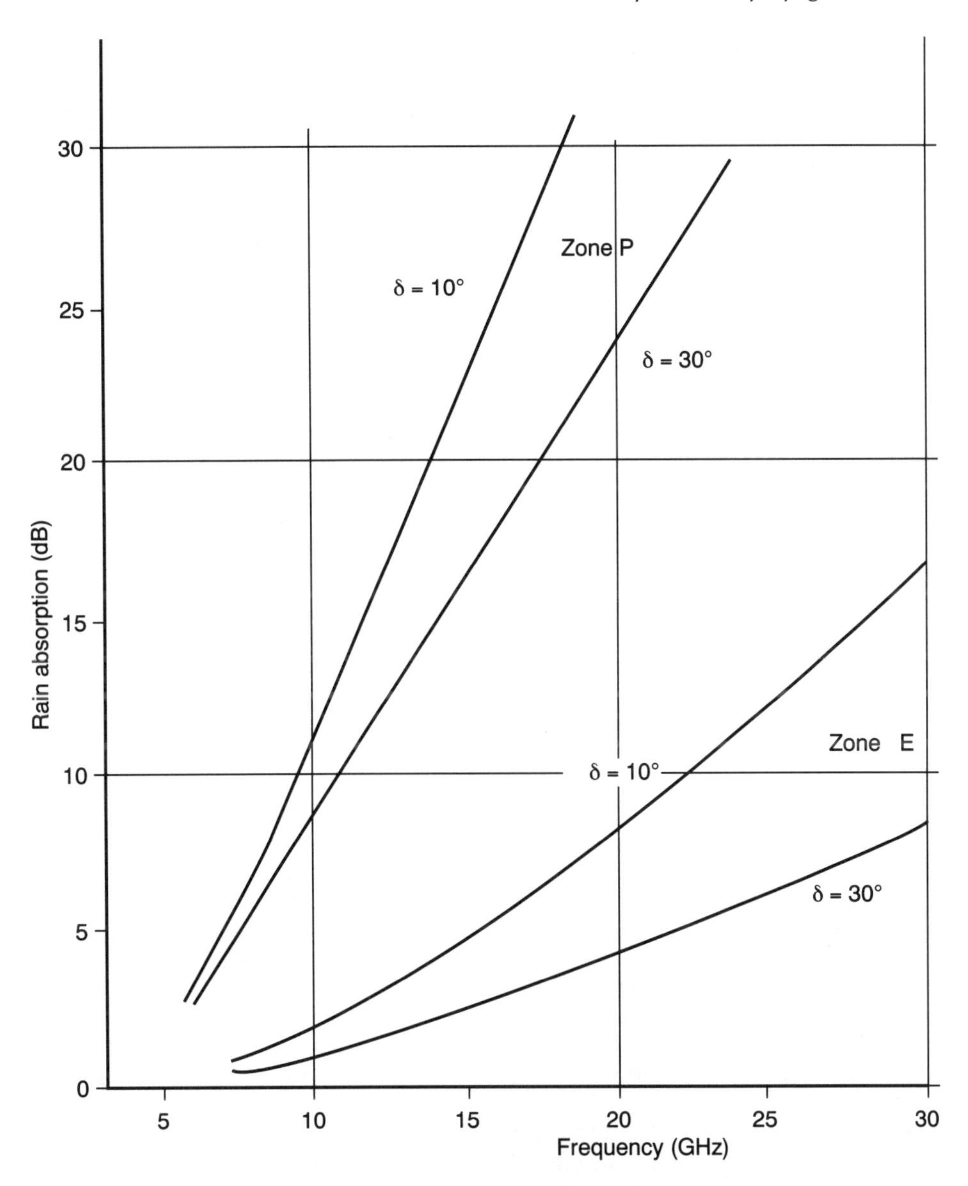

Figure 5.3 *Predicted rain absorption on a space–Earth path for 0.1% of an average year for earth stations in rain climate zones E and P and with angles of elevation of 10° and 30°*

5.1.3 The effects of clouds, snow, etc.

Liquid water in the propagation path, in the form of cloud, mist or fog, may cause significant loss at 100 GHz and above, but the effect can be neglected in the frequency bands in use now for satellite communication.

Hail causes a small loss due to scattering but none due to absorption. Dry snow likewise does not absorb energy from radio waves. However, the water content of wet snow, dropping slowly through the ray path, causes more absorption than would arise if the same amount of water was falling as rain.

5.1.4 Absorption by atmospheric gases

Molecular resonance in the gases of the atmosphere causes loss to radio waves. This loss is most severe in certain specific narrow frequency bands, the absorption bands, although there is quite significant absorption between these bands above 30 GHz. There is a weak absorption band due to water vapour at 22.3 GHz and there are very strong ones at 183 and 324 GHz. There is a very strong and broad absorption band due to oxygen centred on 60 GHz and a weaker, narrow oxygen band at 119 GHz. The specific attenuation γ_G at sea-level due to water vapour and oxygen under typical conditions of temperature and humidity at sea-level is shown in Figure 5.4, taken from ITU-R Recommendation PN.676 [9]. There is no significant absorption due to the other atmospheric gases within the frequency range currently of interest for satellite communication.

The quantity of oxygen in the atmosphere as a whole is unvarying and it is equivalent to a layer about 6 km thick at the partial pressure that oxygen has at sea-level. Thus, to calculate the absorption between an earth station at sea-level and a satellite, due to oxygen, the model of Figure 5.1 and equation (5.7) may be used, with the height of the layer taken as 6 km and the specific attenuation for oxygen taken from Figure 5.4. At 30 GHz, for an earth station at sea-level, seeing the satellite at an angle of elevation of 10°, the absorption due to oxygen would not exceed 0.7 dB, which would be negligible compared with the absorption due to rain in most climates. For an earth station at a site above sea-level, the absorption would be less. In any other frequency range currently of interest for satellite communication, the absorption would be much less.

The specific attenuation for water vapour, shown in Figure 5.4, has been calculated for a water content of 7.5 g m^{-3}. This is equivalent to 50% relative humidity at 16.5°C or 75% relative humidity at 10°C. The total absorption due to water vapour under these conditions, for an earth station at sea-level, can be calculated approximately from the specific attenuation, using the method suggested for oxygen in the previous paragraph, the uniform layer of humid air being assumed to be 2.2 km thick. A more exact calculation, covering earth stations above sea-level, would require information on typical atmospheric water vapour content for the earth station locality; means are provided in ITU-R Recommendation PN.676 [9].

The absorption due to water vapour on a space–Earth path below 15 GHz is negligible. Between 17.3 and 20.2 GHz, a potentially important frequency range, A_G may be as much as 1 to 3 dB, depending on the angle of elevation and the water vapour content of the atmosphere at the earth station, rising to a peak at 22.3 GHz, and remaining significant above that frequency. This loss is small compared with the loss due to rain, but it is likely to be present for much of the time.

Above 50 GHz, A_G ranges with frequency from substantial to very severe. Such losses may place an upper bound on the range of the spectrum that might, in the future, be used for uplinks and downlinks for commercial satellite communication. However, the absorption bands might well be found ideal for intersatellite links.

Ionospheric absorption

The presence of free ions in the ionosphere causes energy to be absorbed from a wave passing through it. In polar regions ionization is, at times, very intense. At 100 MHz the space–Earth loss for an earth station in the polar regions may reach some tens of decibels at a low angle of elevation. However, the loss is proportional to $1/f^2$; it is negligible above 1 GHz, and it can be

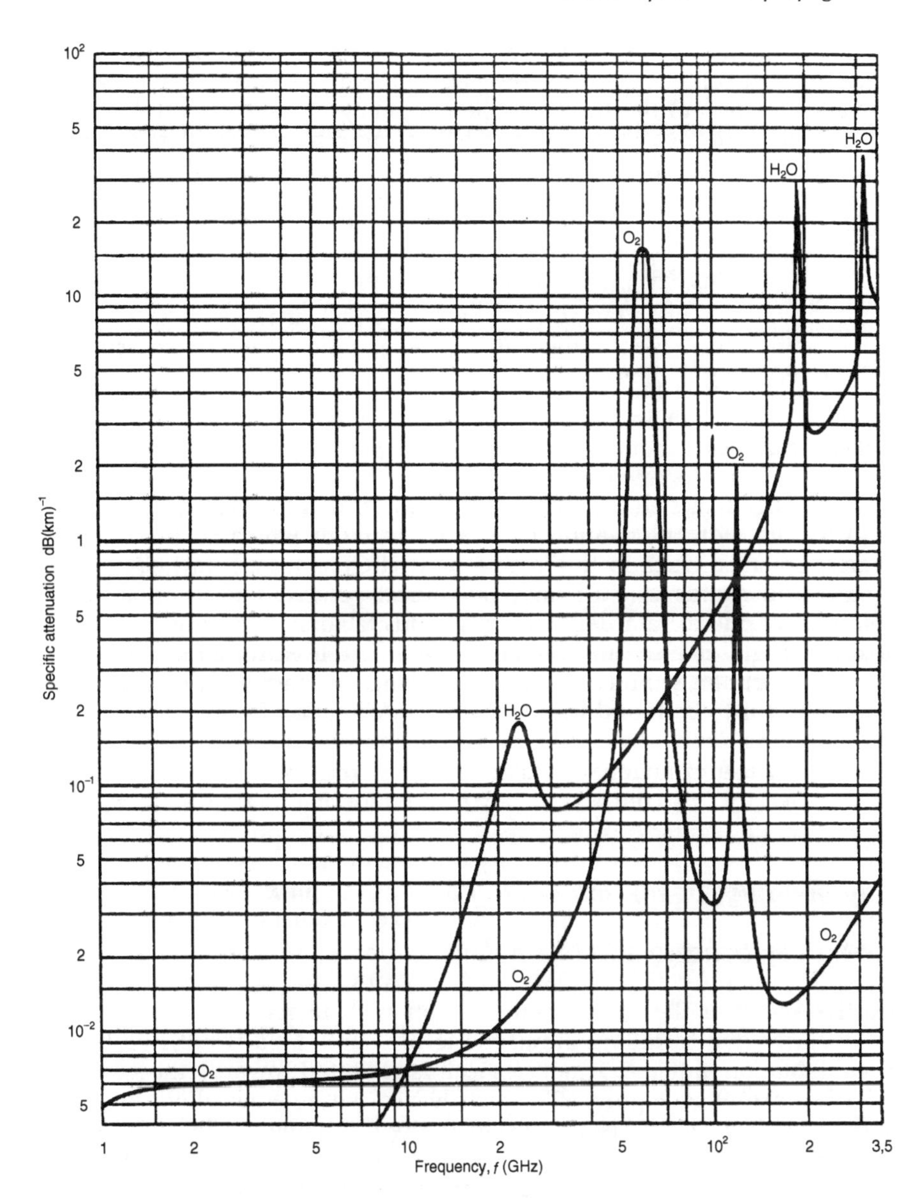

Figure 5.4 *Specific attenuation due to atmospheric gases in the frequency range 1–350 GHz, for an atmospheric pressure of 1013 millibars, a temperature of 15°C and a water vapour content of 7.5 g m^{-3}. (ITU-R Recommendation PN. 676 [9].)*

disregarded, except in polar regions, at any frequency likely to be used for satellite communication.

5.1.5 Polarization effects

Faraday rotation

The simultaneous presence of free electrons and the Earth's magnetic field in the ionosphere causes rotation of the plane of polarization of a wave passing through it. This is called Faraday rotation. The extent of the rotation is a direct function of the total electron content (TEC) of the path through which the wave travels and an inverse function of the square of the frequency.

At VHF the plane of polarization is rotated a total of many complete turns in the course of one journey from space to Earth or vice versa, and the turbulent nature of the ionosphere ensures that the final condition of a linearly polarized wave is continually changing. At 5 GHz, Faraday rotation is small but not negligible. It does not exceed 1° at 10 GHz and can be disregarded for most purposes at higher frequencies. See ITU-R Recommendation PI.531 [10].

At frequencies where Faraday rotation is substantial, it is usual to employ circular polarization to prevent a mismatch in polarization states of an arriving wave and the receiving antenna.

Depolarization by rain

Raindrops are not, in general, perfectly spherical and this asymmetry causes changes to the state of polarization of a beam through which they fall. To a small degree, other particles in the troposphere, such as ice crystals above the melting layer and clouds of dust, may also affect polarization. The extent of the depolarization due to rain depends on various factors: it increases with the frequency, the angle of elevation of the satellite at the earth station and the rainfall rate. An eight-stage procedure for calculating the magnitude of the cross-polarization effect is to be found in ITU-R Recommendation PN.618 [2].

This tropospheric depolarization can cause an increase in interference levels in systems which use orthogonal polarizations for frequency reuse. It may be desirable and feasible to compensate adaptively for it at earth stations; see Section 7.3.4.

5.1.6 Ray bending, beam defocusing and multi-path phenomena

Tropospheric ray bending and beam defocusing

The refractive index of the atmosphere falls off with increasing height above ground, primarily due to falling atmospheric pressure. As a result, the beam from an earth station antenna, if not vertical, is refracted towards the ground. A ray of a beam (*OA* in Figure 5.5) which leaves the antenna along the axis of symmetry, *OB*, will have been deviated through the angle a by the time it has passed through the atmosphere on its way to the satellite. This deviation is significant only at very low angles of elevation. When the direction *OB* is 1° above the horizontal the deviation is about 0.5° but it falls to around 0.1° at an elevation of 5°. This mechanism is operative at all frequencies of interest to satellite communication.

Tropospheric ray bending also has an indirect effect. The ray *OC* in Figure 5.5, leaving the earth station antenna at a somewhat higher angle than *OA*, is bent through a smaller angle, while ray *OD*, leaving at a lower angle, is bent rather more. If the straight portions of these rays are projected backwards, they appear to radiate, not from the earth station antenna *O* but from a point (O'), and the angle between the virtual rays CO' and DO' at O' is greater than the angle between the actual rays *CO* and *DO* at *O*. Thus, the beam is diverging more rapidly than would be expected from the gain of the antenna, so the effective gain is reduced. As would be expected, this defocusing effect is also significant only at low angles of beam elevation; at 1° elevation it is typically equivalent to a reduction of antenna gain of about 0.8 dB, falling to about 0.15 dB at 5° elevation.

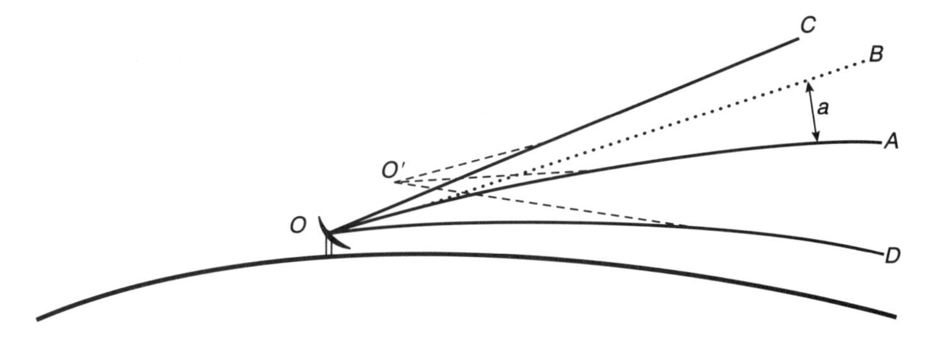

Figure 5.5 *Ray bending and beam defocusing in the troposphere*

Ionospheric ray bending

The ionosphere also causes bending of the rays passing through it, but this can be disregarded in microwave bands and its effects are not very significant at VHF frequencies.

Scintillation

Apart from the relatively coherent effects of a non-uniform atmosphere such as ray bending and beam defocusing, the amplitude and direction of waves arriving at a receiving antenna sometimes vary in an irregular manner as a result of wave interference arising from irregularities in the troposphere or the ionosphere. This so-called scintillation is sporadic.

Scintillation tends to be worse at low angles of elevation. Tropospheric scintillation may be significant at frequencies above 10 GHz and angles of elevation below 10°. Ionospheric scintillation is more likely to be observed at lower frequencies and at earth stations near the geomagnetic equator. Rapid variations of the received amplitude of up to a few decibels are observed in the microwave range and the direction of arrival may vary by a small fraction of one degree. A nine-stage procedure for the prediction of the fade depth due to tropospheric scintillation is to be found in ITU-R Recommendation PN.618 [2].

Ground reflection

Some part of the signal entering or leaving an earth station antenna on land may be reflected from the ground or a nearby building. Similarly, earth station operation on an aircraft or a ship may be affected by reflections from the wings or fuselage of the aircraft, from the superstructure of a ship or from the surface of the sea. These reflections cause multi-path interference when the reflected ray mixes with the direct ray. The effect may be serious if the gain of the antenna is low. This is a principal cause of signal fading at earth stations on ships and aircraft where the angle of elevation of the satellite is low.

5.2 Radio noise

5.2.1 Thermal noise of terrestrial origin

The mechanisms that cause absorption of energy from a wave passing between space and Earth also cause the emission of thermal noise at radio frequencies.

Some of this noise is added to the emission reaching the receiver. And the Earth itself radiates noise which can enter the transmission path via the satellite or the earth station receiving antenna.

The thermal noise power N, available from a 'black-body' having a temperature T (K), measured in bandwidth B (Hz) is given by

$$N = kTB \quad \text{(W)} \tag{5.13}$$

where k = Boltzmann's constant (1.38×10^{-23} JK^{-1}). The spectral power density N_0 of noise from a source is given by

$$N_0 = N/B = kT \quad (\text{W Hz}^{-1}) \tag{5.14}$$

and T is called the 'noise temperature' of the source.

In considering the level of noise received at an earth station or a satellite from external sources it is convenient to identify a brightness temperature T_B for each separate source and a coefficient η which represents the efficiency with which the receiving antenna captures noise from that source. Then t, the noise temperature component due to the identified source, is given by

$$t = \eta T_B \quad \text{(K)} \tag{5.15}$$

and the total noise entering the system from all of these sources, expressed as a noise temperature, can be obtained by summing all the component noise temperatures. The value of η lies between 0 and 1: η approaches 1 if the relevant part of the environment fills the solid angle formed by the main lobe of the antenna and it approaches 0 if that part of the environment lies in the side-lobes.

Earth station receiver noise from the environment

Thus, considering the noise received at an earth station from the ground, the sea, etc., and buildings nearby, T_B typically lies between 100 and 250 K, approaching the lower limit at sea at low angles of elevation and the upper limit on land. This noise temperature component t_{gr} is given by

$$t_{gr} = \eta T_B \quad \text{(K)} \tag{5.16}$$

If the angle of elevation of the main beam and the gain of the antenna are both high and if the antenna design is good, with well suppressed side-lobes and little subreflector spillover, the corresponding value of η will be small and t_{gr} may be no more than 20 K. If, however, the gain of the antenna is very low, as is typical of land mobile earth stations, η may reach 0.5 and t_{gr} may exceed 100 K.

Earth station receiver noise from the atmosphere

The brightness temperature of the atmosphere due to the permanent gases, and rain if it is present, is related to A, the loss in dB which they cause, by

$$T_B = T_m(1 - 10^{-A/10}) \quad \text{(K)} \tag{5.17}$$

where T_m = the mean temperature of the atmosphere, typically about 270 K. This expression may exaggerate T_B by a small amount, since a part of the loss A may be due to scattering, which does not generate noise. Since it may be assumed that this noise source fills the main lobe of the antenna, $\eta \approx 1$ and t_{at}, the noise temperature component due to the atmosphere, is approximately equal to the value of T_B given by equation (5.17). In the strong absorption bands, such as the oxygen band at 60 GHz, t_{at} is equal to the temperature of the atmosphere. At lower frequencies, where gaseous absorption can be disregarded, rain which, for example, causes absorption equal to 3 dB also generates a noise temperature equivalent to about 135 K.

Satellite receiver noise from the Earth

Satellite antennas directed towards the Earth, having sufficient gain for the main lobe to be filled by the Earth, will also receive noise from the Earth with $\eta \approx 1$. T_B will be about 210 K if land occupies a large fraction of the beam footprint. However, except in the atmospheric absorption bands, T will be somewhat less, perhaps as little as 160 K, if sea occupies a large part of the footprint.

5.2.2 Extraterrestrial noise

Galactic noise

Vast amounts of radio noise power are generated in stars and in interstellar gas in the central regions of our galaxy, causing high brightness temperatures in the VHF frequency range and below. However, T_B falls away rapidly with rising frequency. Furthermore, the part of the sky where this broad background of galactic noise radiation is strongest, which is in the constellation Sagittarius, never passes behind the geostationary orbit as seen from the Earth. In the microwave frequency range t_{sky}, the noise component due to galactic noise perceived by an antenna of substantial gain directed towards a geostationary satellite, will not exceed a few kelvins; it can usually be disregarded.

Solar noise

The Sun is a powerful source of radio noise. The Sun is close to the Earth's equatorial plane at the time of the equinoxes, from early March to mid-April and from early September to mid-October. The dates and the times of day when the Sun passes behind a geostationary satellite as seen from an earth station, potentially causing interference at that earth station, depend on the latitude of the earth station, the gain of its antenna and the difference between the longitudes of the earth station and the satellite (angle A in Figure 2.8). Powerful noise from the Sun enters the receiver of an earth station with a high gain antenna and served by a geostationary satellite for a few minutes daily on several days during these equinoctial periods. With an antenna with less gain, interference occurs on more days, for longer periods, but with less intensity. If the satellite is not geostationary, solar interference may arise for brief periods at other times, depending upon the orbit, but it may not arise at all.

The intensity t_{sol} of the solar noise entry depends on the frequency, the state of the Sun and the extent to which the source fills the main lobe of the earth station antenna. At VHF frequencies, the Sun's brightness temperature is normally around 1×10^6 K, falling to about 2.5×10^4 K at 4 GHz, 1.2×10^4 K at 12 GHz and 0.9×10^4 K at 18 GHz. When the Sun is disturbed, its brightness temperature is considerably greater. The coefficient η may be as high as 1 if the beamwidth of the earth station antenna is less than 0.5°. The maximum value of η is less than 1 when the antenna beam is broader and it may be so small for, for example, a VSAT (see Chapter 13) that the effect of t_{sol} can be neglected.

Solar noise interference is severe enough to interrupt service on most systems using high-gain antennas. Loss of communication can be avoided by building earth stations in pairs, separated by sufficient distance to ensure that solar interference occurs on different days or at different times of day, reception being transferred from one earth station to the other when necessary to avoid the interference. However, this is obviously a costly solution, used seldom if ever. It is usually accepted that channels will not be available without any interruptions whatever, and provision may be made for virtually unavoidable interruptions of this kind by including an 'unavailability' budget in the performance specification; see Section 10.4.

Astronomical point noise sources

There are several isolated point sources of radio noise ('radio stars') in the sky, the best-known of which are Casseopeia A, Cygnus A and Taurus A. The planets, other than the Earth, are weak emitters of noise, but much closer; Venus is the strongest planetary noise source. All of these sources have a very small angular extent as seen from the Earth, so η for practical antennas is very small. Furthermore, the more powerful radio stars are remote from the GSO as seen from the Earth and the planets seldom pass close to it. Thus, these noise sources cause no degradation to communication systems. However, for earth stations with fully-steerable high-gain antennas and low-noise receivers, these stellar and planetary sources provide noise of accurately known power which may be used for testing antennas; see Section 7.3.6.

5.3 Transmission time effects

5.3.1 Mean transmission times

The transmission of signals from earth station to earth station via a satellite takes a perceptible amount of time t, given by

$$t = d/c \quad \text{(s)} \tag{5.18}$$

where d = the distance in m, calculable for example from equation (2.8) and
c = the velocity of light in free-space $\approx 3 \times 10^8\ \text{m s}^{-1}$.

For example, for a link between two earth stations via a geostationary satellite, the range of transmission times is 238–274 ms. These transmission times are about ten times longer than an intercontinental connection by wideband terrestrial transmission media.

A connection by satellite between two earth stations may not be limited to one uplink and one downlink. For various reasons, two satellite links might sometimes be connected in tandem at an intermediate earth station; in the future, such connections may be made directly between satellites, close together or far apart, via intersatellite links (ISLs). See Figure 5.6. A short ISL may add little to the transmission time of a connection via a single satellite hop, but a long ISL between geostationary satellites or the connection of two geostationary satellite links in tandem at an intermediate earth station might raise the transmission time to as much as 550 ms.

Furthermore, long transmission delay is not caused solely by the physical length of the radio path. Terminal equipment can introduce substantial delays. High-capacity TDMA systems may introduce delays of several milliseconds and TDMA systems designed primarily for efficient handling of data signals may

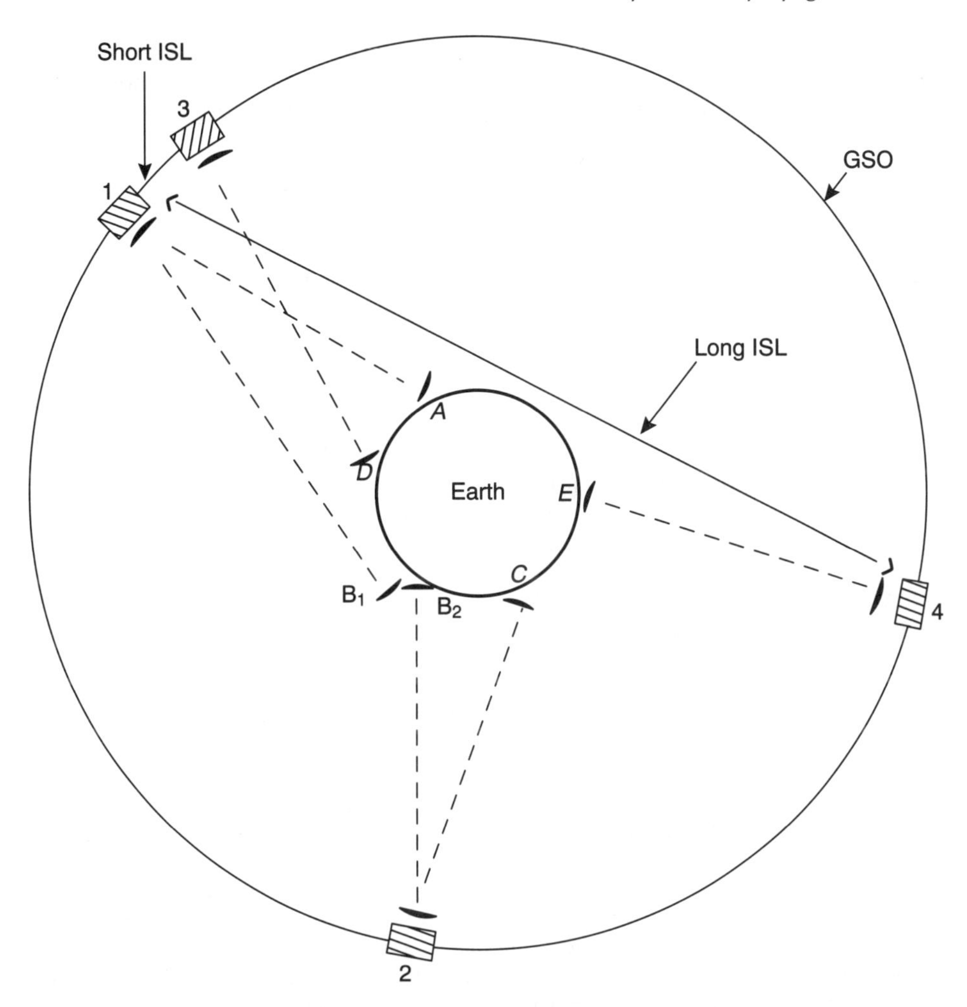

Figure 5.6 *Multi-satellite connections using an intermediate earth station (earth stations A and C via satellites 1 and 2 and earth station B), a short ISL (earth stations A and D via satellites 1 and 3) or a long ISL (Earth stations A and E via satellites 1 and 4)*

cause considerably greater delay. Above all, special codecs that provide substantial economy in the bit rates of digital signals, originally analogue, typically speech signals, may introduce a delay of several tens of milliseconds.

Some telecommunication services are little affected by long transmission time; examples are programme material for television and sound broadcasting and unidirectional data transmissions with forward error correction (FEC). Data transmission systems that correct detected errors in received material by ARQ can be designed to accommodate long transmission times. But problems arise for telephony.

The effects of long transmission time on telephony

Two kinds of difficulty may arise when a circuit having a long transmission time is used for a telephone conversation. The more basic difficulty is subjective.

When people converse politely, they give and receive, consciously or subconsciously, the cues that indicate when one is about to stop talking and the other is invited to start talking. If these cues are not seized upon promptly, conversation does not flow smoothly. With practice, but only within limits, people learn to accommodate delay in their correspondent's response to their cues.

The other problem is speech echo. Long-distance telephone circuits are always operated over most of their length in the 4-wire mode, that is, they have separate channels for the two directions of transmission. But with rare exceptions the user's telephone is connected into the 4-wire network by a bidirectional 2-wire link by means of a hybrid transformer or the equivalent; see Figure 5.7. Ideally the 2-wire/4-wire device is accurately matched for both directions of transmission, so that all of the speech energy received from the distant speaker via the 'return' leg of the 4-wire circuit is directed to the local speaker's telephone. Usually the hybrid transformer is not perfectly balanced and some part of the distant speaker's own speech signal is coupled back to him via the 'go' leg.

When the transmission time for the arrival of this reflected energy back at the distant speaker is short, as it usually is when the transmission medium is terrestrial, the echo, then called the 'sidetone', merely gives a reassuring 'live' impression of the connection. However, when the transmission time is very long, as for a two-way journey via a single geostationary satellite, the reflected signal is clearly perceived as an echo; if the echo is strong, it can disrupt conversation.

An echo suppressor can be used to interrupt the 'go' leg when the distant speaker is talking, so suppressing the echo but perhaps also interrupting a remark of the local speaker. Better results are obtainable by using an adaptive echo canceller to inject into the 'go' leg an appropriately attenuated and phased sample of the distant speaker's signal arriving on the 'return' leg, so neutralizing the echo but not interrupting the signal path for the local speaker's outgoing signal; see ITU-T Recommendation G.165 [11] and Figure 5.7. A centre clipper, for an analogue implementation of the echo canceller, or a non-linear processor for a digital implementation, may be used to suppress low-level remnants of the echo.

Investigations have shown that, given good control of echo, a telephone connection that includes one hop via a geostationary satellite is acceptable to the public, as is acknowledged in ITU-T Recommendation G.114 [12]. The ITU recommends that circuits with a one-way transmission time exceeding 400 ms, such as those involving two hops via geostationary satellites, should be considered unacceptable as a basis for planning, but recognizes that transmission

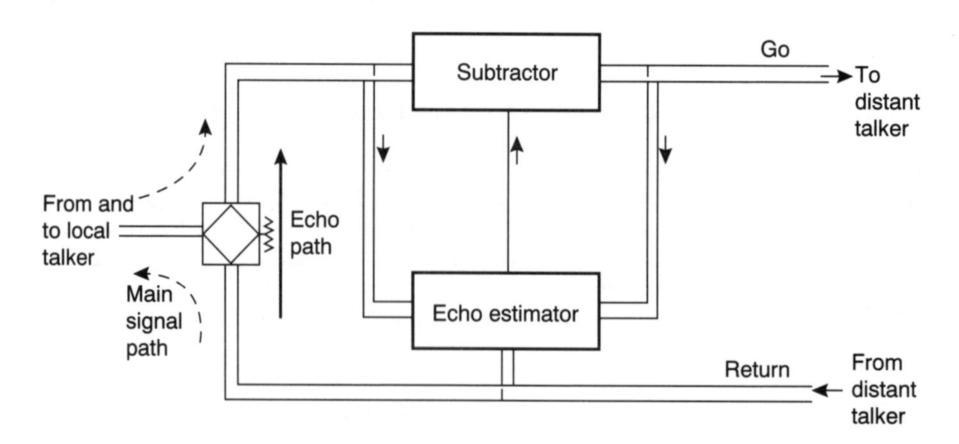

Figure 5.7 *The use of an adaptive digital echo canceller to eliminate echo arising at the 2-wire/4-wire interface*

times longer than 400 ms may arise in exceptional cases, typically where no better option is available.

5.3.2 Short-term variations of transmission time

Ionospheric effects

An Earth–space link is not entirely in free-space and consequently the radio waves that form it do not travel the whole distance at the same velocity as a wave in free-space. In particular, the group velocity in the ionosphere is somewhat less than c. The consequential increase in the transmission time may be several microseconds at VHF and a substantial fraction of 1 μs at 1 GHz. This additional delay varies erratically with time. However, the delay due to atmospheric ionization is inversely proportional to the square of the frequency; it is only a few hundredths of a microsecond at 4 GHz and its effect on communication systems can usually be neglected in comparison with the effect of satellite motion.

Effects due to satellite and earth station movement

Even a satellite which is nominally geostationary has a residual motion relative to fixed earth stations which is significant in some applications. For links with mobile stations, and especially aircraft, the movement of the earth station may have a greater effect.

The ITU Radio Regulations (see Section 6.5, item 3) require most kinds of geostationary satellite to be maintainable within $\pm 0.1^\circ$ of the nominal longitude but put no explicit constraints on inclination of the orbital plane and orbital eccentricity. It is feasible to keep a satellite on station with greater precision than this and some systems aim to keep satellites within $\pm 0.05^\circ$ of their nominal position. Such a satellite would move daily, relative to the Earth's surface, from side to side and from the top to the bottom of a volume of space that measures about 75 km east–west and north–south and about 25 km vertically. The diurnal variation of the distance between an earth station at a fixed location and the satellite would depend on various factors but it might typically be about 30 km. The transmission time from one earth station to another could vary in the course of a day by up to 0.2 ms. Such small movements must be regarded as the best currently attainable; most geostationary satellites move several times as much.

For satellites in 12-hour or 24-hour elliptical orbits, used around apogee, the variation would depend on the length of the orbital arc over which the satellite is used, but it would be very much greater than for a geostationary satellite. A change of the satellite/earth station distance of 2000 km might occur in the course of two hours, equivalent to a mean satellite velocity v_s relative to the earth station of $0.28\,\mathrm{km\,s^{-1}}$. For a medium-altitude circular orbit, v_s would be of the same order. For a low Earth orbit, v_s would be much greater, up to about $5\,\mathrm{km\,s^{-1}}$. A supersonic aircraft travelling at Mach 2 would, under the least favourable circumstances, be moving relative to a geostationary satellite at a rate approaching $0.7\,\mathrm{km\,s^{-1}}$.

The movement of a satellite, relative to its earth stations, causes carrier frequencies to exhibit a Doppler frequency shift δf, such that

$$\delta f / f_c = v_s / c \qquad (5.19)$$

where f_c is the carrier frequency. For the same reason, and to the same degree, the baseband signals of analogue transmissions will be stretched or compressed in the frequency domain and the bit flow of digital signals will be speeded up or slowed down as the physical length of the satellite link changes. The extent of these various effects will change with v_s from minute to minute. It will be different for each of the earth stations that a satellite serves. The aggregate effects for each uplink/downlink pair will be different and there may be a substantial step-change if and when operation is handed over from one satellite to another.

For narrowband transmission systems, these effects will usually be negligible. However, for high-capacity frequency division multiplex systems, the interface equipment which connects the satellite links with the terrestrial network would have to be designed to compensate for the baseband stretch and compression of links via low-orbiting satellites. High-capacity plesiochronous digital systems would have to be specially designed to accommodate the variability of systems using satellites in medium-altitude circular and elliptical orbits. Even the 0.2 ms daily transmission time excursion of a well-controlled geostationary satellite link would be a factor demanding attention in the design of a high-speed digital system which is part of a synchronous network.

These matters are considered further elsewhere, in particular in Chapters 9 and 11.

References

1. Allnutt, J. E. (1989). Satellite-to-Ground Radiowave Propagation. Peter Peregrinus.
2. *Propagation data and prediction methods required for the design of Earth–space telecommunications systems* (ITU-R Recommendation PN.618-3, 1994), ITU-R Recommendations 1994 PN Series. Geneva: ITU.
3. Leitao M. J., Watson P. A. and Brussard G. (1984). Report No. 352, University of Bradford, Interim Report for ESA/ESTEC Contract.
4. *Influence of the atmosphere on interference between radio communications systems at frequencies above 1 GHz* (Final report of COST 210 study, 1991). Commission of the European Communities, Luxembourg.
5. *Characteristics of precipitation for propagation modelling* (ITU-R Recommendation PN.837-1, 1994), ITU-R Recommendations 1994 PN Series. Geneva: ITU.
6. *Specific attenuation model for rain for use in prediction methods* (ITU-R Recommendation PN.838, 1994), ITU-R Recommendations 1994 PN Series. Geneva: ITU.
7. *Worst-month statistics* (CCIR Report 723-3, 1990), Reports of the CCIR, Annex to Volume V. Geneva: ITU.
8. *The concept of 'worst month'* (ITU-R Recommendation PN.581-2, 1994), ITU-R Recommendations 1994 PN Series. Geneva: ITU.
9. *Attenuation by atmospheric gases in the frequency range 1–350 GHz* (ITU-R Recommendation PN.676-1, 1994), ITU-R Recommendations 1994 PN Series. Geneva: ITU.
10. *Ionospheric effects influencing radio systems involving spacecraft* (ITU-R Recommendation PI.531-2, 1994), ITU-R Recommendations 1994 PI Series. Geneva: ITU.
11. *Echo cancellers:* (ITU-T Recommendation G.165, 1993). Geneva: ITU.
12. *Mean one-way propagation time* (ITU-T Recommendation G.114, 1993). Geneva: ITU.

6 Control of radio interference

6.1 Spectrum management for satellite communication systems

National spectrum management

Governments control the use made of radio by stations within their jurisdiction, typically by making spectrum management a function of a civil service department or by setting up an agency for the purpose. These national regulating bodies are known as 'administrations'. A key function in spectrum management is the assignment of carrier frequencies to transmitting and receiving stations, to be used for approved purposes and within stated parameters. In making these assignments, an administration aims to enable radio operating organizations to achieve their objectives without suffering or causing interference while using the spectrum efficiently, so that subsequent applicants will also be able to get access to the radio medium. When it is necessary to take account of cross-frontier interference liabilities, these administrations collaborate with equivalent agencies in neighbouring countries.

The International Telecommunication Union

Where good use of the spectrum requires wider consultation or agreement to permanent policies and procedures, the administrations use the International Telecommunication Union (ITU) as their global forum. Main decisions on policies and procedures are made at periodical World Radiocommunication Conferences (WRCs), held under the aegis of the ITU. These conferences were called World Administrative Radio Conferences (WARCs) prior to 1993. The ITU *Radio Regulations* (RR) [1] bring together the decisions of past WRCs and WARCs, with any subsequent amendments. A major revision of the RR was agreed at WRC-95.

Technical issues are studied internationally and reports and recommendations, generally observed though non-mandatory, are made by periodical ITU Radiocommunication Assemblies and their permanent ITU-R Study Groups. This was a function of the International Radio Consultative Committee (CCIR) prior to 1993. Where technical issues extend beyond the radio field they may become a responsibility of the ITU-T Study Groups of the ITU Telecommunications Standardization Sector, formerly the International Telegraph and Telephone Consultative Committee (CCITT).

Frequency allocation

Radio systems are categorized as 'services' for management purposes; at present there are 37 services. Frequency bands have been allocated for each service by international agreement (see RR [1] Article 8) and frequencies are normally selected from appropriate frequency allocations for assigning to stations. The satellite systems reviewed in this book fall into one or other of the following services:

(1) the fixed-satellite service (FSS), which serves earth stations at fixed and declared locations;
(2) the mobile-satellite service (MSS), which serves earth stations that are mobile, whether they move by land, at sea or in the air;
(3) the broadcasting-satellite service (BSS), which transmits signals from satellites for direct reception by members of the public.

Some of the frequency bands allocated for mobile-satellite use have been given a more specialized role, being allocated to the maritime mobile-satellite service (MMSS), the aeronautical mobile-satellite service (AMSS) or the land mobile-satellite service (LMSS).

In a few cases a frequency band is allocated for more than one satellite service and almost all of the bands which are allocated for satellite services are also allocated for terrestrial services, usually the fixed service (FS) or the mobile service (MS). The FS is made up mostly of line-of-sight radio-relay systems. The MS consists mainly of hand-portable or vehicle-borne radiotelephones below 3 GHz and transportable, temporary line-of-sight links at higher frequencies. This practice of allocating frequency bands for more than one service is called 'band sharing'. In shared bands the different services do not necessarily have equal status; stations of a service with secondary status must not cause interference to a station of a service with primary status, and they have no formal remedy if they suffer interference from the latter.

Some of the ITU frequency allocations are not uniform throughout the world. For frequency allocation purposes the ITU has defined three geographical regions, approximately as follows:

Region 1 consists of Europe, Africa, the CIS, Mongolia and the countries of Asia to the west of Iran.
Region 2 consists of North and South America.
Region 3 consists of Australasia and the countries of Asia which are not included in Region 1.

However, with the exception of the BSS, there are few significant regional differences in the satellite allocations.

Frequency assignment for satellite systems

The management of spectrum for satellite systems differs from that for other kinds of radio system for several reasons, as follows:

(1) Satellite systems are seldom entirely within the jurisdiction of a single administration. Many satellites are owned and operated by international consortia or by companies which lease the right to use the satellite to the earth stations of foreign clients. Only the managers of such satellite systems can design a frequency assignment plan that is satisfactory for the system. The frequency usage that the managers would prefer for each earth station is therefore proposed to the administration responsible for the earth station. If the managers' plan would be compatible with other national spectrum usage, a formal assignment would then be made by the administration.

(2) For most kinds of radio system, sufficient geographical separation between a receiving station and an unwanted transmitting station will prevent interference at any frequency above about 60 MHz, but this is not usually true of satellite systems. A satellite system has little protection from interference from other satellite systems using the same frequency bands except that which is provided by the directivity of the earth station and satellite antennas of the two systems. Thus an administration or a satellite operator acting alone is unable to determine whether the operation of a proposed new system will be compatible with other systems. Protection of satellite systems from interference is, in general, an international function. It is essential that full agreement be reached on how a new system will be harmonized with other systems at an early stage, since little can be done to modify system parameters after the satellite has been launched.

(3) The total number of satellite systems that can operate in a pair of frequency bands without excessive interference may be large, but it is finite. Technical constraints have been placed on system design to enable more satellite systems to use the medium, and in particular the GSO. Notwithstanding these constraints, it is already difficult to get agreement between the interested parties on the entry of new geostationary satellites into the more heavily used frequency bands. The possibility that new systems will be unable to get access to the GSO in the future, in frequency bands providing good radio propagation conditions, has been and remains a matter of great concern to many governments.

As a result of these special factors and concerns, the frequency management of these satellite services has to be carried out very largely at an international level. Two basic methods are used to determine whether and how a new system can be launched and used without causing unacceptable interference to other systems. The two methods are briefly as follows:

(1) The frequency coordination method is used for the MSS and, in most frequency bands, for the FSS. In this method, proposals for launching a new satellite are formally announced. Any administration perceiving a risk of interference with a system for which it is responsible can require the proposing administration to discuss ways of reducing interference to an acceptable level. The consent of these other administrations to significant interference levels, if they are foreseen, is a necessary condition for the registration of the new system by the ITU. And registration is an important defence against subsequent interference.

(2) The frequency planning method is used for the BSS and for certain FSS frequency bands. In this method, a plan is drawn up, usually in advance of any substantial use of the frequency bands concerned, providing a satellite location for every country that foresees a need to launch its own satellite in the future. The system parameters assumed in planning are chosen so that interference will be acceptably small even if the plan should be fully implemented. When the time comes to set a new system up, frequency assignments are made and announced in the usual way, but international clearance should be no more than a formality provided that the parameters of all the systems involved lie within the range of those that were assumed in drawing up the plan.

The frequency coordination method is reviewed in more detail in Section 6.2 and the frequency planning method in Section 6.3.

Since most satellite systems operate in frequency bands that are also allocated with equal status for the FS and some are also shared with the MS, various measures are taken to limit interference between satellite and terrestrial systems; these measures are reviewed in Section 6.4. Finally, the technical constraints on satellite systems which have been agreed, as a means to promoting efficient use of the GSO and the spectrum, are reviewed in Section 6.5.

6.2 Frequency coordination between satellite systems

6.2.1 Regulatory procedures

Several pairs of frequency bands below 15 GHz are allocated to the FSS and managed by frequency coordination. Each pair is already used by a large number of satellite systems. A similar situation, although involving fewer satellites, is developing in the MSS bands around 1.6 GHz.

These bands are allocated for either uplinks or downlinks. In the uplink bands, earth stations do not receive and in the downlink bands they do not transmit. Thus, interference does not occur between one earth station and another. For similar reasons there is no interference between satellites. The interference that has to be brought within acceptable limits by frequency coordination arises from the reception at a satellite of earth station emissions intended for a different satellite and the reception at earth stations of signals intended for a different earth station.

These interference paths are illustrated in Figure 6.1. For example, the uplink signal from earth station A, directed at geostationary satellite A, also illuminates satellite B, although at a reduced level because the ray path is at an angle ϕ_{AB} relative to the axis of the earth station antenna; this interference may become mixed with signals in the process of being relayed to one of the earth stations which satellite B serves. Similarly, downlink interference from satellite A enters earth station B, again somewhat diminished by the off-beam angle ϕ_{BA}. The interference level may be further reduced due to the directivity of the satellite beams, but not if both earth stations are within both satellite beam footprints.

Reverse band working

It is technically feasible for different satellites and their associated earth stations to use a pair of frequency bands in opposite directions, one band being used for uplinks to some satellites and for downlinks from others, and vice versa for the other band. Such an arrangement, called 'reverse band working', might allow a pair of frequency bands to be used more intensively than the conventional mode permits. However, with this configuration, interference could also occur between one earth station and another and between one satellite and another. These interference paths could be coordinated by the application of the same principles as are used for the interference paths shown in Figure 6.1, although with differences of detail. However, earth stations located in the same broad geographical area using frequency bands in opposite directions of transmission would have to be sited with great care, relative to one another, to avoid unacceptable interference.

At present, reverse band working is limited to a few special situations. For example, the band 17.7–18.1 GHz is allocated for downlinks for networks of the FSS and for uplinks (feeder links) to broadcasting satellites. In these special cases it is to be expected that there would be very few earth stations per country,

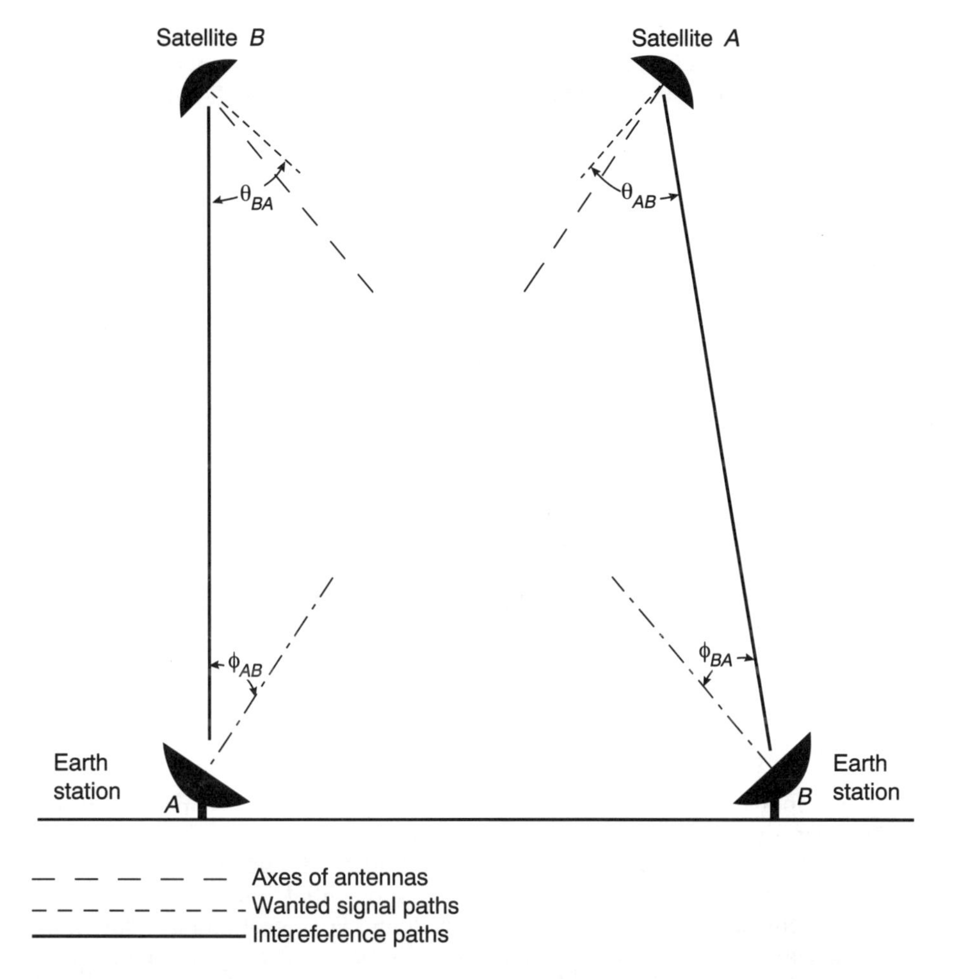

Figure 6.1 *Modes of interference from satellite system A to satellite system B, both systems having transparent transponders and using the same frequency bands in the same directions of transmission*

perhaps only one, for at least one of the two services involved, (the feeder link earth station in this case) making more manageable the earth station siting problem.

Frequency coordination

An administration proposing to make frequency assignments to a planned new satellite must first carry out the frequency coordination process. This process has been designed to ensure that all administrations whose participation in the coordination process would be desirable do in fact have an opportunity to do so. The detailed procedures have been framed to ensure that the process of coordination will not be unduly prolonged by vexatious disputes between participants. A third objective has been to prevent an orbital slot that has been successfully coordinated for a proposed new satellite from remaining indefinitely unused if the proposer is not able to launch and to start operating the satellite in a timely way.

The procedure for frequency coordination, and the subsequent registration of frequency assignments is rather complex but its essence is illustrated in the flow chart in Figure 6.2 and its four phases are set out in more detail below.

Phase 1. Advance publication of information and associated discussions

Frequency coordination starts with the advance publication phase. The procedure is set out in Section I of RR [1] Article 11. Section I of the Annex to RR Resolution 46 [2] may also apply. Proposed new satellite systems are notified to the ITU at an early stage, although not more than six years before the forecast date of bringing into use. The notification gives general information on the proposed new system, such as

(a) the radio service to which it will belong,
(b) the frequency bands that it will use,
(c) the orbit, including a broad indication of the proposed longitude for geostationary satellites,
(d) the geographical coverage of the satellite antenna beams,
(e) the maximum spectral power density of the satellite's emissions, and
(f) basic characteristics of the earth station antennas.

This information is published in the ITU's weekly circular and circulated to all administrations. Any administration that foresees that the new system would cause unacceptable interference to a satellite system, existing or planned, for which it is responsible informs the notifying administration of its concern. Discussions between the interested parties may follow.

At this advance publication stage, it is not likely to be feasible to determine with confidence whether there will be substantial interference between the new system and the various existing systems. For geostationary satellites, much will depend, for example, on the details of the carrier frequency plan proposed for the new system, the parameters of carriers and details of the footprints of satellite antennas. For non-geostationary satellites, details of proposed orbits, the orbital arcs over which satellites would be operated, and additional information on the radio transmission system that is proposed would need to be added to the list of uncertainties applicable to geostationary satellite systems. Such details are unlikely to be finalized until later in the development of the new system.

Nevertheless, discussions that follow advance publication can be of value. They heighten the awareness of all parties to the orbit and spectrum sharing problems that will arise in the main coordination phase that lies ahead. If, as a result of the discussions, it is concluded that these problems are likely to be acute, the operator proposing the new system may decide that it would be desirable to modify his plans or objectives. If it can be foreseen that changes to, say, the footprints or the out-of-beam characteristics of the satellite antennas or to the proposed location in the GSO are going to be necessary, these changes will be made at least cost and with least delay if the decisions are taken at an early stage in the designing and fabrication of the satellite. Moreover, the advance publication procedure creates an opportunity for the owners of other satellites involved to modify their own systems, perhaps when a satellite is replaced, to facilitate the entry into the medium of the newcomer. Any agreements arising from the discussions are reported to the ITU and published.

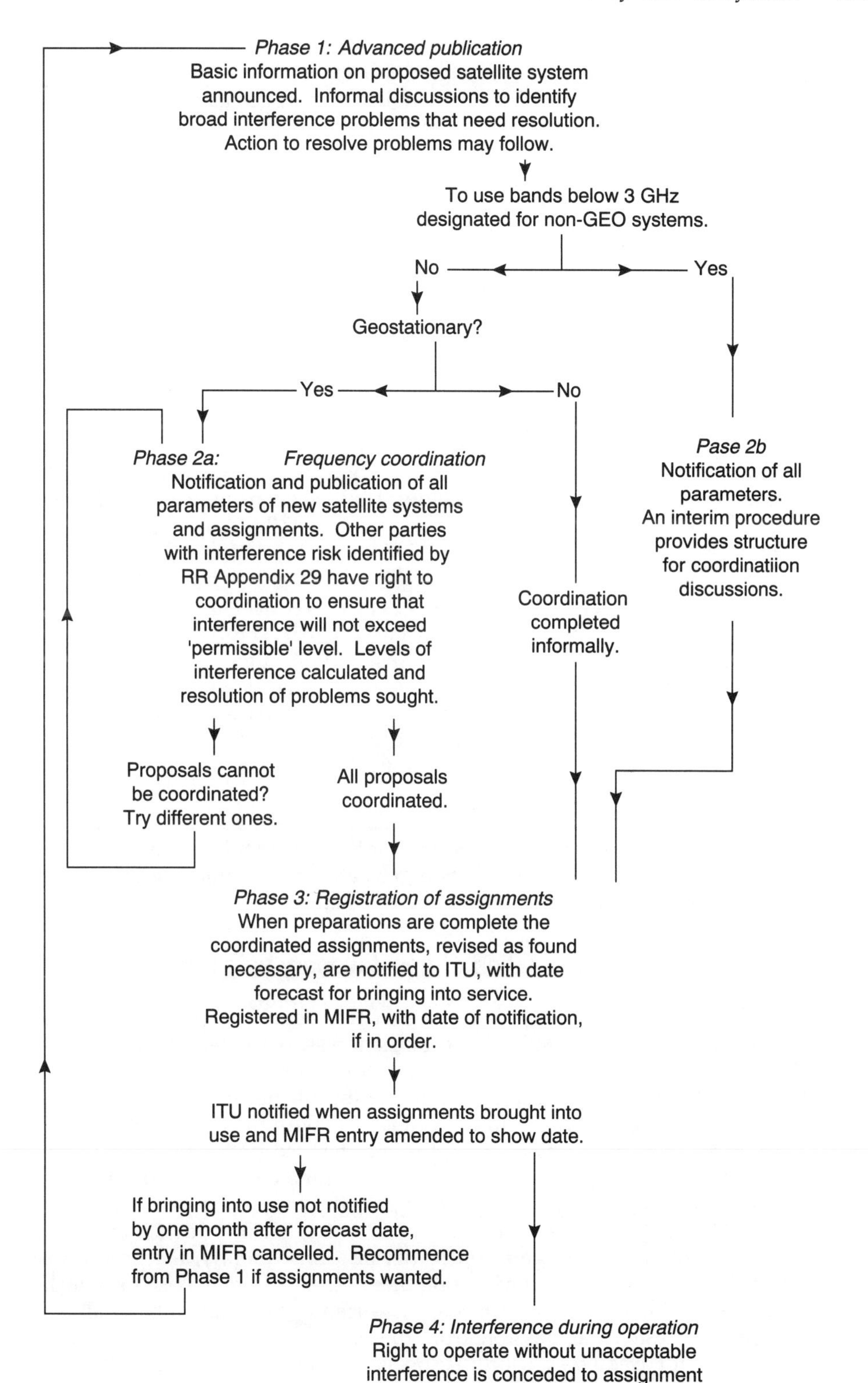

Figure 6.2 *Flow-chart for frequency coordination between satellite systems*

From this point onwards, the procedures for different kinds of new system diverge. Certain frequency bands, all below 3 GHz, were allocated for mobile-satellite services at WARC-92 or WRC-95, with use by non-geostationary satellites of the MSS particularly in view. Systems intended to operate in these frequency bands and the associated feeder link bands are frequency coordinated in accordance with a special procedure outlined under Phase 2b below.

There is no formal procedure for coordinating systems using non-geostationary satellites and operating in other frequency bands. Thus, the managers of these systems must devise their own means for avoiding interference problems to and from other non-geostationary systems. When solutions have been reached, the administration responsible for a new non-geostationary system proceeds directly to Phase 3 of the procedure. This lack of an agreed procedure has been a matter of no practical consequence for the FSS so far, because all commercial non-geostationary satellites have been owned by the USSR or Russia and the ITU is not involved in domestic interference problems.

Geostationary satellite systems for operation in frequency bands other than those identified for the special procedure are coordinated in accordance with the procedure outlined under Phase 2a below. These systems are protected from interference from systems using non-geostationary satellites by the following regulation:

> Non-geostationary space stations shall cease or reduce to a negligible level their emissions, and their associated earth stations shall not transmit to them, whenever there is insufficient angular separation between non-geostationary satellites and geostationary satellites, resulting in unacceptable interference to geostationary-satellite space systems in the fixed-satellite service operating in accordance with these Regulations.
>
> (ITU Radio Regulations, paragraph 2613, as amended at WARC-92)

Exceptionally, it was agreed at WRC-95 that this regulation should not apply to certain designated frequency bands, allocated for the FSS, that had been specially identified for use by FSS systems using non-geostationary satellites or for feeder links for MSS satellites using non-geostationary satellites.

Phase 2a. Frequency coordination of geostationary satellites

When an administration is ready to declare in full detail the parameters of a proposed new geostationary satellite, it becomes necessary to identify other satellite systems, existing or previously announced through the advance publication procedure, that are likely to cause or suffer significant interference. This done, the managers of the proposed system can collaborate with the managers of other systems to clarify and, if possible, solve the interference problems. This is the frequency coordination process. But it is costly and time consuming. There may be dozens of other satellite systems operating in the frequency bands and the arc of the GSO which the newcomer wishes to enter, all of which will suffer and cause some interference, significant or trivial. Consequently, the procedure includes means for identifying, from among these other systems, those for which or from which the interference would almost certainly be acceptably small, so releasing as many systems as possible from the process.

The procedure for frequency coordination is set out in Section II of RR [1], Article 11. The administration responsible for the proposed new satellite sends to the ITU a detailed statement of its intended assignments of transmitting and receiving frequencies. This statement includes emission parameters, the areas where earth stations will be located and the characteristics of the satellite antennas and typical earth station antennas.

A list is then drawn up by the ITU of all the other geostationary satellite systems for which interference might exceed acceptable limits under worst-case conditions. The criterion that is used to identify possible problems is set out in RR [1] Appendix 29 and is outlined below in Section 6.2.2. This criterion is important, not only because it expedites the frequency coordination process but also because it establishes an administration's right to be included in the process if a substantial risk of interference to a system for which it is responsible exists. The details of the proposed new satellite are published in the ITU's weekly circular and the administration responsible for the satellites identified by the Appendix 29 criterion are also directly informed.

The various administrations and system managers concerned, in possession of all the facts, then collaborate to determine whether the interference levels would indeed exceed acceptable limits. This determination has to be made by computation, long before the new system has been built, since by the time that the new system could be in operation, it may be very costly to change its parameters. The methods used in calculating the impact of interference on the performance of channels are outlined in Section 6.2.3. The ITU has recommended that certain levels of interference (called 'permissible interference' levels) between geostationary satellite systems should be accepted as a matter of course in telecommunications channels which may be used for international circuits; see Table 10.1. Higher levels of interference may be found acceptable, but if so, it may be necessary to find ways of reducing channel degradation due to other causes to maintain overall performance objectives.

If it is found that the interference would exceed the ITU's recommended permissible interference levels by a limited amount, that higher level may be agreed between the system managers. If not, ways must be found for reducing it. Typically, reducing interference levels involves changing the transmission plans of the networks involved, and the onus to change usually lies to a large degree with the proposed new system. Carrier frequencies might be changed to avoid interference. Alternatively the system suffering interference might be able to increase the power of its carrier to reduce the effect of interference on signal degradation. More far reaching changes to the equipment may be necessary, such as the use of earth station antennas with better suppression of off-axis gain. Another possible option is a small change in the relative longitudinal positions of satellites, although it is becoming difficult to get benefit from this option as the GSO becomes more completely occupied.

Sometimes it is very difficult to find a configuration for the new satellite network that is simultaneously acceptable to all of the parties with systems coming within the criterion of RR Appendix 29. The ITU may use its good offices to help to resolve major problems and has limited powers to arbitrate if administrations appear to be obstructive without justification. Ultimately, agreement may be reached and the administration responsible for the new satellite system could then move on to Phase 3. In an increasing proportion of cases, however, Phase 2a fails to reach a satisfactory conclusion and it may be necessary to reconsider the project and start frequency coordination again with substantially changed proposals.

Phase 2b. Frequency coordination of non-geostationary satellites

Until recently the ITU provided no formal procedure for the frequency coordination of non-geostationary satellites. However, by the time of the WARC in 1992, the development of MSS systems using non-geostationary satellites was already well under way. WARC-92, which made several new MSS frequency allocations below 3 GHz for such systems, also agreed an interim procedure to provide a regulatory basis for frequency coordination in these bands; see the Annex to RR Resolution 46 [2]. This resolution was revised at WRC-95.

Under this procedure an administration proposing a new system sends details of all relevant system parameters to the ITU, for inclusion in the ITU weekly circular. If these proposals are thought, by another administration, to be likely to cause too much interference to a systems for which it is responsible, the parties discuss possible solutions to the problem. So far the ITU has no permissible interference level recommendations to guide discussions. When the parties have found a basis for agreement, the ITU is advised and the proposing administration is free to progress to Phase 3.

Phase 3. Registration of new frequency assignments

After Phase 1 and, where appropriate, Phase 2 have been completed, firm frequency assignments are notified to the ITU, together with a forecast of the date of bringing into use. The date of bringing a frequency assignment into use should, in principle, be no more than six years after the date of the advance publication of information which initiated Phase 1. It is, however, recognized that setting up a new satellite network is often delayed, not least by problems arising in frequency coordination, and the ITU will consider granting an extension of time of up to three years on request.

The ITU scrutinizes notifications of successful coordination to ensure that the assignments are in full conformity with the RR. If all is in order, the assignments are registered in the Master International Frequency Register (MIFR) and dated.

In due course the notified frequency assignments are brought into use and the fact is notified to the ITU. If the start of use is notified not more than one month after the date given when the assignments were notified, the date of bringing into use is added to the entry in the MIFR. If, however, there is a greater delay, the entry in the register is cancelled. If the frequency assignments are still required, the process must be repeated from Phase 1.

Phase 4. Interference arising in operation

Despite the inter-administration discussions held in Phase 1 and even after formal coordination in Phase 2, interference in excess of what is acceptable sometimes arises in operation. For example, antenna parameters may be found, in use, to be less satisfactory than had been specified. Or the cause may be incorrect operation of the system; for example, a satellite may have drifted away from its planned orbital position or its attitude in orbit may be wrong.

When this happens, the interference is reported to the management of the interfering network, directly or via the administration concerned. If the cause is incorrect operation of one or other of the networks, then the error is corrected if possible. If the error cannot be corrected, or the interference is due to some other cause, it is customary for the frequency assignment with the later registered date of bringing into service to be modified to reduce interference to an acceptable level; if that is not possible, the assignment with less priority is usually withdrawn.

Earth station frequency assignments

The satellite frequency assignments which have been coordinated by the procedures outlined above are associated with earth station locations, stated in approximate terms, in order that interference between the earth stations and other satellites can be calculated. However, the frequency assignments to earth stations must also be registered. This is a simple process in frequency bands which are not shared with terrestrial radio services having equal frequency allocation status. More usually, the procedures outlined in Section 6.4 must be followed to ensure protection from, and for, the radio stations operating terrestrial radio services.

6.2.2 The criterion for frequency coordination

RR [1] Appendix 29 provides a method for determining whether it is safe to disregard the risk that a proposed new geostationary satellite system will cause interference greater than the permitted interference level to an existing geostationary satellite system. The method used for satellites having transparent transponders and using the same frequency bands in the same directions (see Figure 6.1) involves:

(1) Identifying the maximum uplink spectral e.i.r.p. density that earth station A would transmit in an off-axis direction towards satellite B, and converting that e.i.r.p. into an equivalent noise temperature ΔT_s at the output of the receiving antenna of satellite B. This noise would be passed on as an increase in the noise level of the signals relayed by satellite B to earth station B.
(2) Identifying the maximum downlink spectral power density delivered to the transmitting antenna of satellite A and converting that power density into an equivalent noise temperature ΔT_e at the output of the antenna of earth station B.
(3) Calculating the percentage increase that the sum ΔT of these two equivalent noise temperatures will cause to the noise temperature T at the output of the antenna of earth station B due to all other causes under normal conditions.
(4) The RR Appendix 29 procedure provides for account to be taken of polarization discrimination against interfering signals in making these calculations if both parties agree to do so. There are, however, large uncertainties about the degree of protection that polarization discrimination will provide outside the central region of the main lobes of antennas.

If γ is the transmission gain from the output of the receiving antenna of satellite B to the output of the receiving antenna of earth station B, expressed as a numerical power ratio (usually less than unity), then

$$\Delta T = \gamma \Delta T_s + \Delta T_e \quad \text{(K)} \tag{6.1}$$

and the apparent proportionate increase in T due to interference, that is, $\Delta T / T$, can be expressed conveniently as a percentage. RR Appendix 29 rules that there is no requirement for coordination if $\Delta T / T$ is less than 6 per cent.

The effect of an interfering carrier on the performance of the system suffering the interference depends on many factors. The power density of the interference expressed as an equivalent noise level is only one of these. The RR Appendix 29 method is approximate only, and it is applied on a broadband basis, without regard to differences in emission parameters in different parts of a wide frequency band, as is inevitable in a process carried out in the early stages of planning of a system. It is pessimistic, tending to give false positive results; that is, it often indicates a requirement for coordination when the recommended level of permissible interference will not be exceeded in practice. However, it is broadly reliable, giving false negative results in only one set of circumstances. The 6 per cent criterion can be relied on to detect all other cases where the impact of interference would exceed the recommended level of permissible interference.

The procedure fails to take adequately into account the vulnerability of narrowband systems, typically single channel per carrier (SCPC) emissions, when exposed to interference from powerful carriers which are frequency modulated with analogue television signals. These TV carriers usually have an ancillary modulation, a triangular carrier energy dispersal waveform operating at the TV frame rate or field rate, to reduce interference to terrestrial radio systems; see Section 6.4.1. This waveform reduces interference due to the carrier component of the TV emission to most satellite emissions also. However, the dwell time of the video carrier within the pre-demodulator bandwidth (typically 30–50 kHz) of SCPC emissions is so long that the level of interference is exceptionally high; see ITU-R Recommendation S.671 [3]. Various alternative carrier energy dispersal techniques are under study for FM/video carriers. Meanwhile, it is important that system managers ensure that SCPC emissions are not assigned frequencies near the carrier frequency of an analogue TV emission that can cause interference. This policy is called macrosegmentation; see Section 12.4.3.

Antenna gains to be assumed

In carrying out the Appendix 29 procedure it will be necessary to consider the pair of earth stations, one in system A and the other in system B, where the most severe interference conditions would arise. It will also be necessary to make assumptions about the gain of the earth station and satellite antennas that would be appropriate to those interference paths. The topocentric angles ϕ_{AB} and ϕ_{BA} between the satellites (see Figure 6.1) can be calculated from equation (2.10). Ideally the gain of the earth station antennas in these off-beam directions would

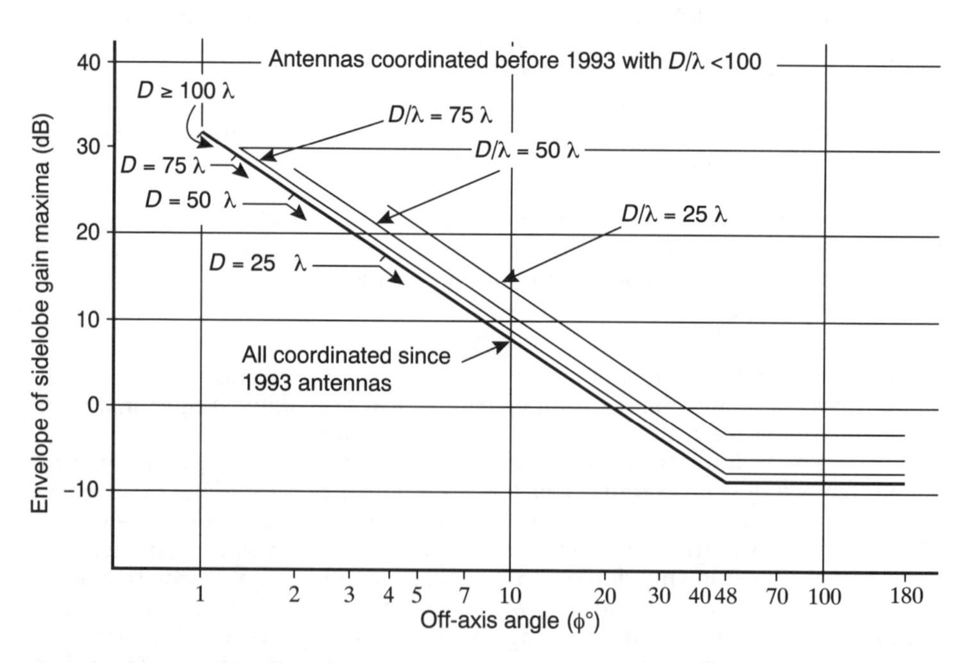

Figure 6.3 *Reference earth station off-beam antenna radiation patterns for use in frequency coordination when measurements of antenna performance are not available. (Based on ITU-R Recommendation S.465 [4].)*

be obtained from gain measurements made on the antennas or from the relevant type approval measurements. More usually, in the absence of data from measurements, ITU-R Recommendation S.465 [4] recommends that the following values should be assumed for the off-beam gain G of an earth station antenna having a primary reflector D metres in diameter, operating at a wavelength of λ metres:

$$G = 32 - 25 \log \phi \quad \text{(dBi)} \tag{6.2a}$$

when ϕ, the angle between the interfering ray path and the axis of the antenna, lies between 1° (or $(100\lambda/D)°$ if greater) and 48°, and

$$G = -10\,\text{dBi} \tag{6.2b}$$

where ϕ exceeds 48°. For small antennas ($100\lambda \geq D$) which were coordinated before 1993, somewhat less stringent assumptions are recommended; see Figure 6.3.

The gain of the satellite antennas in the direction of a specified earth station can best be obtained from the maps of the antenna footprints which form part of the coordination data.

6.2.3 The impact of interference on satellite channels

In the formal coordination stage for geostationary satellites (Phase 2a), it will be necessary to calculate interference levels with greater precision than the method of RR Appendix 29 allows.

Measured data on side-lobe gain levels should be used for calculating the interference which an earth station transmits. If measurements are not available, the reference radiation pattern of ITU-R Recommendation S.465 [4] (see Figure 6.3) will have to be used. For calculating how much interference an earth station receives, this reference radiation pattern should be assumed unless measurements are available showing better performance.

With these earth station antenna data, the interference geometry set out in Figure 6.1 and the data on satellite antenna that will be available to the parties to the coordination process, the wanted-signal-to-interference power density at the input to the demodulator of the earth station suffering interference can readily be calculated. It is then necessary to calculate the impact of the interference after demodulation on a wanted communication channel. For analogue emissions a signal-to-interference ratio is calculated, typically in the form of the equivalent S/N_i ratio, where S is a standard channel test signal and N_i is the noise level equivalent of the interference in the channel. For digital emissions, the impact is expressed as the contribution to the received bit error ratio (BER) due to the interference.

Thus, for example, in determining the post-demodulator impact of interference into an analogue system, the problem is to relate C/I to S/N_i. An interference reduction factor B can be defined by

$$B = 10 \log \frac{S/N_i}{C/I} \quad \text{(dB)} \tag{6.3}$$

But the value of B varies in an intricate way with a large number of factors relating to the wanted and the interfering carriers. For digital systems, the relationship between C/I and BER degradation is even more intractable. For a review of the theoretical bases of such evaluations, and an extensive bibliography, see

CCIR Report 388 [5]. See also ITU-R Recommendations S.741 [6], S.740 [7] and SF.766 [8].

6.3 International planning of spectrum use

6.3.1 The frequency assignment agreement for BSS at 12 GHz

Broadcasting services are more susceptible to interference than other radio services and eliminating the interference is particularly difficult to achieve. There are various reasons for this; among them, broadcasting receivers are numbered in millions, they have a very wide geographical spread and some are likely to have below-standard performance. Consequently, governments take particular care to manage broadcasting frequency bands so that interference seldom arises.

For these and other reasons summarized in Section 15.1.1, detailed international agreements have been drawn up for the use of the main frequency bands allocated for the BSS. FSS uplink allocations that have been set aside for the feeder links which are used to deliver programme material to the satellites are treated similarly. These bands are defined in Table 15.1.

The agreements incorporate planned frequency assignments for every country existing when the planning was done (1977–88), listing in detail the technical parameters of the planned emissions and procedures for taking the plan into use. Provided that the systems that are set up to operate in these bands adhere to these parameters, significant interference between different broadcasting systems should not arise. Countries that ratify the agreements, and most have done so, thereby giving them the status of an international treaty, undertake to use these bands for satellite broadcasting in accordance with the agreements. The frequency assignment plans and the system parameters assumed in planning are reviewed in Section 15.1. The full texts of the agreements are to be found in RR [1] Appendices 30 and 30A.

These uplink and downlink frequency bands are not allocated exclusively for the BSS and the FSS respectively, and interference to broadcasting might arise from these other services. Furthermore, since the BSS and FSS allocations are not the same for the three regions, the possibility exists that interference to broadcasting would arise from a transmitter in another region belonging to a service which has primary status in that region. Accordingly, a complex web of regulations and constraints has been devised to protect planned broadcasting assignments from such interference.

The implementation of the satellite broadcasting plan

When an administration decides to assign for use the frequencies that have been planned for it under the agreements, all the significant parameters intended for the system are notified to the ITU and published. Other administrations are free to challenge the proposal. If it is thought that the intended parameters would be harmful to the interests of another country, whether those interests consist of a working system or a right, established by the plan, to set up a system at any future time, the matter can be discussed between the administrations. If the proposals are not in conformity with the plan, the proposing administration will be told by the ITU to follow a procedure for having the plan amended, involving consultation with all other affected administrations. But in general, if the proposal is in conformity with the plan, in its original form or as it may have been amended, the details will be entered in the MIFR without further question as soon as the assignments are put into operation.

6.3.2 The allotment plan agreement for the FSS

A rigid frequency assignment plan, like the ones used for satellite broadcasting at 12 GHz (see Section 6.3.1) would be inefficient if it were applied to the FSS, which is very varied in its applications. The FSS allotment plan and the associated management procedures aim to provide that necessary degree of flexibility. The text of the agreement is to be found in RR [1] Appendix 30B. This agreement also has treaty status.

The frequency bands that have been used for the plan are listed in Section 14.4.1. All satellites using these frequency bands are to be geostationary. The plan consists of a series of allotments, one reserved for each country (two for a few very large countries that cannot be served from a single location on the GSO) and designed to serve the territory of that country. Each allotment consists of a nominal orbital position and the right to use all five frequency bands included in the plan, subject to technical constraints. Procedures have been included in the agreement for providing allotments for new nations as the need arises and for replacing planned national allotments on request with new allotments for the joint use of a group of countries.

Each nominal orbital position is the centre of a 'predetermined arc' of the GSO, initially $\pm 10^\circ$ wide. The nominal orbital positions have been selected so that, given certain broad assumptions (see Section 12.4) about the technical parameters of the satellites, the earth stations and the emissions used, interference between the national systems will not exceed agreed levels, regardless of the actual parameters of the systems. The parameters do not take the form of constraints on specific equipment characteristics or emission parameters; instead, they are generalized, defining interfering potential and interference susceptibility, leaving a large measure of discretion for the system designer to optimize the system.

Implementation of the agreement

When a country is actively preparing to make use of its allotment, its administration informs the ITU of the more significant elements of the proposed network. These include the service area (which may be smaller than the area assumed in planning), the satellite and earth station antenna characteristics and general information on the power spectra of the emissions (but without details of specific emissions).

If the service area does not extend beyond the limits allowed for in the plan and if the generalized parameters A, B, C and D (see Section 14.4.3) of the proposed network are within the limits assumed in preparing the plan, the network is regarded as fully compliant with the terms of the agreement. If the values of parameters B and D, defining the network's susceptibility to interference, show a prospect of substandard protection, the proposal is accepted but without a commitment to ensure full protection from interference. If the values of A and C, defining the network's propensity to interfere, exceed the values assumed in planning, the proposals are rejected, not being compliant with the agreement.

When proposals received by the ITU have been found compliant with the agreement, the nominal orbital position stated in the plan is reviewed, to take into account the characteristics of other networks already in operation or in the course of implementation. The predetermined orbital arc is then reduced to $\pm 5^\circ$ relative to the new nominal position and the specification of the satellite antennas can be finalized.

In due course the administration notifies the frequency assignments for the satellite to the ITU. The optimum orbital position is reviewed once more, taking into account the most recent developments in the implementation of networks. A final orbital position is chosen, the predetermined orbital arc is reduced to zero, the network is brought into use and the frequency assignments to the satellites are registered in the MIFR.

The right to use these satellite frequency assignments without interference from other satellite networks at the final orbital position is guaranteed by the governments that have ratified the agreement, regardless of the date of bringing into use.

Assignments to earth stations

As with satellite networks which have been coordinated using the procedures outlined in Section 6.2, it is usually necessary to coordinate the frequency assignments to the earth stations with terrestrial radio stations using the same frequency bands. See Section 6.4.

6.4 Sharing spectrum with terrestrial radio services

6.4.1 The general case

Almost all of the bandwidth allocated for satellite communication below 31 GHz is shared to some substantial extent with terrestrial radio services, typically for the fixed and mobile services. It is necessary to have means for controlling cross-frontier interference between the satellite services and the sharing terrestrial services if both have primary allocation status.

As indicated in Section 6.3.1, special provisions have been made in the RR to prevent interference with satellite broadcasting which is in accordance with the plans for the 12 GHz bands. With minor exceptions, the terrestrial allocations sharing with the MSS bands around 1600 MHz have secondary status and therefore should pose no problem for satellite services. And an interim coordination procedure has recently been introduced for use in the frequency bands below 3 GHz that have been designated for use by systems using non-geostationary satellites, typically the MSS; see Section 6.4.2.

However, the main satellite/terrestrial band sharing problem arises in the bands allocated for the FSS. All of the FSS bands below 31 GHz are shared globally with equal allocation status with the FS and most bands have primary allocations for the MS also, although there are large areas where the terrestrial allocations have not been implemented nationally in some bands. The arrangements for limiting the interference, which are applied also to other satellite/terrestrial sharing situations not otherwise provided for, are reviewed in this section.

Most FS systems above 1 GHz are multi-channel line-of-sight radio links at permanent locations. These links are very numerous and widely spread and many of them are formed into radio relay chains, some of them thousands of kilometres long. Most stations of the MS above 3 GHz are transportable but otherwise similar to the FS stations; they are used typically to provide temporary wideband connections into fixed radio relay networks.

The terrestrial services were established in these frequency bands long before satellite services began. At that time quite simple arrangements for liaison between the administrations of adjacent countries were sufficient to prevent cross-frontier interference. Many administrations did not notify frequency assignments to the ITU for registration in the MIFR if there was no evident risk of cross-frontier interference. However, the sharing of frequency bands with satellite communication systems created four new modes of interference between satellite systems and terrestrial stations, some of these modes being operative at great

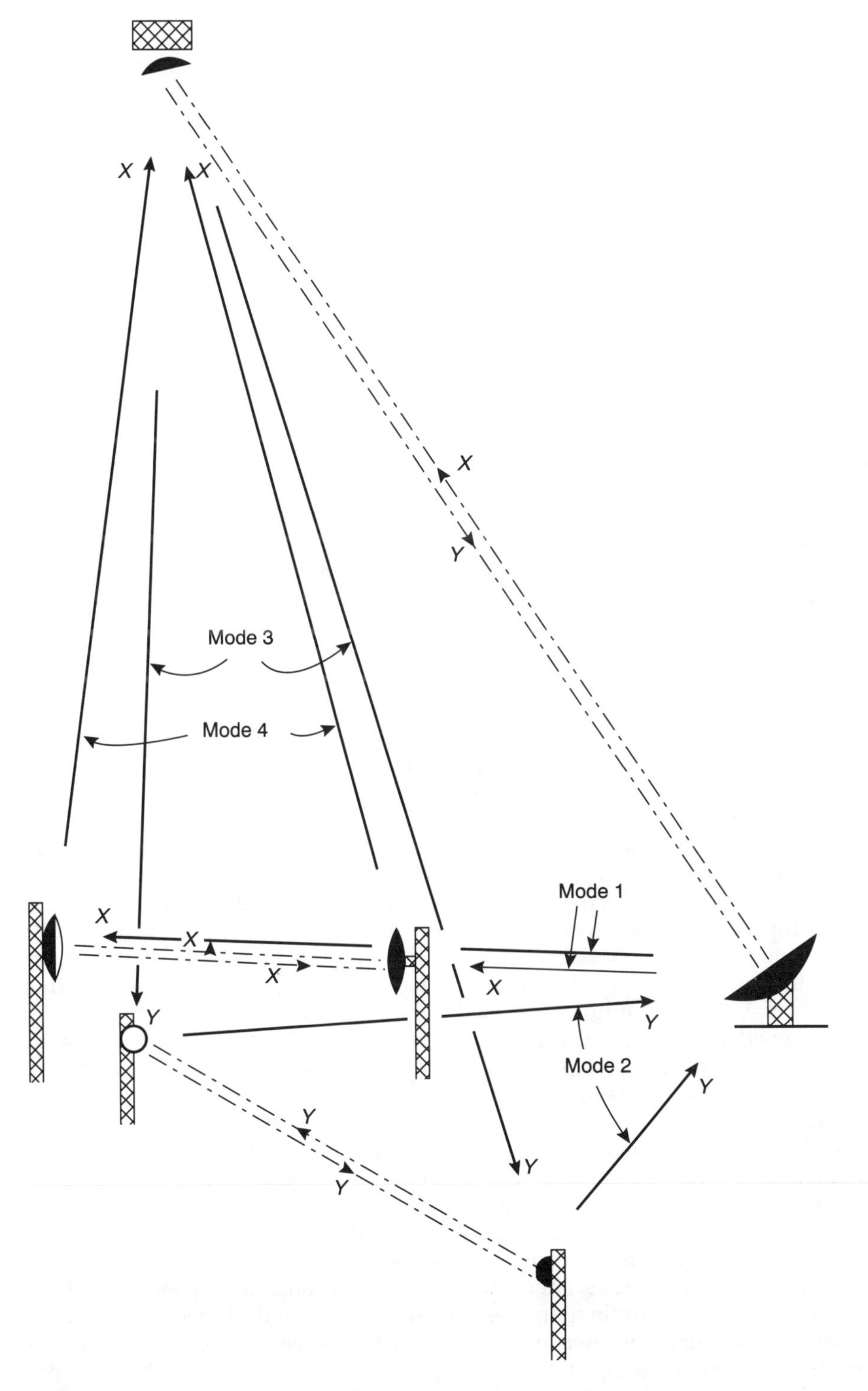

Figure 6.4 *Interference modes between satellite links and terrestrial links using the same frequency bands, band X being used for uplinks and band Y being used for downlinks. Wanted signal paths are chain-dotted and interference paths are shown by bold lines*

distances; see Figure 6.4. Methods have been devised for preventing unacceptable levels of interference to either terrestrial systems or satellite systems through these various modes, without sacrificing the convenient spectrum management methods that were used previously for terrestrial services in areas where there are no earth stations.

Interference modes 1 and 2: earth station/terrestrial station

Most earth station emissions are much more powerful than those of most terrestrial stations of comparable bandwidth. Furthermore, earth station receivers are very sensitive. As a result, strong interference may arise between earth stations and terrestrial stations hundreds of kilometres apart. When both stations are in the same country, the administration responsible must resolve the problem. A mandatory frequency coordination procedure has been agreed to identify and solve these interference problems when the stations are in different countries; the procedure is set out in the RR [1]; see Article 11, Sections III, and IV, Article 12 Sections I, IIE and III, and Article 13.

The procedure starts by identifying all terrestrial stations that, making worst-case assumptions, could possibly interfere with, or suffer interference from, an earth station that is planned to be built in an adjacent country. Then, as many as possible of these stations are released from the process by applying simple criteria to evaluate the likelihood of interference. The remaining earth station/terrestrial-station interference situations are then studied in detail, prospective levels of interference being calculated as accurately as necessary and compared with interference limitation targets; if interference levels are found to be too high, ways must be found for reducing them before the plan for the earth station is implemented and the frequency assignments are taken into use and registered.

Thus, when a site has been provisionally chosen for an earth station, a map is prepared showing the proposed location. A line is drawn on the map round the station marking the limit beyond which significant interference would not arise at the receiver of any conventional terrestrial station operating in the frequency band which the satellite system uses for uplinks, the effects of rain scatter on radio propagation being disregarded. The parameters of the proposed earth station emissions, the radiation pattern of its antenna and any screening of the earth station site are taken into account in drawing the contour. At this stage worst-case assumptions are made about the terrestrial stations; that is, it is assumed that the terrestrial stations will be located at exposed sites with high-gain antennas directed towards the earth station. This 'propagation mode 1' contour is computed using standardized propagation data contained in RR [1] Appendix 28.

Auxiliary contours may be added for propagation mode 1 as an aid to the elimination of terrestrial stations which are less susceptible to interference than the worst-case assumptions, having for example a lower antenna gain towards the earth station or a site that is screened in the direction of the earth station.

A second contour is drawn ('propagation mode 2'), defining the extreme range of interference due to rain scatter. The area which is enclosed by either of these two contours is the coordination area for interference mode 1. See Figure 6.5.

Contours defining the area outside which the emissions of a conventional terrestrial station operating in the satellite downlink frequency band would not cause significant interference to reception at the earth station are also shown on a map for interference mode 2. Auxiliary contours may be added as for interference mode 1. As before, worst-case assumptions are made about the terrestrial stations; it is assumed that these stations would be at unscreened sites, radiating

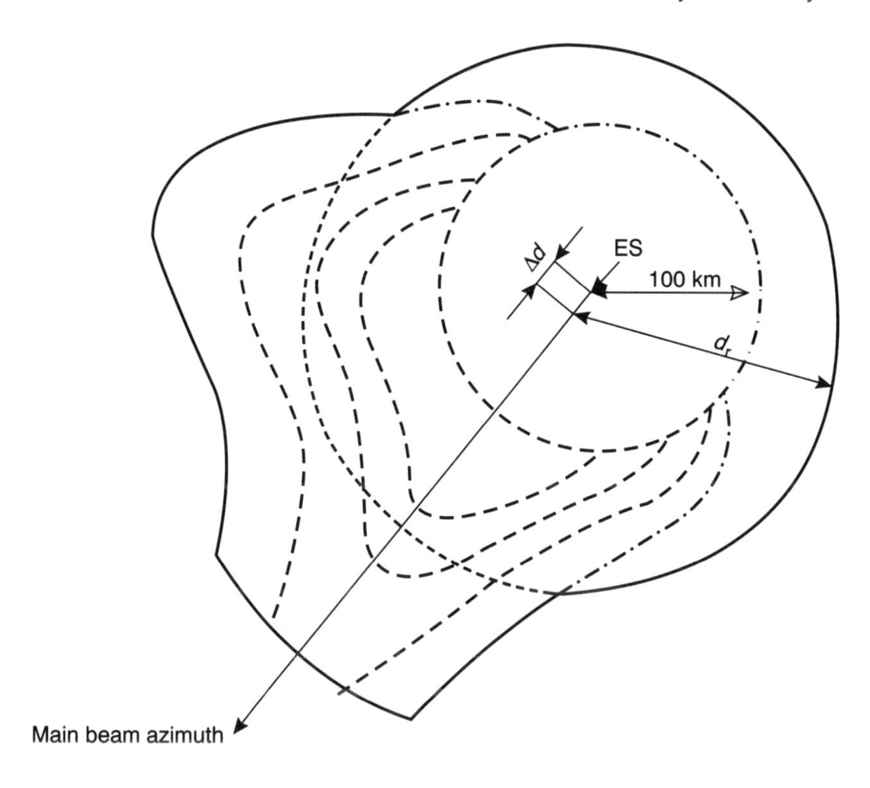

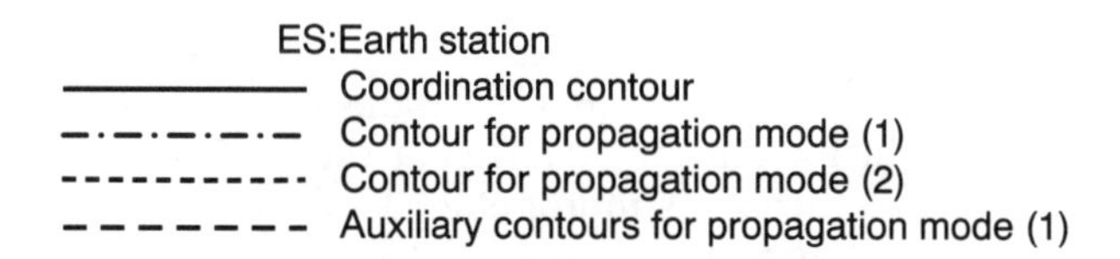

Figure 6.5 *An example of coordination contours surrounding an earth station, showing the two propagation modes for one interference mode. Here d_r is the radius of rain scatter interference risk. (Radio Regulations [1], Appendix 28.)*

the highest e.i.r.p. typically used and with their antennas directed towards the earth station.

Copies of these maps are sent to the ITU and to the administration of countries with territory in the coordination areas, together with details of the proposed earth station frequency assignments. An administration receiving a map will list terrestrial stations within its jurisdiction which are operating within the coordination areas and at the frequencies involved. Taking account of such factors as the terrestrial station's transmitter power, the gain of its antenna in the direction of the earth station and any screening in the direction of the earth station by hills, it will usually be possible to eliminate many terrestrial stations as potential sources of, or sufferers from, significant interference. If potential interference problems remain, the administrations will collaborate to determine what the level of interference is likely to be, using the best available propagation data for the locality.

In the absence of measured data on the off-axis gain of the two antennas involved, the reference radiation pattern of ITU-R Recommendation S. 465 [4]

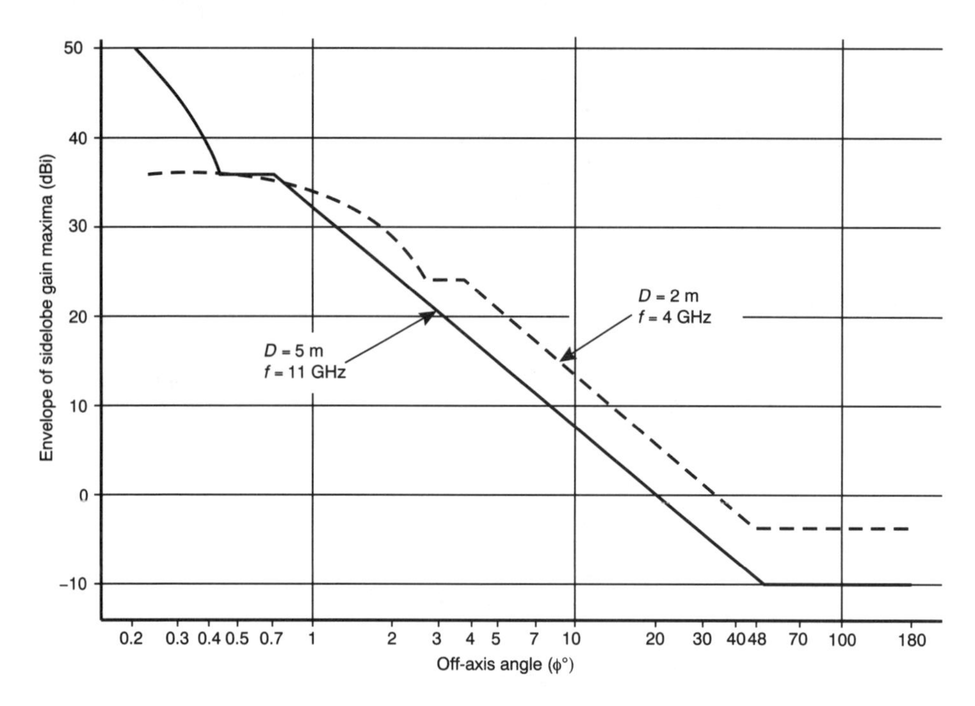

Figure 6.6 *The reference radiation pattern for the antenna of a line-of-sight terrestrial station having the dimensions indicated. (After ITU-R Recommendation F.699 [9].)*

(see Figure 6.3) would be used for the earth station and that of ITU-R Recommendation F.699 [9] for the terrestrial station. Figure 6.6 illustrates two examples of the Recommendation F.699 pattern, for an antenna with a diameter of 5 m operating at 11 GHz and one 2 m in diameter operating at 4 GHz, spanning the range commonly encountered. As indicated in Section 6.2.3 above, the calculation of the impact on system performance of a known C/I ratio is complex. The ITU-R recommendations referenced in that section provide guidance.

In considering whether the computed level of interference in modes 1 or 2 should be accepted, the administrations will be guided by recommendations of the ITU as to the total interference that should be allowed to enter a channel of one service from the emissions of all of the stations of the other. For example, ITU-R Recommendation SF.356 [10] recommends that interference, reckoned as noise, entering an analogue satellite telephony channel from terrestrial systems should be permitted under normal propagation conditions up to a total of 1000 pW0p. ITU-R Recommendation SF.357 [11] recommends a similar level for the total interference from satellites into any telephony channel in an analogue line-of-sight terrestrial link forming part of a radio-relay chain. Both texts also recommend maximum interference levels that should be permitted for brief periods of time when abnormal propagation conditions are present. There are corresponding recommendations for analogue video signals and digital telecommunication signals.

However, interference may enter a satellite channel at an earth station from several terrestrial stations, and interference may enter the same channel in a radio-relay chain from different earth stations at many points along the chain. Furthermore, additional interference enters both satellite and terrestrial channels via interference modes 3 and 4 (see below). Thus, in determining how much interference can be accepted from any one source via modes 1 and 2, these recommendations have to be interpreted with careful judgement.

If it is found that unacceptable interference would arise, some solution must be found. It may be necessary, for example, to choose another site for the earth station, to specify an earth station antenna or a replacement terrestrial station antenna with better side-lobe suppression or to accept some limitation of the frequency band that either station will use.

When all the problems identified in the coordination process have been solved, success is reported to the ITU. The frequency assignments made to the earth stations are notified to the ITU for registration in the MIFR. Thereafter, additional frequencies assigned to either the earth station or to any terrestrial station within the coordination area must first be coordinated between the two administrations.

Interference mode 3: satellite to terrestrial station

Interference from any satellite may enter the receivers of thousands of terrestrial stations spread all over the large part of the Earth that is visible from a satellite. The impact of the interference may be particularly severe at those terrestrial stations that are close to the edge of the Earth's disc as seen from the satellite, since the interference will arrive there from an almost horizontal direction, where the gain of the antenna of the terrestrial station may be high. It is not feasible to limit this interference by coordination, because of the large number of terrestrial stations involved. Instead, a limit has been placed on the spectral PFD (Φ_s) of the signal that a satellite may set up anywhere on the surface of the Earth in any frequency band which is shared with terrestrial services. Limits like these are called sharing constraints.

Details of this constraint are to be found in Article 28 of the RR [1]. For example, in the frequency band 3400–4200 MHz, Φ_s may not exceed $-152\,\text{dB(W}\,\text{m}^{-2})$ in any sampling bandwidth of 4 kHz if the angle of elevation δ is less than 5°, nor $-142\,\text{dB(W}\,\text{m}^{-2})$ if δ exceeds 25°, with a linear transition of 0.5 dB per degree between 5° and 25°. The constraints that apply in other FSS downlink allocations differ in detail; see Figure 6.7. It is calculated that, if the GSO were fully occupied with satellites having full Earth coverage, spaced in orbit sufficiently to avoid excessive mutual interference and operating in the same frequency band as a typical terrestrial station, the latter would not suffer more than half of the interference recommended, for example, in ITU-R Recommendation SF.357 [11]. The other half of the allowance would remain available for entries of interference from earth stations. Paragraph 2585 of the RR permits these spectral PFD limits to be exceeded, subject to the agreement of the administrations of all countries affected.

A sampling bandwidth of 4 kHz was used in determining the sharing constraint below 15 GHz because the frequency division multiplex (FDM) telephony systems using frequency modulation (FM) that are typical of the radio relay systems in this part of the spectrum are susceptible to even relatively small amounts of interference power if it is concentrated within the bandwidth of one telephone channel, nominally 4 kHz. For means of calculating the maximum spectral power density, within a sampling bandwidth of 4 kHz, of an angle-modulated carrier, see ITU-R Recommendation SF.675 [12]. However, it has been assumed that there will be few FDM/FM radio relay systems in bands above 15 GHz; in this part of the spectrum wideband digital systems are typical and the penalty to satellite systems arising from this sharing constraint has accordingly been made less onerous by using a sampling bandwidth of 1 MHz.

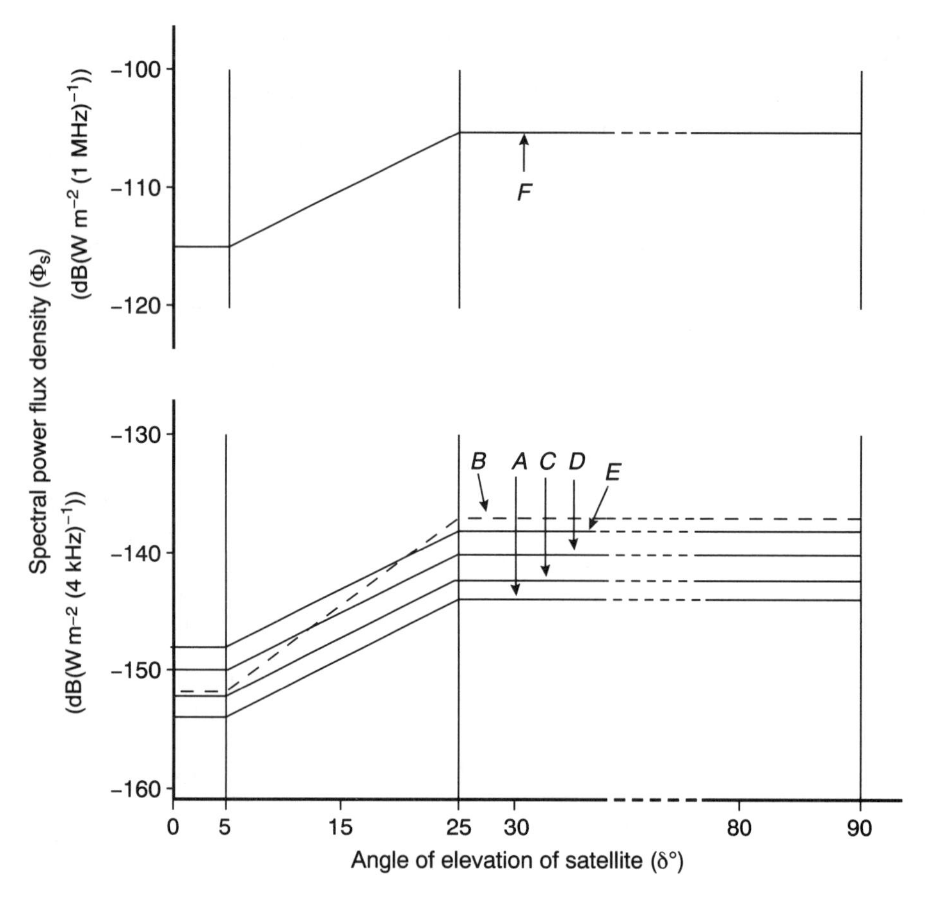

Figure 6.7 *The constraint, as a function of the angle of elevation on arrival, δ, on the spectral PFD, Φ_s, at the surface of the Earth from a satellite transmitter operating in a frequency band shared with a terrestrial service having a frequency allocation of equal status. The limits apply in the following frequency bands: line A, 1.525–2.3 GHz; line B, 2.5–2.69 GHz; line C, 3.4–7.75 GHz; line D, 8.025–11.7 GHz; line E, 12.2–12.75 GHz; line F, 17.7–27.5 GHz*

Interference mode 4: terrestrial station to satellite

Another sharing constraint is used to prevent excessive interference from terrestrial transmitters entering the satellite receiver. The general effect of this constraint, which is set out in detail in RR [1] paragraphs 2505–2511, is that transmitting stations of terrestrial services operating in frequency bands below 10 GHz which are shared with the uplinks of satellite services may not deliver power exceeding 13 dBW to their antennas. This power limit is reduced to 10 dBW between 10 and 29.5 GHz. Furthermore, the gain of the antenna may not cause the e.i.r.p. to exceed 55 dBW in any direction. Below 10 GHz the e.i.r.p. is constrained more severely in the azimuthal directions at which the GSO cuts the local horizon.

Carrier energy dispersal

When the level of signal modulation is low, the spectra of most wideband analogue emissions used in satellite communication show a residual carrier spike with a spectral density that is much more powerful than the power density averaged across the whole bandwidth of the emission. If it were necessary to limit the power of satellite emissions to the degree necessary to ensure that the PFD in the vicinity of the carrier did not exceed the sharing constraint, impracticably large antennas would be needed at the earth station to derive an adequate carrier-to-noise ratio from such weak signals. In an analogous way, digital systems carrying the repetitive bit streams that are typical of light loading exhibit powerful spikes within the spectrum of their emissions.

Techniques called carrier energy dispersal are used to enable more advantage to be taken of the opportunities allowed by the PFD sharing constraints. Typically, for frequency modulated carriers with analogue basebands, a triangular waveform of sufficient amplitude and with a period of a few milliseconds is added to the baseband at the transmitting earth station. This spreads the energy of the carrier over a bandwidth that is much larger than 4 kHz. For digital signals it is sometimes necessary to add a pseudo-random sequence of digits to the modulating signal at the transmitter, thus eliminating concentrations of spectral energy; the same sequence is subtracted at the receiver to restore the signal bit stream to its original state. In systems using extremely small earth station antennas and therefore requiring very powerful satellite emissions, it may be necessary to use spread spectrum modulation to achieve adequate spreading of the signal energy. ITU-R Recommendation S.446 [13] recommends the use of carrier energy dispersal techniques to protect terrestrial services.

The practical consequences of sharing

In an area where earth stations operate, it will usually be necessary to exclude terrestrial mobile stations operating close to frequencies assigned to an earth station.

The methods which have been devised to enable frequency bands to be shared by the FS and the FSS are broadly satisfactory for satellite networks which are used for trunk communication networks. The methods can also be applied, although with rather more difficulty, for thin route satellite networks in areas where the fixed service is not deployed intensively.

However, for VSAT networks (see Chapter 13), sharing with terrestrial services raises three problems. The downlink spectral PFD constraint limits to some degree the technical options that are available for networks using very small earth station antennas. Secondly, satisfactory frequency coordination is difficult to achieve, because the stations of both services may be numerous and the preferred location for both is often at roof-top level in city centres, where opportunities for screening one kind of station from the other are minimal. Finally, frequency coordination is labour-intensive, with a cost that may be high relative to the value of the facilities which a VSAT station supplies.

The internationally agreed frequency allocations do not involve sharing all FSS frequency allocations with terrestrial services in all countries. The main exceptions are as follows:

(1) In most European countries the FSS uplink band 14.0–14.25 GHz is not shared with any other service having equal allocation status and FSS downlinks have higher allocation status than any other service at 12.5–12.75 GHz.
(2) In North and South America the FSS has higher allocation status in the uplink band 14.0–14.4 GHz than any other service. For downlinks, the band 12.1–12.2 GHz is allocated exclusively to the FSS throughout North and South America and the FSS has higher allocation status than any other service in the band 11.7–12.1 GHz in Canada, Mexico and USA.

(3) The allocation status of the FSS in the bands 29.5–30.0 GHz (uplinks) and 19.7–20.2 GHz (downlinks) is also favourable in most countries.

These bands are to be preferred for thin route and VSAT earth stations.

6.4.2 Sharing with non-geostationary systems below 3 GHz

In addition to interim procedures for the mutual coordination of satellite systems using non-geostationary satellites in certain designated frequency bands below 3 GHz (see Section 6.2.1 above), RR Resolution 46 [2] contains an interim procedure for coordinating these satellite systems with terrestrial systems operating in the same frequency bands and having equal allocation status. This latter procedure uses the same basic approach as the general one described in Section 6.4.1 above, but it is simpler. This may be because the procedure is at present an immature one, but perhaps also because the scope for coordinating terrestrial services with satellite systems using non-geostationary satellites and serving mobile earth stations, typically with little antenna directivity or none, may be quite limited.

Under this procedure an administration proposing to assign frequencies in one of the designated bands to earth stations within its jurisdiction produces maps showing the location of fixed earth stations, the area within which mobile earth stations would be authorized to operate, and the coordination areas for both kinds of station. The coordination area for a fixed earth station is to be circular, centred on the station and having a radius of 500 km. The coordination area for land mobile and maritime mobile earth stations extends 500 km in all directions beyond the area in which operation is authorized. For aircraft earth stations the coordination area extends 1000 km beyond the limits of the operating area. A copy of this map and details of all proposed frequency assignments are sent to the ITU and to the administrations of all countries with territory in these coordination areas, with a request for coordination.

An administration receiving this request will consider whether the terrestrial radio systems for which it is responsible, and new systems to be installed within three years, will cause interference to the proposed satellite system or suffer interference from it. If problems are identified, the proposing administration will be sent details of the stations within the coordination area. The administrations will coordinate, seeking solutions. When all the problems have been solved, the ITU will be advised and the proposing administration will be free to notify the satellite system frequency assignments to the ITU for registration in the MIFR.

Later, if any of the administrations with which the satellite system has been coordinated proposes to make a new frequency assignment to a terrestrial station within the coordination area and within spectrum that has been coordinated, then that new assignment is to be coordinated with the administration having the satellite system interest.

6.5 Constraints to enhance satellite communication

In addition to the constraints, reviewed in Section 6.4.1, to facilitate the sharing of spectrum between the FSS and terrestrial services, various measures have been agreed to prevent avoidable interruptions to satellite communication systems and to maximize the number of satellites that can operate in the GSO. Some of these measures have been made mandatory through the RR. Others are recommended by the ITU, are widely implemented, and may one day be made mandatory. A few apply to all satellite systems, but most apply to geostationary systems only. The more important of these constraints are outlined in this section.

Mandatory constraints and requirements for all satellites

Item 1. *Control of satellite transmitters.* It must be made possible to switch off all satellite transmitters by telecommand. (paragraph 2612 of the RR [1])

Item 2. *Spurious emissions* (that is, energy unintentionally radiated at frequencies remote from the wanted signals, such as harmonics and parasitic oscillations). Such emissions from all kinds of radio station are to be limited to the lowest possible value (paragraph 304 and Appendix 8 of the RR [1]). Studies are in progress to define specific limits for spurious emissions from earth stations and satellites. The first concrete result for satellite communications is a recommendation for VSAT earth stations; see Chapter 13.

Mandatory constraints for geostationary satellite systems

Item 3. *Satellite station-keeping.* It must be feasible to maintain a FSS or BSS satellite within $\pm 0.1°$ of its nominal longitude (paragraphs 2615–2627 of the RR [1]). For MSS satellites which make no use of FSS frequency allocations for feeder links, the corresponding limit is $\pm 0.5°$. There are relaxations of these requirements for old or experimental satellites. There is no formal constraint regarding north–south station-keeping except for some broadcasting satellites in the 12 GHz frequency assignment plan; see Section 15.1. However, it is likely that, for regulatory purposes, a geosynchronous satellite will not be regarded as geostationary if its orbital inclination exceeds 5°.

Item 4. *Satellite beam pointing accuracy.* It must be feasible to maintain the pointing of satellite beams within 0.3° of the nominal direction, or within 10 per cent of the −3 dB beamwidth, whichever is the less stringent requirement (paragraphs 2628–2630 of the RR [1]). More stringent constraints apply to satellites operating in accordance with the satellite broadcasting frequency assignment plan and the allotment plan for the FSS; see Sections 15.1 and 12.4. It has recently been agreed that more stringent standards should be adopted, as a design objective, for all FSS satellites; see Item 10 below.

Item 5. *Earth station off-beam radiation.* The level of signal power radiated off the axis of an earth station antenna in directions that would cause interference to geostationary satellites is to be minimized. Stringent assumptions were adopted for off-axis radiation from the antennas of feeder link earth stations in preparing the satellite broadcasting frequency assignment agreement for ITU Regions 1 and 3 and these are to be adhered to in the implementation of the plan; see Section 15.1. For BSS feeder link earth stations in Region 2 the corresponding constraint takes the form of a limit on the gain of antenna side-lobes. In the implementation of the FSS frequency allotment plan, the assessment of conformity with generalized parameter *A* has much the same effect.

Elsewhere the standards given in ITU-R recommendations are to be taken as a guide to what should be achieved (paragraph 2636 of the RR [1]). ITU-R Recommendation S.524 [14], which applies to the earth stations of networks of the FSS, recommends that the off-axis spectral e.i.r.p. density levels shown in Figure 6.8 should not be exceeded at 6 and 14 GHz in directions within 3° north and south of the GSO, although there are relaxations of these recommendations in certain special cases. A separate recommendation has been made for Ku band VSATs; see Section 13.2.2.

Item 6. *Satellite antenna beams.* Constraints on the characteristics of satellite antennas are applied to broadcasting satellite systems under the 12 GHz agreements (see Sections 6.3.1 and 15.1) and are implicit in the FSS allotment plan

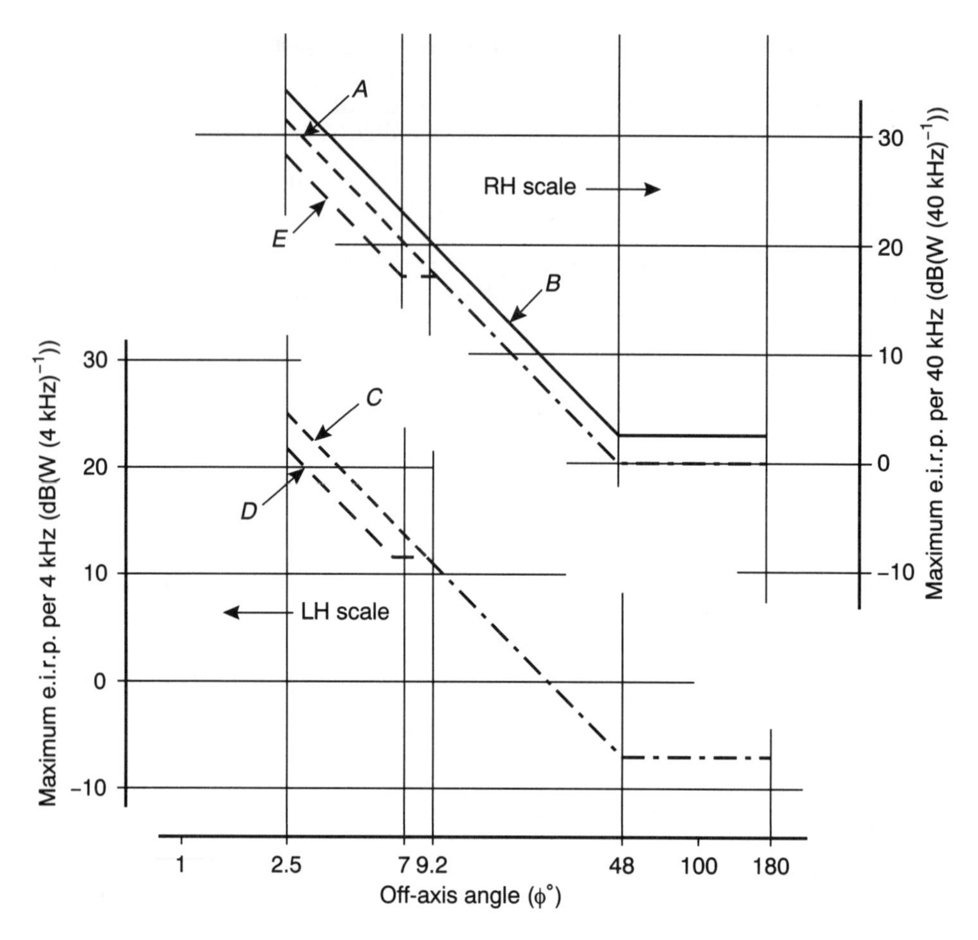

Figure 6.8 *Maximum levels of off-axis e.i.r.p. density from the antennas of FSS earth stations (excluding Ku band VSATs) in directions within 3° of the GSO (after ITU-R Recommendation S.524 [14]). The various curves apply as follows: A 6 GHz, voice-activated telephony, SCPC/FM. B 6 GHz, voice-activated telephony, SCPC/PSK. C 6 GHz, with types of emission other than A and B and an antenna installed before 1988. D 6 GHz, as C but antenna installed since 1988. E 14 GHz, for all types of emission but not including feeder links within the BSS frequency assignment plan*

agreement (see Sections 6.3.2 and 12.4) through the application of generalized parameter C. See also Item 9 below.

Recommended design objectives for geostationary satellite systems

Item 7. *Earth station antenna side-lobes.* Within the solid angle where systems are most likely to cause and suffer interference (see Figure 6.9), ITU-R Recommendation S.580 [15] recommends, as a design objective for new antennas installed after 1995, that the gain G of 90 per cent of the side-lobe peaks should not exceed

$$G = 29 - 25 \log \phi \quad \text{(dBi)} \tag{6.4}$$

That is 3 dB more stringent than the principal pattern given by ITU-R Recommendation S.465 [4]; see Figure 6.3. Outside that solid angle, the objective is to

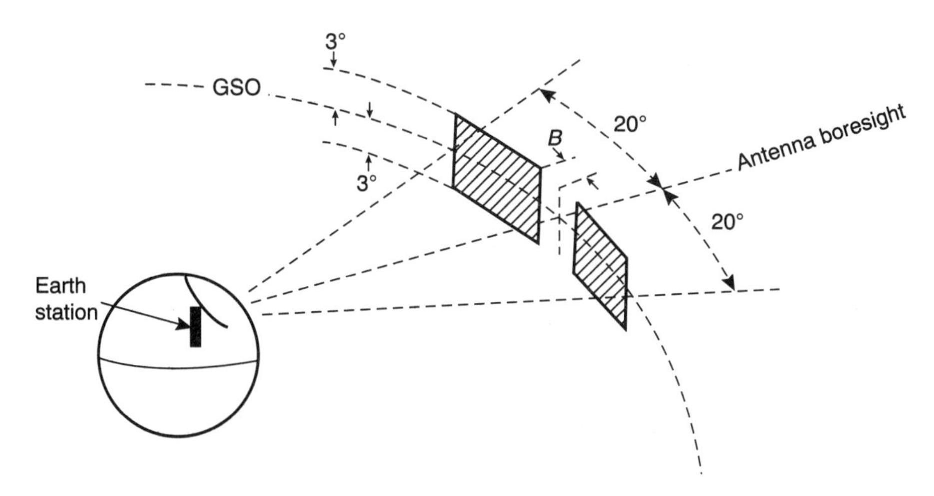

Figure 6.9 *The solid angles over which the constraint on peak side-lobe gain recommended by ITU-R Recommendation S.580 [15] applies (after ITU-R Recommendation S.580). If $D \geq 100$, then $B = 1.0°$. For smaller antennas, $B = 100\lambda/D$ degrees, where D = antenna diameter and λ = wavelength*

ensure that the side-lobe peaks do not exceed the limits given in Recommendation 465. Rather less stringent objectives were set for some categories of antenna before 1995.

Item 8. *Flexibility in satellite stationing.* To facilitate frequency coordination, ITU-R Recommendation S.670 [16] recommends, as a design objective, that the satellites for new FSS networks should be designed so that they can operate up to ±2° in longitude from the nominal orbital position, unless the configuration of the service area makes this impracticable.

Item 9. *Satellite antenna beam overspill.* The ITU is in the process of developing design objectives for the gain of the antennas of FSS satellites outside their coverage area. ITU-R Recommendation S.672 [17] shows the present state of the work. For antennas having a single feed, producing a circular or elliptical beam, the following gain pattern is recommended:

$$G(\theta) = G_m - 3(\theta/\theta_b)^\alpha \quad \text{(dBi)} \tag{6.5a}$$

for values of θ between θ_b and $a\theta_b$ and

$$G(\theta) = G_m + L_n \quad \text{(dBi)} \tag{6.5b}$$

for values of θ between $a\theta_b$ and $b\theta_b$,

where $G(\theta)$ is the gain of the antenna at $\theta°$ off-axis (dBi),
G_m is the maximum gain of the antenna (dBi),
θ_b is one half of the 3 dB beamwidth relative to G_m (degrees),
L_n is the near-in side-lobe level in dB relative to the peak gain required by the system design, and
a, b and α are coefficients related to the value of L_n.

However, many FSS satellite antenna beams are neither circular nor elliptical and many have multiple feeds. Such antennas have complex coverage patterns and the definition of gain objectives outside the coverage area that will promote more efficient utilization of the GSO is proving to be difficult. The work continues.

Item 10. *Satellite beam pointing accuracy*. The ITU Radiocommunication Assembly recommended in 1994 (see ITU-R Recommendation S.1064 [18]) that a design objective more stringent than the regulatory requirements referenced in Item 4 above should be adopted in the FSS. This recommendation calls for satellite beams to be maintained within 0.2° of their nominal direction, or within 5 per cent of the −3 dB beamwidth, whichever is the less stringent objective.

Environmental protection of the GSO

This section would not be complete without a reference to ITU-R Recommendation S.1003 [19], which draws attention to the continuing increase in the number of objects that are drifting, uncontrolled, in the vicinity of the GSO and in transfer orbits intersecting the GSO. These objects range from complete, non-functional satellites and the upper stages of spent launchers to quite small pieces of debris. The danger of collision between operational satellites and these passive objects is growing, and the consequencies of a collision are likely to be major damage to a valuable operational asset at best, and a substantial increase in the number of pieces of debris at worst. The principal recommendation is that a geostationary satellite, near the end of its operational lifetime but before its supply of propellant for orbit adjustment has been completely exhausted, should be boosted into a higher orbit that does not intersect the GSO.

References

1. *Radio Regulations* (1994). Geneva: ITU.
2. *Interim procedures for the co-ordination and notification of frequency assignments of satellite networks in certain space services and the other services to which certain bands are allocated* (RR Resolution 46, 1992), Final Acts of WRC-95. Geneva: ITU.
3. *Necessary protection ratios for narrow-band single-channel-per-carrier transmissions interfered with by analogue television carriers* (ITU-R Recommendation S.671-3, 1994), ITU-R Recommendations 1994 S Series. Geneva: ITU.
4. *Reference earth-station radiation pattern for use in co-ordination and interference assessment in the frequency range from 2 to about 30 GHz* (ITU-R Recommendation S.465-5, 1994), ITU-R Recommendations 1994 S Series. Geneva: ITU.
5. *Methods for determining the effects of interference on the performance and the availability of terrestrial radio relay systems and systems in the fixed-satellite service* (CCIR Report 388-6, 1990), Reports of the CCIR, Annex to Volume IV/IX-2. Geneva: ITU.
6. Carrier-to-interference calculations between networks in the fixed-satellite service (ITU-R Recommendation S.741-2, 1994), ITU-R Recommendations 1994 S Series. Geneva: ITU.
7. *Technical co-ordination methods for fixed-satellite networks* (ITU-R Recommendation S.740, 1994), ITU-R Recommendations 1994 S Series. Geneva: ITU.
8. *Methods for determining the effects of interference on the performance and the availability of terrestrial radio relay systems and systems in the fixed-satellite service* (ITU-R Recommendation SF.766, 1994), ITU-R Recommendations 1994 SF Series. Geneva: ITU.
9. *Reference radiation patterns for line-of-sight radio-relay system antennas for use in co-ordination studies and interference assessment in the frequency range from 1 to about 40 GHz* (ITU-R Recommendation F.699-2, 1994), ITU-R Recommendations 1994 F Series. Geneva: ITU.
10. *Maximum allowable values of interference from line-of-sight radio relay systems in a telephone channel of a system in the fixed-satellite service employing frequency modulation, when the same frequency bands are shared by both systems* (ITU-R Recommendation SF.356-4, 1994), ITU-R Recommendations 1994 SF Series. Geneva: ITU.

11. *Maximum allowable values of interference in a telephone channel of an analogue angle-modulated radio-relay system sharing the same frequency bands as systems in the fixed-satellite service* (ITU-R Recommendation SF.357-3, 1994), ITU-R Recommendations 1994 SF Series. Geneva: ITU.
12. *Calculation of the maximum power density (averaged over 4 kHz) of an angle-modulated carrier* (ITU-R Recommendation SF.675-3, 1994), ITU-R Recommendations 1994 SF Series. Geneva: ITU.
13. *Carrier energy dispersal for systems employing angle modulation by analogue signals or digital modulation in the fixed-satellite service* (ITU-R Recommendation S.446-4, 1994), ITU-R Recommendations 1994 S Series. Geneva: ITU.
14. *Maximum permissible levels of off-axis e.i.r.p. density from earth stations in the fixed-satellite service transmitting in the 6 and 14 GHz frequency bands* (ITU-R Recommendation S.524-5, 1994), ITU-R Recommendations 1994 S Series. Geneva: ITU.
15. *Radiation diagrams for use as design objectives for antennas of earth stations operating with geostationary satellites* (ITU-R Recommendation S.580-5, 1994), ITU-R Recommendations 1994 S Series. Geneva: ITU.
16. *Flexibility in the positioning of satellites as a design objective* (ITU-R Recommendation S.670-1, 1994), ITU-R Recommendations 1994 S Series. Geneva: ITU.
17. *Satellite antenna radiation pattern for use as a design objective in the fixed-satellite service employing geostationary satellites* (ITU-R Recommendation S.672-1, 1994), ITU-R Recommendations 1994 S Series. Geneva: ITU.
18. *Pointing accuracy as a design objective for earthward antennas on board geostationary satellites in the FSS* (ITU-R Recommendation S.1064, 1994), ITU-R Recommendations 1994 S Series. Geneva: ITU.
19. *Environmental protection of the geostationary satellite orbit* (ITU-R Recommendation S.1003, 1994), ITU-R Recommendations 1994 S Series. Geneva: ITU.

7 Radio system technology

7.1 Introduction

7.1.1 The radio system

Satellite systems serve many purposes and the superficial differences between the various kinds of earth station are great. There are much closer similarities between satellites that are used for different purposes. However, all earth stations, large or small and designed for whatever use, have very similar basic architecture. Figures 7.1 to 7.3 show in block schematic form the basic elements of a typical radio path from earth station to earth station through a satellite. The functions of these various units are briefly as follows.

The earth station transmitting chain

The signals entering an earth station take various forms. Most commonly they consist of a narrowband channel or a number of such channels multiplexed together, the channels being designed for telephony but also usable for other kinds of signals of low information rate. Other signals are video channels, complete television signals or data channels. Whatever they are, these signals, after any necessary processing, modulate a carrier, typically at 70 MHz and, after passing through the first intermediate frequency (IF) amplifier, enter a frequency changer, called an up-converter. The modulated carrier leaves the up-converter with the carrier frequency raised, typically to around 1 GHz. A second IF amplifier and a second up-converter follow. The frequency of the local oscillator input to the latter is chosen to change the carrier frequency to f_u, the frequency assigned for the uplink to the satellite. After further amplification, finally in the high-power amplifier (HPA), the modulated carrier is passed to a transmit port of the earth station antenna. See Figure 7.1.

Many earth stations transmit, not one, but several or many modulated carriers to a satellite. As shown in Figure 7.1, the various carriers, at their different frequencies, are combined at some convenient point, typically at the input to the HPA, before being passed to the antenna.

The satellite communications payload

The function of the satellite in the transmission chain is to receive the modulated carriers that earth stations emit as uplinks, amplify them and retransmit them as downlinks for reception at the destination earth station. In the course of this process, the carrier frequency of each emission is moved to a frequency band in which the satellite does not receive, in order that the downlinks should not cause interference to the reception of uplinks.

Satellites of the FSS and the BSS are assigned frequency bands, typically 500 MHz wide, for uplinks and for downlinks. However, the bandwidth of the emissions that they relay is seldom more than 30 MHz and most are much narrower. So it is feasible to subdivide the total bandwidth into a number of

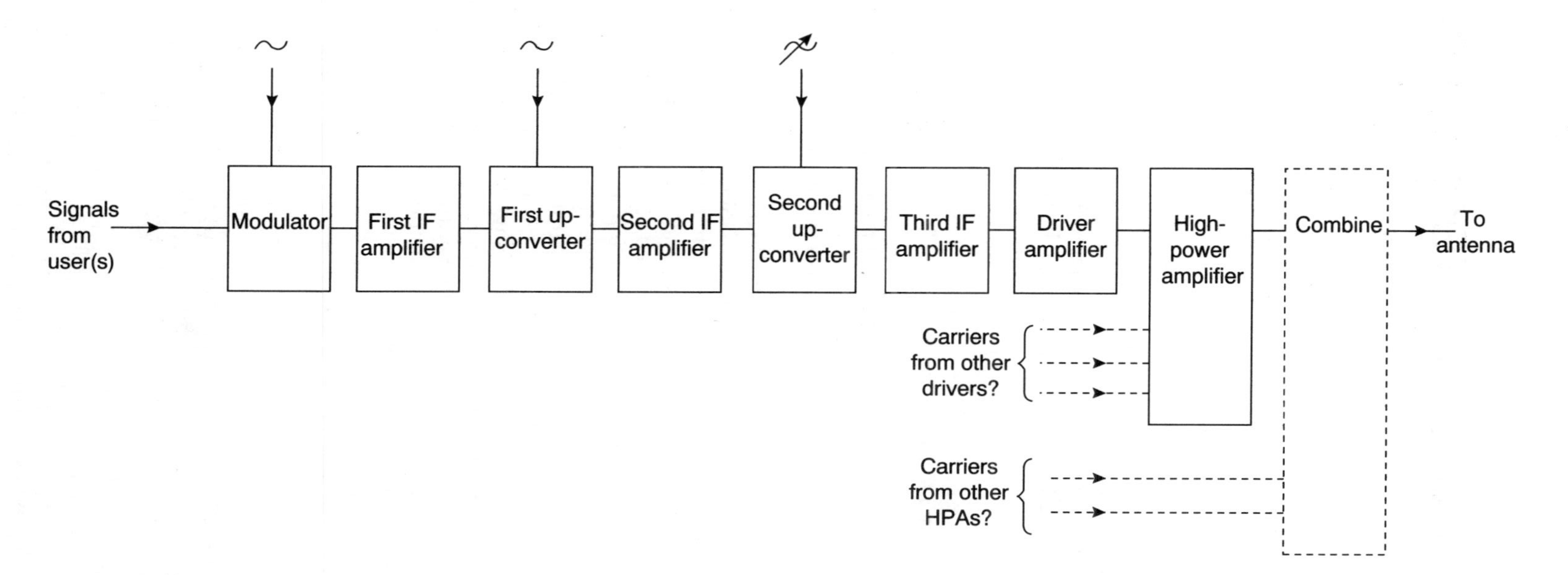

Figure 7.1 *A block schematic diagram of the basic elements of an earth station transmit chain*

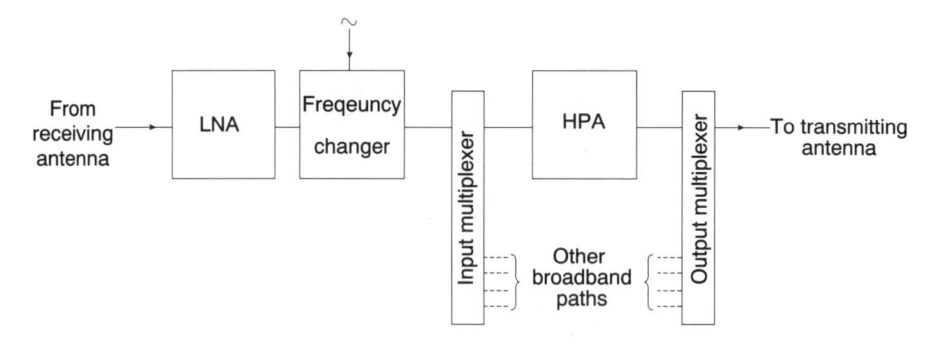

Figure 7.2 *A block schematic diagram of the basic elements of a transparent transponder*

broadband paths through the satellite payload, each typically 36 or 72 MHz wide, each with its own amplifiers, etc. Substantial advantages in efficiency and flexibility are obtainable by doing so.

The essential hardware elements of a broadband path through the payload of a satellite, called a transponder, are shown in Figure 7.2. The satellite receiving antenna collects emissions from earth stations. They are amplified in a low-noise amplifier (LNA), which is typically shared with several broadband paths operating in other bands of frequency. The output of the LNA is passed to a frequency converter where the uplink carrier frequency f_u is changed to the frequency f_d assigned to the emission for the downlink. Then, for each transponder, a band-pass filter located in the input multiplexer selects the emission (or emissions) lying within the frequency band allotted to the transponder and passes it (or them) to an HPA. The output from the HPA is then passed to a transmitting antenna via an output multiplexer, where it is combined with the outputs of any other HPA which are destined for the same transmitting antenna.

The earth station receiving chain

A receive port of the earth station antenna supplies a wideband aggregate of emissions, received from the satellite, to an LNA, followed by a wideband radio frequency amplifier (RFA) operating at the downlink frequency. This aggregate may contain a large number of emissions and some earth stations have interest in only one of them. If so, a double frequency change receiver, its first down-converter tuned to select the frequency of that required carrier, is used to extract the carrier and present it to a demodulator; see Figure 7.3. If the earth station needs to receive several or many of the carriers simultaneously, a splitter following the RFA supplies the wideband aggregate to other down-converters, receivers and demodulators, one being tuned to each carrier of interest.

System technology

Figures 7.1 to 7.3 show that much of the radio technology used in satellite communication is common to other kinds of UHF and SHF radio system used, for example, for terrestrial radio relay or land mobile systems. In this book, for example, it would be inappropriate to consider in detail the transmitting and receiving chains of an earth station. However, the circumstances for which some of this hardware has to be optimized are, to a considerable degree, peculiar

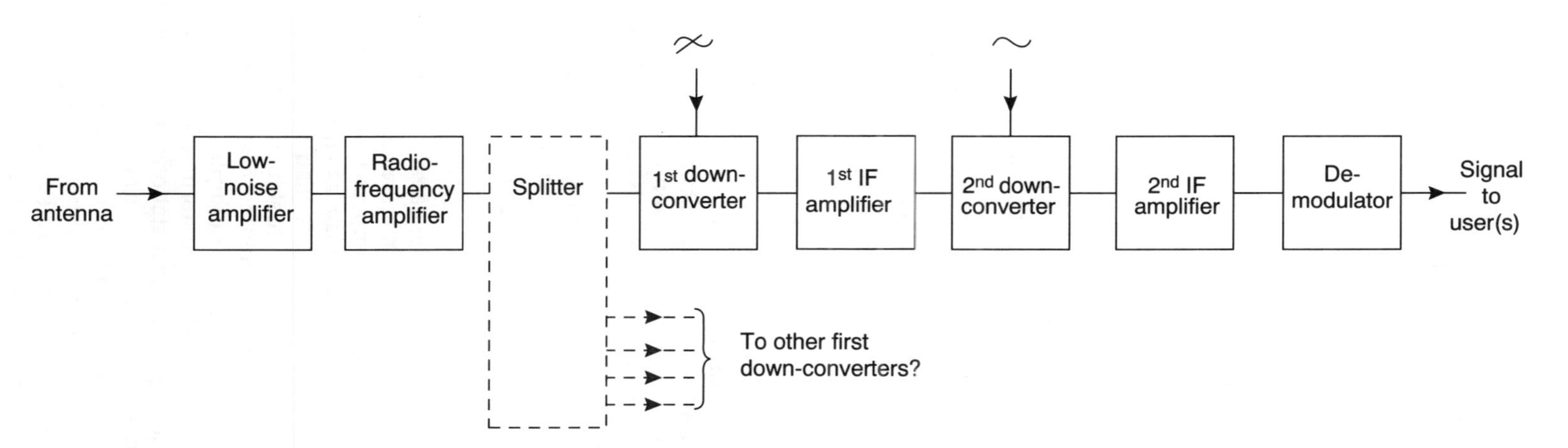

Figure 7.3 *A block schematic diagram of the basic elements of an earth station receive chain*

to satellite communication. This is true, for example, of much of the satellite payload. There are special requirements for earth station antennas, such as stringent side-lobe suppression and, in some cases, satellite tracking facilities. Receiver input stages are designed with special regard to the minimization of thermal noise. Power amplifiers, both on Earth and in space, are often required to amplify several or many modulated carriers simultaneously, raising problems in systems which are very susceptible to the performance degradations that amplifier non-linearities cause in these circumstances. These various topics are considered further in Sections 7.2 to 7.5.

There now follows a brief introduction to various system techniques which, while not exclusive to satellite communication, nevertheless play a particularly important part in it.

Multiplexing techniques

The signal delivered to the input terminal of the modulator in Figure 7.1 may be a single telephone channel, in analogue or digital form; this is usual in MSS systems and is often found in the FSS at earth stations with a small traffic load. But heavily loaded FSS earth stations may have hundreds or even thousands of telephone channels to transmit and receive. Large numbers of channels will be combined into multiplexed aggregates, analogue channels by frequency division multiplex (FDM) and digital channels by time division multiplex (TDM). These multiplexing techniques are considered in Chapters 8 and 9 respectively.

Modulation techniques

The transmission loss between earth stations and satellites is always high and at the present stage in the development of satellite communication there is seldom enough power from satellite transmitters to permit the use of modulation methods that would support a high ratio of information to bandwidth. To enable acceptable signal to noise ratios to be obtained, analogue signals are always transmitted using FM and with relatively wide deviation. Digital signals are always transmitted using phase shift keying (PSK) or a closely related technique, with seldom more than four significant phase conditions.

Multiple access

Many earth stations may make use of the same satellite at the same time. In the BSS one earth station usually has exclusive use of one or more transponders. However, although it occasionally happens that one FSS earth station needs the whole capacity of a transponder to relay its emissions, it is more usual for several or many earth stations to share one, and in the MSS a transponder is always shared by a large number of earth stations. The techniques by which a transponder can provide continuous communication simultaneously between several or many earth stations are called 'multiple access'.

In the simplest and most widely used form of multiple access, the emissions from the various participating earth stations are assigned different uplink carrier frequencies within the passband of the transponder, the frequency spacing between the carriers being sufficient to prevent overlap of the spectra of the emissions. This is called frequency division multiple access (FDMA).

In another form of multiple access, called time division multiple access (TDMA), short bursts of emission from different earth stations pass through the transponder, not simultaneously but in a rapidly recurring sequence, each burst being modulated by time-compressed digital signals. A series of bursts transmitted from one earth station, when received at another earth station, can be expanded in the time dimension to provide continuous digital communication channels.

A third form of multiple access is coming into use, in particular for the LMSS. Signals which overlap in frequency are transmitted by participating earth stations using wideband spread spectrum modulation. An earth station receiver extracts wanted signal elements from the aggregate by recognizing the signature code of the wanted transmitter. Used in this way, the technique is called code division multiple access (CDMA).

These various multiple access techniques are reviewed in Chapters 8 and 9.

Frequency reuse

These various multiple access techniques are not to be confused with 'frequency reuse' techniques, by means of which the same band of frequencies may be used twice or several times over, each use being made with little or no regard to the others. Using geostationary satellites, frequency reuse can be obtained in three basic ways.

Figure 7.4(a) shows two satellites with a substantial angular separation in orbit but using the same frequency bands and having overlapping service areas on Earth. The two systems do not interfere provided that the off-axis gain of the earth station antennas, in the direction of the unwanted satellite, is sufficiently low. (Note that the area on the Earth's surface that lies within a specified gain contour of a satellite beam (typically −3 dB relative to the peak gain of the beam) is called the 'footprint' of the beam. This concept is usually meaningful only in the context of geostationary satellites.

Similarly Figure 7.4(b) shows two satellites which serve well-separated footprints. The two systems do not interfere provided that the off-axis gain of the satellite antennas towards any earth station in the unwanted footprint is sufficiently low. This may hold true even if the two satellites have the same orbital location. By extension, a single satellite having antenna beams which have well-defined footprints illuminating well-separated service areas may be able to use the same frequency bands in both beams.

Thirdly, satellite antennas may transmit and receive orthogonal polarizations. These dual-polarized antennas may serve two different groups of earth stations, one transmitting and receiving in one polarization mode and the other operating in the orthogonal mode. Alternatively, a satellite with dual-polarized antennas may serve a single group of earth stations which, like the satellite, are equipped for transmit and receive in both polarization modes, thereby effectively doubling the width of the assigned frequency band.

Where the directivity of earth station antennas or satellite antennas, or the polarization discrimination of either, are not separately sufficient to enable the spectrum to be used without regard for other uses, a combination of two of them, or all three, may often succeed.

7.1.2 Satellite system terminology

Chapter 6 shows that satellite communication is a closely regulated medium and regulatory terminology permeates the literature. Many formal definitions of relevant terms, mainly of a regulatory nature, are to be found in Article 1 of the RR [1]. Technical terms particularly relevant to the FSS and the BSS are defined in ITU-R Recommendations S.673 [2] and BO.566 [3] respectively. A

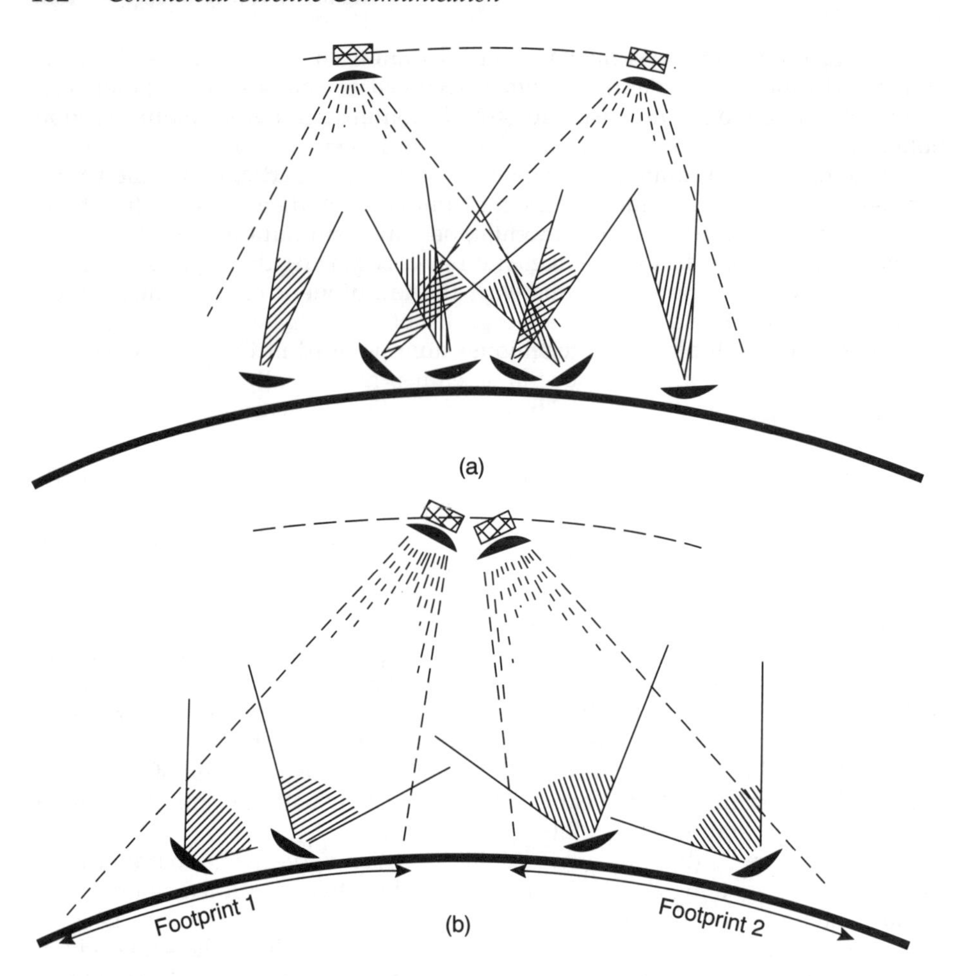

Figure 7.4 *Frequency reuse by two satellites and their earth stations. (a) involves the use of earth station antenna directivity and orbital separation; the service areas may overlap. (b) involves the use of satellite antenna directivity and discrete service areas; earth station antenna directivity may be small*

multilingual vocabulary covering orbits and systems is to be found in the International Electrotechnical Vocabulary, Chapter 725 [4]; a revised edition, extending the coverage to communications technology, is in preparation. These four authorities are harmonized, one with another, and the usage of terms in this book conforms with these sources with few exceptions. The main exceptions are as follows:

1. These official vocabularies define three terms, namely 'spacecraft', 'satellite' and 'space station'. A spacecraft is a vehicle to be launched into space. When a spacecraft has been put into a stable orbit it is called a satellite. A space station is the complement of an earth station; it is a communications payload installed on a satellite for relaying emissions received from earth stations. These distinctions are valid and logical, but they are rather cumbersome and 'space station' has come to have a different meaning in everyday usage. The word 'satellite' is used for all three concepts in this book.

2. A 'satellite network' is formally defined as one satellite plus all its cooperating earth stations. The term 'satellite system' is applicable to a single satellite network, as defined, but it also covers a group of satellite networks; the satellites of such a group might be functionally independent but under common management (as, for example, the Intelsat system) or all of them might have to be used together to provide a single facility (as in a typical non-geostationary satellite system). But the development of satellite communications since these terms were defined has made 'satellite network' rather ambiguous; a single satellite may now support many networks that have no common element other than the satellite itself. In this book a group of mutually supportive satellites is called a 'constellation' and 'satellite system' is applied to either one satellite and its earth stations or several satellites, functionally independent but under common management, with or without the cooperating earth stations; the context will make clear what is intended.
3. To define a frequency band precisely, the upper and lower limiting frequencies must be identified. But precision is not always necessary and it is often convenient to have a simple, more generalized, way of indicating a frequency band or an order of frequency. Terms like VHF, UHF, SHF and EHF are standardized and widely understood but the frequency ranges that they cover are very broad. However, terms like

   'L band' for the MSS allocations near 1.6 GHz,
   'S band' for the MSS allocations near 2 GHz,
   'C band' for the FSS allocations between 3.4 and 7.075 GHz,
   'Ku band' for the FSS allocations between 10.7 and 14.5 GHz, and
   'Ka band' for the FSS allocations between 17.7 and 30 GHz

survive from Second World War usage in radar. The rationale for these terms has now disappeared and they are not precise enough for many purposes but they are adequate and very convenient for other purposes and are widely used in satellite communication literature. They are used where convenient in this book.

7.2 The communications payload of a satellite

7.2.1 Transponders

The transponder described in outline in Section 7.1 and Figure 7.2, usually serving one of a number of broadband paths through the satellite payload, receives an emission (or several) from Earth at frequency f_u, changes the carrier frequency to f_d, raises the power level, and transmits the emission back to Earth. Inevitably the emission suffers some distortion in the process and some noise is added, but no intentional changes, apart from the frequency change, are made to it. Such a transponder is said to be transparent. Figure 7.5 is a more complete schematic diagram of a satellite payload using transparent transponders. It is convenient to divide a transponder into two parts, a receiver and a transmitter, interfaced by the input multiplexer.

There are various ways in which a signal might be modified on board the satellite, increasing the effectiveness of the satellite system as a whole, but making the transponder not fully transparent. These options tend to be limited to emissions using digital modulation. For example, digital carriers may be demodulated and regenerated before modulating a second carrier for onward transmission, eliminating noise and signal distortion that has occurred to the uplink. Other options arise for digital carriers if the satellite has several antennas

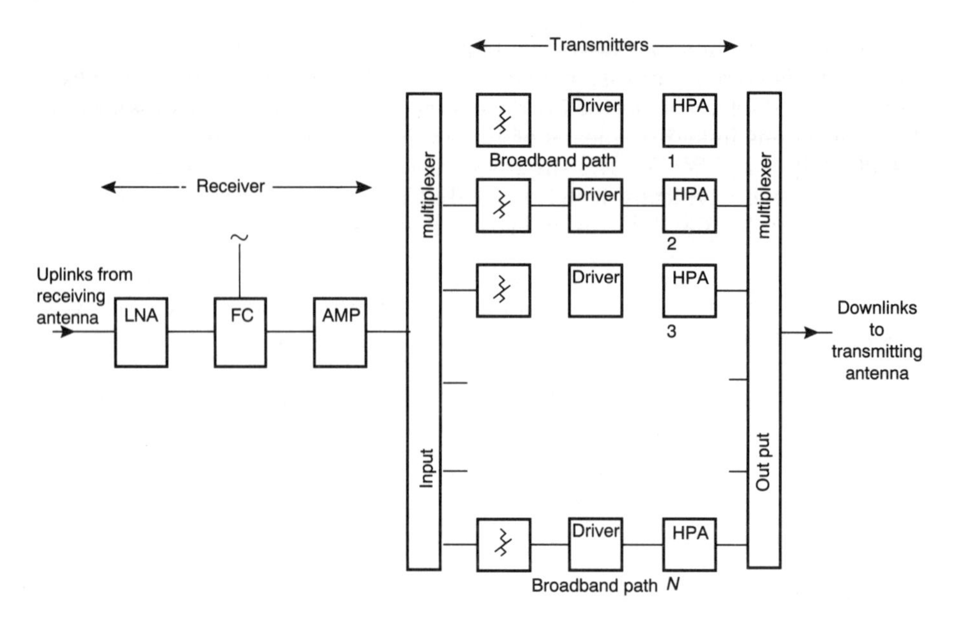

Figure 7.5 *Block schematic diagram of a satellite payload with transparent transponders using single-frequency conversion; provision for redundancy is not shown*

or if it operates in various parts of the spectrum. These options are considered further in Sections 7.2.3 and 9.4.6. However, almost all transponders now in use in commercial satellite systems are transparent.

The receiver

The LNA is a wideband stage, usually serving several broadband paths, often amplifying the whole aggregate of signals from one receiving antenna output port. In the early days of satellites, tunnel diode amplifiers were used in LNAs. Schottky barrier field effect transistors made from gallium arsenide (GaAs FETs) and high-mobility electron transistors (HMETs) are now mainly used in C and Ku bands, providing noise temperature contributions in the range 450 to 600 K. Parametric amplifiers are used in Ka band, with a similar noise contribution.

Low-noise GaAs FET amplifiers are based on a three-port device in which the gate controls the flow of electron current from the source to the drain by varying the electric field. This causes the formation of a depleted carrier region in the active layer beneath the gate.

The active element of a parametric amplifier is a varactor (variable capacity) diode which presents a negative resistance at the signal frequency, thus providing amplification. A varactor diode is a p-n junction device in which the depletion layer acts as the dielectric of a capacitor. In order to maintain the depletion layer, the junction is reverse biased; by increasing the bias, the layer width is increased, thus decreasing the capacitance. A bias is applied to the varactor diode, varying at the pump (oscillator) frequency, causing the capacitance to vary at the same frequency and creating the negative resistance effect.

The frequency converter comprises a mixer, a local oscillator and filters. The local oscillator runs at a frequency which is the difference between the centre frequencies of the uplink and downlink bands respectively. For example, for the frequency bands commonly assigned to FSS satellites at C band, the difference frequency is 2.25 GHz. State of the art mixers employ Schottky diodes with beam leaded connections linked by means of coplanar slotted lines.

A multi-stage amplifier follows the frequency converter, to make up the total gain of the receiver, typically to 60 or 70 dB. This gain must be constant over the specified bandwidth (e.g. the gain ripple should be less than 0.5 dB over 500 MHz).

The input multiplexer

An input multiplexer consists of a chain of three-port circulators feeding a set of bandpass filters which define the frequency bands allotted to the transmitters that the multiplexer serves. The design of the filters to meet stringent performance requirements is made simpler by dividing the transmitters into two groups, ensuring that filters operating in adjacent frequency bands are not tightly coupled to one another through the circulator chain. For example, if there are N transmitters operating in adjacent frequency bands to be fed from the same receiver, they might be split into two groups, odds and evens, the two groups being fed by different chains of circulators, the feed from the receiver being divided between them by means of a directional coupler; see Figure 7.6. The third port of the final circulator of each chain is matched to a dissipative load, to stop emissions that are not accepted by any of the bandpass filters being reflected back along the chain, causing a mismatch at its input. For the same reason it may be desirable to insert a similarly terminated circulator to act as an isolator at points marked X in the figure at the beginning of each chain.

Each bandpass filter passes emissions to its transmitter if their frequencies lie within its designated frequency band, rejecting the emissions proper to other transmitters. Low loss within the passband is not of prime importance since, with substantial gain in the receiver, loss in the filter can be made good without significantly degrading the carrier-to-noise ratio. However, it is essential for the loss in the filter to rise very rapidly at the edge of the passband, so that there should be little wastage of bandwidth where the passbands cross over. The variation with frequency of loss and group delay within the usable width of the passband must be small and free from abrupt changes, since these would cause signal distortion. These filters are usually realized in the form of high-Q cavity resonators, the materials used being chosen to ensure that the electrical characteristics of the filters vary very little indeed within the temperature range to which the filters will be exposed.

It may not be feasible to meet extremely stringent filter specifications, such as may arise for example in the MSS, at the downlink frequency. For such satellites a double frequency conversion architecture is sometimes used; see Figure 7.7. The received signal is down-converted to an intermediate frequency for filtering and amplification. It is then up-converted to the downlink frequency and amplified by a transmitter. See, for example, the description of the Inmarsat 2 satellite in Section 14.2.3. Dual frequency conversion architecture may also be used for satellites operating in frequencies above Ku band, since technical constraints may limit the gain that can be obtained at the downlink frequency. Double frequency conversion has been used in the Ku Band payload of Intelsat VI satellites (see Section 11.2.2) for a different purpose; there the use of an IF at 4 GHz provides a convenient interface for 'cross-strapping' between the Ku Band and the C Band transponders.

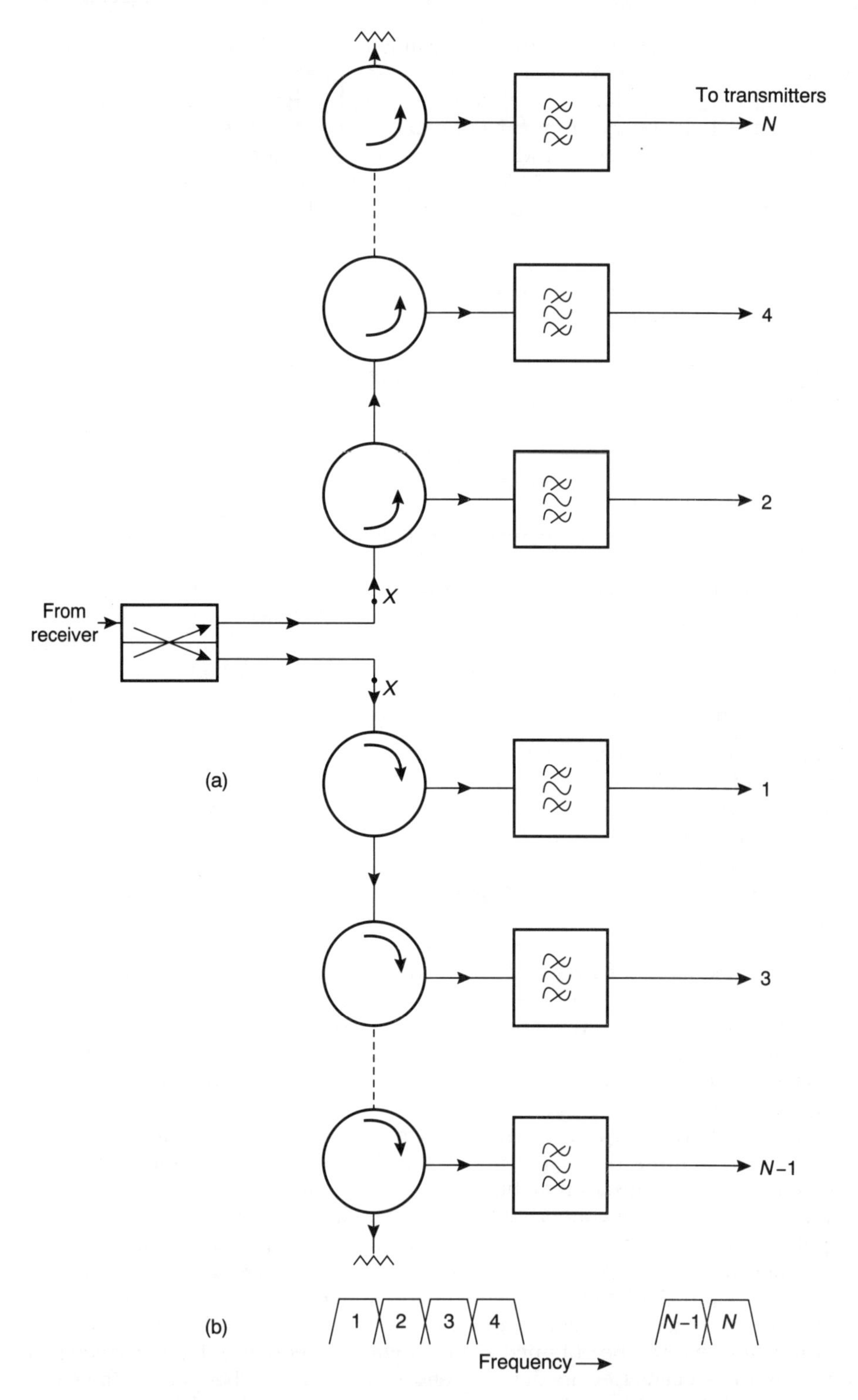

Figure 7.6 *An input multiplexer suitable for operating with adjacent radio channels*

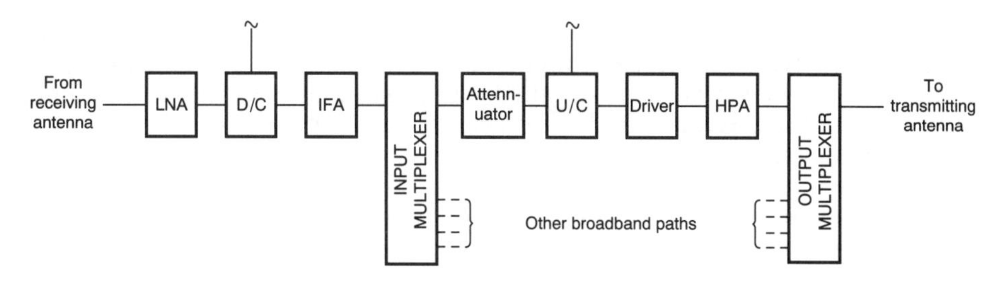

Figure 7.7 *Block schematic diagram of a transparent transponder with dual-frequency conversion*

The transmitter

The telecommandable variable attenuator, typically adjustable in steps of 1 dB over a range of 10 or 15 dB, serves two functions. First, the gain of HPAs tends to decrease with time, and this can be compensated for by a reduction of attenuation. Secondly, the attenuator permits adjustment of the gain of the transponder for the purpose of optimizing the parameters of the earth station-to-earth station link as a whole, having regard to the nature of the emission(s) which the transmitter is to carry. This is necessary, especially for the FSS, to enable uplink thermal noise (and therefore the transmitting earth station e.i.r.p. requirement) to be traded off against intermodulation noise generated in the transponder. The process is outlined in Section 10.4.

A driver amplifier is used to boost the output of the input multiplexer to the level necessary to drive the HPA; typically a gain of 25 to 50 dB is required. Field effect transistors and hybrid technology enable a compact and lightweight physical design to be realized.

Two types of output amplifier are used for satellite HPAs, namely travelling wave tube amplifiers (TWTAs) and transistor solid state power amplifiers (SSPAs).

A TWT is a vacuum tube device consisting of an electron gun and an associated permanent magnet system that fires an electron beam through a slow wave structure to a collector. The slow wave structure typically takes the form of a metallic helix. The radio emission requiring amplification is induced into the slow wave structure near the source of the beam. The radio wave and the electron beam travel along the vacuum tube at about the same speed and electromagnetic interaction between them induces a velocity modulation of the beam. Bunching of the electrons of the beam follows, which increases with the distance travelled along the beam, and this feeds energy back into the wave, to be extracted near the collector. TWTAs are available providing a maximum output power for a single emission of up to several hundred watts with gains of about 55 dB and an efficiency of up to 40 to 50 per cent, efficiency being defined as the ratio of the radio emission output power to the electric power input.

Solid state power amplifiers are constructed using field effect transistors. They have some limitations of output power and operating frequency but provide high reliability, compactness and low mass. As their limitations are being overcome by technological advances, SSPAs are increasingly replacing TWTs, especially in the lower frequency bands and where low to medium power outputs are required. SSPAs, typically using transistors in parallel, are currently being used to provide a maximum output power for a single emission of around 10 and 30 W at 4 and 12 GHz respectively with a gain of about 50 dB and efficiency between 20 and 35 per cent.

TWTAs distort in various ways the emissions that they amplify. The effects of distortion are worse when several emissions are amplified simultaneously in the same tube, and worst of all when the tube is operated close to its point of maximum output. Solid state devices also distort in the same circumstances, although the effects tend to be less severe. In receivers, where the number of emissions present is large but their level is low, the consequences of distortion need not be severe because these low level signals can be amplified at a linear part of the amplitude transfer characteristic of the amplifier without unacceptable cost in terms of power or mass. However, HPAs must be operated at a point in their characteristics where efficiency is fairly high. If more than one emission is present at any instant in an HPA, the trade-off between efficiency and distortion becomes crucial.

These distortion phenomena, which also affect earth station HPAs, are considered in more detail in Section 7.5.3.

The output multiplexer

The bandpass filter sections which form the output multiplexer have four main functions. They combine the outputs of several transmitters into a single waveguide to feed a transmit antenna. They reduce the radiation of unwanted emissions arising from distortions in the HPA but falling outside the passband of the transmitter that produced them. They reduce the radiation of spurious emissions, far outside the band allotted to the satellite, such as harmonics of emissions and even-order intermodulation products. They also ensure that the output port of each HPA in the group presents a high impedance to the output port of every other HPA in the group.

It is important that the insertion loss within the passband of a filter should be low, since any loss at this point in the system reduces the e.i.r.p. available for downlinks. Accordingly, multiplexer designs avoid the use of circulators. As with the input multiplexers, filter sections take the form of cavity resonators, coupled to a waveguide manifold which is short circuited at one end and is coupled to a transmitting antenna feed port at the other. Here again the design can be considerably simplified if odd and even frequency bands are combined separately and connected to different feed ports.

Circuit fabrication techniques

The units of the transponder are typically of modular construction using hybrid circuits containing unencapsulated chips. The process used in forming the hybrid circuits depends on the operating frequency. Hybrid circuits operating at IF are constructed using thick film technology. Resistors, capacitors and interconnections are formed by depositing layers of suitable materials on an alumina substrate, the chips of the active devices being bonded directly on to the substrate. Hybrid circuits operating at radio frequencies use the same process, except that the microstrip circuits that provide the conducting paths are formed using thin film technology. In this case, the substrate is first covered with a thin layer of gold which is then subjected to a photolithographic process to create the required conducting paths. Printed circuit boards are used for devices operating at low frequencies.

Reliability through redundancy

Operational payloads are designed for reliability by duplication of the devices, the failure of which would render the payload inoperable. In particular, reliability through redundancy is applied to those devices that may be expected to wear out or are foreseen to be liable to sudden failure. For example, there are usually two identical receiver units for each receiving antenna output, of which one is active and the other is on standby, providing 2-to-1 redundancy; either unit can be activated by telecommand. The driver amplifiers and HPAs are also replicated to provide the required end of life reliability.

7.2.2 Satellite antennas

The Earth subtends an angle of 17.4° from the GSO, filling only 0.4 per cent of the sky. All communications satellites have their attitude stabilized rather precisely with respect to the Earth while in normal operation and it is clearly wasteful to use satellite antennas that do not concentrate on to the Earth much of the energy that they radiate. An antenna on a geostationary satellite, directed towards the Earth's disc, can provide a gain of at least 17 dBi towards all parts of the Earth's surface that are visible to it; such an antenna is said to have global coverage, although the term is obviously not precise. If a smaller footprint is sufficient, higher antenna gain can be used. Very high gain beams are called spot beams.

Currently, commercial communications satellites are equipped with horn, array or reflector antennas. Each of these types of antenna can be used in the dual polarization mode, given appropriate feeds. It may be found desirable to introduce the use of lens antennas in the future when growing congestion in the GSO demands more effective reduction of radiation outside the required footprint. In the sections that follow, antennas are described in the transmitting mode, but the same structures also function in the receiving mode.

Horn antennas

The horn antenna has the advantages of simple construction and small size. Its side-lobe characteristics are poor but they can be improved by corrugation of the internal surface of the horn. Its small size and wide beamwidth also make the horn ideal as the primary source for a reflector antenna.

Horn antennas can be divided into two main types, pyramidal and conical, according to the shape of the aperture. The pyramidal horn has the advantage that it transmits waves with no cross-polarization products. This characteristic, together with the fact that its gain can be calculated accurately from its physical dimensions, make it a useful gain measurement tool.

Conical horns are used as satellite global beam antennas. For example, a horn with 0.3 m diameter aperture provides a beamwidth of 17.5° at 4 GHz. This type of antenna can be classified as single-mode, multiple-mode or hybrid mode depending on the wave propagation mode in the waveguide that feeds it. The single-mode conical horn antenna is tuned to the predominant mode of the circular waveguide, the TE_{11} mode. It is the most basic of the three types. The multi-mode type is tuned to the TE_{11} waveguide wave propagation mode together with the TM_{11} mode (one of the higher propagation modes). The hybrid-mode type as the name implies, is tuned to a hybrid mode which offers low side-lobes over a particularly wide beamwidth and a symmetrical beamwidth along its axes.

Array antennas

An array antenna consists of a radiating aperture which is formed by a number, sometimes a large number, of radiating elements placed side by side over the physical area that constitutes the aperture. The phase and amplitude of the feed to each element can be adjusted by the use of controllable phase shifters and power dividers to ensure that the combination of individual radio waves from each radiating source produces the required beam pattern. Dipoles, horns and printed elements are all used as radiating elements in array antennas.

Reflector antennas

Reflector antennas are a convenient means of obtaining high gain and an efficient match between the beam footprint and a complex geographical distribution of earth stations. The simplest reflecting antenna consists of a paraboloidal reflecting surface illuminated by a radiating source placed at the focus of the paraboloid; a conical horn is often used as the source. Offset front-fed and dual reflector geometry (Cassegrain or Gregorian) are also used. These configurations are similar in principle to typical earth station antennas and their characteristics are discussed in more detail in the context of that application in Section 7.3.

An elliptical beam can be generated by a reflector antenna with a single radiating source at the focus of the primary reflector by the use of a reflector that is itself elliptical; the beam is narrower in the plane of the major axis of the reflector. This provides a simple means of adjusting the shape of the footprint.

More complex shaped beams can be obtained by using an array of radiating sources in the antenna focus, thereby generating an overlapping pattern of footprints. See, for example, Figure 11.3, which shows one of the arrays of horns that is used to form Hemi and Zone beams for Intelsat VI satellites, and Figure 11.2 showing the footprints, matching the required service areas, that these arrays generate.

7.2.3 Multi-beam satellites

Many satellites of the FSS operate in two pairs of frequency bands, most usually C band and Ku band. Others, such as Eutelsat II, have downlinks in different bands within the frequency range called Ku band. The Intelsat satellites provide global coverage in C band using global beams from horn antennas but they also have C band reflector antennas providing beams of higher gain to the geographical areas where most earth stations are located. Furthermore, most satellites operate with dual polarizations, which provide in effect pairs of beams with the same coverage.

It is usually necessary to use separate antennas for operating in different parts of the spectrum. However, it is also feasible to use a reflector system, with two or more horns or arrays of horns offset from the focus, to generate more than one beam, each with its own footprint. Under favourable circumstances, in which the gain of one beam in the direction of any earth station served by another beam is sufficiently small, these various beams may use overlapping frequency bands, a very valuable form of frequency reuse.

When a satellite has more than one uplink and one downlink beam, it will often be necessary to receive an uplink in one beam and retransmit it as a downlink in another beam with different coverage. It will be desirable to have flexibility in coupling transponders to beams and this can be provided by switches which can be operated by telecommand. In some satellites these switching facilities are elaborated into a completely flexible matrix; see, for example, the Intelsat VI facilities described in Section 11.2.2.

However, the availability of multiple beams, flexibly accessed, provides the means for various alternatives to the simple transparency of the transponders described in this section. One, permitting bursts of a TDMA emission uplinked to the satellite to be distributed between several downlink beams according to the destination of the signals carried in each burst, is already in use in Intelsat; it is described in Section 11.4.5. Other options, involving a greater degree of signal reconstruction on board the satellite, remain for the future; some possibilities are outlined in Section 9.4.6.

7.3 Earth station antennas

7.3.1 Axisymmetric Cassegrain antennas: the main lobe

Almost all earth stations at fixed locations have antennas with paraboloidal primary reflectors. The very high gain antennas at some FSS earth stations use Cassegrain geometry. These antennas have a large circular primary reflector fed from a feed assembly close to its pole via a hyperboloidal subreflector, the whole being symmetrical about the axis of the primary reflector; see Figure 7.8. Other antenna configurations, currently used with smaller primary reflectors, are described in Section 7.3.3.

For the axisymmetrical Cassegrain antenna, using simplifying assumptions, the ray path geometry is shown in Figure 7.8. Rays radiating from a phase centre

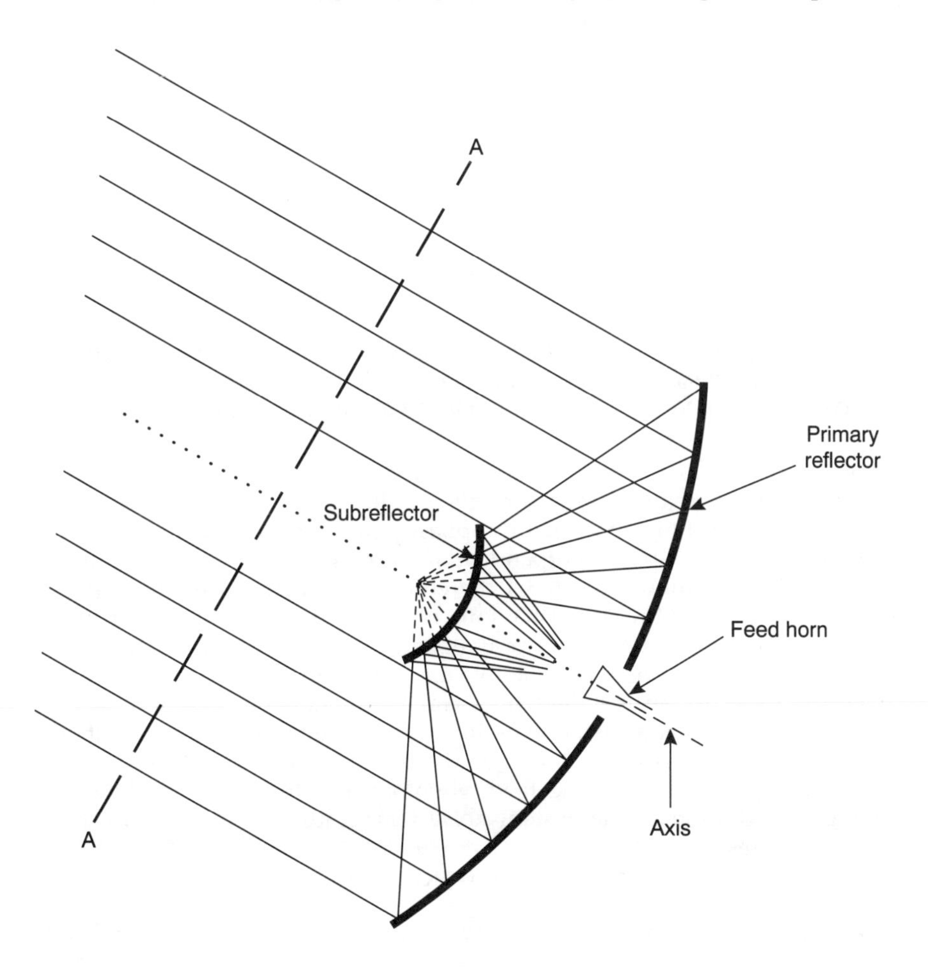

Figure 7.8 *An axisymmetrical Cassegrain antenna in cross-section, showing typical ray paths*

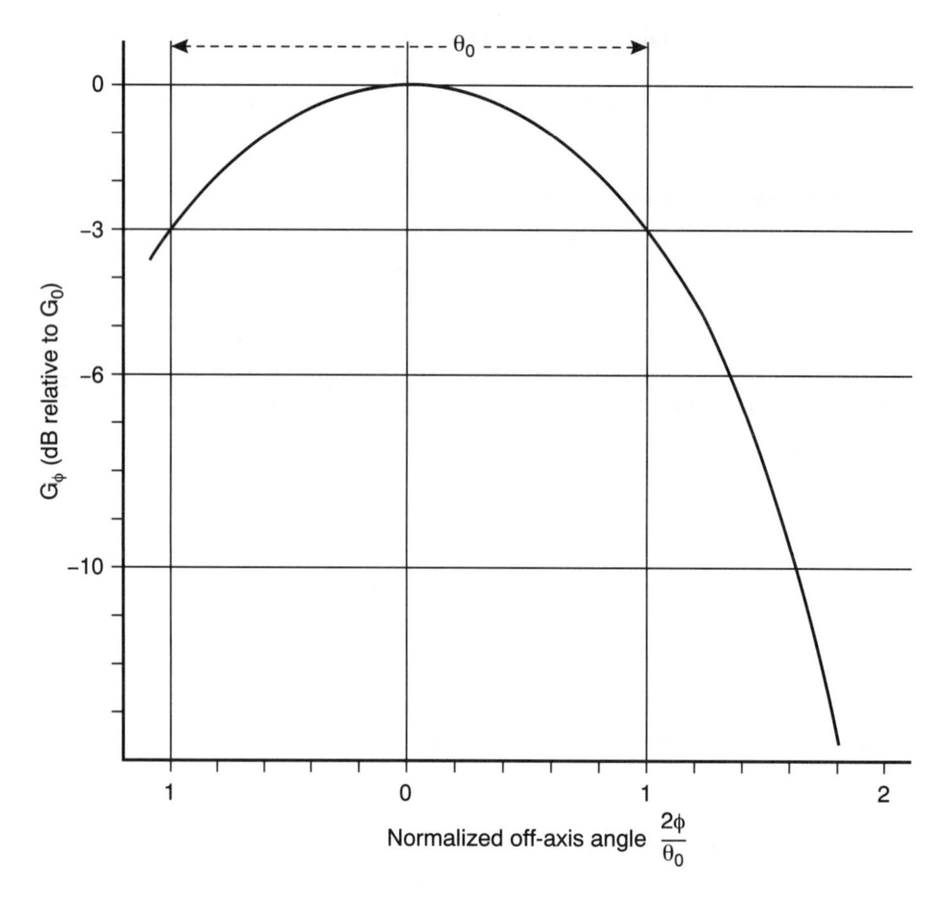

Figure 7.9 *The gain G_ϕ of the main lobe of an antenna at an angle ϕ from the axis, relative to the maximum on-axis gain G_0, plotted on rectangular coordinates, the ordinate for ϕ being normalized to the half-power beamwidth θ_0*

located at the mouth of the feed horn are reflected at the subreflector and are reflected again at the primary reflector as a parallel beam along the axis of the main reflector. The distance between the phase centre at the feed horn and the point where a ray cuts a plane perpendicular to the beam (*AA* in the figure) is the same for all rays; the beam at *AA* therefore has a plane wavefront. (Here, as elsewhere in this section, the behaviour of an antenna is considered in the transmit mode, but the conclusions apply equally to the receive mode.)

This configuration is particularly suitable for large high-gain antennas, because it enables the LNA to be installed in a cabin, immediately behind the feed assembly, that can be made easily accessible for maintenance while the antenna is in operation without imposing limits on the pointing direction of the antenna. The short waveguide between the feed assembly and the LNA has low loss, minimizing the addition to the system noise temperature in the receive mode that this link causes.

The maximum gain G_0 of the antenna, which is along the axis, is given by

$$G_0 = 10 \log 4\pi A_e/\lambda^2 \quad \text{(dBi)} \tag{7.1}$$

where A_e is the effective aperture of the antenna (m^2), and
λ is the wavelength (m).

A_e is related to the physical area A of the main reflector by

$$A_e = \eta A \quad (\text{m}^2) \tag{7.2a}$$

$$= \tfrac{1}{4}\eta\pi D^2 \quad (\text{m}^2) \tag{7.2b}$$

where D is the diameter of the main reflector (m) and
η is the aperture efficiency.

Thus,

$$G_0 = 10 \log \eta(\pi D/\lambda)^2 \quad (\text{dBi}) \tag{7.3}$$

The aperture efficiency η is usually expressed as a percentage and a typical value for a large Cassegrain antenna is 65 per cent. Using this approximation,

$$G_0 \approx 8.1 + 20 \log D/\lambda \quad (\text{dBi}) \tag{7.4}$$

In directions which do not coincide with the axis of the antenna, the gain declines symmetrically, forming the main lobe of radiation. The beamwidth θ_0 at which the gain is -3 dB relative to G_0, is a key parameter of an antenna, called the half-power beamwidth. It is given with an accuracy sufficient for most purposes by

$$\theta_0 = k\lambda/D \quad (\text{degrees}) \tag{7.5}$$

where k is about 65° for large efficient antennas.

The gain G_ϕ of the antenna at an angle ϕ from the axis but within the main lobe is represented well by the function $(\sin^2 \phi)/\phi^2$. With this assumption, the variation with ϕ of G_ϕ relative to G_0 is shown in Figure 7.8, normalized for the angle where $G_\phi = G_0 - 3$ dB, that is, where $\phi = \frac{1}{2}\theta_0$.

For values of ϕ between 0 and $\frac{1}{2}\theta_0$, the gain is given approximately by

$$G_\phi \approx G_0 - 12(\phi/\theta_0)^2 \quad (\text{dBi}) \tag{7.6}$$

The gain falls to a very low value, the 'first null', when ϕ is of the same order as θ_0. At greater angles from the axis of the antenna, there are side-lobes; see Section 7.3.2.

'Shaped' reflector profiles

Figure 7.10(a) shows how the feed horn illuminates the subreflector of an axisymmetrical Cassegrain antenna. The side-lobes of the feed horn radiation pattern miss the subreflector altogether and the main lobe delivers the most intense flux to the central part of the subreflector and a rather weaker flux to the outer parts of the subreflector.

The energy reflected from the central part of the subreflector reaches a part of the primary reflector that is obstructed by the subreflector. This energy does not join the radiated beam; after further reflections from one or both reflectors, the rays emerge in directions that are not parallel with the axis of the antenna and with random phase, as shown in Figure 7.10(b). The energy reflected by the outer parts of the subreflector illuminates the primary reflector, but not uniformly, tapering towards the edge. Moreover, as may be seen from Figure 7.8, the core of the beam carries no direct flux while it is close to the antenna, being screened from the primary reflector by the subreflector. (This empty core is filled in by diffraction downstream.) These are some of the reasons why the aperture efficiency of axisymmetrical Cassegrain antennas is not more than about 65 per cent.

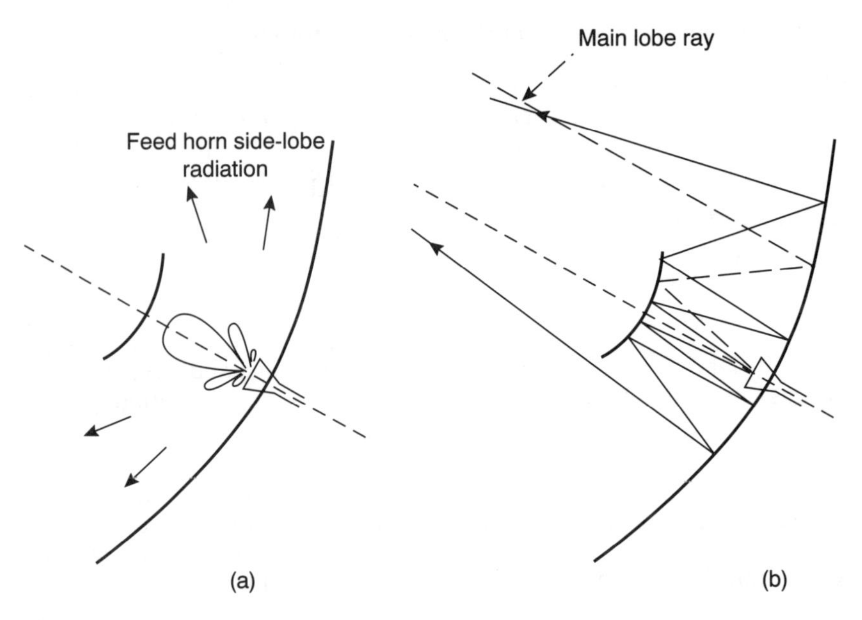

Figure 7.10 *An axisymmetrical Cassegrain antenna in cross-section. (a) shows a typical feed horn radiation pattern, with side-lobes that are not reflected by the subreflector. (b) shows two ray paths that are reflected close to the centre of the subreflector and, after multiple reflections, fail to join the main beam*

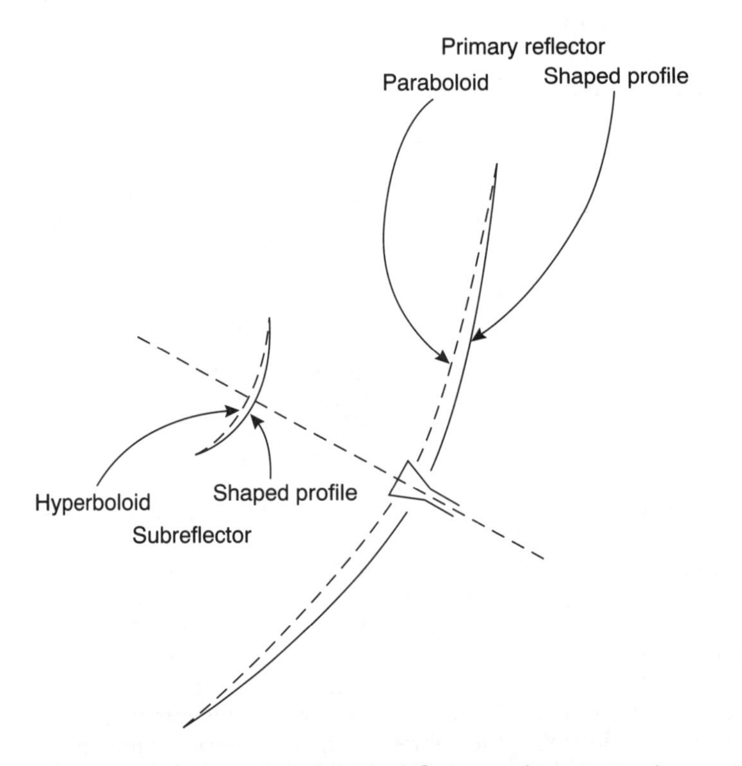

Figure 7.11 *An axisymmetrical Cassegrain antenna in cross-section using 'shaped' reflectors. Unshaped paraboloidal and hyperboloidal profiles are also shown for comparison, the differences being greatly exaggerated*

The aperture efficiency can be improved to some extent by modifying the profile of the subreflector. The hyperbola which generates the surface of the subreflector is distorted by increasing the curvature near the axis and reducing the curvature near the edge. In consequence, the illumination of the central part of the main reflector is made less intense, whereas at the outer part of the main reflector it is increased; see Figure 7.11. Having modified the subreflector in this way, it is necessary to modify the primary reflector to equalize the distances travelled by all rays that join the beam, thus maintaining an equiphase wavefront at the aperture. This 'shaping' of reflectors, which increases both η and G_0, is often adopted for large antennas.

7.3.2 Axisymmetric Cassegrain antennas: the side-lobes

The plane wavefront at the aperture of an ideal antenna sets up a main lobe of radiation and also a diffraction pattern of side-lobes extending in all directions. The peak gain of the side-lobes declines rapidly to a low level as the angle ϕ relative to the axis increases. The number and width of the side-lobes depend on the ratio of the diameter of the aperture to the wavelength. For the largest antennas in current use at 14 GHz, D/λ is around 800; a cross-section through the radiation pattern would show thousands of side-lobes, each about one-tenth of a degree wide. A cross-section through a small VSAT antenna, with $D/\lambda = 30$, would show one or two hundred side-lobes, each one or two degrees wide.

However, the gain of some side-lobes of practical antennas is much greater than diffraction theory indicates, for several reasons. Figure 7.10 shows how side-lobe radiation from the feed horn and also the radiation from the feed horn main lobe which strikes the subreflector too close to its centre are both lost from the main beam and add to the energy in the side-lobes. In addition:

(1) Large-scale imperfections in the profiles of the reflectors cause a significant increase in the gain of side-lobes close to the main lobe.
(2) Small-scale reflector profile imperfections scatter energy in all directions.
(3) Even small errors in the relative locations of feed horn, subreflector and primary reflector may increase the gain of the side-lobes close to the main lobe, as well as distorting the shape and reducing the gain of the main lobe.
(4) A structure, typically consisting of four struts, is required to support the subreflector at its correct location in front of the primary reflector. These struts are exposed to the energy reflected by the primary reflector, and they scatter some of it.
(5) The main lobe of the feed horn will have significant gain in the direction of the edge of the subreflector, and likewise the subreflector will reflect significant flux in the direction of the edge of the primary reflector. Energy that spills over the edges of the reflectors is radiated in the side-lobe field, directly or after diffraction at the reflector edge. See Figure 7.12.

Thus, a practical earth station antenna has powerful side-lobes adjacent to the main lobe; typically, the gain of the first side-lobe is within 20 or 25 dB of G_0. The peak gain of the outer side-lobes declines with increasing remoteness from the axis but in an erratic manner. Usually the most significant side-lobes are those due to subreflector spillover.

The energy that goes into the side-lobes is lost from the main lobe, reducing η, A_e and G_0. Energy radiated through the side-lobes causes interference to other satellite systems and to terrestrial radio systems; for geostationary satellites the most serious interference arises from energy radiated in the solid angles identified in Figure 6.9. Each of these transmit mode effects has its counterpart in the receive

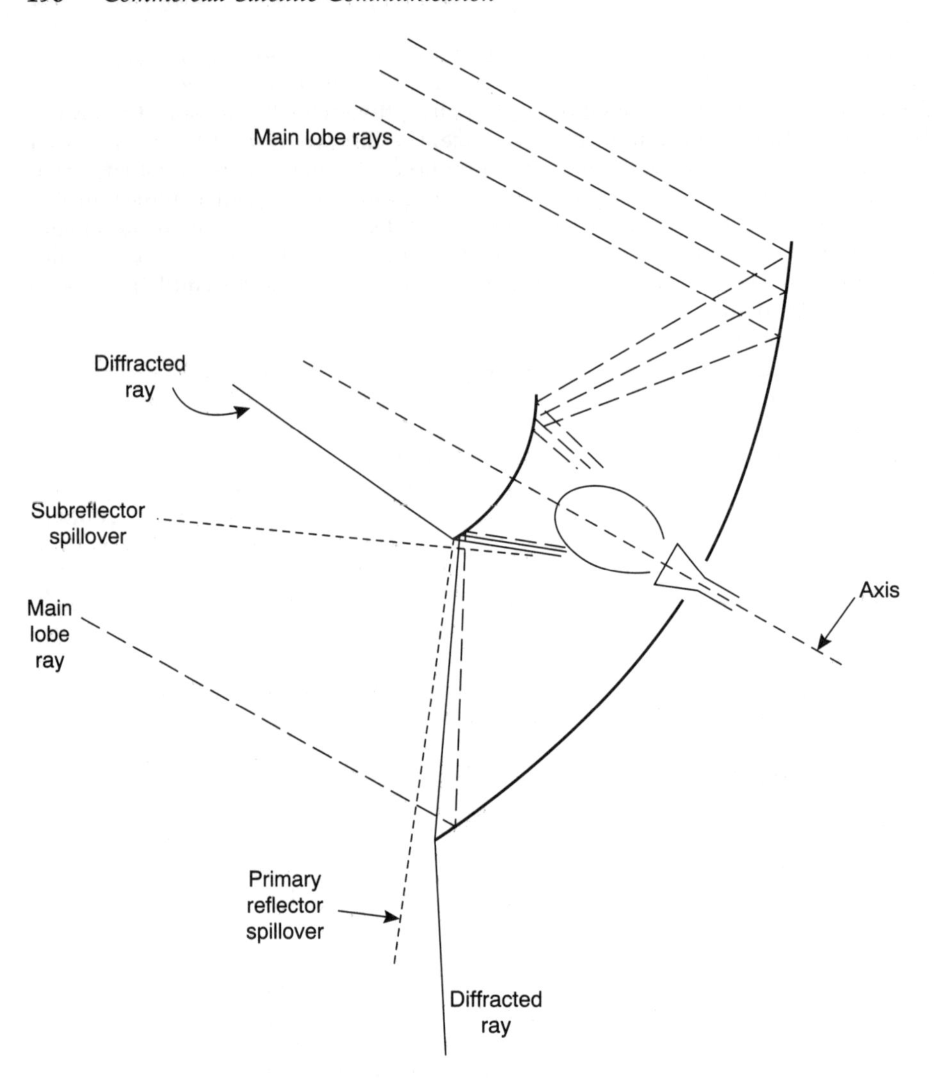

Figure 7.12 *An axisymmetrical Cassegrain antenna in cross-section showing spillover at the edge of the subreflector and the primary reflector*

mode, permitting the entry of interference from other satellite systems and of thermal noise radiated by the ground, nearby buildings, etc.

Side-lobe levels can be reduced by using reflectors with accurate profiles; profile errors not exceeding a fraction of a millimetre are usually specified. Great care must be taken to position the feed horn and subreflector in their correct positions relative to the primary reflector and the structural members that support them must be robust enough to maintain these relative positions under extreme weather conditions. Lossy material that absorbs radio wave energy, placed at the edge of the subreflector and the primary reflector, can reduce spillover.

An antenna designer must find an acceptable compromise between aperture efficiency and side-lobe suppression, since maximizing η demands uniform illumination of the primary reflector, which leads to high spillover at the edges of both reflectors, whereas the best side-lobe suppression is achieved by a tapered illumination of the primary reflector and minimization of spillover.

Regulatory constraints

Good side-lobe suppression could significantly increase the number of satellites that can operate simultaneously in the GSO. The side-lobe gain levels actually being achieved by large earth station antennas have been studied for many years by the CCIR and subsequently by Study Group 4 of ITU-R. The original purpose of this work was to provide a basis for recommendations on the antenna gains that should be assumed, in the absence of specific measured data, when calculating prospective interference levels; see Section 6.2.2 and equation (6.2) for the current recommendation. CCIR Report 391 [5] shows that this recommendation is realistic for existing large axisymmetrical Cassegrain antennas and better results are being achieved. However, CCIR Report 998 [6] indicates that great care is necessary in the design of small axisymmetrical antennas if this standard is to be achieved.

However, as indicated in Section 6.5, item (7), the ITU-R recommends, as a design objective for new earth station antennas, that the gain of 90 per cent of side-lobe peaks within the solid angles of greatest concern should lie below the curve given by equation (6.5), which is, in general, 3 dB more stringent than equation (6.2). In addition, limits are recommended for the spectral e.i.r.p. density radiated through earth station side-lobes in the general direction of the GSO; see Section 6.5, item (5). These objectives are readily achievable with large well-designed earth station antennas and using carrier energy dispersal but for smaller antennas the objectives may demand a degree of side-lobe suppression that is not feasible with axisymmetrical Cassegrain antennas.

7.3.3 Low side-lobe antenna configurations

There are reflector antenna configurations which have much lower side-lobe gain levels than the axisymmetrical Cassegrain design. The primary reflectors of these antennas have a parabolic profile, but, as shown in Figure 7.13, the profile is offset

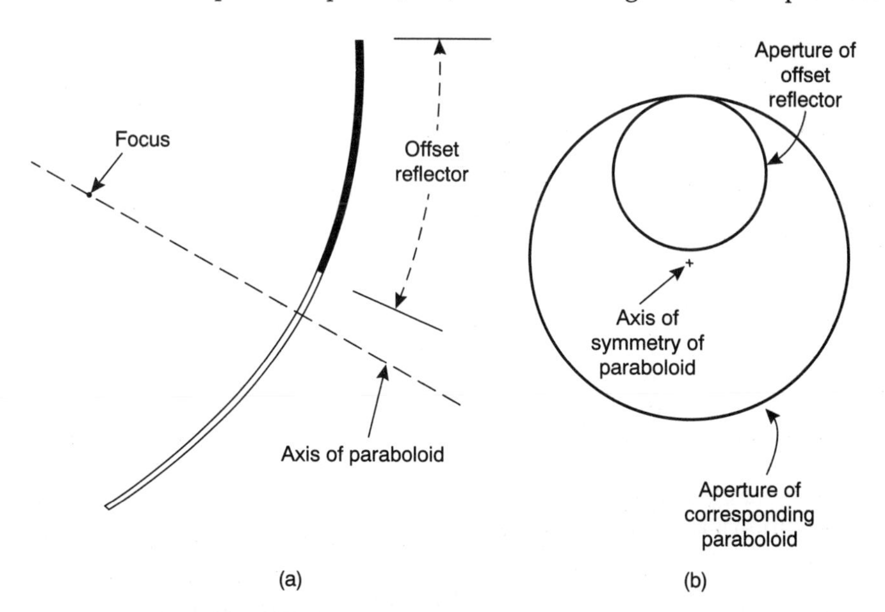

Figure 7.13 *The offset paraboloidal reflector: (a) in cross-section; (b) the aperture of a circular offset reflector, showing its relationship with a symmetrical paraboloid*

from the axis of symmetry of the paraboloid. The aperture is usually circular, although an elliptical aperture has advantages in some applications. Offsetting the profile from the axis of symmetry of the paraboloid makes possible ways of feeding the reflector that do not obstruct the aperture.

The offset front-fed antenna is the simplest of these configurations; see Figure 7.14. The feed is located in front of the reflector but to one side, at the focus of the reflector. There is no subreflector, no obstruction of radiation close to the main beam, and therefore no reinforcement of side-lobes due to scattering, apart from what is due to imperfections of the reflector profile.

The principle of the offset Cassegrain antenna is shown in Figure 7.15. This has an hyperboloidal subreflector placed between the offset paraboloidal primary reflector and its focus, but the subreflector and its supports are located where they do not obstruct the main beam. A variant of the offset Cassegrain antenna uses offset Gregorian geometry. In this, a concave ellipsoidal subreflector is placed beyond the focus of the primary reflector; see Figure 7.16.

In each of these configurations there is some spillover of energy from the feed which passes by the edge of the reflectors; feeds must be designed so as to limit this spillover to avoid radiation hazards in the vicinity of the antenna. The

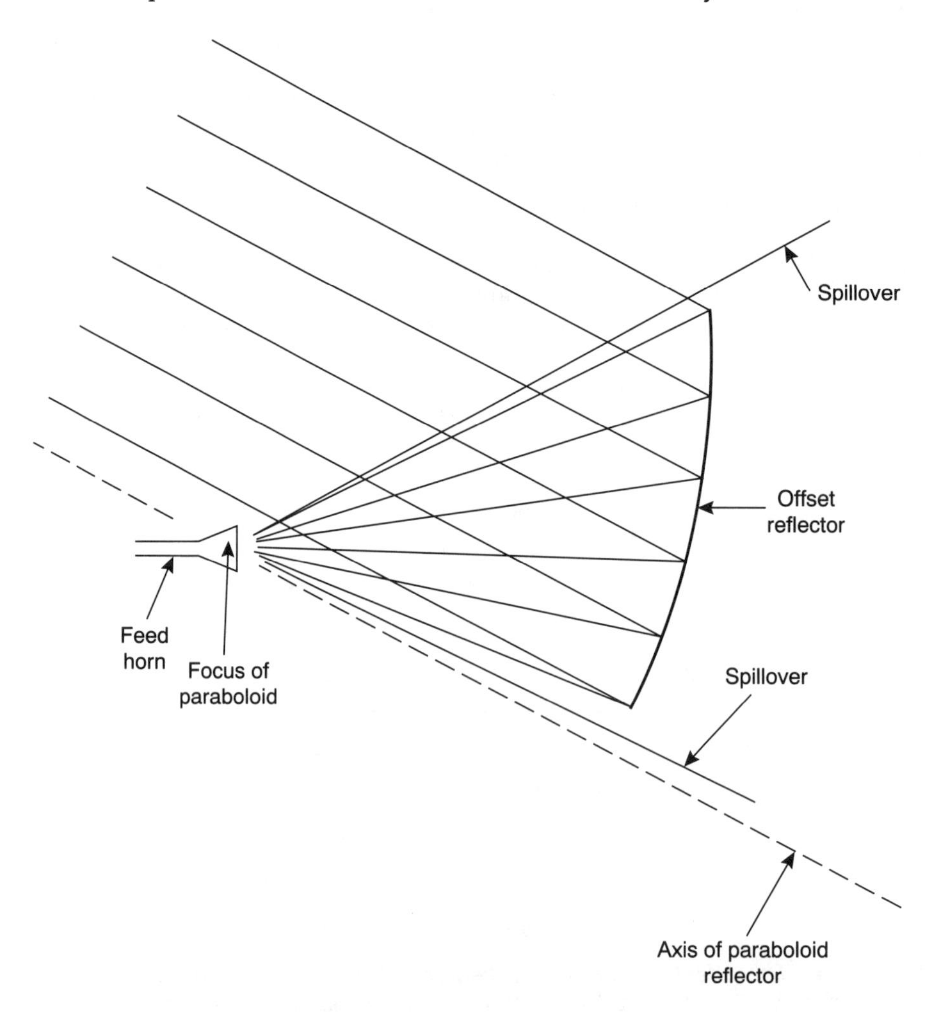

Figure 7.14 *The cross-section of an offset front-fed antenna*

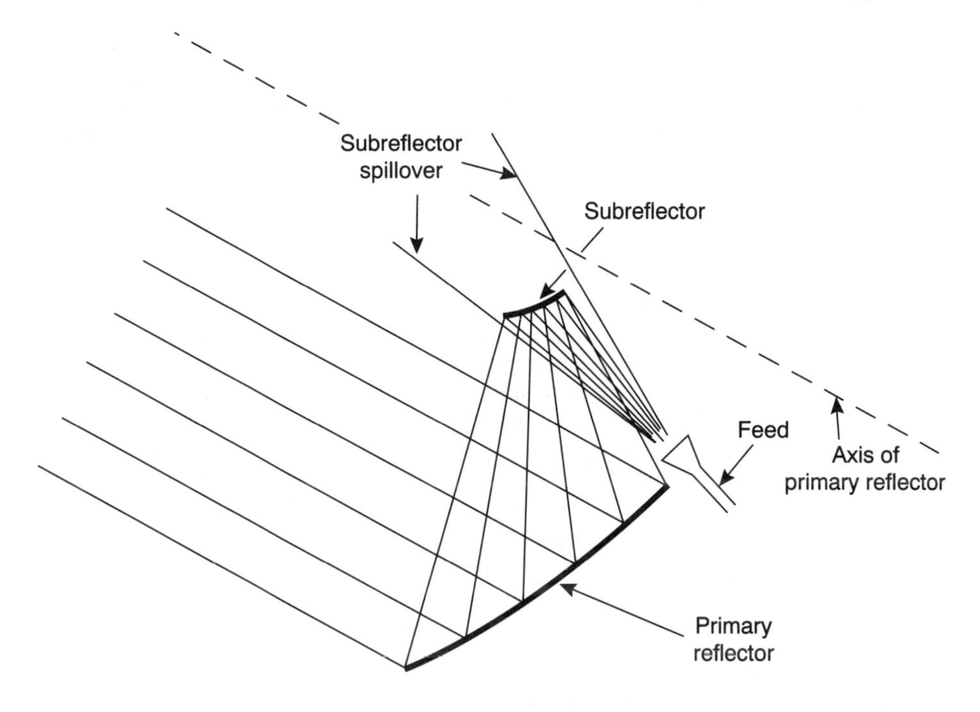

Figure 7.15 *The cross-section of an offset Cassegrain antenna*

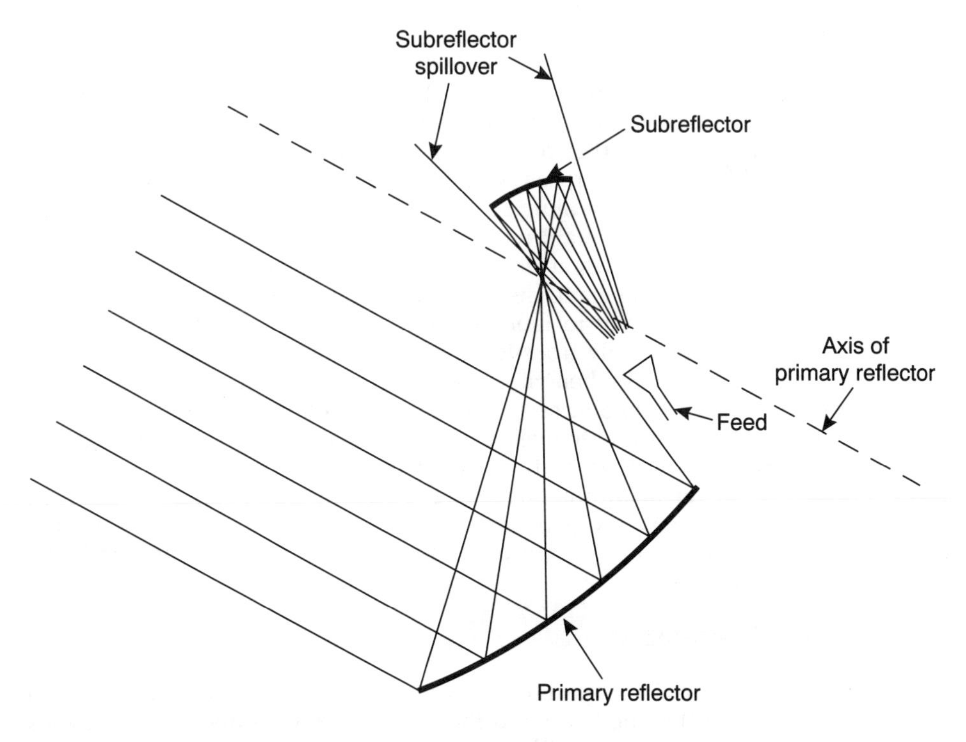

Figure 7.16 *The cross-section of an offset Gregorian antenna*

spillover is not likely to cause much interference to other satellite systems, nor should much interference enter the system in the receive mode by this route, since with proper orientation of the system these spillover lobes will not be directed towards the GSO. If necessary the angular separation between spillover lobes and the GSO can be increased for dual reflector antennas by arranging for the feed to illuminate the subreflector through a hole in the primary reflector.

As with the axisymmetrical Cassegrain configuration, the profiles of the reflectors of an asymmetrical antenna can be 'shaped' or the properties of the feed can be modified to make more uniform the illumination of the aperture and thereby to increase the aperture efficiency.

Offset antennas can be designed to have excellent radiation characteristics. Aperture efficiencies η exceed 75 per cent, although it should be noted that the product ηA, which gives the effective aperture of the antenna (equation (7.2)) and thus the gain G_0, is to be calculated from the area of the aperture projected into a plane perpendicular to the axis of the system, not the physical area of the main reflector. Side-lobe peak levels in the direction of the GSO about 10 dB better than the standard recommended in ITU-R Recommendation 580 (see equation 6.5) can be obtained; see CCIR Report 998 [6].

Some offset reflector antennas have an elliptical or rectangular aperture. The main lobe of such an antenna in the plane corresponding to the longest dimension of the primary reflector is narrower than that obtained with a circular aperture of the same gain. This asymmetrical directivity can be arranged to generate a narrower main lobe in the east–west direction at the GSO, providing better protection against interference between satellite networks. This asymmetry also facilitates the mounting of an antenna on a road vehicle without folding the main reflector, an important consideration for a transportable earth station.

Another offset antenna configuration using front-fed geometry and a non-circular reflector, typically rectangular, is the torus antenna. With this design the curvature of the reflector in the direction of the shortest dimension is parabolic but the curvature is circular in the longest dimension. The antenna is oriented to take advantage of its higher directivity along the GSO. However, the circular profile in this plane ensures that the loss of gain due to defocusing for a feed located to either side of the true focus point is less than it would be if the reflector were paraboloidal. Thus, by installing several independent feeds along the image of the GSO, as focused by the main reflector, it is possible to get access with the one reflector to several satellites at different longitudinal stations.

These offset configurations have disadvantages. The primary reflectors, being asymmetrical, are more costly to fabricate, and if access is required by waveguide to the feed of a steerable front-fed offset antenna, the waveguide may be long and lossy. However, the benefit of low side-lobes in the critical directions amply compensate for the higher cost. Offset front-fed antennas are very widely used for small receive-only antennas. Offset Cassegrain and Gregorian geometry is often preferred for earth station antennas with diameters of 7 m or less, especially where powerful uplink spectral e.i.r.p. density makes low side-lobe gain essential. Offset geometry is relatively more costly to implement for large antennas, but the large on-axis gain of these antennas makes better side-lobe suppression less necessary.

7.3.4 The feed assembly

The feed assembly provides the link between the earth station transmitter(s) and/or receiver(s) on the one hand and the antenna on the other. The functions that the feed assembly must provide depend very much on the mission of the earth station and the design of the satellite with which it works. Two examples follow, spanning the range of the FSS.

A simple receive-only feed assembly

For many receive-only earth stations with low-gain antennas, typically VSATs and satellite broadcasting receivers, the feed assembly consists of a feed horn and, usually, a polarizer. A polarizer is often required to reject carriers transmitted by the wanted satellite in the polarization mode orthogonal to the wanted mode. The feed horn usually operates in a wideband mode, delivering the wanted signal, together with a spectrum of other signals transmitted by the wanted satellite with the wanted polarization, to the LNA and a wideband down-converter. These latter units are usually located adjacent to the feed assembly, with a coaxial cable link to the rest of the receiver.

The asymmetry of offset front-fed antennas introduces a cross-polar signal component, but this can, to a large extent, be compensated for by a modification to the design of the feed. This problem does not arise to a significant degree with dual-reflector systems.

A transmit/receive dual-polarization feed assembly

An earth station used for one of the major FSS systems has a complex feed assembly, realized in waveguide and carrying out a variety of functions. The principal units of one typical arrangement are shown in block schematic form in Figure 7.17 and an outline description of the role each plays follows. Dual polarization in both directions of transmission is provided. The outputs from the HPAs, to be transmitted to the satellite with orthogonal polarizations, are delivered by waveguide to an orthomode transducer (OMT), which combines the two waves with their electric vectors accurately perpendicular, and passes them to a polarizer. If orthogonal circular polarizations are to be radiated, a quarter-wave polarizer is used; this converts the two linearly polarized signals into the required circularly polarized state. If orthogonal linear polarizations are to be radiated, no conversion of the state of polarization is required but a half-wave polarizer is used to enable the angle of polarization to be adjusted to match the state of the polarizer of the satellite receiving antenna. The signal to be radiated

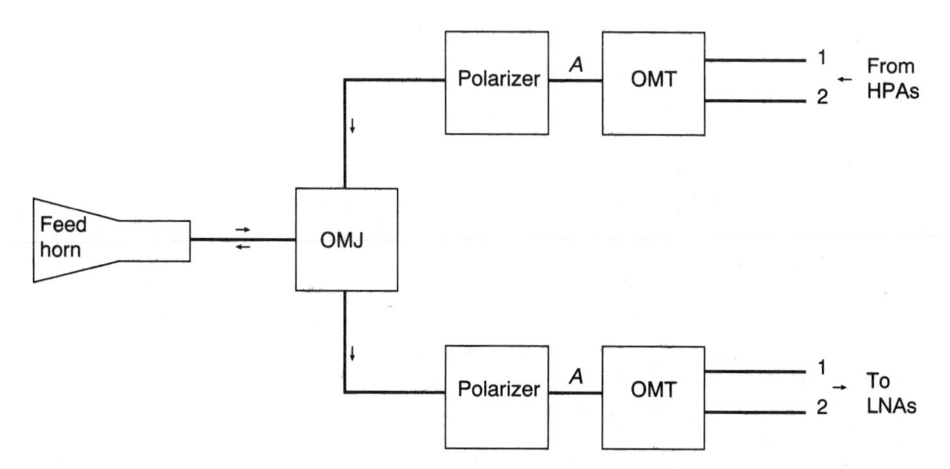

Figure 7.17 *A feed assembly for a dual-polarized, transmit/receive earth station antenna. Half-wave polarizers might be inserted at points A and A for adaptive compensation of differential phase shift due to rain in the radio path*

then passes via an orthomode junction (OMJ) to the feed horn, which delivers the required pattern of illumination to the antenna.

The OMJ acts as a duplexer, passing uplink signals to the feed horn and accepting downlink signals from the feed horn for passing to the LNAs, while ensuring that no significant amount of the uplink signals passes through to the earth station's LNAs. In the receive path, the polarizer and the OMT function in a way that is complementary to those of the transmit path, delivering downlink signals of the two polarization states to the twin LNAs.

The feed horn is a critical component of the antenna system. It must function over a wide frequency range, in both transmit and receive modes, the gain of the horn side-lobes being closely controlled to avoid degradation of the side-lobe performance of the antenna. Above all, it must preserve, over as wide an angle as possible in the vicinity of the axis of the antenna, the polarization characteristics of waves delivered to it by the OMJ for transmission or received from the satellite. The corrugated conical horn has outstandingly good qualities in these respects. Even so, while the axial ratio of a signal radiated along the axis of the antenna may be almost as good as that of the signal delivered from the feed assembly to the horn, the cross-polar discrimination (XPD) is likely to be significantly less in a direction only a small fraction of a degree away from the axis of the antenna.

Brief notes on the functioning of the polarizers and on the effects of Earth–space propagation on the polarization of signals follow. Satellite tracking is an additional function sometimes performed by the feed; see Section 7.3.5.

The quarter-wave polarizer

The electric field of a circularly polarized wave can be resolved into two linearly polarized components, each of peak amplitude E, in quadrature in time and space. If one of the components were shifted in time by $1/4f$ seconds (where f is the frequency in hertz) or by $\frac{1}{2}\pi$ radians, which is the electrical equivalent, the two components would combine to form a linearly polarized wave, its amplitude being equal to $\sqrt{2}E$. Conversely, if a linearly polarized wave were split into two equal parts and one part were shifted by $\frac{1}{2}\pi$ radians in time and space, the wave on recombination would be circularly polarized.

This is the basis of operation of the quarter-wave polarizer. The polarizer typically takes the form of a circular waveguide containing a quarter-wave plate of dielectric material; see Figure 7.18. If a linearly polarized wave, $E \sin \omega t$, enters the polarizer, the angle between the electric vector and the plate being ϕ, then the resolved component of the wave which is perpendicular to the plate, $E \sin \phi \sin \omega t$, is not affected by the plate but the component parallel to the plate is delayed by $\frac{1}{2}\pi$ radians, becoming $E \cos \phi \sin(\omega t - \frac{1}{2}\pi) = E \cos \phi \sin \omega t$. The wave leaving the polarizer is elliptically polarized, and the polarization is circular if the angle ϕ is 45° or 135°; the sense of circular polarization depends on whether ϕ is 45° or 135°. If the input wave consists of a pair of orthogonal linearly polarized components, meeting the quarter-wave plate at 45° and 135°, the outgoing wave will be a pair of orthogonal circularly polarized components.

The process is reciprocal; a pair of orthogonal circularly polarized waves is converted by the polarizer into a pair of orthogonal linearly polarized waves, the orientation of their polarization being determined by the angle of the plate.

The half-wave polarizer

The half-wave polarizer is a similar kind of device, typically using a plate of dielectric material in a circular waveguide to cause a phase delay of π radians to

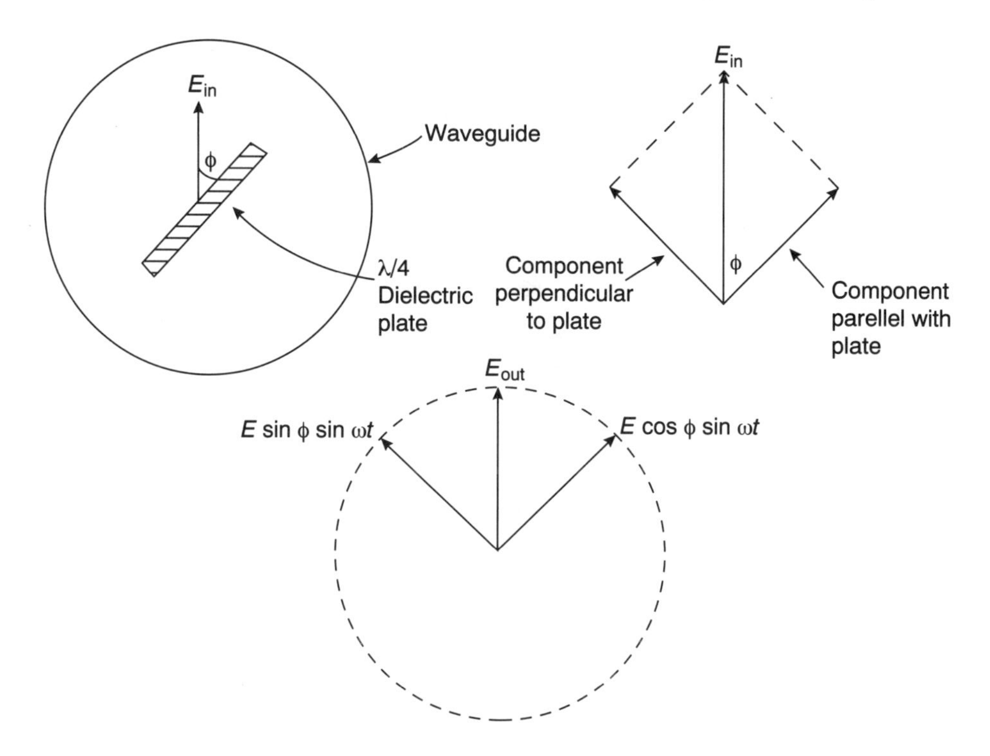

Figure 7.18 *A polarizer using a quarter-wave plate*

the component of the electric vector of a linearly polarized wave which passes parallel to the plate. On recombination with the perpendicular component, which would not be affected by the plate, the wave remains linearly polarized but the orientation of the electric vector is changed by $\pi - 2\phi$ radians, where ϕ is the angle between the electric vector of the incoming wave and the plate; see Figure 7.19. Thus, when the plate is rotated through an angle $\delta\phi$, the plane of polarization is rotated through an angle $2\delta\phi$.

Depolarization and its compensation

As a radio wave is propagated between an earth station and a satellite, its polarization state may be affected, possibly leading to a significant reduction of the cross-polar discrimination (XPD) that can be achieved over links where dual polarization is used. See Section 5.1.5.

At C band, Faraday rotation of the plane of polarization, arising in the ionosphere, causes rapid, unpredictable variations of the orientation of a linearly polarized wave. It is for that reason that circular polarization is used at C band. Faraday rotation is negligible above about 10 GHz.

A radio wave passing through rain or wet snow suffers attenuation and phase delay due to the presence of liquid water in the propagation path. Both effects increase with the rain rate and with frequency. The effect on the polarization state of the wave would not be significant if raindrops were symmetrical in shape. But raindrops are not symmetrical; as they fall through the air their spherical form becomes oblate. Furthermore, this asymmetry is not random; the orientation of raindrops tends to be aligned. For a given angle of elevation of the ray path at the earth station (the 'tilt angle'), the nature of the depolarization is in general

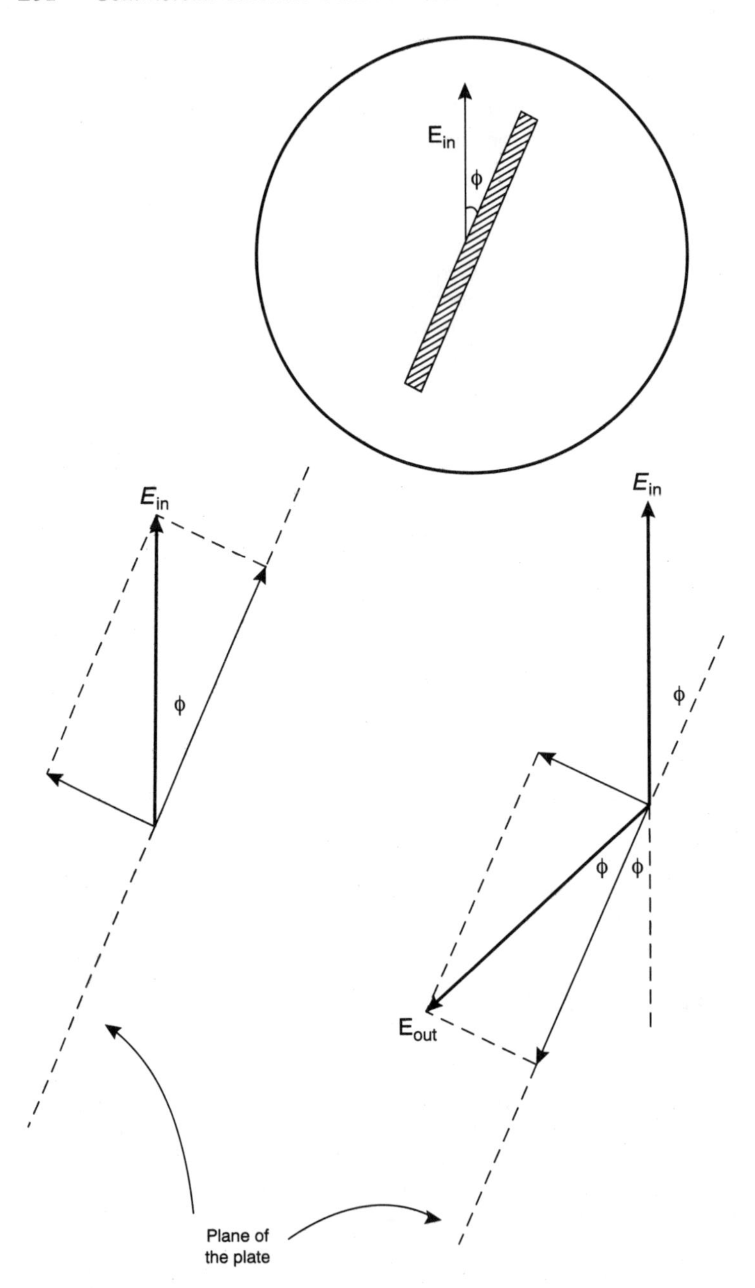

Figure 7.19 *A polarizer using a half-wave plate*

a function of the size of the raindrops and the angle between the minor axis of the raindrops and the vertical (the 'canting angle').

As a result, the magnitude of the attenuation and the phase delay which the wave suffers is a function, among other factors, of the angle of the electric vector in the plane of the wavefront. This differential phase shift and differential attenuation causes circularly polarized waves to become somewhat elliptical in polarization. Linearly polarized waves also become elliptically polarized. This leads to a reduction of XPD in both cases. The extent of this degradation varies from second to second with the amount of rain and the canting angle of raindrops in the path traversed by the wave. The impact of the reduction in XPD that this causes on cross-polar interference is quite severe in heavy rain in Ku band, although the effect is less at C band.

In principle, the loss of XPD due to rain can be considerably reduced by adaptive compensation for the differential phase shift and differential attenuation. However, the precise nature of any depolarization event is peculiar to each earth station where rain is falling, so it is not feasible to carry out compensation at the satellite. Moreover, once a depolarized signal has entered the satellite transponder, it is not possible for an earth station to eliminate the cross-polar interference that had occurred in the uplink. Thus, to eliminate loss of XPD due to rain, each earth station experiencing rainfall must compensate for depolarization that has occurred in the downlink and also precompensate for the depolarization that is going to occur in the uplink.

The depolarization occurring to a downlink carrier can most readily be assessed at the earth station involved by measuring the polarization state of a beacon signal radiated by the satellite at a frequency close to that of the downlink in question. The uplink depolarization can also be deduced, although with considerable uncertainty, by applying a frequency scaling factor to the measurements on the beacon signal transmitted from the satellite.

Compensation for differential phase shift, uplink and downlink, can be implemented by inserting half-wave polarizers in the transmit and receive paths of the feed assembly, at the points marked *A* in Figure 7.17, the phase shift introduced by the polarizers being controlled adaptively by the beacon monitor. The introduction of gain or loss into the receive and transmit chains to implement differential attenuation compensation is, however, technically much more difficult and costly and has not yet been reduced to commercial practice.

7.3.5 Satellite tracking

Nominally geostationary satellites are not perfectly stationary as seen from an earth station. As indicated in Section 6.5 item 3, the ITU RR require geostationary satellites to be kept within ±0.1° of the longitude of their nominal position. This is generally achieved and station-keeping is often better than these regulations require. There is, however, no corresponding regulatory requirement for limiting the movement of FSS and MSS satellites in the north–south direction which arises from inclination of the plane of the orbit. More stringent regulations apply to the BSS.

There is advantage in limiting the satellite's diurnal north–south excursion to a small fraction of one degree of latitude. Nevertheless, towards the end of the operational life of a satellite, too little fuel may remain available for the thrusters that correct orbital inclination. Depletion of the fuel supply may arise because an unexpectedly high proportion of the initial load had been used to make good an initial orbit that was not fully satisfactory due to imperfect functioning of the launcher. A more usual cause is that a satellite remains, in most respects, fit for use long after the end of its design lifetime. For example, Intelsat V satellites were designed for seven years' life in orbit and were launched with a load of hydrazine sufficient for orbital inclination correction throughout that period. However, the first of the series, launched on 6 December 1980, was fully serviceable seven years later although the supply of hydrazine was severely depleted by that time. Accordingly, a decision was taken in 1988 to discontinue orbital inclination corrections and to use the remainder of the fuel mainly for longitude corrections. As a result, the satellite was still in service for selected purposes in December 1994, 14 years after launch, and with the prospect of operation for three further years, although the inclination of the plane of the orbit had built up considerably [7]. For reasons such as these, it is not unusual for the satellite movement as seen from an earth station to reach several degrees per day.

Equations (7.5) and (7.6) confirm that there will be an unacceptable increase in space–Earth path losses due to satellite movements of this magnitude with even the smallest earth station antennas unless the antenna is steered to track the moving satellite. Indeed, for earth station antennas with the largest value of D/λ in current use (namely Eutelsat TDMA), a pointing error of 0.01° would be unacceptable. Two basic techniques are used with high-gain earth station antennas to provide automatic tracking of satellite movement. These are the monopulse and step-track techniques.

Monopulse tracking

When a satellite beacon signal enters the feed horn from an off-axis direction, high-order waveguide modes are set up which are absent when the source is on-axis. These modes are detected at ports in the horn in modern monopulse systems, enabling the presence and the sense of the pointing error to be determined. Earlier monopulse designs used auxiliary horns, offset in azimuth and elevation, to achieve a similar result. When a pointing error is detected, the antenna drive motors are activated to steer the antenna to the correct pointing angle.

The step-track system

The basic step-track system uses the main beam of the antenna to explore variations with pointing direction of the received power level of the satellite beacon signal. The antenna is displaced by a small angle in an arbitrary direction; if the beacon level rises, it is assumed that there had previously been a pointing error and that this error was reduced by the exploratory displacement of the antenna. The effect of a further displacement in the same sense would then be tried. If, on the other hand, the initial displacement is followed by a reduction in beacon level, then it is assumed that the pointing error, if there had been one, has been made worse by the displacement, and the direction of exploration would be reversed. This exploratory process is carried out in both elevation and azimuth, the tracking system continuing to hunt for a direction from which an exploratory move in any direction causes a reduction in beacon level. The process is often called 'hill-climbing'.

Current practice

Monopulse systems are capable of very precise tracking but they are costly and their ability to track rapid, unforeseen, movements in the direction of arrival of signals, important in the radar systems in which they originated, is not needed in satellite communication. Step-track systems have adequate performance and cost less; they are currently the usual choice for high-gain antennas.

However, basic step-track systems have disadvantages. The antenna may be driven off the correct direction by a fluctuation in beacon signal level caused, for example, by a brief deflection of the primary reflector by a gust of wind, and frequent activation of the steering mechanisms of massive antennas causes severe wear. More advanced step-track systems maintain adequate pointing precision with less frequent explorations of the direction of arrival of the beacon signal, and reduce the risk of responding to false indications of beacon direction, by taking advantage of the highly repetitive nature of diurnal satellite movements. Such systems typically use a basic step-track system to provide periodic corrections to a stored steering programme based on precise knowledge of the parameters of the satellite orbit.

For earth station antennas of relatively low gain it may be sufficient, and much less costly, to draw up a daily programme of steering corrections, based on the satellite's orbital parameters and to drive the steering mechanism from that alone. This is called programme tracking. Very small antennas may not need any form of tracking; see Section 12.2.1.

7.3.6 Earth station antenna testing

Antenna gain

It is essential to be able to measure the on-axis gain of an earth station antenna and important to measure the gain of the inner side-lobes, to verify that the antenna is functioning correctly. G_0 can be measured using a monitoring station within line of sight of the antenna and several kilometres distant, usually by comparison with a standard horn antenna of known gain. A monitoring station like this is called a 'boresight' station. The polar diagram can also be measured, although the measurements made of low-gain side-lobes with a boresight station are unreliable because of the presence of signals reflected from the ground nearby or from buildings. These measurements may be made anywhere within the frequency range over which the antenna is to operate.

However, measurements made at boresight stations are necessarily made with the axis of the antenna approximately horizontal. When the antenna is tilted to the angle of elevation used in normal operation, the geometry of the antenna may change significantly. As an alternative, main lobe and side-lobe gain measurements can be made at another earth station via a satellite, preferably the one with which the earth station antenna is to operate. Such measurements are more dependable, but the signal-to-noise ratio at the cooperating earth station is likely to be too low to allow low-level side-lobes to be measured, and use of the satellite for measurements may be limited to a single frequency.

Antenna figure of merit (G/T)

For many purposes, the performance of an earth station in the receive mode can best be expressed by a figure of merit consisting of the quotient obtained when the maximum gain G_0 of the antenna at the frequency in question, expressed as an arithmetical ratio relative to the gain of an isotropic antenna, is divided by the system noise temperature T_s in kelvin. The figure of merit is usually expressed logarithmically, thus:

$$\text{Figure of merit} = 10 \log G/T_s \tag{7.7}$$

The figure of merit is denoted by G/T and its units are, by convention, expressed as $\text{dB}\,\text{K}^{-1}$.

T_s can be measured by comparing the received noise level with the noise level arising from a noise source of known absolute power level connected to the LNA in place of the antenna. G/T can then be calculated, using the measured value of G_0.

A direct measurement of G/T is also feasible for high-performance antennas using radio noise from a radio star or a planet, provided that the antenna mount allows the main beam to be steered to the appropriate point in the sky. The antenna is directed on to the astronomical noise source and the noise level P_{st}

at the receiver output is measured. Then the antenna is steered away from the noise source and the noise level P_n is measured again. Then

$$G/T = 10 \log \frac{8\pi k(r-1)}{\lambda^2 \Phi(f)} \quad (\text{dB K}^{-1}) \tag{7.8}$$

where $k = 1.38 \times 10^{-23}\,\text{J K}^{-1}$ is Boltzmann's constant
$r = (P_n + P_{st})/P_n$
λ is the wavelength (m)
$\Phi(f)$ is the radiation spectral flux density of the source at frequency f ($\text{W m}^{-2}\,\text{Hz}^{-1}$)

The radio stars commonly used at 4 GHz are Cassiopeia A, Taurus A and Cygnus A. For measurements at 11 GHz and higher frequencies, one of the planets may be preferred since they provide a more powerful signal; Venus is usually chosen. Details of the method and the necessary data are in ITU-R Recommendation 733 [8].

7.4 Earth station receiver input stages

System noise temperature (T_s)

The system noise temperature of a radio receiver is reckoned as an equivalent noise temperature at the antenna output port. It includes noise reaching the demodulator input from sources other than the antenna, chiefly the waveguide connecting the feed to the LNA, the LNA itself and the connection between the LNA and the rest of the earth station receiver. See Figure 7.20. In estimating the total effect of these various noise contributions, it is necessary to make allowance for gains and losses arising between the point of reference (the antenna port) and the source.

The noise power that enters the receiver from the antenna under clear sky conditions, t_a, is made up of noise from extraterrestrial sources, the atmosphere, the environment (the ground, buildings, etc.) and noise due to losses in the antenna and the feed. Below 15 GHz and under clear-sky conditions, noise from

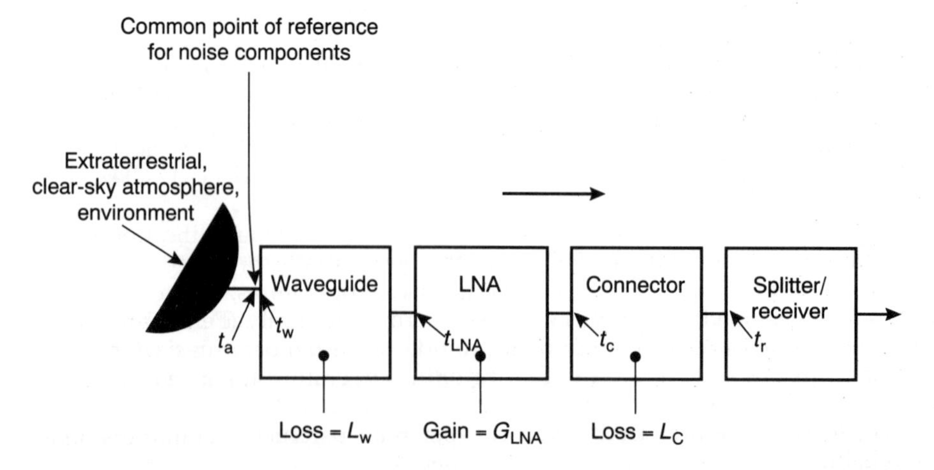

Figure 7.20 *Sources of noise contributing to the system noise temperature of an earth station*

astronomical sources (see Section 5.2.2) and the atmosphere can be disregarded in comparison with noise from the environment (see Section 5.2.1). The latter will depend on the gain of the antenna, its angle of elevation and the effectiveness of side-lobe suppression; it will range from about 20 K for a high-gain FSS antenna operating at a high angle of elevation to over 100 K for a typical LMSS user's antenna of almost zero gain. An addition might need to be made to these figures to cover noise due to losses in the antenna and the feed assembly.

The importance of the noise due to the waveguide between the antenna port and the LNA, t_w, is made clear by the fact that every 0.1 dB of loss in the waveguide, L_w, generates 6.6 K of equivalent noise temperature (assuming an ambient temperature of 290 K) referred to the common point of reference; see equation (5.17). Every effort must be made, in designing a high-performance earth station, to minimize L_w and a loss as small as 0.2 dB, equivalent to 13 K, is achievable.

LNAs with noise temperatures, t_{LNA}, ranging upwards from around 15 K are available, wideband and highly linear if necessary, at costs that range correspondingly. An economic choice between the various options is one of the key decisions in design; it is considered further below. In referring t_{LNA} to the common point, allowance should in principle be made for the loss in the waveguide; for example, if $L_W = 0.2$ dB, a power ratio of $10^{L/10} = 1.05$, the effective value of t_{LNA} is increased to that extent. This is, however, a trivial amount and is disregarded here.

The noise due to loss in the connector, t_c, will depend on the design of the installation; perhaps 6 dB may be taken as typical, leading by equation (5.17) to $t_c = 217$ K. However, in referring t_c to the antenna output port, allowance must be made for the intervening gains and losses. If G_{LNA} equals 30 dB, and disregarding L_W, the power ratio involved = 0.001 and t_c is diminished accordingly, becoming a small part of 1 K at the antenna terminal. Likewise t_r, the noise generated in the receiver, although likely to be of the order of 1000 K, becomes quite small when referred to the common point.

To sum up, given good design,

$$T_s \approx t_a + t_w + t_{LNA} \quad \text{(K)} \tag{7.9}$$

where t_a relates to clear-sky conditions.

Low-noise amplifiers

In the early years of satellite communication, the downlink e.i.r.p. available from FSS satellites was very small and great efforts were made, at great expense, to minimize T_s by minimizing t_{LNA}, using cryogenically cooled masers and parametric amplifiers. Values of T_s around 60 K were achieved. This is no longer essential. However, the situation differs as between 4 GHz and 11–12 GHz, the downlink allocations of main interest.

At 4 GHz in the FSS, the LNA must provide highly linear amplification over an instantaneous bandwidth of 500 MHz; this is currently provided by parametric or GaAs FET amplifiers. The lowest noise temperatures, about 30 K, are obtained from parametric amplifiers, cooled thermoelectrically (Peltier effect) to about −50°C, but GaAs FET amplifiers provide noise temperatures which are little higher, around 40 K, and GaAs FET amplifiers operating at ambient temperature have noise temperatures around 70 K. The solid state option, cooled or uncooled, tends to be preferred, being cheaper than parametric amplifiers and more reliable.

At 11–12 GHz an instantaneous bandwidth of 750 MHz may be required, although 250 MHz or less would be sufficient in many systems. In this frequency range the superiority in terms of noise temperature of cooled parametric amplifiers relative to cooled GaAs FET LNAs is greater, about 90 K compared with 120 K, with ambient temperature GaAs FET amplifiers offering t_{LNA} around 160 K. However, radio propagation factors play a part in choosing between the options in this frequency range. Figure 10.3 shows that, whereas a C band satellite link for international telecommunications will be designed to meet its objectives under clear-sky conditions, a link operating in a higher band must meet its objectives when it is raining. In such rain, t_a will be significantly increased, negating much of the benefit of attaining the lowest feasible value of t_{LNA}, so in this frequency range also the economic choice tends to favour solid state LNAs.

Earth station antennas with low gain do not justify economically the use of costly high-performance LNAs; a higher G/T would usually be obtainable at less cost by using a bigger antenna. Less costly solid state LNAs, with higher noise temperatures and/or narrower bandwidths, are available, suitable for many applications of satellite communications.

7.5 Power amplifiers

7.5.1 Non-linear amplification

Satellite transponders and earth stations use TWTs, klystrons and various kinds of solid state device as amplifiers but none of them has perfectly linear amplitude transfer characteristics. Thus, if a single unmodulated carrier is being amplified in a TWT, the total output power increases as shown by curve A in Figure 7.21 as the drive level is raised, the amplifier becoming saturated at a high level. Furthermore, the level of the component of the output that is at the carrier frequency (curve B) passes through a maximum at the point S, becoming less if the drive level is increased beyond that point; the difference between curve A and curve B for a given drive level is made up of output at harmonics of the carrier frequency, mainly the odd harmonics.

If the input to the amplifier consists, not of a single unmodulated carrier wave, but of several or many at different frequencies, a new situation arises. Components at the input frequencies and their harmonics are present in the output as before but intermodulation products are also generated at frequencies which are equal to the sum of, and the difference between, the frequencies of the various input carriers and their harmonics. The derivation of expressions for the amplitudes and frequencies of the intermodulation products is lengthy (see [9]) but two generalizations can readily be made.

Firstly, the amplitude of each of the first-order output products (that is, the wanted products, at the same frequencies as the input carriers) is a direct function of the amplitude of the corresponding input carrier wave and an inverse function of the amplitudes of all of the other input carrier waves. Thus, if the level of one of the input carriers is increased, the output level of all of the other carriers is reduced. This is called signal compression.

Secondly, with many input carriers there will be a very large number of intermodulation products. However, knowledge of the distribution in frequency of the intermodulation products can be derived from consideration of the results for three input carriers. The intermodulation products generated by input carriers at frequencies f_1, f_2 and f_3, fall into several groups, the most important of which is the group of third-order products at frequencies such as $2f_1 - f_2$ and $f_1 + f_2 - f_3$. These third-order products are important because they are relatively strong and their frequencies are in the vicinity of f_1, f_2 and f_3 (see Figure 7.22), and consequently they are potential sources of interference to the reception of f_1, f_2

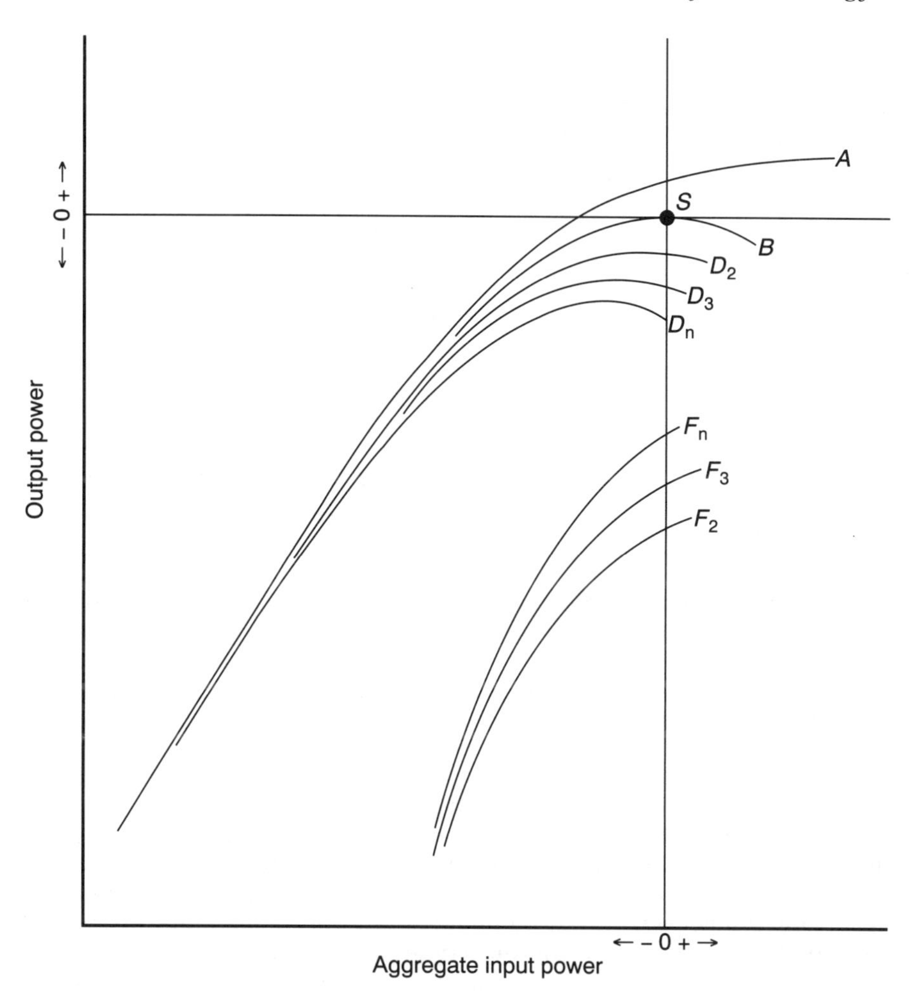

Figure 7.21 *Sketch of the amplitude transfer characteristic of a TWT amplifier. The scales are logarithmic but arbitrary, with their origin at the point S of single carrier saturation. For a key to the labelling, see the text*

and f_3. Fifth-order products at frequencies such as $3f_1 - 2f_2$ and $3f_1 - f_2 - f_3$ are also important since many of them are also in the vicinity of f_1, f_2 and f_3 but their levels are lower than those of the third-order products. Still higher-order distortion products are even lower in level and they tend to be more remote in frequency; their presence reduces the power available for the wanted signals but they can often be suppressed by filters and for most purposes they can be disregarded.

The aggregate output power of the wanted products declines somewhat as the number of carriers increases and the maximum output occurs at a lower level of aggregate input power, as shown by curves $D_2, D_3, \ldots, D_n$ in Figure 7.22. Correspondingly, the aggregate power of the in-band intermodulation products rises; see curves $F_2, F_3, \ldots, F_n$.

If the input carriers are modulated, their spectra have sidebands; the various intermodulation products have broader sidebands, increasing the probability that interference will occur. Fortunately, this interference does not retain the characteristics of the modulation of the carriers which produced it; in telephony

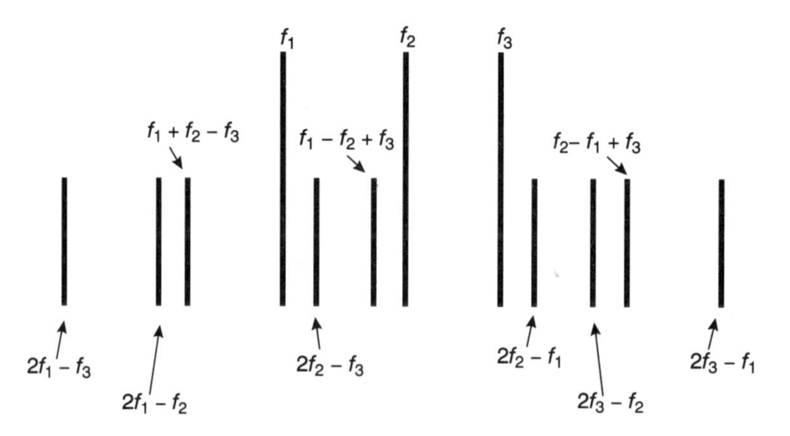

Figure 7.22 *The spectrum of third-order intermodulation products generated by the non-linear amplification of three unmodulated carriers of frequencies f_1, f_2 and f_3*

terms it is unintelligible cross-talk. Intermodulation product power can be treated as noise, although, unlike thermal noise, the noise power density spectrum is not uniform.

In receivers and in the low-power stages of transmitters and transponders, these consequences of amplitude non-linearity raise no major problems; amplifying devices with a sufficiently high output power rating can be used, ensuring that the device can be operated at the low-level linear part of its characteristics. In the output stages of transmitters which are used to amplify a single carrier only, these non-linearities do not raise serious problems, given good filtering of the output signal to suppress harmonics. But earth station and transponder HPAs may be used to amplify more than one carrier at a time. The need for economic solutions to the problems raised by amplifier non-linearities in multi-carrier situations is a basic one for the FSS and the MSS.

Backoff

The output carrier-to-intermodulation-noise ratio, $(C/N)_{\mathrm{im}}$, when n carriers are being amplified, is represented by the distance separating curves D_n and F_n for a given input power level in Figure 7.21. Clearly $(C/N)_{\mathrm{im}}$ increases as the input power level is reduced. The ratio between the carrier level and noise from all the other sources affecting a satellite link, N_{R}, will deteriorate as C is reduced. Nevertheless, there is an optimum value of input power level for an amplifier. If the input power level is raised above that optimum, the reduction of $(C/N)_{\mathrm{im}}$ will exceed the increase of C/N_{R} and the carrier-to-total-noise ratio $(C/N)_{\mathrm{r}}$ will become less. And if the input power level is reduced below the optimum, the increase in C/N_{R} will exceed the reduction in $(C/N)_{\mathrm{im}}$, and $(C/N)_{\mathrm{T}}$ will again deteriorate. See Figure 7.23.

Thus the best operating conditions for a non-linear amplifier are not obtained at the drive level where the wanted carrier output power is greatest but at some lower drive level. An optimum ratio is to be found for the carrier-to-total-noise ratio, leading to a larger potential information throughput or an improved C/N ratio for the end-to-end link. Reduction of the drive level in this way is called 'backoff'. The degree of backing-off which is employed in an operational situation can be defined in terms of the conditions at either the input port or the output port of the amplifier, the datum for either convention being the operating point

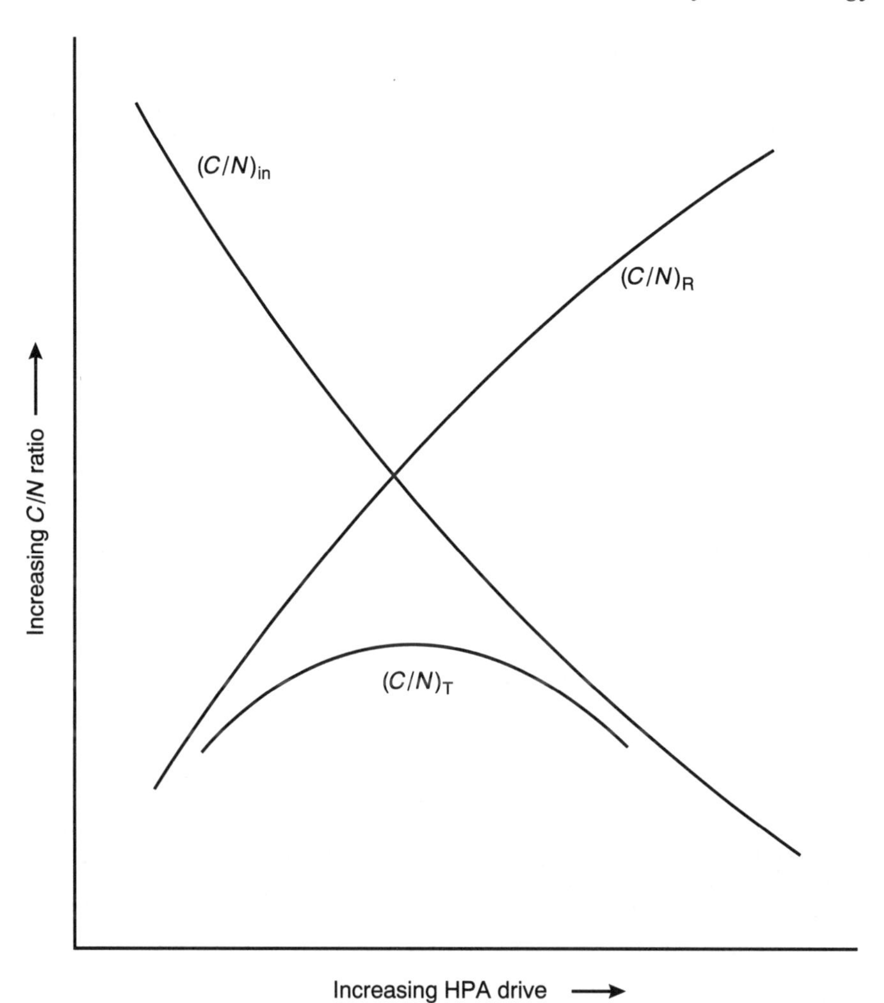

Figure 7.23 *A sketch of the variation with HPA drive level of the ratio of carrier level C to intermodulation noise N_{im}, all other noise N_R and total noise N_T for an end-to-end link*

where maximum output is obtained from the amplifier for a single-input carrier (that is, point S in Figure 7.21). Thus,

$$\text{input backoff} = 10 \log \frac{P_{1i}}{P_{ni}} \quad \text{(dB)} \tag{7.10}$$

$$\text{output backoff} = 10 \log \frac{P_{1o}}{P_{no}} \quad \text{(dB)} \tag{7.11}$$

where P_{1i} = the input power level for a single carrier (watts) which would produce maximum output (P_{1o} watts),
P_{ni} = the aggregate input power level of n carriers producing an aggregate output power level of P_{no} watts,

the power of distortion products in the output being disregarded in both cases.

Strategies for minimizing the effects of intermodulation in satellite transponders used for FDMA can sometimes be applied (see Section 7.5.2) and the amplitude characteristic of amplifiers can sometimes be linearized to some degree by predistortion (see below). However, in the absence of these aids, an input backoff of 6 to 10 dB is typical for FDMA satellite transponders, corresponding to an output backoff of 3 to 4 dB. Larger values of input backoff, perhaps 15 dB, are likely to be required for earth station HPAs amplifying more than one carrier, since although intermodulation is deleterious to the system wherever it occurs, the cost of providing a given reduction of noise through backoff is less at the earth station.

Linearization by predistortion

It is feasible to reduce the intermodulation noise generated in a non-linear amplifier operating in the multi-carrier mode to some useful extent by predistorting the input waveform in a way which is the inverse of the distortion that the amplifier will impose. With predistortion, the degree of output backoff required to ensure an acceptable level of intermodulation noise may be reduced by 2 or 3 dB. However, linearizers of this type may not function effectively outside a rather narrow range of operating levels.

Other aspects of non-linear amplification

Non-linear amplification as described above affects the design and operation of power amplifiers both at earth stations and in satellite transponders. Another phenomenon arising in these amplifiers that must sometimes be taken into account is a variation with the instantaneous drive level of the phase shift that they introduce. This variation of phase shift and other aspects of non-linearity raise different problems for HPAs in transponders and at earth stations and these are reviewed, together with other related issues, in Sections 7.5.2 and 7.5.3 respectively.

7.5.2 Satellite power amplifiers

Strategies for avoiding intermodulation noise

The impairment of performance caused by intermodulation noise generated in transponders operated close to saturation is not always severe. For example, where two wideband emissions, utilizing all or most of the available power are to be amplified, having carrier frequencies f_1 and f_2, the only intermodulation products likely to be significant are centred at $2f_1 - f_2$ and $2f_2 - f_1$; see Figure 7.24. A part of the sidebands of these intermodulation products, having significant spectral energy, may lie within the passband of the transponder but the level of these components is likely to be relatively low. Most of the energy of these intermodulation products lies outside the transponder passband, not affecting reception of the wanted emissions at all. Interference from these out-of-band components may impair reception of emissions from transponders adjacent in frequency but it may be controlled by the filters of the output multiplexer and by differences of gain, polarization or footprint between the various satellite transmitting antennas involved.

Where FDMA involves more than two emissions per transponder there are strategies for reducing the intermodulation noise density in the vicinity of the wanted emissions. For example:

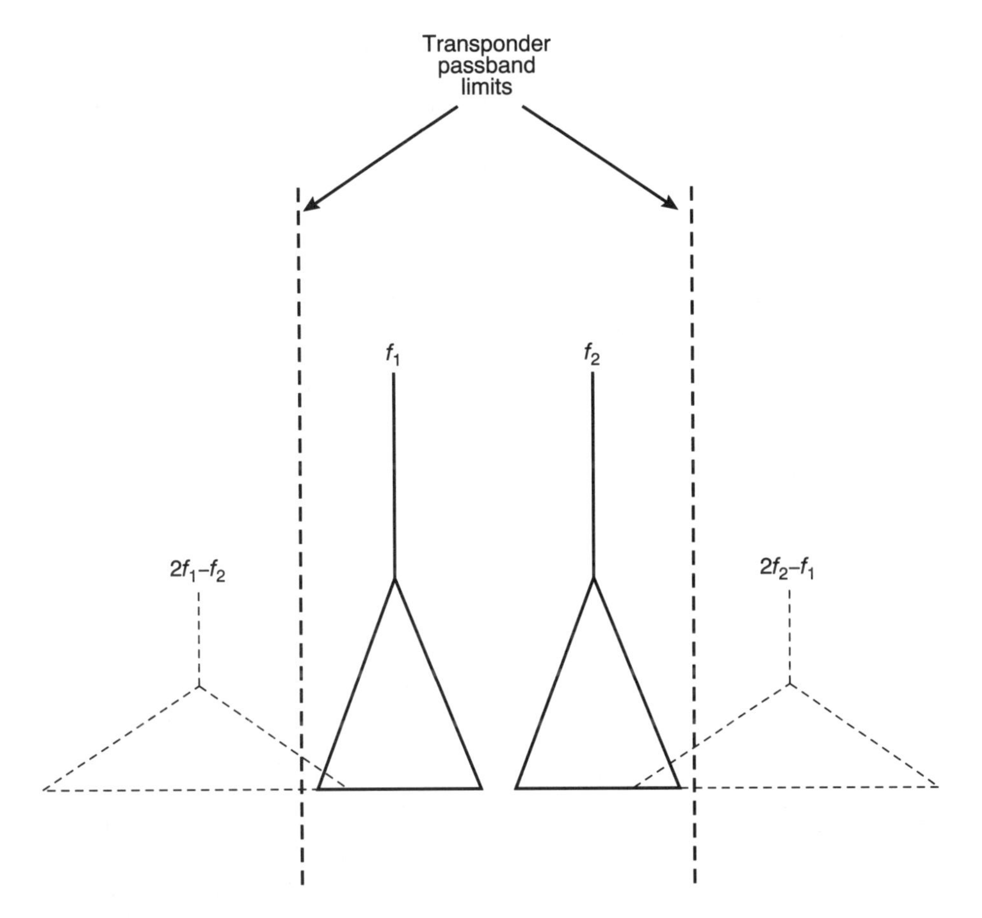

Figure 7.24 *The spectra of two modulated carriers, amplified non-linearly in a transponder, and of their main intermodulation products*

(1) If a transponder of moderate e.i.r.p. rating transmits emissions to earth stations with low G/T, it is likely to be severely 'power-limited', that is, all of the available output power may be required for a group of emissions that occupy only a fraction of the transponder bandwidth. In such circumstances it is often feasible to select carrier frequencies for the emissions in a pattern that ensures that much of the intermodulation noise falls in between the bands occupied by the emissions. See [10–12].

(2) If the bandwidth of an FDMA transponder is wholly occupied by emissions (that is, if the transponder is 'bandwidth-limited') a computer simulation study can be carried out to determine which permutation of the possible carrier frequency assignment plans will give the optimum distribution of intermodulation noise, taking into account the distribution of assignments. That is usually the arrangement that allows the greatest useful information flow to be obtained from the available transponder power and bandwidth.

Phase shift variation with drive level

Most carriers received at satellites are frequency modulated (FM) or phase shift keyed (PSK). Such emissions have relatively constant amplitude on arrival at the satellite, although there may be amplitude variations, caused, for example, by insufficiently equalized post-modulator filtration at the transmitting earth station or by imperfections in the characteristics of transponder input filters. However, when two or more emissions are present simultaneously in a transponder, the aggregate waveform always varies in amplitude; the emissions 'beat' together. The envelope of the aggregate waveform varies in amplitude at the sum and difference frequencies of the various components, with components that include the signal modulation on the carriers, and the peak-to-trough ratio is large in some circumstances.

In transponder HPAs and especially in those using TWTs, the phase shift between the input port and the output port varies with the input power level, particularly when the amplifier is close to saturation. Thus, when a group of emissions is amplified, a phase modulation is imposed upon all of the emissions, the modulating waveform being related to the envelope of the multi-carrier aggregate. This form of signal degradation, called amplitude modulation to phase modulation (AM/PM) conversion, adds to the post-demodulator noise level or the BER at the receiving earth station. AM/PM conversion causes intelligible crosstalk in analogue telephony systems.

Group delay distortion

An ideal band-limiting filter at the input to a transponder would have uniform loss within the passband, no variation of phase delay with frequency and a sudden transition to very high loss outside the passband. Practical transponder filters show small loss ripples in the passband, a roll-off to high loss as the edge of the passband is approached that is fast but not instantaneous and variations of phase delay with frequency, $\delta\phi/\delta f$, that are significant. All the other frequency-sensitive components in the satellite link are also liable to contribute amplitude and phase shift irregularities, although all such components, by design and equalization, are made as free from these irregularities as possible.

The variations of loss with frequency contribute to AM/PM conversion; see above. The variations of phase delay with frequency, $\delta\phi/\delta f$, cause group delay distortion, which adds to the post-demodulator noise level in multi-channel analogue telephone emissions and distorts the waveform of video and digital emissions, causing degradation of picture signals in the former and increasing the risk of bit errors in the latter. The scale of these degradations depends a great deal on the nature and bandwidth of the emission affected and on the way in which $\delta\phi/\delta f$ varies within the band occupied in the transponder by the emission.

Group delay irregularities of transponder filters are minimized by design and by equalization before the satellite is launched, but some irregularities remain. In order to avoid an unacceptable group delay noise level or waveform distortion, it may be necessary to apply further equalization at the transmitting earth station; this is most readily done by inserting equalization networks, made up for the specific application, at the first IF amplifier stage.

7.5.3 Earth station high-power amplifiers

The power per uplink carrier that earth stations must deliver to their antennas varies from a fraction of a watt to several hundreds of watts, and more rarely up to several kilowatts, depending on such factors as antenna gain, emission bandwidth and satellite orbit. Solid state, TWT and klystron amplifiers are used, often with only a single carrier per HPA but sometimes with several carriers amplified by a backed-off high-power stage. At the larger FSS earth stations, the uplink requirements are so great that the output of several HPAs has to be combined.

Transistor HPAs

Transistor amplifiers, typically GaAs FETs, are available with output power ratings up to about 100 W at 6 GHz and up to several tens of watts at 14 GHz and these maximum power ratings are rising as the technology develops. Transistor amplifiers are intrinsically wideband and more reliable, more linear and less costly than either TWTs or klystrons.

TWT and klystron HPAs

In both TWTs and klystrons, used as amplifiers, the input signal modulates the velocity of electrons near the source of an electron beam. These velocity variations evolve into electron density variations in a drift space downstream of the modulation point, and a more powerful replica of the input signal can be extracted from the beam before the beam ends at a collector electrode. Gains in the range 35 to 50 dB are obtained in this way. But the means used for coupling signals into and out of the beam are different for the two devices. The functioning of a TWT is outlined in Section 7.2.1. In a klystron, a chain of resonant cavities, typically three or four, takes the place of the TWT's slow-wave structure, the first and last cavities providing input and output coupling respectively. The cavities may be stagger-tuned to broaden the bandwidth of the device.

With their aperiodic slow-wave structure, TWTs have a wide instantaneous bandwidth. A bandwidth of 500 MHz is obtainable without unacceptable variations of gain or group delay at both 6 and 14 GHz. The cavities of klystrons can be tuned over a wide band, typically 500 or 600 MHz but the instantaneous bandwidth is no more than about 80 MHz, and rather less for tubes with output power ratings in the lower part of the range.

Both TWTs and klystrons are used with output power ratings ranging up to 3 kW, the more powerful tubes being forced-air cooled. Considerably more powerful devices of both kinds are also available, but liquid cooling of the collector electrode is necessary at these very high power ratings, adding substantial complexities and cost to an installation, and 3 kW tends to be treated as the economic limit of power per tube.

The choice between amplifying and combining devices

Transistor amplifiers are preferred where the available output power rating is sufficient. For higher power ratings, while the instantaneous bandwidth advantage of TWT amplifiers is important for some applications, klystrons are preferred in most other respects, being more efficient, more linear, more robust, less costly and having longer life in service. The power supply system requirements for TWTs are particularly complex.

The amplitude transfer characteristics of all of these amplifiers show non-linearity. This leads to the generation of harmonics of the emission when only one carrier is present, especially if the amplifier is operated close to single carrier saturation; see Section 7.5.1. Furthermore, most PSK modulation techniques produce variations in the amplitude of the modulated carrier envelope (see Figure 9.6) and these variations may be aggravated by stringent post-modulator band-limiting filtration. Non-linear amplification and the associated signal compression effects cause the spreading of emission spectra and modulating waveform distortion of digital emissions. For these reasons it may be desirable to operate with a small backoff when amplifying a powerful digital carrier, even in the absence of any other carrier.

All three types of amplifier can be used to amplify more than one carrier, subject to precautions being taken to limit the various distortions that arise. When more than one carrier is present, there will be intermodulation, AM/PM conversion and signal compression. A minimum output backoff of 6 to 8 dB, corresponding to an input backoff of about 13 to 17 dB, is likely to be required for a multi-carrier group of FDM/FM or TDM/PSK emissions. However, some reduction of these backoff requirements can be obtained by the use of linearization by predistortion.

Signal compression takes a particularly severe form when a powerful TDMA emission shares an HPA with other emissions, since the TDMA carrier is switched on and off for each signal burst, perhaps thousands of times per second, and on each occasion the output level of the other emissions rises or falls abruptly. This sharing arrangement should be avoided.

Multi-carrier uplink transmission systems

Some earth stations transmit several or many uplink emissions. There are two basic techniques for providing high-power amplification in these circumstances. The modulated carriers can be combined at the output of the driver amplifiers, then amplified as a multi-carrier aggregate in the HPA; for systems involving substantial numbers of SCPC channels, the carriers may be combined into an aggregate earlier in the transmission chain. Or a separate HPA can be provided for each carrier, the outputs being combined before delivery to the antenna feed port. Or both techniques can be used, one HPA being used perhaps for an aggregate of low power carriers, its output being combined with the higher-powered carriers, each from its own HPA.

Carriers must be combined at the output of the HPA without impairing the matching of sources and loads. Two kinds of device are used for combining carriers without mismatch, namely directional couplers and combinations of filters and circulators. Each has its particular field of application and each has its disadvantages. Directional couplers are aperiodic but lossy; assuming a symmetrical coupler, there is a loss of 3 dB at each coupler and these losses may become large if there has to be a cascade of couplers to combine a number of HPA outputs. Filter/circulator combination involves little loss but the constraints imposed by selectivity may be inconvenient.

In practice, directional couplers are usually preferred for pre-HPA combiners and for combining the outputs of TWT HPAs. The frequency limitations of circulator systems may not be disadvantageous when used for combining the outputs of klystrons, which are themselves band-limited, and their low loss makes circulators preferred in this situation.

References

1. *Radio Regulations* (1994). Geneva: ITU.
2. *Terms and definitions relating to space radiocommunications* (ITU-R Recommendation S.673, 1994), ITU-R Recommendations 1994 S Series. Geneva: ITU.
3. *Terminology relating to the use of space communication techniques for broadcasting* (ITU-R Recommendation BO.566-3, 1994), ITU-R Recommendations 1994 BO Series. Geneva: ITU.
4. *Space Radiocommunications* (1982), Chapter 725 (published separately) of the International Electrotechnical Vocabulary. Geneva: International Electrotechnical Commission.
5. *Radiation diagrams of antennas for earth stations in the fixed-satellite service for use in interference studies and for the determination of a design objective* (CCIR Report 391-6, 1990), Reports of the CCIR, Annex to Volume IV-1. Geneva: ITU.

6. *Performance of small earth station antennas for the fixed-satellite service* (CCIR Report 998-1, 1990), Reports of the CCIR, Annex to Volume IV-1. Geneva: ITU.
7. Intelsat News Release 94–34 of 7 December 1994.
8. *Determination of the G/T ratio for earth stations operating in the fixed-satellite service* (ITU-R Recommendation S.733-1, 1994), ITU-R Recommendations 1994 S Series. Geneva: ITU.
9. Westcott, R. J. (1967). Investigation of multiple f.m./f.d.m. carriers through a satellite t.w.t. near to saturation. *Proc IEE,* **114**(6).
10. Babcock, W. C. (1953). Intermodulation interference in radio systems. *BSTJ,* **XXXII**(1).
11. Fang, R. J. F. and Sandrin, W. A. (1977) Carrier frequency assignment for nonlinear repeaters. *Comsat Tech. Rev.,* **7**(1).
12. *Frequency plans for satellite transmission of single channel per carrier (SCPC) carriers using non-linear transponders in the mobile–satellite service* (ITU-R Recommendation M.1090, 1994), ITU-R Recommendations 1994 M Series Part 5. Geneva: ITU.

8 Analogue transmission

8.1 Frequency modulation

Frequency modulation (FM) is used when analogue signals are to be transmitted by satellite. When a carrier wave is frequency modulated by a signal of constant amplitude at a single frequency, the modulated carrier voltage $M(t)$ can be expressed in the general form

$$M(t) = A_c \cos(\omega_c t + m \cos \omega_m t) \tag{8.1}$$

where A_c = the peak amplitude of the carrier wave,
$\omega_c = 2\pi f_c$ and f_c is the carrier frequency (Hz),
$\omega_m = 2\pi f_m$ and f_m is the modulating frequency (Hz),
m = the peak phase deviation (radians) caused by modulation and is given by

$$m = kA_m / 2\pi \tag{8.2}$$

where k = the modulating signal voltage which produces an instantaneous frequency deviation of 1 rad s^{-1}.
A_m = the peak amplitude of the modulating signal,

Bandwidth occupied by an FM emission

A frequency modulated carrier can be resolved into a carrier vector and sideband vectors. The power of the wave, whether modulated or unmodulated, is constant. Thus, as the depth of modulation rises, increasing the power of the sideband components, the power of the carrier component falls. If the phase deviation m is less than about 0.5 radians at its peak, a fairly close equivalent of the modulated carrier defined by equation (8.1) is obtained by adding two sideband vectors, at $f_c + f_m$ and $f_c - f_m$ respectively, of the appropriate amplitude and phase, to the carrier vector.

It may be noted that a double sideband AM wave also has these spectral components, although with different phase relationships. Whatever processes an FM wave undergoes that disturbs the phase relationship between its spectral components, such as variations of group delay within the passband of a filter, is likely to produce amplitude variations of the wave. Conversely, if the amplitude of an FM wave is made non-uniform, say by the addition of noise or interference, then any subsequent process that affects non-linearly the amplitude of the wave, for example amplitude limitation, adds sideband components to the wave which reflect the amplitude variations that had been present.

An analysis of equation (8.1) that is valid for all values of m involves an infinite series of sideband components above and below f_c, spaced by f_m. The amplitude of the components, which can be evaluated using Bessel functions, falls rapidly with distance from the carrier frequency. The number of sideband components in the vicinity of the carrier becomes very large if the modulating waveform is more complex than a single sinusoid and the calculation of their amplitudes and their distribution in frequency becomes intractable.

In 1939 J. R. Carson developed an empirical rule for calculating the approximate bandwidth B that would contain all the spectral components of an FM emission that were necessary for satisfactory transmission of the signal, namely

$$B \approx 2(f_u + F_p) \quad \text{(Hz)} \tag{8.3a}$$

$$\approx 2f_u(F_p/f_u + 1) \quad \text{(Hz)} \tag{8.3b}$$

where $f_u =$ the highest modulating frequency (Hz),
$F_p =$ the peak frequency deviation (Hz).

Here F_p/f_u is called the deviation ratio. The results given by Carson's rule are sufficiently accurate for most practical purposes over a wide range of values of m and for a variety of types of modulating signal. The rule is widely used.

Noise in FM systems

The block schematic diagram at Figure 8.1 shows the essential elements of FM demodulation. The filters in the final IF amplifier determine the bandwidth B_{IF} of noise that reaches the limiter, along with the emission, and the filter passband must be wide enough to pass all the significant sidebands, typically dimensioned by Carson's rule.

Consider an unmodulated carrier at frequency f_c, having power equal to C W, accompanied by a uniform spectrum of noise with a power density of N_oW Hz^{-1} and band-limited by an ideal IF filter of bandwidth B_{IF} centred on f_c, the total noise power within the passband being small compared with C. Let this noisy carrier, after amplitude limiting, be passed into an ideal FM demodulator. Then at the output of the demodulator, at frequency f in the baseband, the noise power density n_{of} is given by

$$n_{of} = (4\pi^2 N_o f^2)/(k^2 C) \quad (\text{W Hz}^{-1}) \tag{8.4}$$

where k is defined as in equation (8.2). Thus, in the demodulated baseband of an FM system, the noise power density increases with the square of the frequency, or the noise voltage density increases linearly with frequency, the well-known FM triangular noise distribution.

By integrating equation (8.4) between f_l and f_u, the lower and upper frequency limits of the baseband, the total noise power in the baseband is found to be

$$N_{bb} = 4\pi^2 N_o (f_u^3 - f_l^3)/(3k^2 C) \quad \text{(W)} \tag{8.5}$$

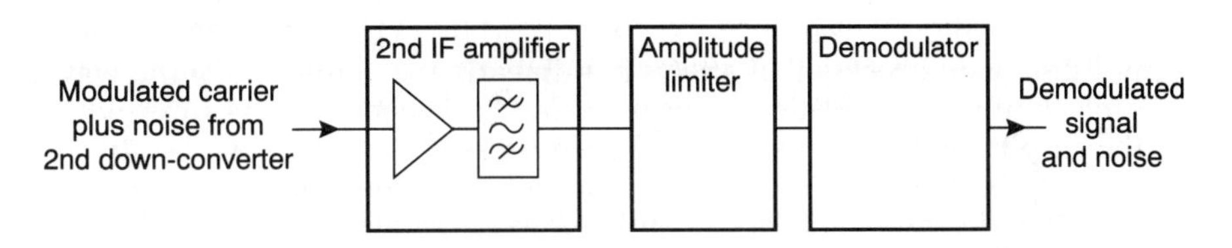

Figure 8.1 *The essential elements for demodulation of a frequency modulated emission*

If $f_u \gg f_l$, equation (8.5) simplifies to

$$N_{bb} = (4\pi^2 N_o f_u^3)/(3k^2 C) \quad \text{(W)} \tag{8.6}$$

Then, if the carrier wave were modulated by a signal of voltage S r.m.s., producing an r.m.s. deviation of f_r Hz, it can be shown that

$$S/N_{bb} = (3Cf_r^2)/(N_o f_u^3) \tag{8.7}$$

But $B_{IF}N_o$ is the total pre-demodulator noise power N_{IF} and $C/(B_{IF}N_o)$ is the pre-demodulator carrier-to-noise ratio. Thus, $C/N_o = (B_{IF}C)/N_{IF}$. Replacing C/N_o by $B_{IF}(C/N_{IF})$ in equation (8.7), rearranging and rewriting in logarithmic notation, we obtain

$$10\log(S/N_{bb}) = 10\log(C/N_{IF}) + 10\log(3B_{IF}f_r^2)/f_u^3 \quad \text{(dB)} \tag{8.8}$$

Thus, for the parameters for which this conclusion holds good (that is, the total pre-demodulator noise power is small compared with the carrier power and $f_u \gg f_l$), the post-demodulator S/N ratio is related to the pre-demodulator C/N ratio by the factor $3B_{IF}(f_r/f_u)^2$. Here f_r/f_u is called the modulation index.

Demodulator threshold

It might seem that S/N could be increased indefinitely for a given carrier power by increasing the modulation index, although at a cost in bandwidth occupied. However, as the deviation increases, and with it the required passband, the total pre-demodulator noise power increases to the point where C/N falls below a threshold value, after which S/N declines rapidly. This demodulator threshold occurs when the peaks of noise voltage begin to exceed the carrier voltage. It is convenient to define the threshold as the value of C/N for which the output S/N is some arbitrary small amount, such as 1 dB, less than it would have been if equation (8.8) still applied; see Figure 8.2. The threshold is at about $C/N = +10$ dB for well-designed conventional demodulators.

Figure 8.2 sketches the relationship between C/N and S/N for a typical FM demodulator and a modulation index of 5. The straight-line section above the threshold is a plot of equation (8.8), the pre-demodulator passband being given by equation (8.3). It is usual to design systems to operate at a point on this curve which is somewhat above the threshold. This provides a threshold margin that allows the link to continue to function without an unacceptable degradation of performance despite small reductions, for whatever reason, in the power of the modulated carrier reaching the receiver. Figure 8.2 shows the operating point with a margin of 6 dB, although smaller margins are usual at C band, where radio propagation conditions are relatively good.

However, demodulator designs are available which extend the range of usable C/N ratios downwards to about +7 dB. One of these is the FM feedback demodulator, in which the passband of the final IF filter is made considerably less than the Carson's rule value, so reducing the noise level at the demodulator. The output of the demodulator frequency modulates the local oscillator which changes the frequency of the carrier to its final IF. The deviation of the wanted carrier is thereby reduced, enabling it to pass through the filter; see [1]. In the dynamic tracking filter demodulator an electronically tunable IF filter, narrower than the passband width given by Carson's rule, is made to track the low frequency components of the modulating signal by means of feedback from the demodulator output. Phase-locked demodulators can also be used to demodulate FM signals at C/N ratios somewhat below the conventional FM demodulator

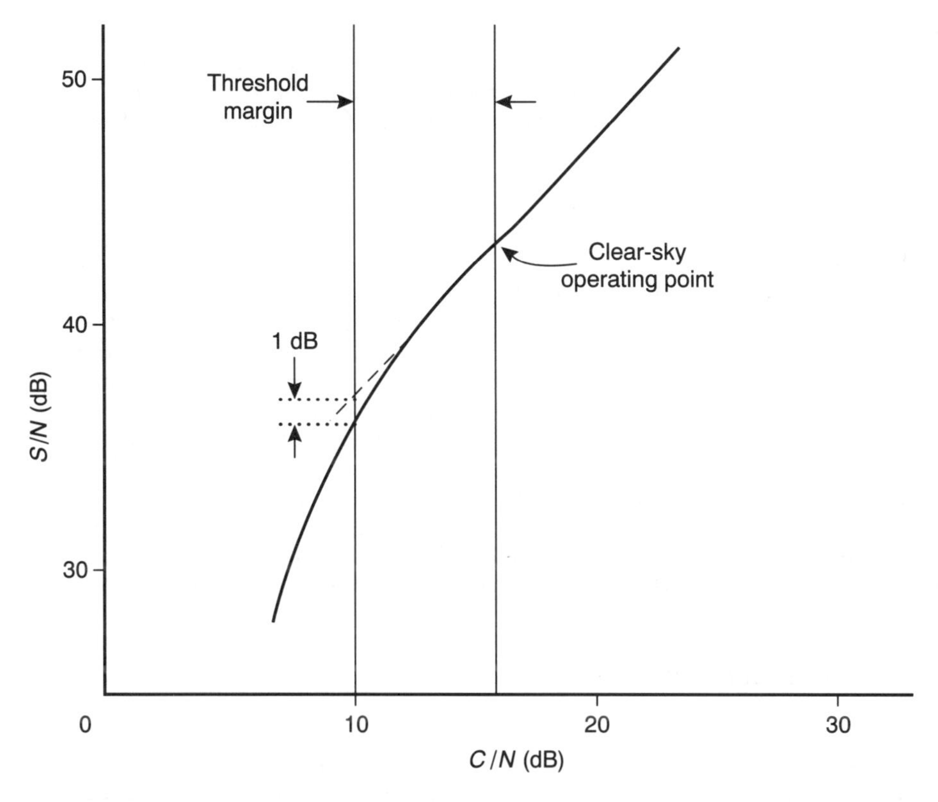

Figure 8.2 *The post-demodulation S/N ratio from a typical FM demodulator for a sinudoidal modulating signal that fills the IF passband, the modulation index being 5*

threshold; see [2]. Extended threshold demodulators are used where the circumstances of a satellite network make the use of minimal carrier power unavoidable.

8.2 FM systems

8.2.1 A single analogue telephone channel

For many links by satellite, it is convenient to use a carrier wave frequency modulated with a single analogue telephone channel; this is called an FM single channel per carrier (SCPC) emission. Consideration of the optimum parameters of such emissions involves knowledge of the level of the telephone signal, the distribution of signal power in the frequency domain, the noise level that users will regard as acceptable and the means that are available for obtaining that channel quality while making efficient use of satellite power and bandwidth.

Telephone signal level

Close to the point of origin (that is, before losses in the telephone network have had a significant effect on the level) the signal power while an average speaker is talking is somewhat less than 1 mW. The mean signal level observed over a sampling period of one minute, and therefore taking into account periods when the local speaker is not speaking but listening to the distant correspondent or considering a reply, may be about 10 dB below 1 mW but on the other hand there may be peaks of speech power substantially exceeding the 1 mW level.

But the loudness of individual speakers' voices differ considerably and any one speaker's loudness varies, for example, with emotion. If the speaker is in a noisy environment, the level of his speech rises. If the sidetone level happens to be high, the speaker will speak more softly. The transmission loss between the point of origin and the point of entry into the four-wire trunk network of the PSTN (within which losses are small and closely controlled) varies considerably from line to line. When allowance is made for all of these factors, a mean signal power of −15 dB relative to 1 mW (that is, −15 dBm0) per channel can be expected within the trunk network at a point of zero relative level while a call is in progress; see CCITT Recommendation G.223 [3]. However, the range of variation of signal levels is large, the standard deviation of values being about 6 dB. Some of these uncertainties apply also to signals that do not have their origin in the PSTN, although the range of variation in level is likely to be considerably smaller.

Despite these wide variations, a basis for system design is required that will combine satisfactory performance for almost all users with economical use of satellite power and bandwidth. Typically it is assumed, for an SCPC link that is part of the public network, that signal peaks will exceed 1 mW but by not more than a few decibels.

The audio-frequency spectrum

Speech signals, after transmission through the PSTN, have a limited spectrum. Almost all of the energy is contained between 300 and 3400 Hz. Little of the acoustic energy in the human voice outside those limits is converted to an electrical signal by the microphone, and that small amount is usually discarded by filters or codecs in the network. Figure 8.3 sketches the spectral distribution of speech signal power for a typical PSTN call and the conventional symbolic representation of the frequency band occupied by an analogue speech channel.

Pre-emphasis and de-emphasis

As indicated in equation (8.4), the noise power density in the received channel of an SCPC FM system is a function of frequency squared, noise being most intense at the top of the audio band. The human ear is near to its maximum sensitivity at 3 kHz and the acceptability of noise in a telephone channel is considerably increased by the use of a post-demodulator filter that reduces the level of the high-frequency components of the noise, the consequential reduction of the level of the speech components at high frequency being compensated for by the application of a filter with the reverse characteristic before transmission; see Figure 8.4(a).

Pretransmission signal spectrum distortion, called 'pre-emphasis', at a rate of +6 dB per octave across the whole spectrum from 300 to 3400 Hz, followed by 'de-emphasis' at −6 dB per octave after demodulation would leave the speech signal unchanged and the noise distributed uniformly. In practice it is found better to use a pre-emphasis characteristic which involves rather less relative reduction in the level of speech signals at the lower end of the audio-frequency spectrum; see, for example, Figure 8.4(b). The pre-emphasis and de-emphasis characteristics cross over at 1 kHz; thus the level of the standard test tone, which is also at 1 kHz, is unaffected. An improvement in the S/N ratio, called the pre-emphasis advantage (P_e), of about 6 dB is obtained by emphasis used in this way with an SCPC transmission.

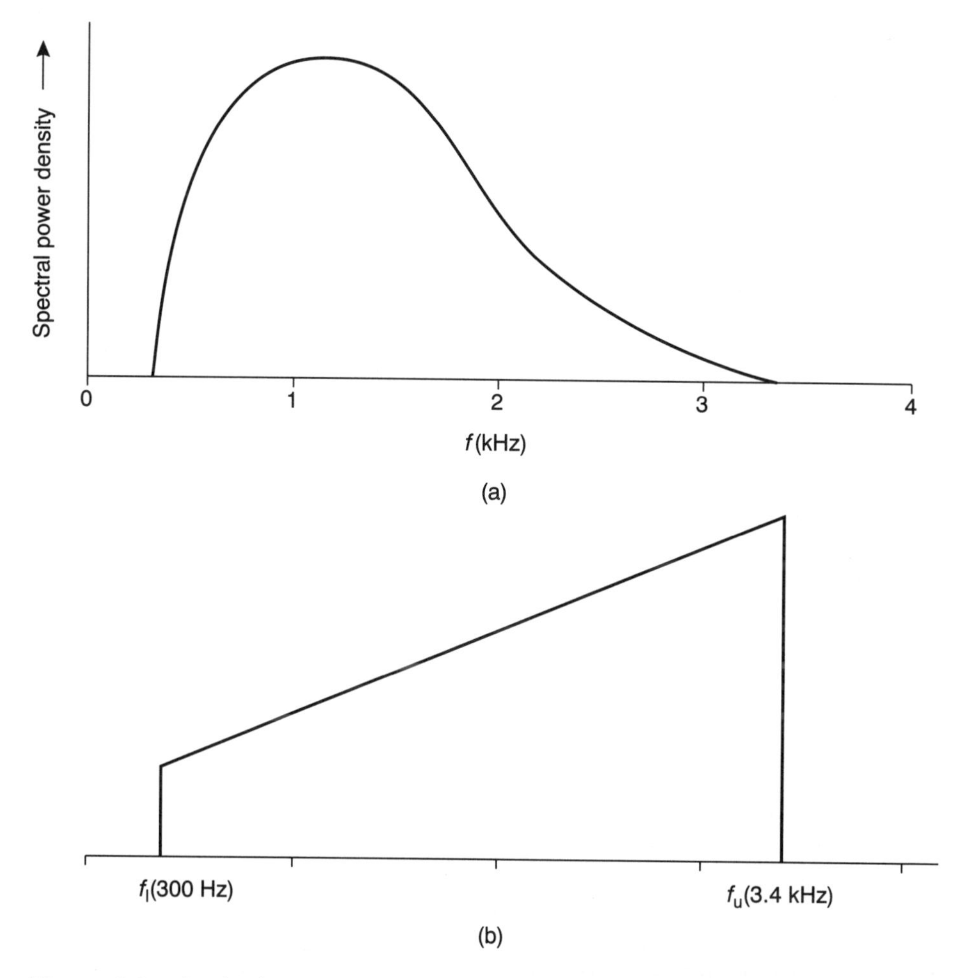

Figure 8.3 *A telephone channel in the PSTN: (a) is a sketch of the spectral distribution of power in a typical speech signal and (b) is the conventional symbol for the channel*

Increasing the level of the higher speech frequency components by pre-emphasis does not cause excessive carrier deviation because the spectral power density of speech at the higher frequencies is low. However, SCPC satellite channels are also used with modems for other kinds of signal, such as facsimile, low-speed data or groups of tones carrying telex channels. Such signals have a spectral power density at the upper end of the frequency band that is much higher than that of a speech signal. Thus channel loading levels and the characteristics of pre-emphasis networks for channels used in this way must be chosen so as to avoid over-deviation of the carrier.

Psophometric weighting of noise

The subjective effect of noise cannot be assessed efficiently by a simple power measurement; in measuring noise levels in telephone channels it is usual to make allowance for differences of the auditory sensitivity of a typical human ear in the various parts of the transmitted spectrum. A standard noise power

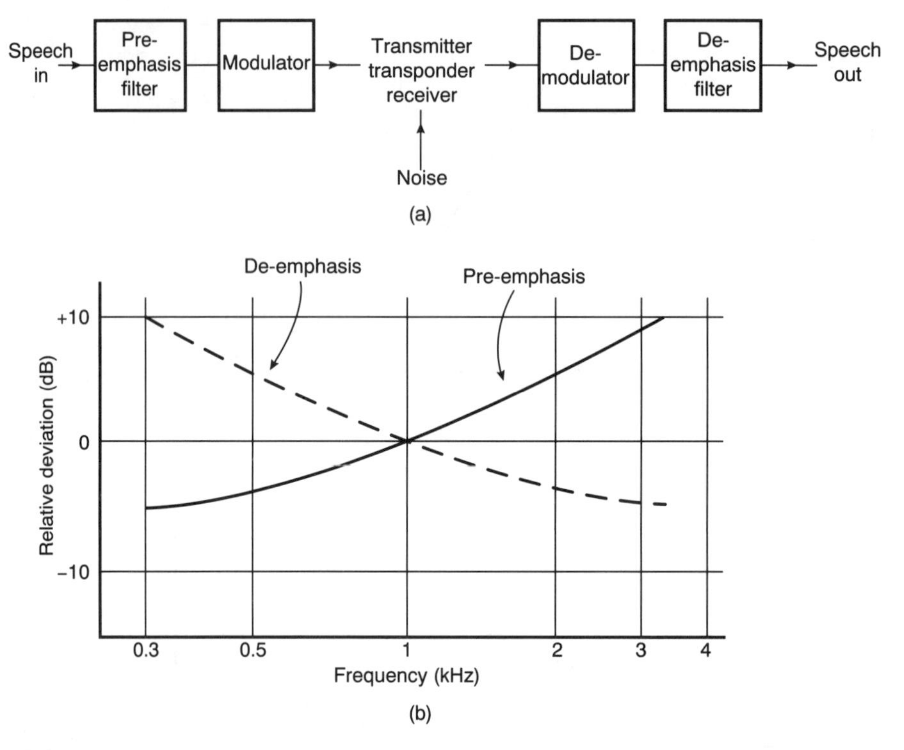

Figure 8.4 *Pre-emphasis and de-emphasis for an SCPC FM emission. (a) shows the block schematic diagram, (b) shows typical characteristics for speech signals*

meter, fitted with a filter, called a weighting network, which reflects these differences is called a psophometer and the modification of measurements of noise level that this network produces is called 'psophometric weighting'. Uniformly distributed noise in a channel limited to 300–3400 Hz, measured through the weighting network recommended by the ITU, has a level 2.5 dB less than an unweighted measurement would give; this is called the psophometric weighting advantage (P_s).

Signal processing: clipping and companding

It is not desirable for signal amplitude peaks to deviate the frequency of the carrier beyond the limits of the assigned frequency band or the IF passband of the receiver. This is because the former may cause interference in adjacent frequency assignments and the latter causes noise bursts in the received signal, called 'truncation noise'. However, the peak-to-mean power ratio of a speech signal is large, as is the standard deviation of the mean levels of different callers. If the drive level at the modulator is set so low that over-deviation by peaks never occurs, the modulation index f_r/f_u is low for most of the time, resulting in a need for a high C/N ratio and a wide IF bandwidth to ensure that the channel quality is acceptable.

The simplest way to enable the mean value of f_r/f_u to be raised without causing over-deviation is to clip the peaks of speech amplitude before the modulator, but severe clipping produces a degree of signal distortion that is unacceptable for many applications. It is usual to clip the highest peaks of speech signals in SCPC systems, but only as a last resort when other measures have ensured that clipping will seldom occur. A simple automatic level control device could produce an improvement in f_r/f_u, not necessarily accompanied by unacceptable distortion, but might cause repeater overloading or loss of circuit stability if the satellite channel were part of the PSTN. A syllabic compandor provides economy in C/N and is compatible with telephone network use.

A compandor consists of a compressor located before the modulator, which reduces the dynamic range of the signal to be transmitted, and an expander located after the demodulator which makes compensating adjustments to the signal; see Figure 8.5. No gain or loss is perceived in the channel if measured between points marked *A* and *B* in the figure, so the stability of the network is not impaired and the amplitude structure of the signal is not affected. However, between points *C* and *D*, the level of speech peaks exceeding the 'unaffected level' is reduced and the mean level, being below the unaffected level, is raised. The S/N ratio is not noticeably affected while speech is in progress, but the noise level is reduced when the speaker is silent, producing a subjective improvement in the S/N ratio. The variation of speech levels as the signal passes through the system is shown in Figure 8.6.

The compander functions as follows. The compressor measures the level of the incoming signal and leaves it unchanged if the signals are at the unaffected level. If the level is x dB above the unaffected level, a loss of kx dB is introduced into the signal path, where k usually equals 0.5. Likewise, if the incoming level falls y dB below the unaffected level, gain is introduced into the signal path equal to ky dB. The response time of these level adjustments is of the order of a few milliseconds, producing compression that operates at the syllabic rate. The reverse process occurs at the expander. See Figure 8.6. The unaffected level is chosen according to the characteristics of the system and the terrestrial network; a value between 0 and −10 dBm0 is usual and 0 dBm0 is used in some SCPC systems. Subjective improvements in the S/N ratio between 10 and 20 dB are obtainable, and few high peaks of speech remain to be clipped after compression.

The specification in CCITT Recommendation G.166 [4] has been drawn up for compandors to be used on high-capacity multiplexed systems but it is also suitable for other applications.

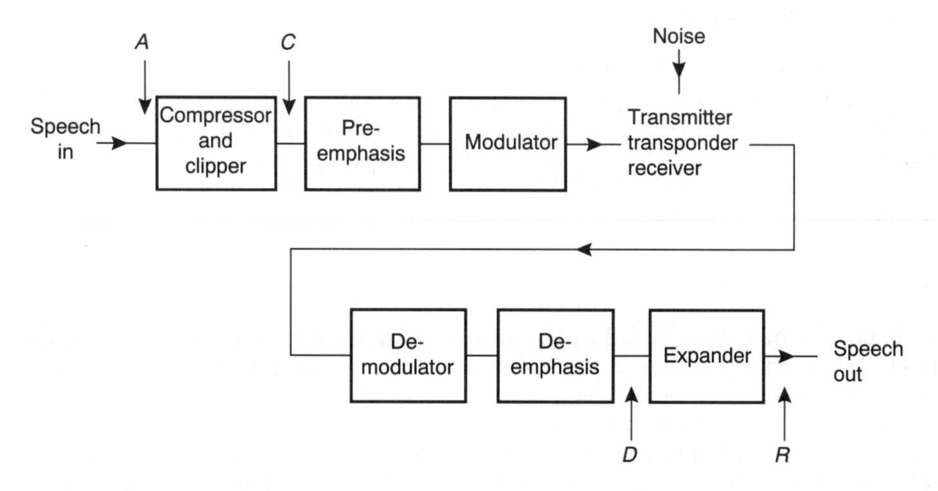

Figure 8.5 *A block schematic diagram of a syllabic compandor system for an analogue telephone channel*

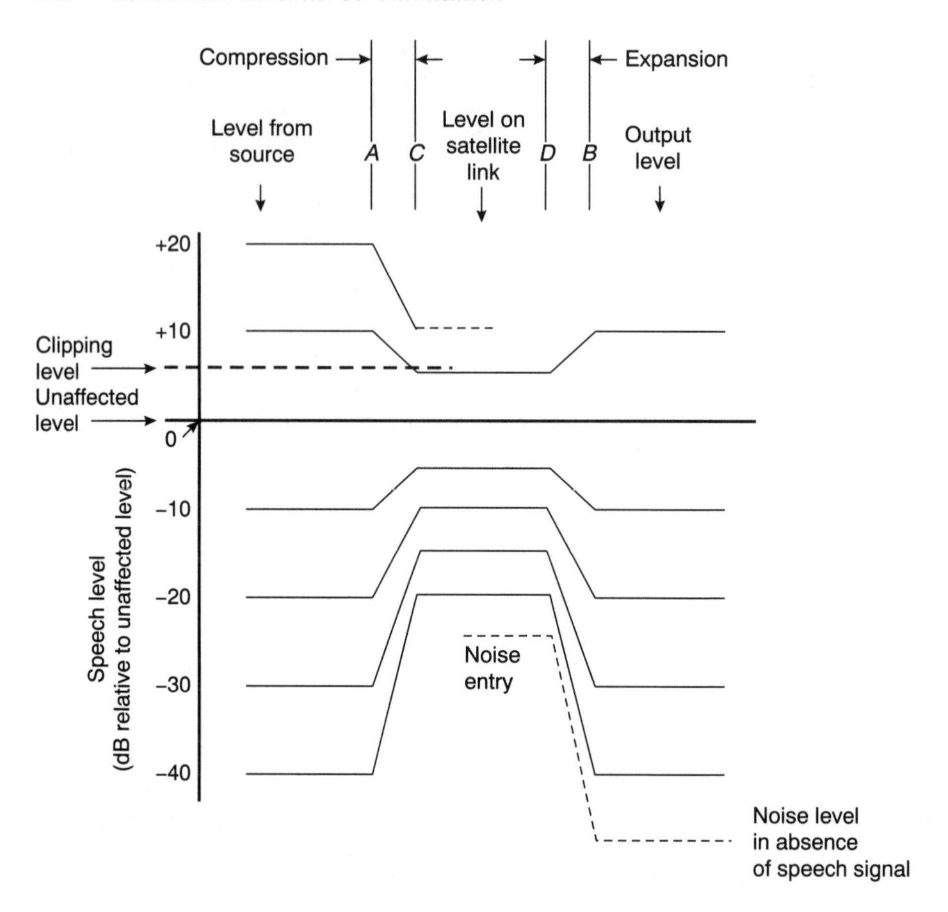

Figure 8.6 *Speech level variations in a companded channel, relative to the 'unaffected level', as a speech signal passes through the system; A, B, C and D relate to points identified in Figure 8.5*

The noise level objective

Section 10.4 shows that the agreed noise objective for international analogue telephone channels in multiplexed systems transmitted by geostationary satellites leads to a budget of around 8000 pW of in-system noise, psophometrically weighted, at a point of zero relative level (= 8000 pW0p). No noise objective has been agreed internationally for SCPC channels but this figure may be taken as indicative of what users would consider desirable.

Carrier parameters

Let it be assumed that $f_u = 3.4\,\text{kHz}$ and that signal peaks, after compression, are clipped if they exceed +3 dBm0. Equations (8.3) and (8.8) enable the down-link post-demodulator noise level, as a function of B_{IF} and C, to be calculated, assuming that B_{IF} is fully occupied when the signal peaks occur. Then if the psophometric weighting, pre-emphasis and companding advantages are 2.5, 6 and 15 dB respectively, Figure 8.7 shows the post-demodulator noise level as a function of relative carrier power for values of B_{IF} between 16 and 28 kHz and pre-demodulator C/N ratios between 11 and 20 dB.

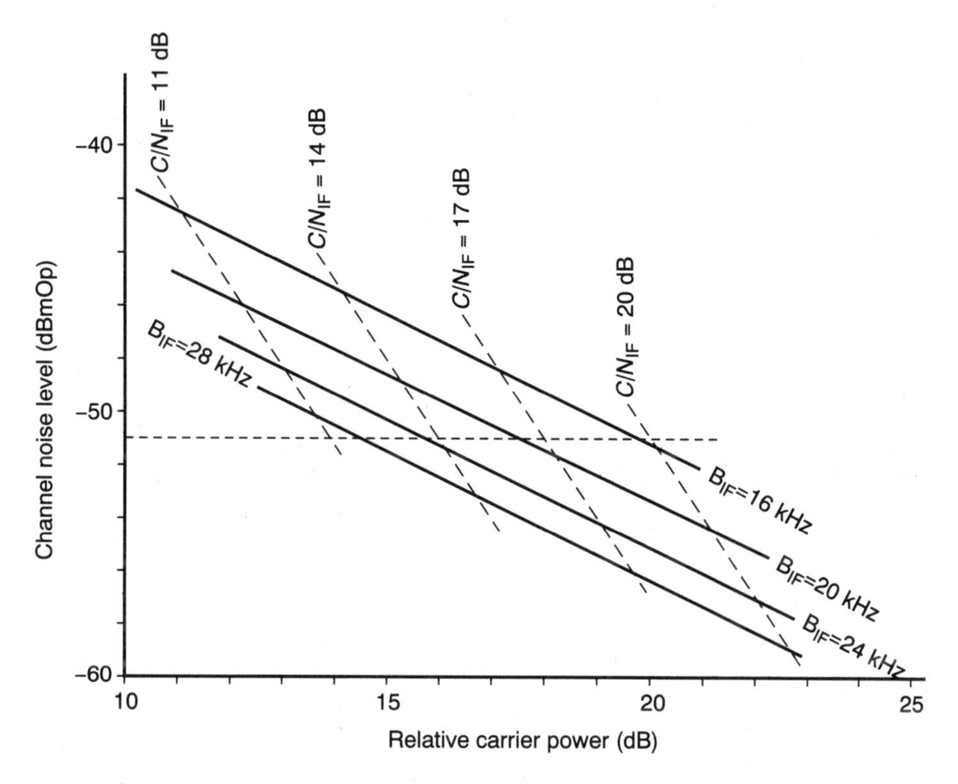

Figure 8.7 *The post-demodulator noise level N_{bb}, as a function of the carrier power used, in SCPC channels carrying speech signals peaking up to +3 dBm0, for received IF bandwidths between 16 and 28 kHz*

If a noise level of 8000 pW0p (= −51 dBm0p) is taken as the performance target, Figure 8.7 shows that this standard could be achieved with a range of sets of carrier parameters. At one extreme the option requiring least downlink power gives $B_{IF} \approx 31$ kHz but it would provide a margin relative to the demodulator threshold (assumed to be at $C/N = 10$ dB) of only 1 dB. Perhaps the other extreme would be represented by the option using B_{IF} equal to only 15 kHz and providing a threshold margin of 10 dB but requiring about 6 dB more carrier power.

The determination of the optimum parameters in any particular situation will clearly be influenced by the need to provide sufficient demodulator threshold margin but other important factors would also have to be taken into account. Do such factors as the gain of satellite and earth station antennas, the power of the satellite transmitter and radio propagation conditions cause the transponder to be power-limited or bandwidth-limited? With power limitation, will power saving techniques such as voice switching of the carrier be used? With bandwidth limitation, are bandwidth saving techniques such as demand assignment of satellite channels to be used? Such methods are considered further in Section 12.3.2 and in Example 5 in Section 10.5; those references are in the context of digital transmission but much is equally relevant to analogue transmission.

8.2.2 Frequency division multiplex using FM

When a satellite link is required to carry several or many channels of analogue telephone traffic, the channels are usually transmitted as an FDM aggregate which frequency modulates the carrier. Satellite FDM/FM systems vary widely in capacity; a large one might have 960 channels in a baseband almost 4 MHz wide and a small one might have 12 channels in a 60 kHz baseband.

The baseband structure

The basic building block of FDM aggregates, used alike in analogue cable systems, radio relay systems and satellite systems, is the group of 12 channels, each limited in bandwidth to 300–3400 Hz and assembled modularly at intervals of 4 kHz between 12 and 60 kHz; see Figure 8.8. There is a pilot tone at 60 kHz which is used for various supervisory purposes, such as automatic level and automatic frequency controls and to raise an alarm in the event of an operational failure.

Links needing larger numbers of channels have a group at 12–60 kHz plus a succession of 12-channel groups or 60-channel supergroups, the latter made up of five 12-channel groups. The first supergroup is located at 60–300 kHz, the second at 312–552 kHz, and so on. The whole aggregate would contain enough complete groups to meet the channel requirement of the link in the least total baseband bandwidth. For more information on FDM baseband structure, see CCITT Recommendations G.341 [5] and G.343 [6].

The baseband signal which modulates the radio carrier consists mainly of this FDM aggregate of channels formed by one or more 12-channel groups. However, one or two more channels are sometimes added to the aggregate between 4 and 12 kHz and used for service communication between the earth stations and between the gateway stations involved. The band below 4 kHz is usually occupied by a triangular waveform for carrier energy dispersal; see below. Figure 8.9 shows a baseband providing 36 channels.

Emphasis in FDM/FM

From equation (8.4), the noise density in the received baseband, after demodulation, increases with frequency. Thus, the highest channel in the baseband is the noisiest. As in the SCPC case, pre-emphasis and de-emphasis can be used to reduce the range of noise levels from the top to the bottom of the baseband, so enabling adequate performance to be obtained in the worst channel with less expenditure of satellite power. The pre-emphasis characteristic recommended by CCIR Recommendation S.464 [7] (see Figure 8.10) and the inverse of the characteristic that is used for de-emphasis enable the noise level in the top channel to be reduced by 4 dB but the top channel remains the noisiest, so its performance is the criterion against which the acceptability of system performance is measured.

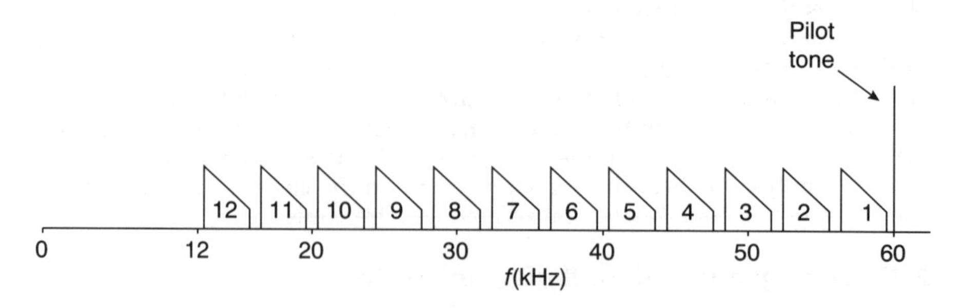

Figure 8.8 *The arrangement of telephone channels in a 12-channel FDM group*

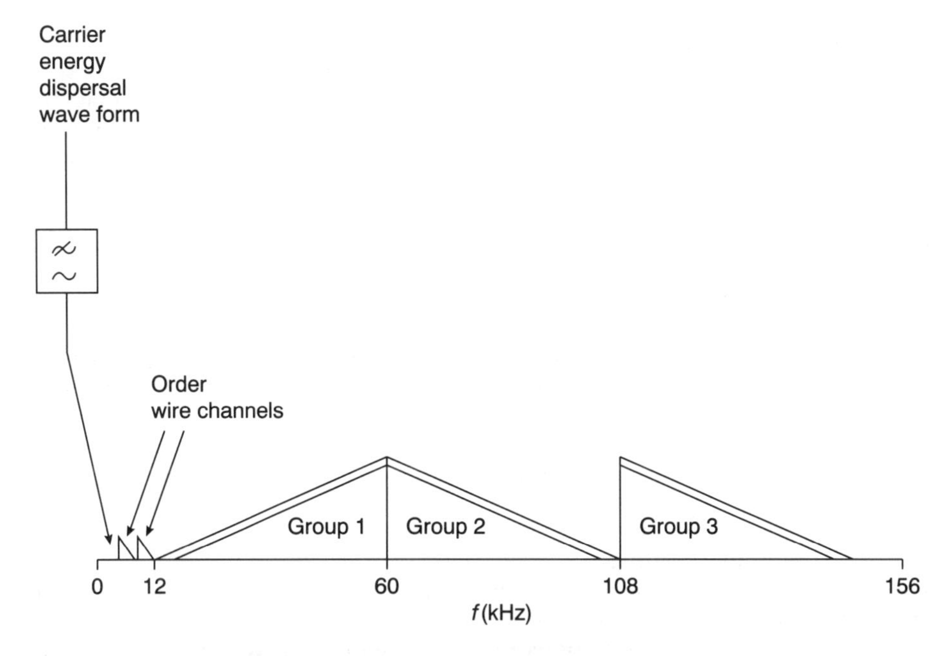

Figure 8.9 *An FDM baseband containing three 12-channel groups, service channels and a carrier energy dispersal waveform*

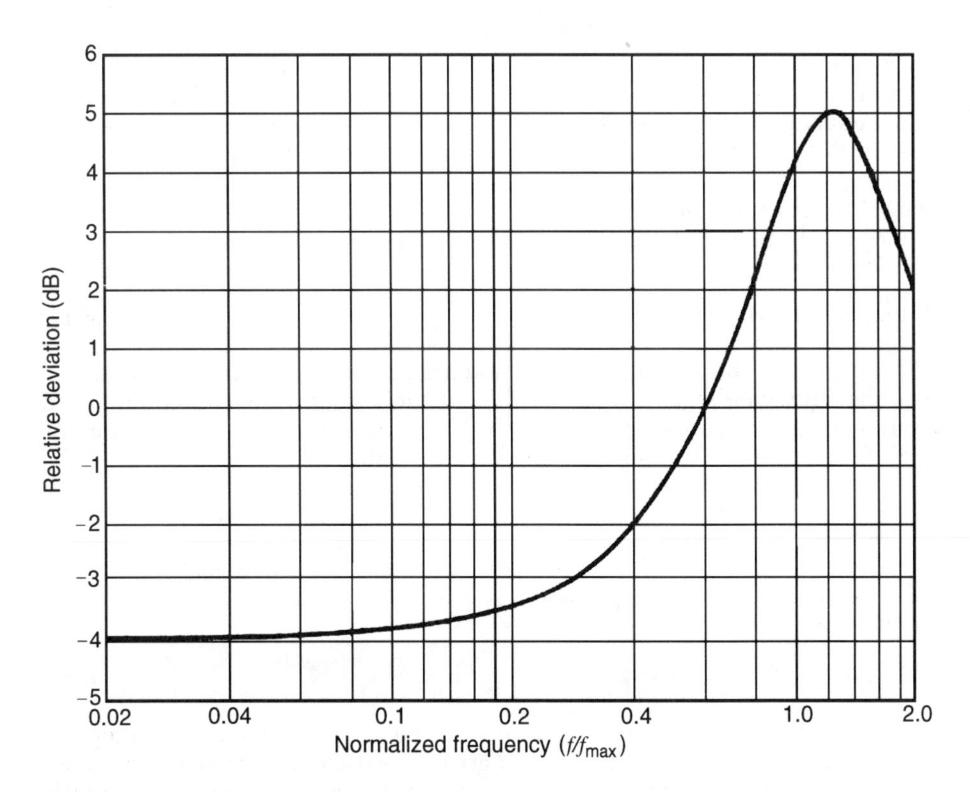

Figure 8.10 *The pre-emphasis characteristic for FDM telephony. (ITU-R Recommendation S.464-2 [7].)*

Signal and noise levels in FDM/FM systems

In any FDM baseband except one with only a very few channels, it may be assumed that the highest modulating frequency f_u is the mid-band frequency of the top channel. Then, the noise level N_{tc} in the top channel is the post-demodulator noise power density at f_u multiplied by the bandwidth b of the top channel, where $b = 3.4 - 0.3 = 3.1$ kHz. Replacing N_o/C by N_{IF}/CB_{IF} in expression (8.4), we get

$$N_{tc} = 4\pi^2 N_{IF} b f_u^2 / k^2 C B_{IF} \quad \text{(W)} \tag{8.9}$$

If a 1 mW test tone were transmitted in the top channel, producing an r.m.s. carrier deviation of f_{tt} Hz, then by a process analogous to the derivation of equation (8.8), we get

$$10 \log \frac{S_{tt}}{N_{tc}} = 10 \log \frac{C}{N_{IF}} + 10 \log \frac{B_{IF} f_{tt}^2}{b f_u^2} + P_s + P_e \quad \text{(dB)} \tag{8.10}$$

where S_{tt} is the power of the test tone (= 1 mW).

The relationships between the amplitude characteristics of the multi-channel signal, the IF bandwidth B_{IF} and the test tone deviation f_{tt} that will allow efficient use to be made of satellite power and bandwidth but without causing the carrier to deviate too often beyond the limits of the IF filter, thereby generating truncation noise, must now be considered.

The relationships between the number of channels, n, and the other parameters are not determinate in nature, but it is found that the following assumptions, derived ultimately from the work of Holbrook and Dixon [8], give acceptable results in most situations. See also CCITT Recommendation G.223 [3]. The multi-channel equivalent signal power due to n telephone channels in FDM, relative to the 1 mW test tone, is assumed to be given by loading factor l, where

$$20 \log l = -15 + 10 \log n \tag{8.11}$$

if $n \geq 240$, and by

$$20 \log l = -1 + 4 \log n \tag{8.12}$$

if $240 > n \geq 12$. It should, however, be noted that these equations may underestimate the aggregate level if many of the channels in an FDM system are used for purposes other than telephony. Also the multi-channel peak factor for the aggregate signal is g. The most commonly used value for this factor is given by

$$20 \log g = 10 \quad \text{(dB)} \tag{8.13}$$

but $20 \log g = 13$ dB has also been used and may be more appropriate for FDM systems with relatively small capacity. Then, applying Carson's rule (equation 8.3),

$$B_{IF} = 2(f_u + g l f_{tt}) \quad \text{(Hz)} \tag{8.14}$$

Using these relationships, equation (8.10) can be used to identify the range of emission parameters that would enable a given worst-channel noise performance objective to be achieved. Figure 8.11 shows how the required carrier power varies with B_{IF} and the pre-demodulator C/N ratio for a top channel noise level of −51 dBm0p. It will be seen for example from the diagram that, for a 72 channel

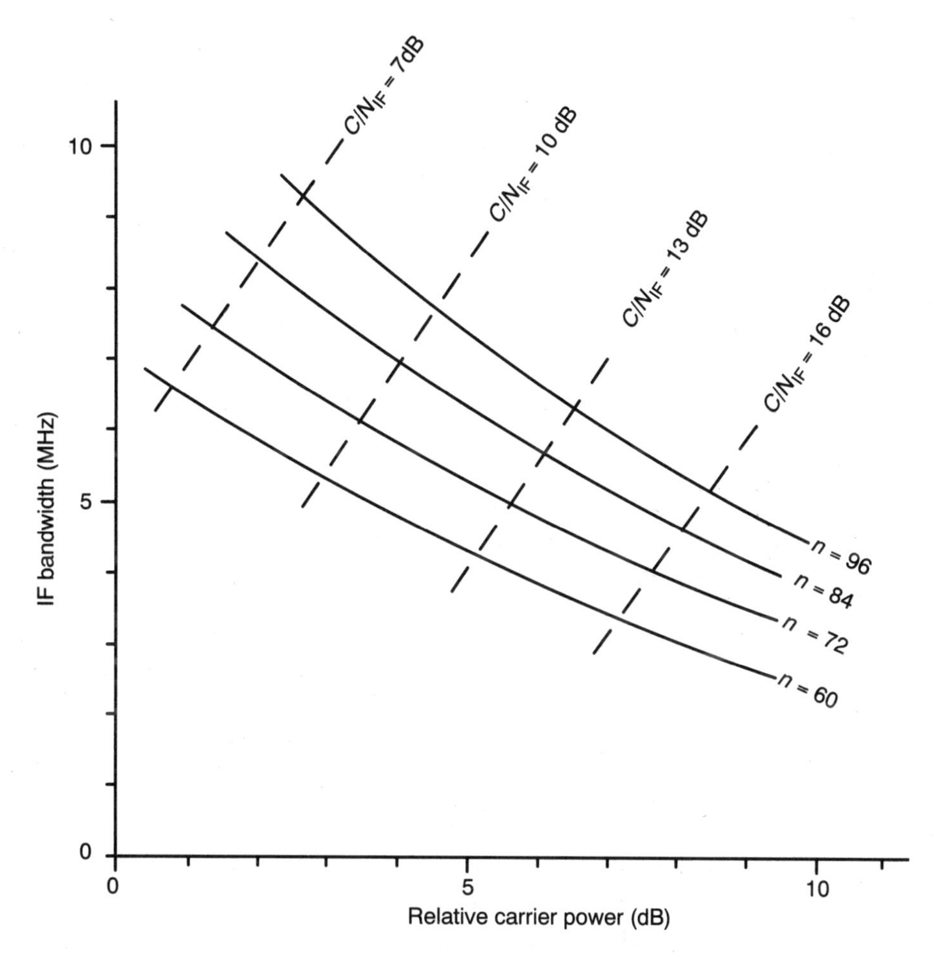

Figure 8.11 *IF bandwidth, carrier power and pre-demodulator C/N ratio for FDM/FM emissions providing a noise level of −51 dBm0p in the top channel of basebands having from 60 to 96 channels*

carrier and a threshold margin of 3 dB with a conventional demodulator, an IF bandwidth of about 5 MHz would be required to achieve that objective.

However, in a bandwidth-limited transponder or for reception with worse radio propagation conditions, it might be preferable to increase the power of this 72-channel carrier by about 2 dB, enabling the same performance to be achieved (in clear-sky) with a 6 dB threshold margin in an IF bandwidth of only about 4 MHz. Further bandwidth economy can be obtained with further expenditure of carrier power, leading towards what is called a 'high-density' carrier. Conversely, in a power-limited transponder the power required for a 72-channel carrier might be reduced by several decibels at the cost of an increase in the IF bandwidth, although the associated reduction in pre-demodulator C/N ratio might make necessary an extended-threshold demodulator.

Carrier energy dispersal with FDM/FM

Some of the energy of an FDM/FM emission is transferred by modulation from the carrier into the sidebands. However, although one or more FDM pilot tones are always present, the mean level of the telephony signals in the baseband is never high and may be negligible outside the daily period of heavy traffic. Unless measures are taken to prevent it, most of the energy of the emission is concentrated for much of the time within a small bandwidth around the carrier frequency. This would often break international agreements limiting the downlink spectral PFD at the Earth's surface; see Section 6.4.

ITU-R Recommendation S.446 [9] reviews the methods available for keeping downlink PFDs within the regulatory constraints. For FDM/FM emissions the simplest method, and the one in general use, involves adding to each FDM baseband at the transmitting earth station a triangular waveform having a period typically between five and twenty milliseconds. The amplitude of the waveform is sufficient to produce a carrier deviation, typically of several hundreds of kilohertz peak-to-peak, depending on the power of the carrier. The waveform is filtered out of the demodulated baseband at the receiving earth station. The carrier deviation due to the waveform increases, to some degree, the IF bandwidth required in receiving the emission.

Although the carrier component of the satellite emission may enter radio relay system receivers at its full power level, it is sweeping rapidly in frequency as a result of modulation by the triangular waveform. As a result, its dwell time in any one telephone channel of a FDM system is short, perhaps only 50 microseconds. The transient response of the FDM channel filters is too slow for unacceptable levels of interference energy to build up in the channel.

As is shown in Figure 8.9, the carrier energy dispersal waveform is injected into the baseband below the telephone channels. Interference from the waveform to the lowest channels in the FDM aggregate is prevented by a low-pass filter. This filter suppresses the higher-frequency components of the waveform, and thereby rounds off its angular peaks; see Figure 8.12. Consequently, whereas an ideal triangular waveform would produce a uniform spread of the energy of the carrier component, the rounded-off waveform allows small peaks to form at a higher energy level at the upper and lower limits of the frequency deviation that the waveform causes. The height of these peaks above the plateau obtained while the waveform is rising or falling at a constant rate varies with the period of the waveform and the cutoff frequency of the filter; at worst, if the filter cuts

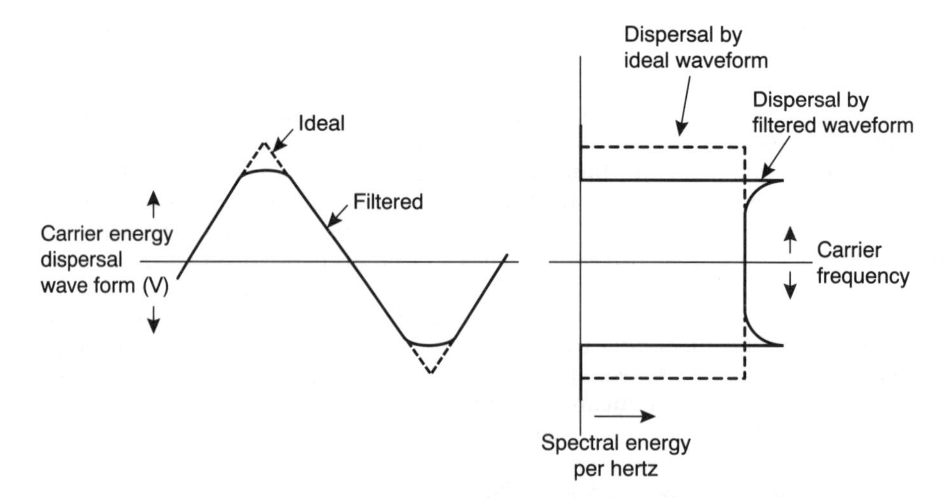

Figure 8.12 *Carrier energy dispersal for an FM emission using a triangular waveform, showing the dispersal produced by an ideal triangular waveform and the effect of waveform filtration*

off below 4 kHz to protect two speech channels below 12 kHz, the height of the peaks above the plateau will typically be between 5 and 10 dB. An allowance can be made for this effect in determining the amplitude of the waveform to be used. However, modulation due to signals and tones in the baseband tends to disperse these peaks.

The minimum deviation due to the waveform that will prevent the PFD of a downlink emission from exceeding the regulatory limit is given by

$$10 \log \tfrac{1}{4} f_{pp} = w_t - S - \Phi_{max} + A \tag{8.15}$$

where f_{pp} is the peak-to-peak deviation due to the waveform (kHz)
w_t is the carrier power of the emission (dBW)
S is the wave spreading factor $= 10 \log 4\pi d^2$ (see equation (5.5)) (dB)
Φ_{max} is the PFD constraint (see Figure 6.7) (dB(W m^{-2} (4kHz)$^{-1}$))
A is an allowance for the peaks due to filtering the waveform, minus an allowance for minimal baseband modulation (dB)

However, there are circumstances in which a higher degree of carrier energy dispersal may be beneficial, because it reduces interference from one satellite system to another, facilitating coordination between systems.

Economy in the use of transponder power and bandwidth with FDM/FM

Most of strategies which are used to save transponder power and bandwidth in SCPC systems are inappropriate with FDM/FM. FDM/FM is intrinsically more efficient than SCPC if the number of channels required by each earth station is sufficiently large, leaving few opportunities for further economy. For example, the proportion of the time in the busy hour when a circuit, one of many on a major route, is in use is much higher than the maximum feasible loading on a route consisting of few circuits. Consequently, demand assignment of channels would give less benefit. Moreover, an FDM multi-channel aggregate can accommodate speech amplitude peaks on its many channels without over-deviating the carrier, because the peaks are unlikely to be simultaneous.

In general, the means that are used to ensure that FDM/FM carriers do not use transponder capacity inefficiently lie more in system management than system design. In particular, the number of carriers amplified by a transponder should be minimized and the number of active channels per carrier should be maximized. Ideally an earth station would transmit only one carrier, multiplexed with the GO legs of all its circuits to all of the other earth stations which the satellite serves. This carrier would be received simultaneously at all earth stations to which any of these GO channels are addressed and each earth station would extract the channels which are proper to it from the aggregate. Each of these distant earth stations would repond with RETURN legs via a multi-destination FDM carrier of its own.

However, as indicated above, carriers with relatively few channels use transponder capacity less efficiently than high-capacity carriers, and channel-by-channel companding is sometimes used, for example, in the Intelsat system to boost the efficiency of low-capacity carriers. See also [10].

8.2.3 Analogue video transmission

This section considers the transmission of analogue video signals from point to point in the FSS. Such transmissions take three typical forms. Some provide brief connections between a place where an event, news, sporting or cultural, is occurring and a location where this topical material can be inserted into a broadcasting programme. Others, with long daily schedules, supply complete programmes to terrestrial broadcasting stations, cable head stations and so on. The third provides video links for non-broadcasting purposes.

At the time of writing almost all terrestrial television broadcasting uses 625/50 or 525/60 picture standards. These systems have 625 or 525 lines per frame, structured as two interlaced fields, and 25 or 30 frames per second respectively. The format waveforms vary in detail from country to country, but these differences are not important in the present context. Analogue luminance signals, which incorporate the synchronization pulses (see Figure 8.13), some of which may be replaced by teletext signals, occupy a bandwidth from about 10 kHz to an upper limit between 4.2 and 6 MHz. A chrominance subcarrier at a frequency between 3.5 and 4.5 MHz is added to the luminance signal, and overlaps with it in frequency, modulated with colour information encoded by one or other of three encoding techniques, PAL, SECAM and NTSC. This baseband signal modulates the video carrier by vestigial sideband (VSB) amplitude modulation; in almost all cases the upper sideband is radiated at full power and the lower sideband is, to a large degree, suppressed.

Sound programme signals are transmitted on separate carriers, adjacent to the video emission and between 4.5 and 6.5 MHz above its carrier. Typically there is one sound carrier, frequency modulated with a monophonic signal, but digital stereophonic sound is sometimes used, or two carriers or more to provide a choice of languages.

Frequency modulation is used when an analogue video signal is to be transmitted via an FSS link. Sound programme signals are sometimes transmitted

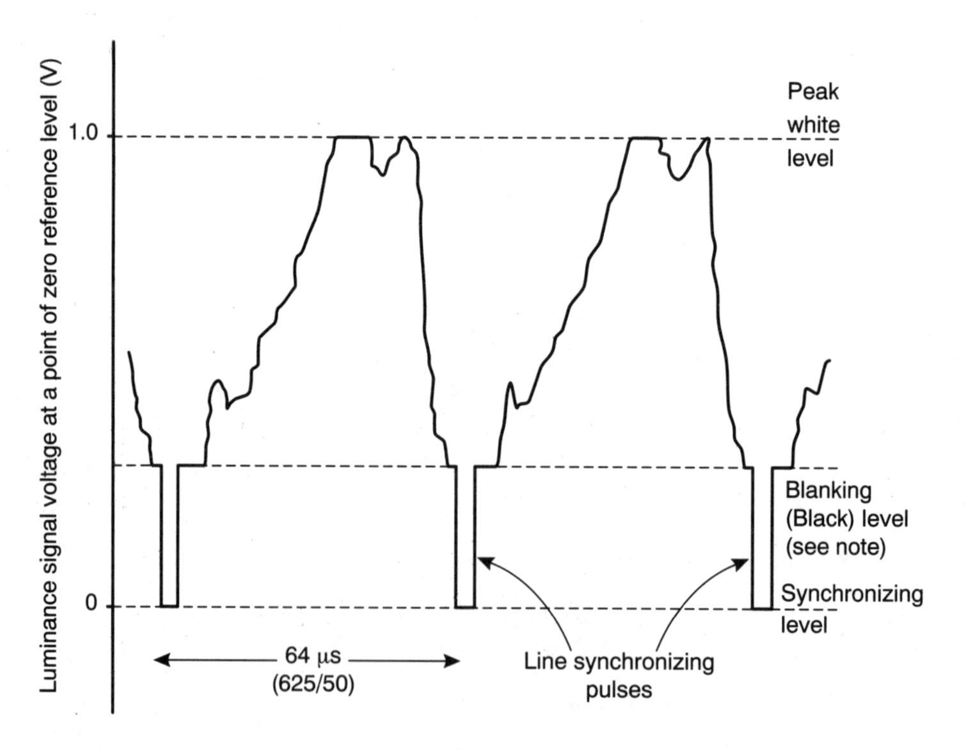

Figure 8.13 *A sketch of the luminance waveform of a standard definition television signal, showing two consecutive lines. Note: the blanking (black) level is at 0.3 V for 625/50 systems and at 0.286 V for 525/60 systems*

via a subcarrier in the same baseband as the video signal, at a frequency above the chrominance carrier. Nevertheless, it is desirable to find some way of transmitting the sound programme signals without using subcarriers which share a transponder with the video carrier, thus making more bandwidth and transponder power available for the FM emission used for the video signals. Often sound programmes are transmitted via a different transponder using the SCPC technique or a wideband 'programme circuit' channel may be used in a multiplexed emission, where it occupies the place of three or four conventional telephone channels. Alternatively a sound programme channel may be PCM-encoded and transmitted in time-division multiplex within the synchronization pulses of the luminance carrier, 20 bits per pulse providing a high-quality 300 kbit/s sound channel; this is the sound-in-sync (SIS) system.

Emphasis characteristics and noise weighting

Noise weighting in video transmission makes allowance for the lessening sensitivity of the human eye to the appearance of noise on the television screen as the frequency of the noise rises. The optimum weighting characteristic varies, depending on the picture standards of the signal to be transmitted, as do the optimum emphasis characteristics. However, the use of optimum characteristics, specific to each kind of TV signal, would raise complexities and provide little benefit in the operation of FSS point to point links. Accordingly, compromise 'unified' characteristics have been agreed for standardized performance calculations and measurements for PAL/SECAM/NTSC video signals in FSS channels.

The unified characteristics are as follows. CCIR Recommendation 568 [11] calls for baseband noise measurements to be limited to the band between 10 kHz and 5.0 MHz. CCIR Recommendation 567 [12] provides a unified noise weighting characteristic; see Figure 8.14. CCIR Recommendation 405 [13] provides two unified pre-emphasis characteristics, one for 625/50 picture standards and one for 525/60 picture standards, see Figure 8.15. The de-emphasis characteristics are the inverse of these pre-emphasis characteristics. The combined advantage of noise weighting and emphasis $P_w + P_e$ total 13.2 and 14.8 dB for 625/50 and 525/60 signals respectively. The performance objective of $S/N = 53$ dB that is quoted in Table 10.1 assumes the application of these unified characteristics.

The signal to noise ratio

It is customary to define the signal-to-noise ratio of a video signal in the form

$$(S/N)' = 20 \log \frac{\text{black/white luminance voltage range}}{\text{r.m.s. noise voltage}} \quad \text{(dB)} \qquad (8.16)$$

The symbol $(S/N)'$ is used here for a signal/noise ratio calculated in accordance with this convention. To incorporate this convention into equation (8.8) it is necessary to replace the r.m.s. frequency deviation (f_r) with an expression for the peak-to-peak deviation of the luminance waveform, excluding the synchronization pulses.

Figure 8.13 shows that, at a point of zero reference level, the luminance waveform measures zero volts at the tip of the synchronizing pulses and 1.0 V at peak white, the blanking or black level being at 0.3 V for 625/50 signals and at 0.286 V for 525/60 signals. Levels in a video link are lined up, end to end, using a test tone with a peak-to-peak amplitude equal to the peak-to-peak amplitude of the luminance signal plus the synchronizing pulses (that is, 1 V at a point of zero reference level).

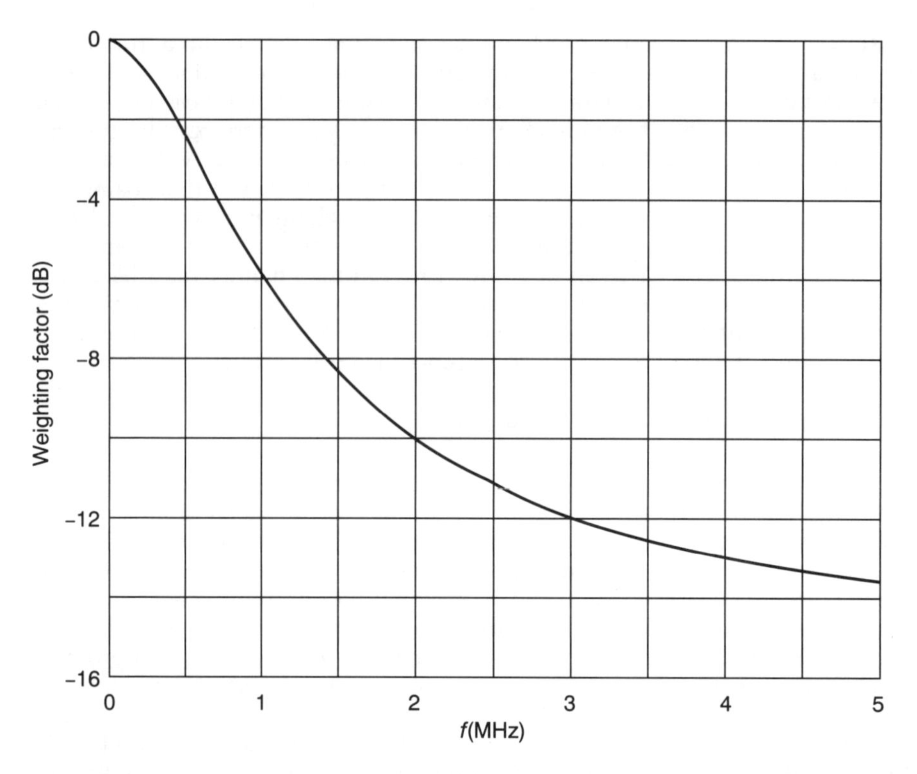

Figure 8.14 *The 'unified' noise weighting characteristic for standard definition FM television. (CCIR Recommendation 567-3 [12].)*

It is convenient to assume for the moment that the frequency of the test tone is at the crossover point of the emphasis characteristics (that is, at about 1.5 MHz for 625/50 systems and about 0.75 MHz for 525/60 systems), although this is not in fact true and it will be necessary to make a compensating adjustment later. With the test tone at the crossover frequency, F_{tt}, the peak-to-peak deviation due to the test tone would be equal to the peak-to-peak deviation due to the luminance signal, including the synchronization pulses. Thus the peak-to-peak deviation of the luminance component, from black to white, would have an amplitude of about 0.7 F_{tt}.

To accommodate the 'unified characteristics' convention it is assumed that the upper frequency limit for the baseband, f_u, is 5 MHz when calculating S/N, although in calculating the bandwidth B_{IF} from Carson's rule it is, of course, necessary to use the actual value of f_u.

Thus, adapting equation (8.8) to the video signal case, assuming that the demodulator is operating above its threshold, applying the conventions and assumptions indicated above, allowing for noise weighting and pre-emphasis advantage and making minor rearrangements, we have

$$10\ \log(S/N_w)' \approx 10 \log C/N_{IF} + 10 \log \frac{3B_{IF}F_{tt}^{z}}{(2 \times 5 \times 10^6)^3} + X + P_w + P_e \quad \text{(dB)} \tag{8.17}$$

where N_w is the weighted noise level and

X is the adjustment required to compensate for the fact that the test tone frequency is 15 kHz (dB).

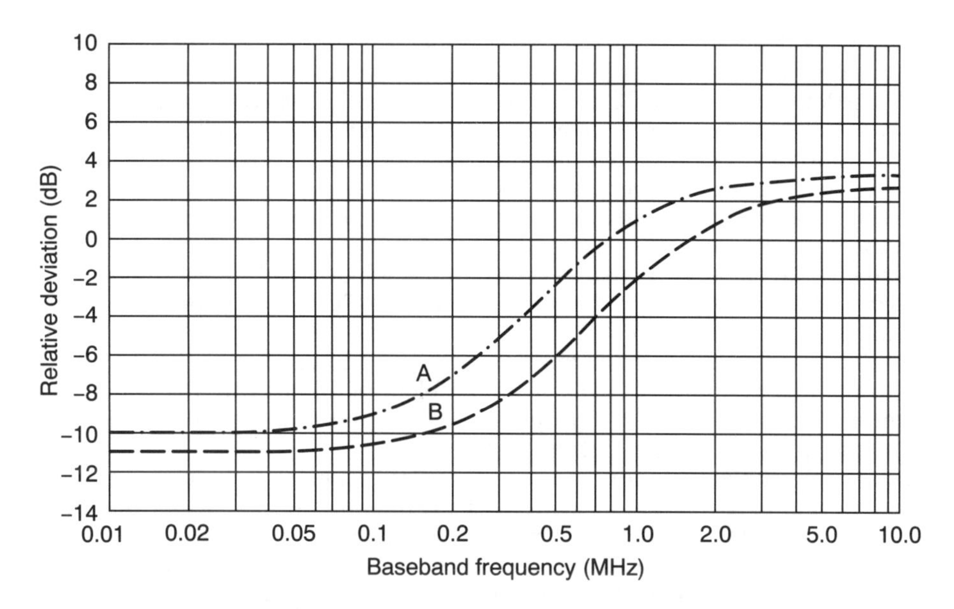

Figure 8.15 *The pre-emphasis characteristics for standard definition FM television. Curves A and B are for 525-line and 625-line systems respectively. (CCIR Recommendation 405-1 [13].)*

The pre-emphasis characterisics (see Figure 8.15) attenuate a 15 kHz tone by 11 dB in 625/50 systems and by 10 dB in 525/60 systems. Thus, the value of F_{tt} used in solving equation (8.17) understates the peak-to-peak deviation corresponding to the luminance waveform by these amounts. Thus the value of X is +11 or +10 dB, depending on the system. After substituting for X, P_w and P_e, equation (8.17) is reduced, for 625/50 systems, to

$$10\ \log(S/N_w)' \approx 10\ \log(C/N_{IF}) - 175 + 10 \log(B_{IF}F_{tt}^2) \quad \text{(dB)} \tag{8.18}$$

0.6 dB should be added for 525/60 systems.

Carrier energy dispersal with FM/TV

For FM/TV satellite connections with downlinks in frequency bands in which the power flux density limits discussed in Section 6.4 apply, it is necessary to use some means to disperse the carrier energy of the powerful emissions that such links require, to protect terrestrial systems from unacceptable interference. A symmetrical triangular waveform having a period equal to the frame period of the television picture (that is, 40 or $33\frac{1}{3}$ ms for 625/50 and 525/60 picture standards respectively) and with its points of inflection synchronized with the beginning of each field causes little degradation of picture quality and is commonly added to the baseband. A saw-tooth waveform with the same period as the field and synchronized with it may also be used. A peak-to-peak deviation of the order of 1 MHz is usually necessary.

A triangular waveform with a period of 40 or $33\frac{1}{3}$ ms can provide the mandatory protection required for telephone channels 3.1 kHz wide in terrestrial radio relay systems. However, carrier energy dispersal affects, not only the level of interference from satellite transmissions to terrestrial systems but also the interference between different satellite networks. Narrowband SCPC emissions being relayed by another satellite are very susceptible to interference from an FM/TV emission, despite carrier energy dispersal of this kind, since the dwell time of the TV carrier within an SCPC IF bandwidth, typically between 20 and 45 kHz, is likely to be of the order of 1 ms, giving sufficient time for the full impact of the interference to develop. It would, in principle, be possible to reduce interference arising in this way to an acceptable level by increasing the orbital separation between the satellites, but at the cost of severely reducing the efficiency with which the geostationary orbit is used.

Alternative carrier energy dispersal techniques which would meet the mandatory requirements but also reduce interference to SCPC emissions are being studied; see, for example, ITU-R Recommendations 446 [9] and 671 [14]. One possibility that has been studied would use a saw-tooth waveform with its period equal to the line period ($\approx 64\,\mu s$) but the problem of eliminating the waveform at domestic receivers without picture degradation and at a cost acceptable in a mass market product has not been solved.

Pending the implementation of carrier energy dispersal techniques that are effective in reducing this kind of inter-system interference, the rising numbers of satellites operating in the GSO make it essential to find ways of ensuring that interference of this kind does not arise. The FSS allotment plan encourages system managers to assign frequencies for high-density carriers, such as FM/TV, in the upper 60 per cent of each frequency band, leaving the lower 40 per cent of bandwidth for low-density carriers, a concept called macrosegmentation; see Section 12.4.3. Macrosegmentation has not been found acceptable for general application, if only because many satellites use much, or all, of their bandwidth for FM/TV. The alternative is for system managers to use the frequency coordination process (Section 6.2) to ensure that spectrum within a few megahertz of the carrier frequencies of FM/TV emissions from other satellites is used for more robust types of emission, or for none at all.

Economy in the use of transponder power and bandwidth for FM/TV

The performance objectives set for satellite analogue video links are demanding. For example, CCIR Recommendation 567 [12] calls for a weighted S/N ratio of 53 dB for international connections for at least 99 per cent of the time, S/N being defined as in expression (8.16). Furthermore, the operational circumstances in which FSS links are used for television signals often involve satellite antennas with wide geographical coverage and low gain and transmitting earth stations with limited e.i.r.p. capabilities. These factors make more difficult the achievement of high performance. It would often be technically desirable to make use of all of the power and most of the bandwidth, perhaps 30 MHz, of one transponder for each television connection, but this is costly and the uplink power required to drive the transponder to maximum output would be too great for some earth stations. There is therefore a need for means of obtaining acceptable signal quality with less power and bandwidth for FM/TV.

An often-used economy technique is called 'over-deviation'. If B_{IF} is made substantially less than the Carson's rule bandwidth (equation (8.3)) which would be appropriate for the peak deviation which is to be employed, some of the spectral energy of the emission will be driven beyond the limits of the receiving IF filters when the frequency deviation caused by the picture signal is at its highest value. This causes noise bursts to occur in the wanted channel. However, the spectrum of the television baseband signal is such that these bursts of noise seldom occur, and within limits they may be tolerated. The power and bandwidth used at earth station and satellite with over-deviation can be considerably less than the full Carson's rule bandwidth would require and the weighted S/N

ratio that is obtainable for most of the time is considerably higher than the same bandwidth and power could provide without over-deviation.

References

1. Enloe, L. H. (1962). Decreasing the threshold in FM by frequency feedback. *Proc. IRE*, **50**, 18–30.
2. Acampora, A. and Newton, A. (1996). Use of phase subtraction to extend the range of a phase-locked demodulator. *RCA Review*, **27**, 577–599.
3. *Assumptions for the calculation of noise on Hypothetical Reference Circuits for telephony* (CCITT Recommendation G.223, 1988), CCITT Blue Book, Fascicle III-2. Geneva: ITU.
4. *Characteristics of syllabic compandors for telephony on high capacity long distance systems* (CCITT Recommendation G.166, 1988), CCITT Blue Book, Fascicle III-2. Geneva: ITU.
5. *1.3 MHz systems on standardized 1.2/4.4 mm coaxial cable pairs* (CCITT Recommendation G.341, 1988), CCITT Blue Book, Fascicle III-2. Geneva: ITU.
6. *4 MHz systems on standardized 1.2/4.4 mm coaxial cable pairs* (CCITT Recommendation G.343, 1988), CCITT Blue Book, Fascicle III-2. Geneva: ITU.
7. *Pre-emphasis characteristics for frequency modulation systems for frequency division multiplex telephony in the fixed-satellite service* (ITU-R Recommendation S.464-2, 1994), ITU-R Recommendations 1994 S Series. Geneva: ITU.
8. Holbrook, B. D. and Dixon, J. T. (1939). Load rating theory for multi-channel amplifiers. *BSTJ*, **18**, 624–644.
9. *Carrier energy dispersal for systems employing angle modulation by analogue signals or digital modulation in the fixed-satellite service* (ITU-R Recommendation S.446-4, 1994), ITU-R Recommendations 1994 S Series. Geneva: ITU.
10. Krishnamurthy Jonnalagadda (1980). Syllabic companding and voice capacity of a transponder. *RCA Review*, **41**, 275–295.
11. Single value of the signal-to-noise ratio for all television systems (CCIR Recommendation 568, 1990), Recommendations of the CCIR 1990, Volume XII. Geneva: ITU. (Soon to be republished as ITU-T Recommendations J.62.)
12. Transmission performance of television circuits designed for use in international connections (CCIR Recommendation 567-3, 1990), Recommendations of the CCIR 1990, Volume XII. Geneva: ITU. (Soon to be republished as ITU-T Recommendation J.61.)
13. *Pre-emphasis characteristics for frequency modulation radio-relay systems for television* (CCIR Recommendation 405-1, 1990), Recommendations of the CCIR 1990, Volume IX. Geneva: ITU.
14. *Necessary protection ratios for narrow-band single-channel-per-carrier transmissions interfered with by analogue television carriers* (ITU-R Recommendation S.671-3, 1994), ITU-R Recommendations 1994 S Series. Geneva: ITU.

9 Digital transmission

9.1 Introduction

It has long been recognized that digital transmission is potentially more efficient than analogue transmission for satellite systems, especially when TDMA is used. The development of experimental TDMA systems began very early; see, for example, [1]. Digital transmission was taken into limited use in the Intelsat system as early as 1972 [2], although not in the form of TDMA. At that time almost all of the PSTN used analogue transmission and electromechanical switching. Where digital transmission was used in the terrestrial network it provided additional telephone channels on existing short junction cables in situations where it would be costly to lay additional cables and where the big bandwidth requirement of digital transmission was of little consequence. Digital satellite links were not seen as an economical means of interconnecting analogue terrestrial networks.

However, by 1980, electromechanical telephone switching was giving way to digital switching. In areas where most telephone exchanges had been re-equipped with digital switches, digital terrestrial transmission networks became economically attractive, and the coming of optical fibre cables greatly reduced the cost of wide bandwidths in the terrestrial network. In addition, most innovative non-speech applications for telecommunications are digital. These developments are making digital transmission predominant in satellite networks also.

A digital signal consists of a sequence of signal elements of equal duration, called 'binary digits' or 'bits'. During a signal element, the signal has one of two states, usually represented by '1' and '0'. Each bit by itself can convey a single piece of information; it may, for example, indicate a choice between two possible responses to a question. More usually a bit is one of a group of bits, called a 'byte' in computing but a 'word' in telecommunications. A word can convey a choice between more than two options; for example, a 5-bit word may select one character from a set of 32 alphanumeric characters and an 8-bit word can identify which voltage level, from a predetermined set of 256 possible levels, best represents the instantaneous voltage of a speech signal in analogue form. A word of N_b bits is required to specify which of M possible options is being chosen, where

$$N_b = \log_2 M \qquad (9.1)$$

A carrier wave can be modulated in various ways to transport digital signals over a radio link. For example, each bit may be indicated by one or other of two different phase or frequency states of the carrier, the difference being recognizable at the receiver. These successive carrier states, each having the same duration as a single bit, are called 'symbols'. The modulation technique may be capable of putting the carrier into, not just two, but four, eight or more recognizable states, so conveying not one, but two, three or more bits per symbol. The number of bits per symbol, N_{bs}, is related to the number of recognizable carrier states per symbol, M_s, available by

$$N_{bs} = \log_2 M_s \qquad (9.2)$$

The reception of digital signals without an unacceptably high proportion of incorrectly recognized bits depends mainly on the availability of an adequate C/N ratio at the receiver. However, there are error correction techniques that may enable a signal of acceptable quality to be recovered from a received signal that has a high BER. The techniques for transporting digital signals over a radio link are reviewed in Section 9.2.

A small but growing proportion of satellite traffic, typically for machine-to-machine communication, is digital in form from end to end but most traffic starts and finishes in the analogue form. Before an analogue signal can be transmitted via a digital link, terrestrial or satellite, it must be converted into the digital form and it must eventually be converted back to the analogue form for delivery to the user. Conventions used, in particular for telephony and video signal transmission by satellite, are reviewed in Section 9.3. And the more important kinds of communication system that are used to enable digital transmission to use satellite communication systems efficiently are discussed in Section 9.4. See [3].

However, this chapter can be no more than an elementary introduction to a complex subject. Comprehensive theoretical and practical treatments of digital transmission by satellite are to be found in textbooks such as [4] and [5].

9.2 Bit stream transport

9.2.1 Phase shift keying

The modulation technique generally used for digital transmission in satellite communication is one or other of the forms of phase shift keying (PSK) and its variants. Quadriphase PSK (QPSK) provides a balance between carrier power requirement and information capacity that is suitable for many current satellite systems, although biphase PSK (BPSK) is sometimes used where the downlink e.i.r.p. that is available is not sufficient to ensure that the transponder will be bandwidth-limited for QPSK carriers. Higher-order PSK modulation systems are also feasible and they provide a higher information capacity for a given bandwidth occupied. For example, eight-phase PSK is sometimes used. However, the relatively high carrier-to-noise ratio that such emissions require for satisfactory reception limits their application.

Coherent PSK

A BPSK modulator takes a carrier wave and passes it to the earth station transmitter with constant amplitude and with its phase either unchanged or shifted by π radians, depending on the state of the digital modulating signal. See Figure 9.1. The two carrier phase conditions can be identified as states A and B.

The demodulator at the receiving earth station recovers a modulation-free carrier wave from the received emission, locked in phase with one or other of the phase states of the received carrier. This may be done, for example, by rectifying the incoming carrier, extracting the second harmonic which exhibits little trace of phase shift, and dividing the frequency of the second harmonic by two. The phase of the recovered carrier is then compared with that of the received emission. Given an adequate C/N ratio, the latter will be found to be approximately in phase or in antiphase with the recovered carrier, thus allowing the modulating signal to be recovered by the detector.

However, there is ambiguity, since the A state of the received carrier may have become interpreted as either the 1 state or the 0 state of the original modulating signal. The ambiguity can be resolved by injecting into the modulating

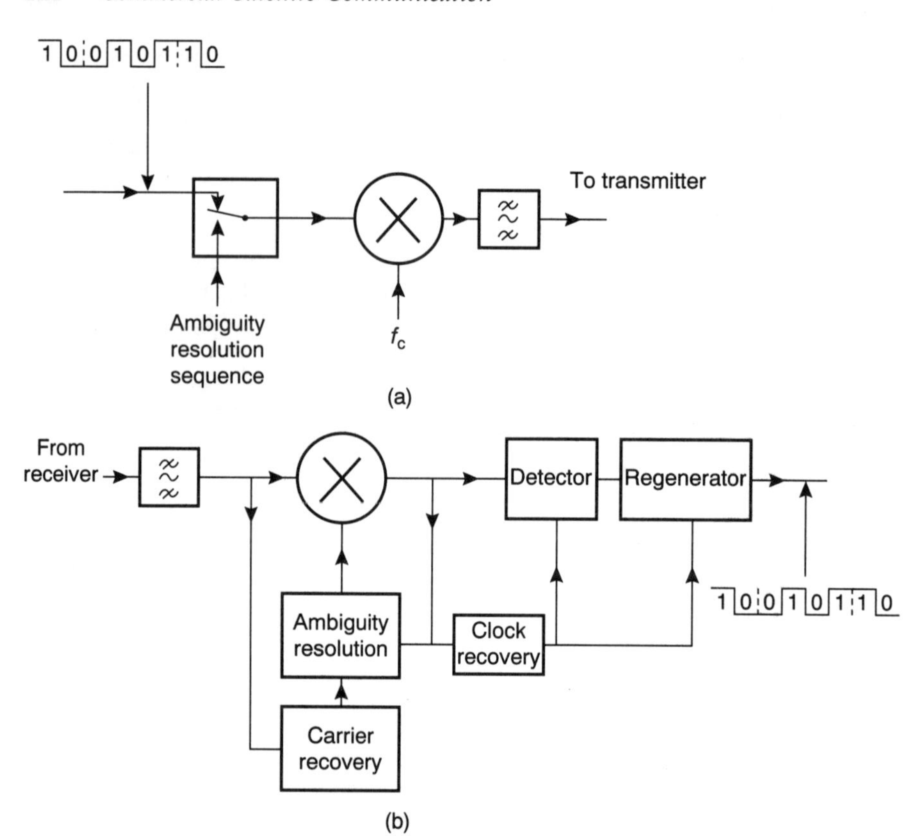

Figure 9.1 *Block schematic diagram for coherent BPSK showing (a) the modulator and (b) the demodulator*

signal at the transmitting earth station, typically in the form of a preamble to a non-continuous traffic burst or part of the synchronizing bits framing a continuous traffic signal, a predetermined sequence of bits of recognizable polarity, the pattern being chosen so that it will not be confused with traffic words. If, when this pattern is recognized at the receiver, it is found to have been inverted in polarity, the recovered carrier is phase-inverted before it is fed to the demodulator. This technique of demodulating with a recovered carrier which is coherent with the original carrier, is called coherent PSK (CPSK).

A QPSK emission is generated and can be coherently demodulated in ways that are analogous to those for coherent BPSK. Two streams of bits, I (the in-phase stream) and Q (the quadrature stream), are fed to the modulator. These two streams may consist of alternate bits of the same digital signal, with their bit period doubled and one stream time-delayed by half of the new bit period to bring it into synchronism with the other stream. Alternatively the input streams may be independent in origin, though synchronized. The four phase states may be identified as A, B, C and D respectively, their phase relationships being as given in Table 9.1.

QPSK modulation is typically achieved by using two BPSK modulators, the I bit stream modulating a carrier in phase with the reference carrier and the Q bit stream modulating a carrier which is in phase quadrature with the reference carrier; the two BPSK carriers are then summed to produce the QPSK carrier; see Figures 9.2 and 9.3.

Table 9.1 *QPSK phase relationships*

Input I	*Input Q*	*Relative phase state of modulated carrier (radians)*
1	1	A
0	1	$B = A - \frac{1}{2}\pi$
0	0	$C = A - \pi$
1	0	$D = A - \frac{3}{2}\pi$

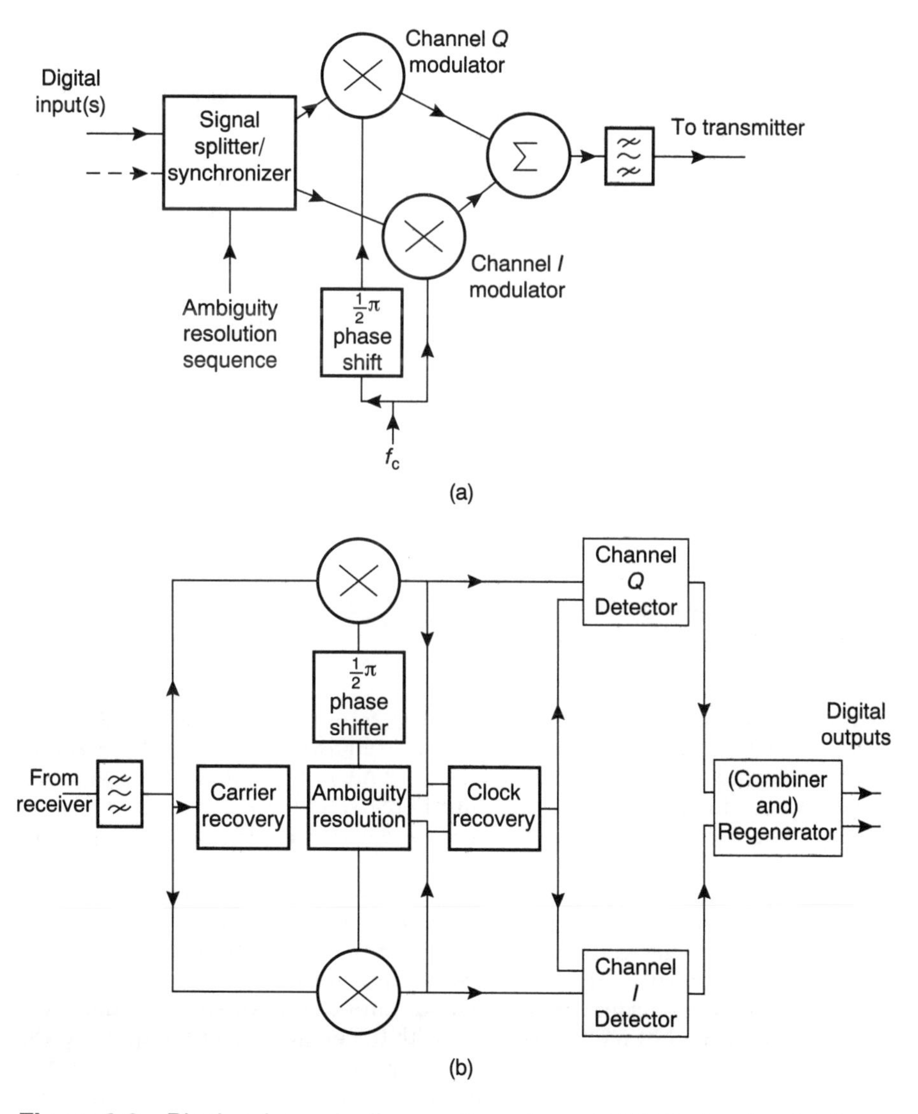

Figure 9.2 *Block schematic diagram for coherent QPSK showing (a) the modulator and (b) the demodulator*

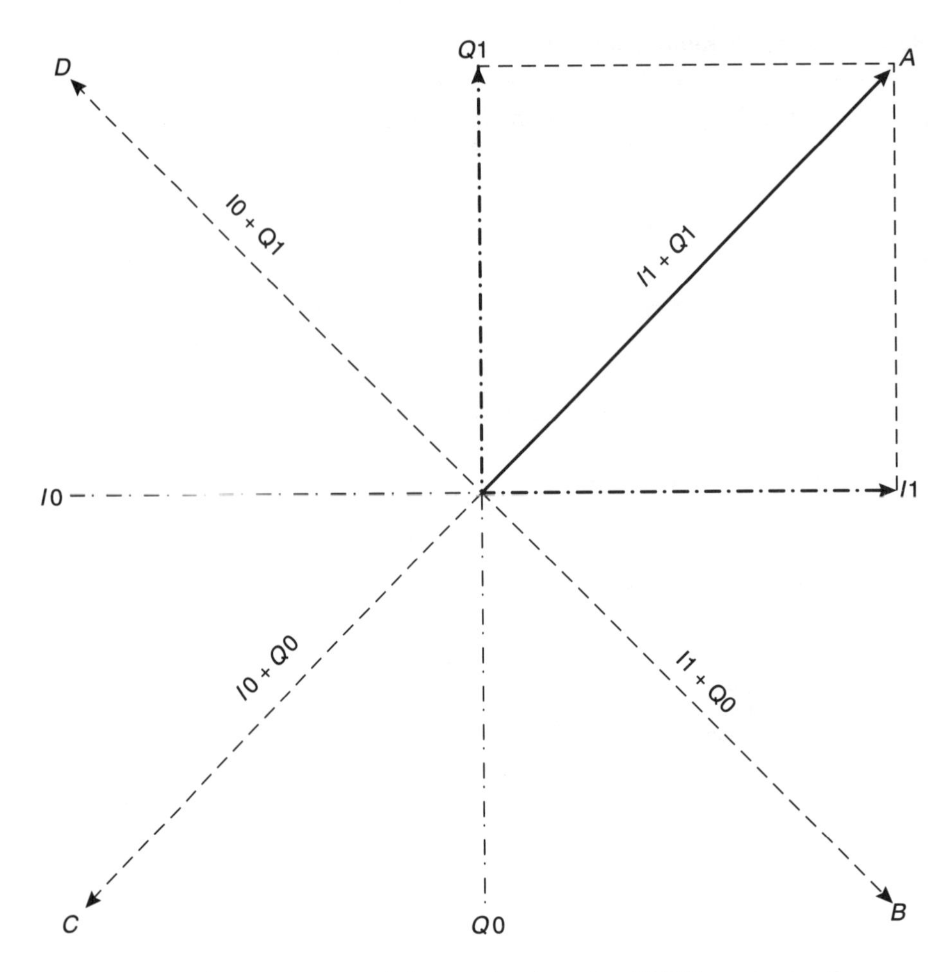

Figure 9.3 *Vector diagram of the production of a QPSK carrier by the combination of two BPSK carriers, I and Q*

Carrier recovery and ambiguity resolution is achieved for coherent QPSK in ways similar in principle to the methods used for BPSK, although they are rather more complex, since it is necessary to identify which of four possible recovered carrier phase conditions is the one that directs the received signals correctly into the I and Q channels, each with the correct polarity. See the block diagram in Figure 9.2(b).

One important advantage of digital transmission, compared with analogue transmission, lies in the feasibility of regenerating at the receiver a perfect facsimile of the digital signal delivered to the modulator, provided that the noise, interference and signal distortion added to the carrier in transit are not so severe that the polarity of symbols leaving the demodulator cannot be correctly detected.

The polarity of a symbol will be detected with the greatest likelihood of success if it is assessed at or near its midpoint, its phase condition being least affected at that instant in time by the various distortions that arise in the transmission path. It is therefore necessary to recover symbol timing ahead of the detector. Symbol timing recovery, more usually called clock recovery, may be done for example by generating at the receiving earth station a regular square wave with the same period as the original signal, the timing of its transitions being advanced or retarded by small amounts until the perceived transitions of the demodulated signal are occurring equally often before and after those of the locally generated waveform. The local waveform is then used for timing the detection and regeneration processes.

Differentially coherent PSK

Coherent PSK is efficient, in that it can provide link performance of the required quality with the lowest C/N ratio, but demodulation is rather complex. Other PSK transmission techniques are somewhat less efficient but simpler to implement and differentially coherent PSK (DCPSK) is the most widely used of these.

With binary DCPSK it is necessary to convert the digital signal into a transitional code before it modulates the carrier. A bit of polarity 1 is conveyed, not by a symbol of phase state A or B, but by transition from either of these states to the other. Conversely, a bit of polarity 0 is conveyed by a symbol during which the carrier remained constant in either phase state. This coding process involves a transmission delay equal to half the symbol period.

At the receiver, the carrier is split between two paths and one component is time delayed by the duration of one symbol before being compared in phase with the other component by means of a coherent demodulator. See Figure 9.4. If, on comparison, the two components are found to be in antiphase throughout a symbol, the polarity of the original modulating signal bit is assumed to be 1. If the two components are in phase, the polarity of the modulating signal is assumed to be 0. Thus, with Binary DCPSK it is not necessary to recover the carrier or resolve carrier phase ambiguities, although it is necessary to recover the symbol clock to optimize the timing of the detection process. However, if one or other of two symbols being compared has been so badly affected by noise or interference that its phase state is misinterpreted, then not only that bit, but also the subsequent bit, will be detected incorrectly.

Similar methods can be used to implement quadriphase DCPSK. However, Ogawa [6] notes that biphase DCPSK yields BERs for relatively high values of C/N which are little inferior to biphase CPSK, but as the number of significant carrier phase states is increased, the inferiority of performance relative to the corresponding CPSK mode becomes greater. High-order DCPSK systems are not used much in satellite communication.

Staggered or offset QPSK

In a variant of QPSK called staggered QPSK (SQPSK) or offset QPSK (OQPSK), the timing of the bits modulating the Q channel is delayed by half of the QPSK

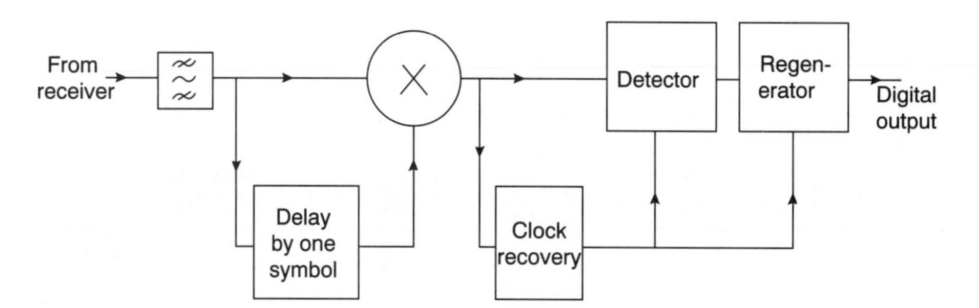

Figure 9.4 *Block schematic diagram of a differentially coherent biphase PSK demodulator*

symbol period after any necessary synchronization of the I and Q channels. As a result, the phase state of the SQPSK carrier changes in steps of $\frac{1}{2}\pi$ radians, never taking a step of π radians. This reduces the distortion that arises in the earth station transmitter and the satellite transponder (see Figure 9.6).

Minimal shift keying

Minimal shift keying (MSK) can be regarded as a variant of BPSK. However, whereas with BPSK the phase of the carrier may be changed abruptly by π radians at the start of a symbol, there being no further phase shift during the rest of the symbol period, with MSK the phase changes, forwards or backwards, throughout every symbol period at a uniform rate of $\pi/2t_s$ radians per second, where t_s is the symbol period in seconds. Thus, in MSK, during any one symbol, the carrier phase is advanced or retarded by a total of $\frac{1}{2}\pi$ radians.

An MSK emission is usually demodulated as PSK, coherently. However, the sustained, uniform, phase shift of a carrier by $\pm\pi/2t_s$ radians, beginning at the start of a symbol, is exactly equivalent to an abrupt change of frequency, from f_c to $f_c \pm 1/4t_s$ Hz, where f_c is the frequency of the unmodulated carrier. Thus MSK can also be treated as frequency shift keying (FSK).

The spectrum of a PSK-modulated carrier

The spectrum of a carrier, PSK-modulated by a random stream of bits, is sketched at Figure 9.5. Most of the energy of the carrier is located within $\pm 1/t_s$ Hz of f_c, the frequency of the unmodulated carrier, where t_s is the symbol period, but low-level modulation products extend well beyond those limits. Approximately

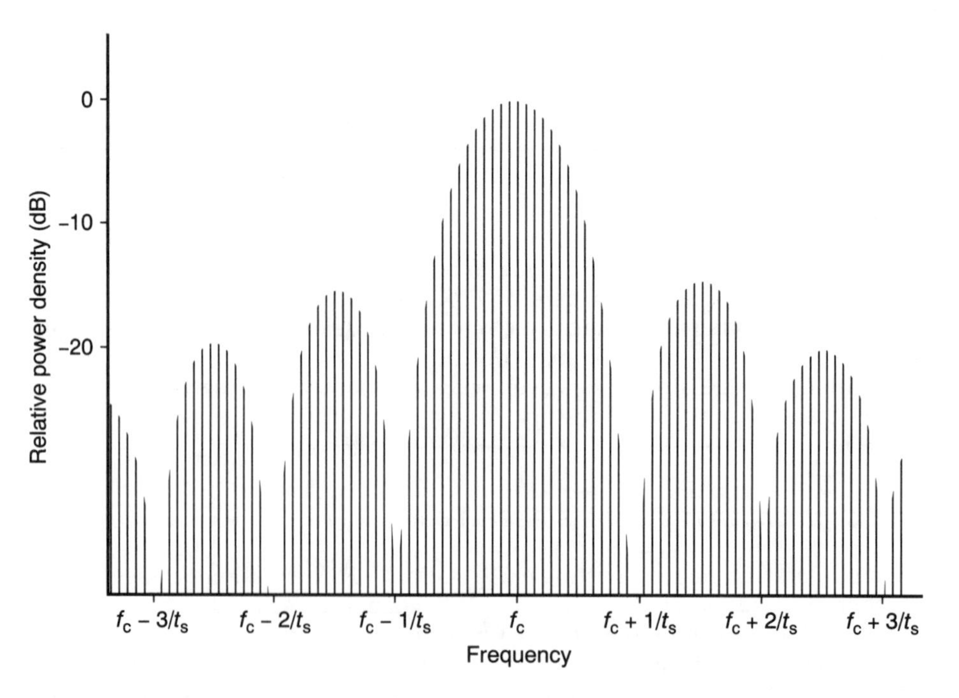

Figure 9.5 *Sketch of the spectrum of a carrier, PSK-modulated by a random digital signal*

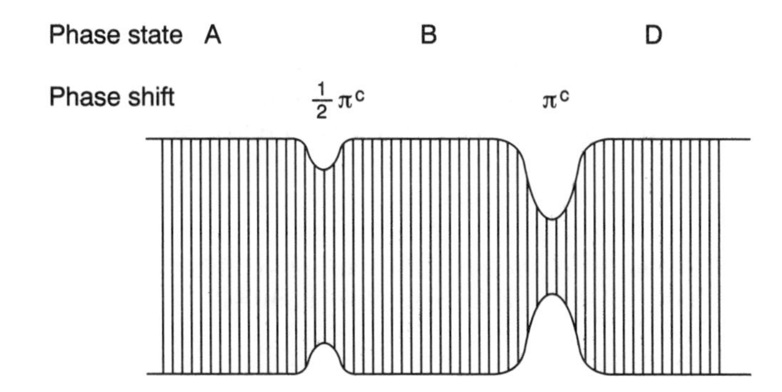

Figure 9.6 *A sketch of a QPSK emission envelope distorted by post-modulator filtration*

90 per cent of the power of BPSK, QPSK, SQPSK and MSK emissions having the same symbol rate is contained within the same bandwidth. The phase continuity of MSK ensures that the power in its higher-order modulation products is least. However, it is found in practice that the performance of links is not seriously degraded if the pre-demodulator bandwidth is limited to $\pm 0.6/t_s$ Hz.

Carrier energy dispersal

If the modulating signal can be regarded as a random bit stream, the energy within the spectral envelope shown in Figure 9.5 will be distributed among a very large number of components. This does much to distribute the energy of an emission over its bandwidth.

However, if the modulating signal contains a repetitive pattern of bits, as commonly occurs when channels are lightly loaded, a number of discrete spectral lines emerge, some of which may have a spectral energy density which is considerably higher than the smooth spectral envelope produced by random signals. To prevent such lines from exceeding regulatory power limits (see Section 6.4), it may be necessary to use artificial spectral energy dispersal. This is usually done by the modulo-2 addition of a pseudo-random bit stream to the digital signal before transmission, an identical pseudo-random bit stream being generated at the receiving earth station to allow the process to be reversed after demodulation.

Exceptionally, where the earth stations which are to receive the emission have a very low G/T ratio, it may not be feasible to keep a PSK downlink signal within the regulatory limits even with artificially randomized modulation. In such cases it may be necessary to use, not PSK, but spread spectrum modulation.

Signal degradation in the transmission path

Once modulated at the transmitting earth station, a PSK emission passes through a number of stages where the carrier and the signal it carries may be distorted, before reaching the demodulator at the receiving earth station. The most important of these distortions are due to the filter at the output of the modulator, the HPA of the earth station transmitter, the band-defining filter at the input to the satellite transponder, the power amplifier of the transponder, and the filter at the input to the demodulator.

The effects of filters

The purpose of the filter at the input to the demodulator, which typically has a passband of the order of $\pm 0.6/t_s$ Hz, is to reject noise and interference adjacent in frequency to the emission. The main function of the filter at the output of the modulator is to suppress inessential sideband components of the modulated signal, thereby reducing interference to emissions with adjacent frequency assignments. However, in addition to these important intended effects, these filters have effects on the wanted emission which are undesirable.

Suppression of high-order sidebands distorts the signal waveform that is recovered at the receiver. Also, the emission, deprived of its high-order sidebands, ceases to have an envelope of constant amplitude; at the output of the modulator filter the amplitude is reduced while the carrier shifts phase by $\frac{1}{2}\pi$ radians and the dip is deeper for a phase shift of π radians. See Figure 9.6. SQPSK avoids the deeper dips by eliminating phase shifts of π radians and MSK is more effective in the same respect.

Secondly, non-linear variations of group delay across the passband of a filter, and especially near the edges of the passband, adds to the distortion of the signal waveform and of the carrier envelope variations caused by the suppression of the high-order sidebands. It is necessary to use a type of filter that has low group delay non-linearity for these purposes and to improve group delay characteristics further by equalization. Additional equalization may be required at the transmitting earth station to pre-equalize group delay in the satellite transponder input filter that has been incompletely equalized on board.

Thirdly, energy from each symbol is spread, through the response of the filters to the impulsive nature of the carrier modulation, into the leading edge of the subsequent symbol. To minimize this 'inter-symbol distortion', the in-band amplitude responses of the modulator and demodulator filters must be carefully matched to one another and to the characteristics of the signal. The presence of significant amounts of inter-symbol distortion makes it particularly important that the detector should assess the phase state of the received carrier at the middle of the symbol period, where the level of inter-symbol interference can be expected to be least.

Power amplifier distortion

To maximize the information throughput of a satellite without raising the earth segment costs, the transponder HPA is operated as near to saturation as the good functioning of the network permits; in consequence, amplification is usually somewhat non-linear. This leads to signal degradation.

When a PSK emission, band-limited for transmission and therefore having an envelope that varies in amplitude, is amplified non-linearly, it re-acquires some of the high-order sideband products which the modulator output filter had suppressed. This re-establishes a source of interference for adjacent frequency slots.

Secondly, intermodulation products are generated if the transponder is also amplifying other emissions at the same time. Some of these products may fall within the spectrum required for other emissions, reducing the C/N ratio.

Thirdly, AM/PM conversion occurs in shared amplifiers by two mechanisms. The most substantial effect is likely to arise from variations in the amplitude of the envelope produced by the summation of all of the emissions simultaneously present in the amplifier. However, AM/PM conversion also arises from variation of the amplitude of the envelopes of individual PSK emissions, caused by the suppression of high-order sideband products by modulator filters. The phase modulation of other emissions carrying similar signals that this latter phenomenon causes may be particularly troublesome because, in the terminology of telephony, it causes intelligible crosstalk.

Finally, the sudden changes of loading conditions in the HPA of a transponder, which occur at the beginning and end of each burst of a high-capacity TDMA system, change the level of any other carrier that is sharing the transponder (signal compression).

It is necessary to ensure that these various distortions occurring in the transponder HPA are kept to an acceptably low level by the use of an appropriate degree of backoff. Some earth station HPAs also amplify two or more emissions at once and the same kinds of distortion tend to arise there also. However, whereas there are strong economic reasons for limiting transponder HPA backoff to the minimum acceptable value, the cost of operating earth station HPAs with a large backoff is much less. High-capacity TDMA emissions should not share earth station power amplifiers with other emissions and it is preferable to avoid sharing with high-capacity digital emissions which do not operate in the burst mode. Where earth station amplifiers are shared, a high degree of backoff is essential. A lesser degree of backoff is desirable even for amplifiers that are used for a single PSK emission, to limit the interference caused by the reintroduction of high-order sideband products.

Bit error ratio and C/N ratio

The quality of a digital link is measured by the ratio between the number of bits sent and the number received incorrectly. This criterion is the bit error ratio (BER) and it is usually expressed as P_e, the probability that any received bit will be in error. Thus, if 12 bits are received in error in a block of data containing one million bits, $P_e = 1.2 \times 10^{-5}$. Performance targets in the range 1×10^{-4} to 1×10^{-7} are typical, depending on the nature of the signal and the purpose for which it is sent.

The waveform of the demodulated signal is distorted by the various degradations reviewed above. System design must ensure that this waveform distortion is not so great that the likelihood of symbols being misinterpreted by the detector, because of this, is significant. However, thermal noise enters the communication channel at various points, and this will make false some of the detector's decisions if the level of noise is high enough. Intermodulation products and various other kinds of distortion product may also be added to the emission at several points along the route from the modulator to the demodulator, together with interference from other links. If the level of this unwanted energy is not too high, it can be treated as if it were noise of the same total power.

Then, given ideal system design, P_e depends primarily on the ratio, at the input to the demodulator, between the power of the emission, C, and the power of noise plus interference and distortion products that can be treated as noise, N. For BPSK and QPSK emissions, using coherent demodulation, with pre-demodulator filtering matched to the signal parameters and the noise bandwidth limited to $\pm 0.6/t_s$, the theoretical BER is given by

$$P_e = \tfrac{1}{2}\,\mathrm{erfc}(E_b/N_o) \tag{9.3}$$

$$= \tfrac{1}{2}\,\mathrm{erfc}(CB/NR) \tag{9.4}$$

where $\mathrm{erfc}(x) = (2/\sqrt{\pi}) \int_x^{\infty} \exp(-t^2)\,\mathrm{d}t$

$E_b =$ energy per bit $= C/R$,

$R =$ the bit rate (bit/s)

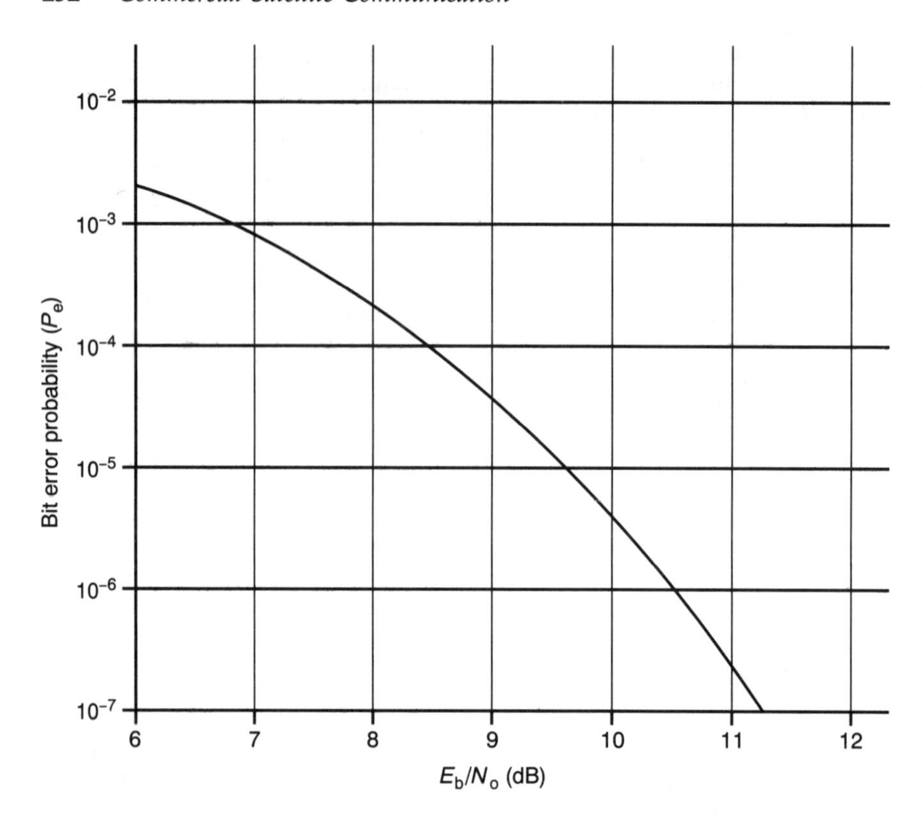

Figure 9.7 *The bit error probability P_e for a BPSK or QPSK carrier with coherent demodulation, an ideal demodulator input filter matched to the signal parameters and an ideal demodulator, in the presence of noise of density level E_b/N_o*

N_o = spectral power density of noise, etc. ($\mathrm{W\,Hz^{-1}}$)
B = the pre-demodulator bandwidth (Hz)

This function for P_e is plotted against E_b/N_o in Figure 9.7. The theoretical P_e for SQPSK and MSK emissions is approximately the same. For an ideal differentially coherent PSK system, E_b/N_o must be about 0.75 dB higher than that for an ideal coherent BPSK system over a P_e range from 1×10^{-4} to 1×10^{-7} if the same BER is to be achieved. An implementation margin of 1 or 2 dB will usually be needed for any of these systems to allow for suboptimal performance of equipment.

9.2.2 Frequency shift keying

In frequency shift keying (FSK) a digital signal is caused to change the frequency, but not the amplitude, of a carrier so that frequency f_A represents condition 1 of the signal and frequency f_B represents condition 0. The binary emission so produced can be represented by

$$V \cos \pi(f_c \pm \delta f)t \tag{9.5}$$

where $f_c = \frac{1}{2}(f_A + f_B)$ (Hz)
$\delta f = \frac{1}{2}(f_A - f_B)$ (Hz).

Here $f_A - f_B$ or $2\delta f$ is called the frequency shift, which may be related to t_s, the bit period of the modulating signal, by the modulation index h, defined by

$$h = 2t_s\,\delta f \tag{9.6a}$$

$$= t_s(f_A - f_B) \tag{9.6b}$$

Four-condition and higher-order FSK modulation is also feasible but it is seldom, if ever, used in satellite communication.

An FSK emission may be generated in several ways. A modulating signal may be caused to switch between sources of f_A and f_B; see Figure 9.8. Such a signal lacks phase coherence. A frequency modulator may also be used. An emission having phase continuity throughout transitions from one frequency condition to the other, called continuous phase FSK (CPFSK), may be produced by combining two orthogonal BPSK carriers offset-keyed by the modulating signal. CPFSK having a modulation index h equal to 0.5 is equivalent to minimal (phase) shift keying (MSK).

A CPFSK emission is usually demodulated coherently, a carrier being recovered at the receiving earth station in ways similar to those used for a coherently demodulated PSK carrier; see Section 9.2.1. A non-coherent FSK emission may also be demodulated coherently, after recovery of carriers from the sources of both f_A and f_B. However, this technique is unattractive because it is more complex and needs a higher C/N ratio for a given BER than a corresponding coherent BPSK system. Instead, a non-coherent FSK emission is usually demodulated non-coherently by means of a frequency discriminator, by a pair of filters tuned

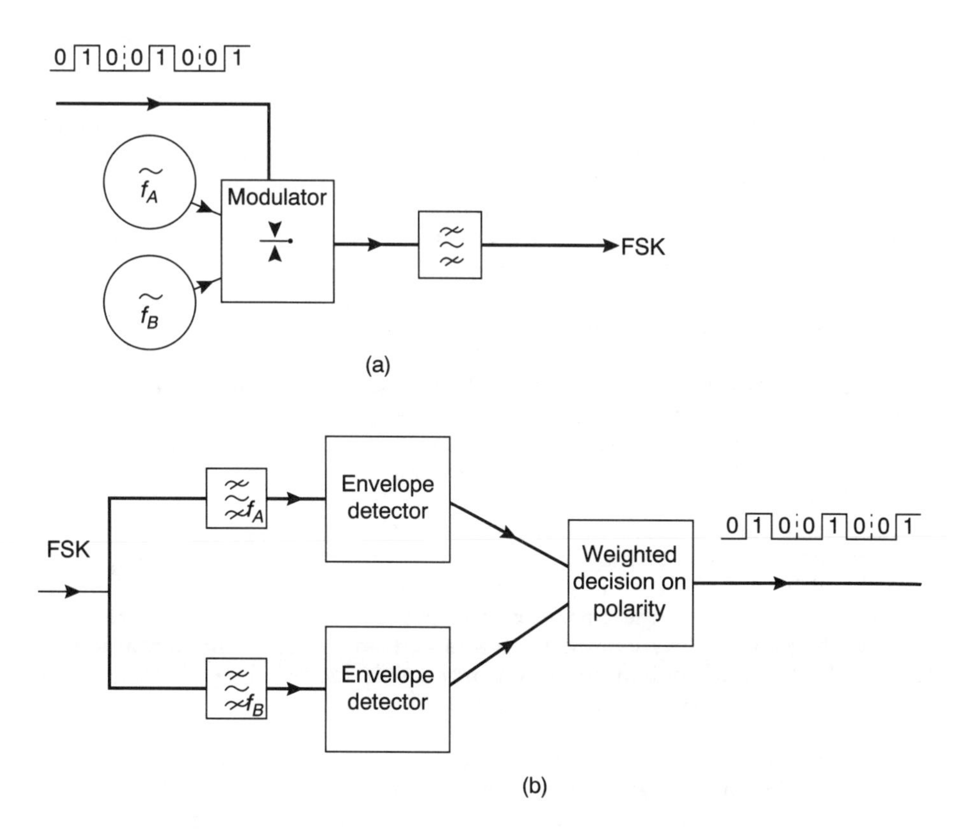

Figure 9.8 *A typical modulator (a) and demodulator (b) for non-coherent FSK*

respectively to f_A and f_B and feeding envelope detectors, or by some equivalent technique. See Figure 9.8. For most non-coherent demodulator configurations, polarity ambiguities do not arise.

Coherently demodulated CPFSK is equivalent to BPSK or QPSK in terms of the bandwidth required for a given information flow or the BER for a given C/N ratio and the level of the outer sideband components due to modulation is low. This technique is attractive where economy in the use of spectrum is important. Non-coherent FSK needs a C/N ratio at least 4 dB higher than BPSK for a given BER but it is simpler to implement and may find application where the cost of receive-only earth stations is of critical importance.

9.2.3 Spread spectrum modulation and CDMA

The term 'spread spectrum modulation' is applied to techniques in which, for some practical advantage, the bandwidth of the modulated emission is made much wider than the information bandwidth of the modulating signal. This section reviews two such techniques, namely direct sequence and frequency hopping spread spectrum modulation. Other techniques, involving, for example, variation in the timing of information bursts and hybrid spread spectrum modulation, involving elements of both direct sequence and frequency hopping modulation, are also used, particularly when privacy is an objective.

Direct sequence spread spectrum modulation

Figure 9.9 shows a basic block schematic diagram for a direct sequence spread spectrum (DSSS) modulator. A digital information signal $S(t)$ is multiplied by a pseudo-random sequence $C(t)$ and the product modulates a carrier wave of frequency f_t, typically by BPSK. The bit period t_s of $S(t)$ is usually 100 to 1000 times as long as t_c, the elemental period of $C(t)$. Thus, each bit of the signal to be transmitted modulates the carrier in the form of a very large number of elements, called 'chips'. The emission has a wideband spectrum, although the information rate of $S(t)$ may be quite small.

The basic receiver arrangement is also illustrated in Figure 9.9. The local oscillator is tuned to heterodyne f_c to the intermediate frequency of the receiver. A pseudo-random sequence is generated at the receiver, the same sequence $C(t)$ as was generated at the transmitter, and this phase modulates the output of the receiver local oscillator. The received carrier and the modulated local oscillator waveform are compared in the correlator. If the pseudo-random sequences from the two generators are in synchronism at the receiver, the correlator delivers a carrier, modulated by a reconstituted $S(t)$ waveform, to the IF amplifier, where it can be separated by narrowband filters from almost all of the unwanted wideband products from the correlator.

If the pseudo-random sequence generator at the receiver is out of synchronism with the generator at the transmitter, the output from the correlator will be wideband and noise-like. Thus, when communication over a DSSS link is being newly established, the phase of the sequence generated at the receiver is allowed to slip until synchronism is attained and a coherent digital signal enters the IF amplifier, whereupon synchronism is locked. The IF filters must be wide enough to pass the carrier modulated by $S(t)$, that is, approximately $\pm 0.6/t_s$ Hz.

Interfering emissions having spread spectrum modulation, on reaching the correlator, will remain spread unless there is correlation between the pseudo-random sequences of the wanted and unwanted emissions. Interfering carriers with conventional modulation that may enter the correlator with the wideband

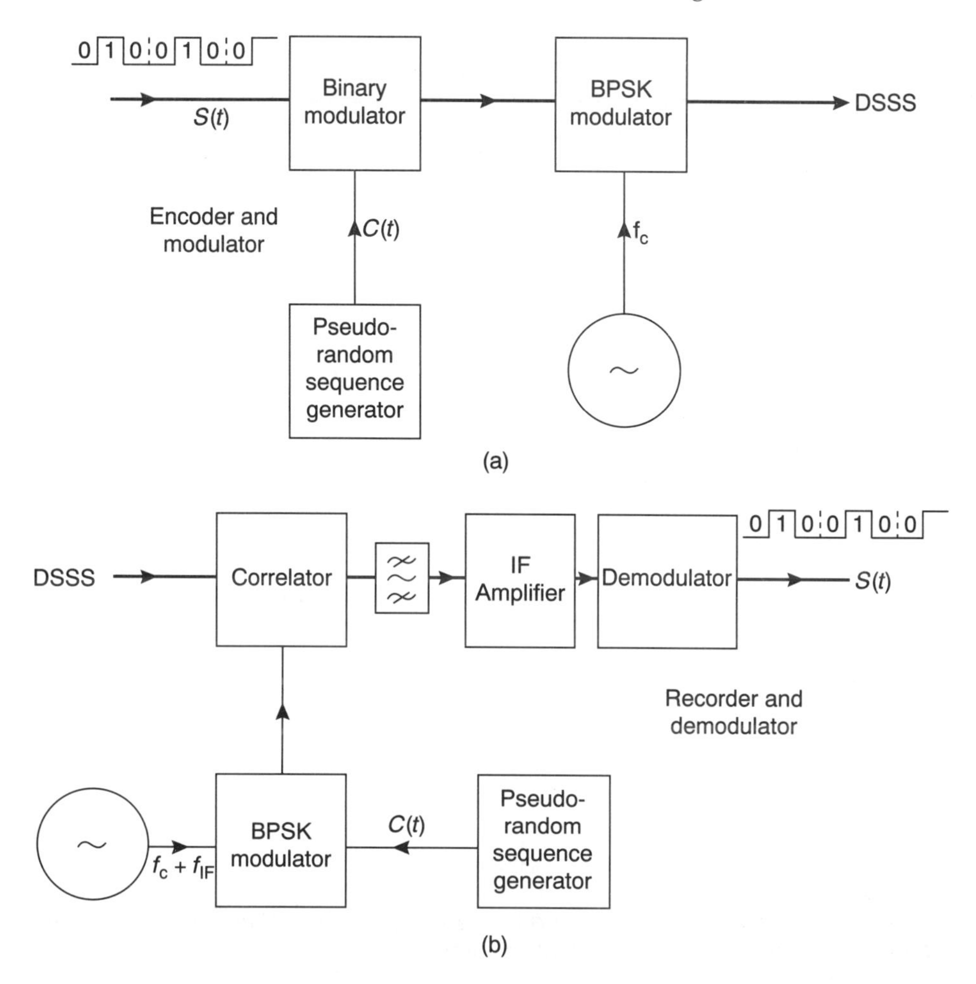

Figure 9.9 *Block schematic diagram of a direct sequence spread spectrum encoder/modulator (a) and decoder/demodulator (b)*

wanted emission are remodulated by the recovered pseudo-random sequence via the local oscillator, their energy being spread out over a wideband. Thus, neither kind of interference adds more than a small part of its power to the coherent wanted emission.

The ratio of the carrier power reaching the receiver from the wanted transmitter, relative to noise and interference, must be sufficient to provide an adequate BER in $S(t)$ after demodulation. However, the necessary power level of the wanted carrier will be very small relative to the total power of the noise and interference in the wideband occupied to the spread spectrum emission.

Frequency hopping spread spectrum modulation

In a typical frequency hopping spread spectrum (FHSS) emission, the carrier is switched at very short intervals from one frequency to another, selected pseudo-randomly from a large but finite complement of frequencies. This frequency hopping carrier is modulated by a digital signal, typically by BPSK. Figure 9.10 is a block schematic diagram of a basic encoder/modulator and

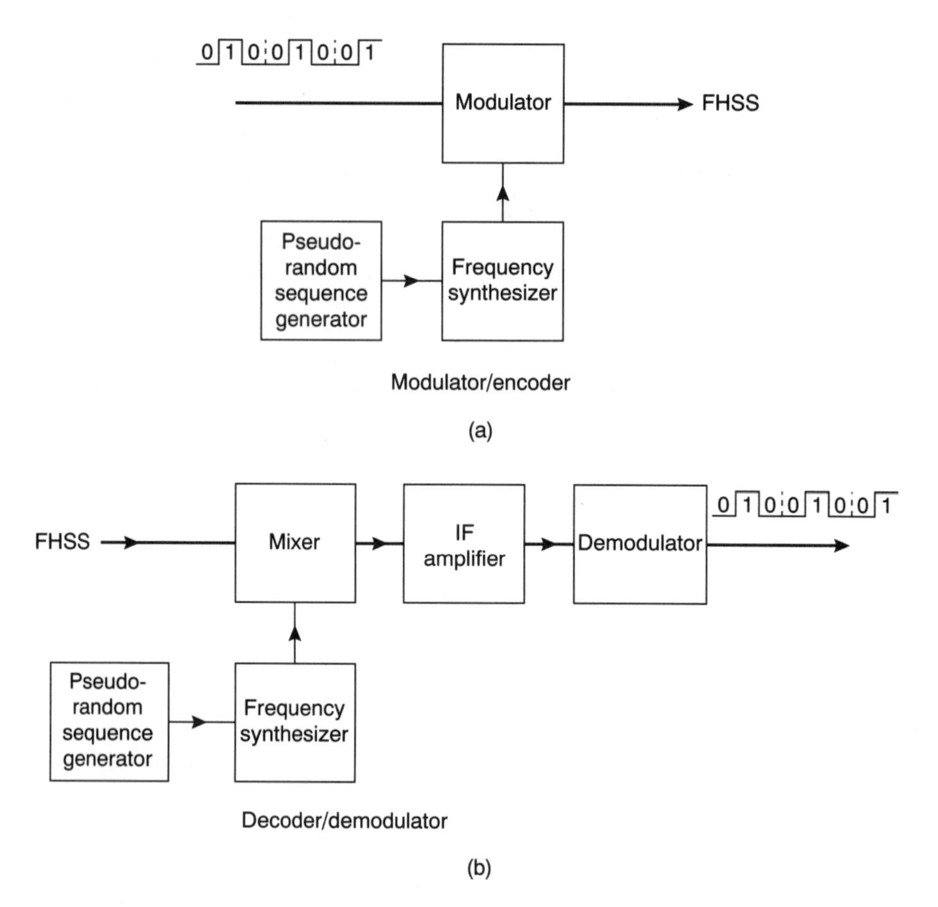

Figure 9.10 *Block schematic diagram of a frequency hopping spread spectrum modulator/encoder and decoder/demodulator*

decoder/demodulator. The principle involved in tracking the frequency of the carrier is akin to the decoding of DSSS and will be sufficiently clear from the figure. The modulation and demodulation of the carrier with the signal is conventional.

In a typical FHSS system as used in commercial satellite communication, the interval between frequency hops is of the same order as the period of the signal bits. The bandwidth spanned by the complement of frequencies is likely to be several megahertz and the number and spectral distribution of the frequencies of the complement is such that, over a small fraction of a second, the spectral energy of the emission is spread fairly uniformly over the band covered by the complement. Other systems use much longer frequency hopping intervals to provide privacy, but these are less effective as a means of spreading carrier spectral energy.

Applications for spread spectrum modulation

A key characteristic of spread spectrum systems is that they can operate with acceptable performance when the downlink carrier C at the receiving earth station has a spectral power density which is much less than that of the noise N and/or interference I in the frequency band in which the system is operating. This ability can be expressed in a numerical form as the processing gain (PG) of the system, where

$$10 \log \mathrm{PG} = 10 \log \frac{C}{(I+N)_{\mathrm{out}}} - 10 \log \frac{C}{(I+N)_{\mathrm{in}}} \tag{9.7}$$

where $C/(I+N)_{\mathrm{in}}$ is measured at the input to the receiver in the bandwidth required by the spread spectrum emission, and

$C/(I+N)_{\mathrm{out}}$ is measured at the input to the demodulator in the bandwidth that would be required if the carrier were modulated only by the wanted signal.

This processing gain is obtained even when the interference consists of a number of similar spread spectrum emissions, provided that the pseudo-random sequences of the wanted and unwanted emissions are uncorrelated.

Thus, in commercial satellite communication, DSSS and FHSS provide multiple access to transponders. When spread spectrum modulation is used in this way it is usually called code division multiple access (CDMA). A CDMA system needs none of the apparatus of TDMA and little of the careful management that both FDMA and TDMA require. CDMA is coming into use for particular applications in land mobile satellite systems, in particular using FHSS, and especially in LEO systems, where more conventional multiple access systems would be difficult and costly to implement. However, CDMA makes relatively inefficient use of power and bandwidth and the information capacity per access may be quite limited.

In addition, spread spectrum modulation provides privacy; indeed the very presence of a spread spectrum emission may be masked by conventionally modulated carriers sharing the same frequency band. Also, it can provide a very high degree of carrier energy dispersal, which can be of great value where earth stations having very low G/T are used.

9.2.4 Error detection and correction

A BER requirement of 1×10^{-7} for almost all of the time is not unusual for satellite channels and more stringent performance objectives are sometimes set. Figure 9.7 shows that a BER of 1×10^{-7} requires a C/N ratio of about 11.3 dB in an ideal BPSK system. It may often be less costly to accept a BER objective around 1×10^{-4} for the radio link between earth stations and to eliminate most of those bit errors by means of an automatic error correction system. Such systems make use, at the receiving earth station, of information about the pattern of signal bits being transmitted that had been injected into the bit stream (typically at the transmitting earth station) to enable symbols that have been incorrectly interpreted at the demodulator to be identified and replaced with bits of the correct polarity. This process is called forward error correction (FEC). By using FEC it is feasible to achieve high performance using several decibels less carrier power from the satellite, despite an increase in the occupied bandwidth to accommodate the additional information that has been injected into the signal.

Another requirement for a data channel may be that no data at all should be passed on to the ultimate destination during periods when the BER is high and the integrity of the link is consequently in doubt, and that no data should be lost in transit. Occasions arise, for example, when there is sporadic interference from another radio link or when solar noise increases greatly at the receiving earth station because the satellite is transiting the Sun as seen from the receiving earth station, and at such times the BER may be too high for an FEC system to be effective. To achieve these objectives, a device is required at the receiving earth station that detects the presence of bit errors, withholds the garbled data from the ultimate destination and directs the transmitting earth station to hold back the flow of new data until channel conditions improve. When conditions are once more satisfactory, the flow of data is restarted, beginning with the retransmission of the data that had originally been received garbled. This process is called automatic repeat on request (ARQ).

FEC can bring benefit to all kinds of digital communication, including those that function in real time, like telephony. ARQ can be used effectively only when information is being transmitted out of a store. ARQ is also virtually limited to situations where reception is required at only one earth station, and that station has a return channel to the transmitting earth station for control messages.

A variety of sophisticated systems of both types have been developed. Some FEC systems are capable of identifying and correcting substantial numbers of isolated errors in a block of data. Others are specifically designed to function efficiently when a burst of consecutive errors occurs. It may be desirable to use two FEC codes with different characteristics, one operating within the loop of the other, to provide a particularly powerful FEC action; in such combinations, codes are said to be concatenated. For some purposes it may be necessary to use both FEC and ARQ, the FEC providing adequate protection of the data without imposing delays when transmission conditions are fairly good and the ARQ maintaining the integrity of the data, albeit with delays, when transmission conditions are worse. There is an extensive literature; see, for example, [7].

The performance of these systems is defined by various criteria.

(1) The proportion of bits that are added to the information stream in order to enable errors to be detected and, in FEC systems, corrected (called the overhead) is defined by comparing the total bit flow with the information bit flow. Thus, if k overhead bits are added to each block of j information bits, causing the total number of bits transmitted per block to become $j + k = n$ bits, the system would be said to have a (n, j) code or a j/n code. Effective FEC systems are available where $j/n = 3/4$ and very powerful $1/2$ codes are available.

(2) The coding gain is defined as the increase in E_b/N_o that is required to give a specified BER without FEC, relative to value of E_b/N_o that provides the same BER when FEC is used. In this calculation, E_b is the energy per information bit, disregarding overhead bits. The coding gain varies with the code but also with BER. It is high, typically up to 3 or 4 dB, when the unprotected BER is as low as $P_e = 10^{-5}$, but it falls considerably when transmission conditions are worse and eventually becomes negative, typically when the unprotected BER is as high as $P_e = 10^{-2}$.

(3) A code capable of correcting isolated bursts of up to q bit errors is called a q-burst error correcting code.

Soft decision detection

When the ratio of carrier power to noise-plus-interference power at the demodulator is high, the post-demodulator signal waveform exhibits little distortion, and clock recovery is efficient. There is therefore little uncertainty about the decisions that the detector takes as to the polarity of the bit or bits which a symbol carries. When conditions are worse, there may be considerable ambiguity, but, in conventional systems, the detector is designed to come to an unequivocal decision, making the best use of the information presented to it. The decision may, of course, be wrong. When FEC is used, the final decision on the polarity of a data bit is based on the detection, not of one symbol, but of several. If there is conflict among the polarities indicated by the several symbols that determine the final decision, a simple majority of these unequivocal decisions may be taken.

However, under these conditions a more trustworthy decision can be obtained if the detector comes to decisions which are qualified by an assessment of their dependability. All of the qualified decisions, called soft decisions, relating to the identity of a bit can then be given an appropriate weighting before being used, jointly, to reach the final decision. The use of this technique, associated in particular with Viterbi [8], raises the coding gain attainable with convolutional coding under typical operating conditions to about 6 dB.

9.3 Digital encoding of analogue signals

9.3.1 Signal encoding for telephony

Analogue signals are encoded for transmission over digital channels by assessing the voltage of samples of the waveform at short time intervals, in absolute terms or by comparison with the assessments of previous samples, and defining the result in the form of a compact digital signal. These digital signals are used at the receiving end of the channel to reconstruct the original waveform.

Usually the analogue signal is sampled by the coder at intervals rather shorter than half of the period of the highest frequency of the analogue signal spectrum. Each sample is compared with a scale of voltage steps, called quantization levels, each of which can be identified by a binary number, and the number of the lowest voltage step which exceeds the voltage of the sample is transmitted to the decoder. At the decoder the voltage of the quantization step can be identified from the number, allowing the analogue waveform to be reconstructed. This process is called pulse code modulation (PCM).

A somewhat different principle is used in differential PCM (DPCM), in which samples of the voltage to be encoded are compared in the coder with the voltage of previously transmitted samples. The difference voltage is expressed as a digital word and transmitted to the decoder so that the waveform can be reconstructed. Delta modulation is an extreme form of DPCM, in which each code word consists of only one bit.

With both techniques the reconstruction of the original waveform is imperfect. The differences between the original waveform and the reconstructed waveform can be treated subjectively as the addition to the analogue signal of a form of noise, called quantizing noise. The level of quantizing noise is determined, not by radio channel conditions but by the number of quantizing steps and the frequency of sampling. Consequently, account is taken of quantizing noise when performance objectives are set and this source of noise is disregarded when the radio parameters are determined for a channel.

Pulse code modulation

Encoding standards for PSTN telephony were developed separately in Europe and North America, primarily for the terrestrial PSTN. Both standards use PCM with 8-bit encoding and 8000 samples per second. In both standards the voltage steps between successive quantizing levels are non-uniform. In both standards the steps are made small for low-level signals and larger for high-level signals, in order that the quantizing noise at low speech levels should not be excessive. This device has an effect similar to companding in analogue systems. However, the laws of these variations are not the same. The European and non-European laws are called the A law and the μ law respectively.

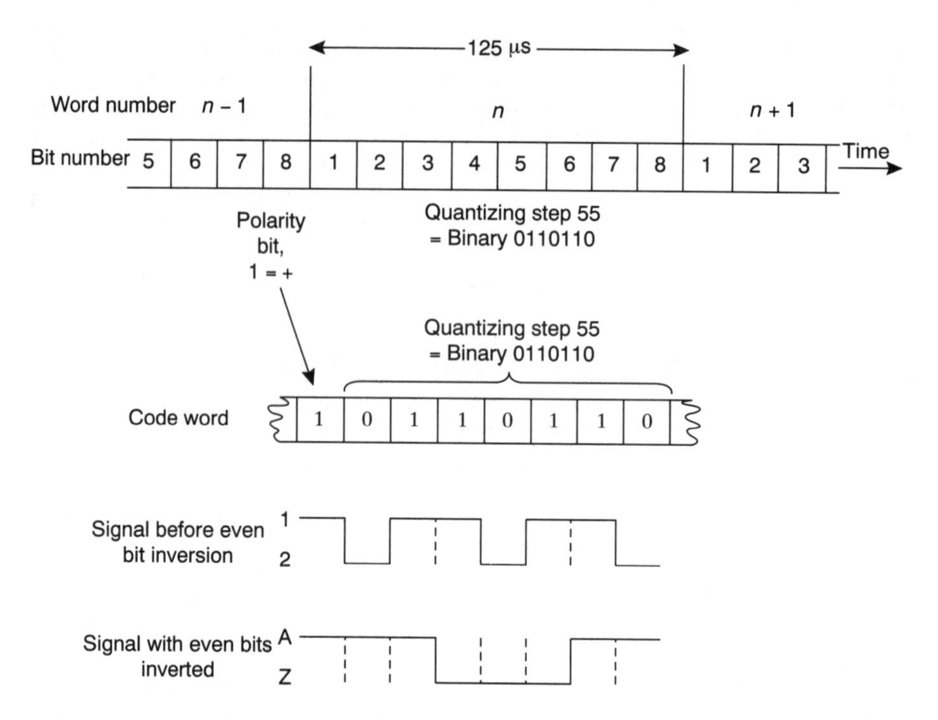

Figure 9.11 *8-bit PCM encoding of a sample (word n) from an analogue waveform with positive voltage at quantization level 55*

In most other main characteristics the two standards are similar. The most significant bit in the sample word, defining the sign of the sample, is the first bit and the remaining seven bits give the number of the quantization level in binary notation, the least significant bit being the last. The even bits of telephony channels are inverted before transmission to reduce the incidence of long periods of the same polarity. See Figure 9.11. One or other of these digital standards has been implemented in national terrestrial networks world-wide. Both standards are defined in CCITT Recommendation G.711 [9], which recommends that the A law should be used on international links where the terminal countries concerned use different companding laws domestically.

An analogue telephone channel that has been encoded as 8-bit PCM will transmit satisfactorily the other kinds of signal for which analogue telephone channels are used, such as facsimile and data transmission using speech-band modems.

The subjective speech quality of channels that have been encoded by these standards is good, but a transmission capacity of 64 kbit/s is required. This substantial bandwidth is quite acceptable for most terrestrial transmission links, many of which are quite short; the cost of digital transmission by terrestrial links is low and falling. A more complex codec would not be cost-effective for most terrestrial applications, even if its bandwidth requirement were less. However, a system with a lower bit rate would offer significant economic advantages for the satellite medium.

Perhaps the simplest way to reduce the bit rate for satellite links without substantially increasing the cost of interfacing with standard 64 kbit/s terrestrial networks would be to omit the least significant bit of 8-bit encoding, thereby reducing the bit rate to 56 kbit/s. This has been done, for example, for SCPC telephony channels in the Intelsat system. The saving in transmission cost is significant and the loss of transmission quality is small. If this practice were extended by omitting the two least significant bits per word, the reduction of transmission quality might not be accepted. More effective means of bit rate reduction are, however, available; see below.

Delta modulation

A delta modulation system transmits a series of 1-bit words, positive or negative, that can be integrated and filtered at the demodulator to reconstruct the analogue signal. In a typical implementation, the coder which generates the bit stream also has an integrator and filter which reconstruct the analogue signal from the output of the coder. The signal that has been reconstructed at the transmitting end is compared with the original analogue signal and the voltage difference is sampled at a high sampling rate. If the original signal voltage is found to be higher than the reconstructed signal voltage, a positive bit is generated and sent, giving a positive increment to the reconstructed signals at both the transmitting and the receiving terminals. A negative bit and a negative increment follows if the original signal voltage is found to be lower at the comparator. Figure 9.12 shows a block schematic diagram of one realization of the transmitting and receiving terminals.

When the input signal to a delta modulator is zero or changing slowly, the reconstructed signal consists of square wave alternations at the sampling rate, producing a background of quantization noise. When the input analogue signal is changing rapidly the reconstructed signal may fail to track the changes accurately, causing the signal reconstructed at the receiver to be distorted. With a sufficiently high sampling rate, a small voltage increment per bit and effective filtering of the receiver output, these problems can be overcome. A sampling rate between 24 and 40 thousand samples per second, used for a speech signal, produces a quality of transmission which is fairly acceptable for high-level inputs, although it is less satisfactory for low-level inputs. There is no good way to interface such a system with a conventional network using PCM. Delta modulation is, however, simple to implement.

Differential PCM

DPCM for speech signals has been developed in various forms to overcome the limitations of delta modulation, providing acceptable channel performance with a bit rate lower than that of PCM. By using code words of several bits, with quantifying intervals which adapt to the level of the signal or its rate of change, and by predicting future sampling results from the trend of past samples, it becomes possible to reduce the sampling rate and the final bit rate while improving the transmission quality. Used for telephony, DPCM at 32 kbit/s is virtually equal in quality to 64 kbit/s PCM, and good intelligibility can be obtained at lower bit rates, down to 9.6 kbit/s. Interfacing problems remain.

Adaptive differential PCM

Adaptive differential PCM (ADPCM) is a specific form of DPCM, having the particular advantage that it can readily be interfaced with the PSTN designed around 64 kbit/s PCM channels. Like other DPCM systems, ADPCM predicts the amplitude of future samples by extrapolation from recent samples and transmits information to correct those predictions when the new samples are received.

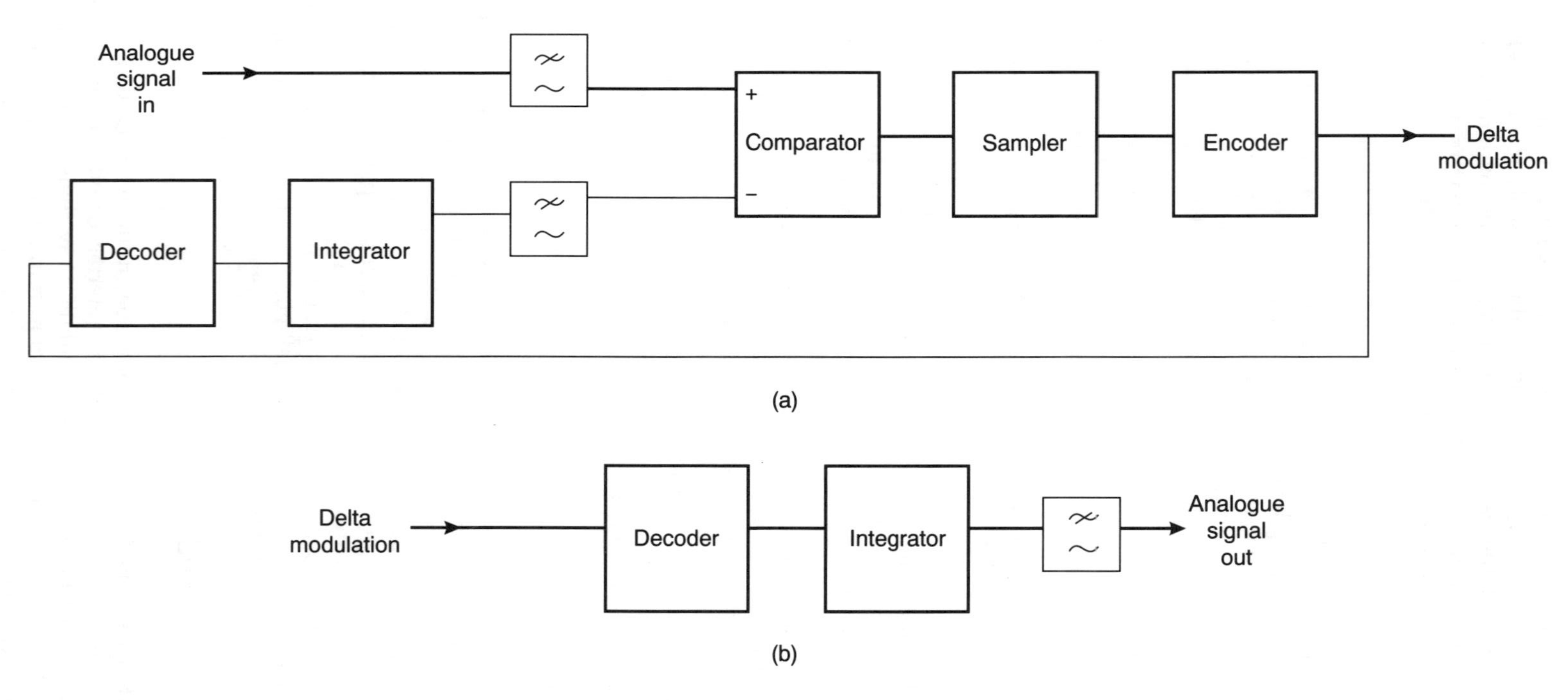

Figure 9.12 *Block schematic diagram of Delta encoder (a) and decoder (b)*

Successive samples of most analogue signals are fairly well correlated, one with another, and so the maximum size of correction is likely to be much less than the full range of possible sample voltages arising with standard PCM encoding. Consequently, the error voltage can be encoded with using five or fewer bits per sample and smaller quantization level intervals than those used in 8-bit PCM. Devices which enable a number of signals, originally digitized using 64 kbit/s, to be routed through a smaller number of 64 kbit/s transmission channels, are called digital circuit multiplication equipment (DCME).

The transmission quality for speech obtained with 4-bit ADPCM, requiring 32 kbit/s per channel is as good as that provided by 64 kbit/s PCM and this form of ADPCM is widely used in the FSS. However, some degradation of performance is to be expected for speech-band modems used for 4.8 kbit/s data if routed through 32 kbit/s ADPCM channels and the use of such channels for 9600 bit/s data transmission may not be feasible.

Internationally agreed DCME specifications of the algorithms that can be used to convert either A law or μ law 8-bit PCM to and from globally standardized ADPCM are to be found in CCITT Recommendations G.726 [10] and G.727 [11]. It will be noted that an incidental by-product of the use of ADPCM in this way is a ready means for interconnecting terrestrial networks using different companding laws.

9.3.2 Speech analysis/synthesis

Reduction of the information rate required for intelligible and recognizable speech to much below 9.6 kbit/s, may not be feasible using systems that digitize the analogue waveform. The way to greater bit-rate reduction probably lies with vocoder systems. These analyse in real time the sounds of speech and define in digital form such parameters as the pitch of vowel sounds, the phonetic envelope of syllables and the level of the signal. The digital description is transmitted over the link to enable a replica of the original speech to be synthesized; see [12]. Vocoders are complex, but they are now being produced in VLSI technology at a price low enough to permit use in commercial mobile–satellite telephone systems. Current systems providing intelligible communication and recognizable speech require less than 4 kbit/s per channel; see [13].

9.3.3 Encoding FDM assemblies of telephone channels

Where the terrestrial link to the earth station uses analogue transmission of telephony with FDM and the satellite link is digital, one option is to interface the two links at the channel level, demultiplexing and remultiplexing as necessary. There are likely, however, to be advantages in interfacing at a multiplex level. There are two ways of doing this.

It will sometimes be convenient to encode directly an FDM supergroup, containing five 12-channel groups, for transmission over the satellite link. With the supergroup occupying the frequency band 12–252 kHz, 9-bit encoding at a sampling rate of 576 000 samples per second would be appropriate. However, this involves a bit rate of 5.184 Mbit/s; that is, 86.4 kbit/s per channel, compared with 64 kbit/s per channel when interfacing at channel level.

Alternatively, a transmultiplexer can be used to convert directly from one 60-channel FDM supergroup to two 30-channel time division multiplex (TDM) aggregates (see Section 9.4.3) and vice versa, requiring 4.096 Mbit/s; that is, 68.3 kbit/s per channel.

9.3.4 Video signal encoding

A colour video signal at its source consists of three analogue components: red, blue and yellow. These are processed into a luminance signal and red/yellow and blue/yellow colour difference signals. These component signals are PCM-encoded for in-studio distribution. The encoding standards used for 625-line 50-fields-per-second systems require sampling the luminance component at 13.5 MHz and the two colour difference components at 6.75 MHz, with uniform quantization and 8 bits per sample. This yields a bit rate of 216 Mbit/s. Similar factors arise for 525/60 systems.

Before being broadcast in analogue form to the public, the three components are combined into a composite signal in which the two colour difference components modulate the chrominance subcarrier in accordance with PAL, SECAM or NTSC standards and are then combined with a luminance signal. This baseband signal can also be PCM-encoded. The sampling frequency is typically 3 or 4 times the frequency of the chrominance subcarrier and, using uniform quantization and 8 bits per sample, this yields bit rates of the order of 150 Mbit/s.

The use of either of these options for 625-line or 525-line video transmission by satellite is not economically attractive. However, there are various ways of processing video signals encoded at either the component or the composite stage to reduce the bit rate. A bit rate reduction of about 20 per cent can be obtained simply by replacing the conventional frame and line synchronizing pulses with digital synchronizing signals and slowing down the remaining bit stream to close up the time slots where the synchronizing pulses had been.

Much more bit rate economy can be achieved by taking advantage of the redundancy that is present in the picture signal. There is a high degree of correlation between successive picture elements along a scanning line. Similarly, most lines closely resemble the previous line and the subsequent line in the same field, and the adjacent, interlaced lines in the previous and subsequent fields. These correlations invite predictive methods to be applied. DPCM with fewer bits per sample can be used to transmit the difference between the present value of a picture element and a prediction of its value based on the previous elements in the same line and the corresponding element of the previous line, already transmitted. Alternate fields may be suppressed, being restored at the receiving terminal by interpolation. Further reductions can be achieved by optimizing sampling patterns and by taking advantage of the subjective characteristics of human visual perception.

Bit rate reduction systems operating a 34 Mbit/s and employing methods like these provide a picture quality which is acceptable for the chain feeding the final, broadcasting link. For the time being, however, analogue transmission is used for most links in FSS satellites providing programme feeds to broadcasting systems; see Section 8.2.3. No doubt digital links will be preferred when digital video broadcasting, satellite and terrestrial, becomes better established; see Section 15.3.

Systems which cause some degradation of picture quality but which provide substantial economy in bandwidth and earth station e.i.r.p. requirements, of great value for satellite news gathering (SNG), operate down to about 8 Mbit/s; see Section 14.5.1.

A typical conference television picture differs greatly from a picture intended for entertainment broadcasting. The conference picture contains much less movement and, in particular, the background changes very little. Using a system that identifies at the source the areas where movement is occurring and updates those areas, the static parts of the picture being maintained from a store at the destination, the required bit rate can be reduced below 1 Mbit/s. Viewphone systems are already in service requiring only 64 kbit/s.

9.4 Digital systems for satellite communication

9.4.1 Digital SCPC radio links

When a single digital speech channel modulates a carrier it is often necessary to add housekeeping bits to the bit stream carrying speech words to enable carrier phase, the digital clock and word timing to be recovered at the receiving earth station. Furthermore, economy in the use of satellite power often dictates that the carrier be suppressed at the transmitting earth station when the near-end speaker is silent; consequently, these housekeeping functions may have to be repeated very often. The following description of an SCPC system for telephony is based on the one used in the Intelsat system. The SCPC carriers are modulated by a 64 kbit/s signal using QPSK and are demodulated coherently. The carrier is suppressed if no speech is present. See Figure 9.13.

Speech signals are delivered to the speech detector at the transmitting earth station as 7-bit PCM at a bit rate of 56 kbit s^{-1}, the least significant bit of 8-bit PCM signals having been suppressed. When speech signals begin to arrive at the transmitting earth station, they are put into a store, the carrier is switched on and a preamble is sent. The first 20 symbols have polarity 1 in both the I and Q channels; that is, for 625 μs the carrier is radiated continuously in phase state A (see Figure 9.3). The receiving earth station recovers carrier and phase during this period, with no ambiguity. The next 40 symbols transmitted have alternate 101010 ... polarity in both channels; that is, for 1.25 ms the carrier is in phase states A and C for alternate symbols. The receiving earth station recovers the transmitting symbol clock timing during this period. The system is then ready to transmit a succession of 256-bit frames containing speech signals.

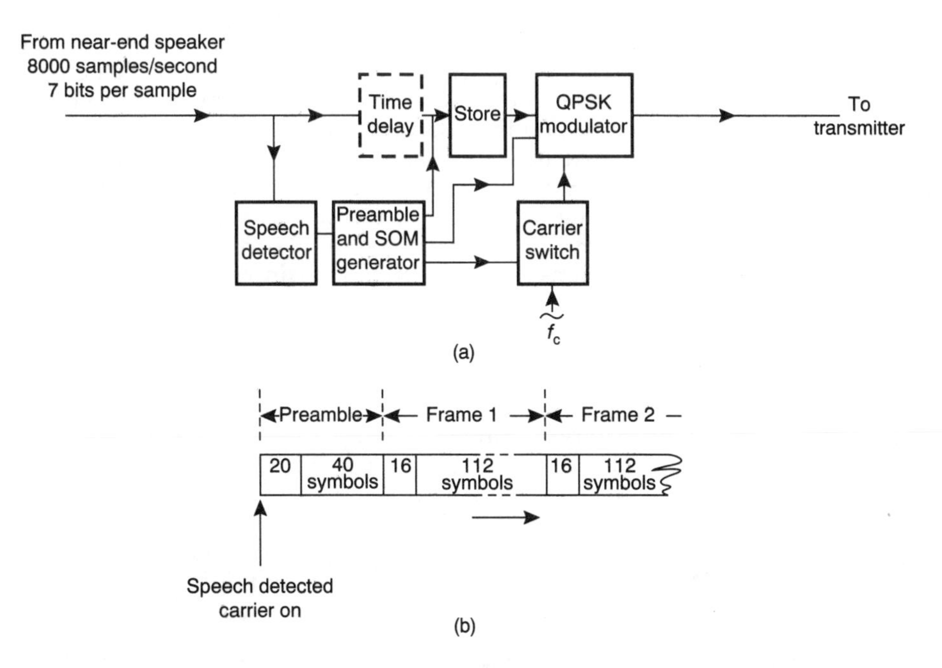

Figure 9.13 *A digital SCPC channel, showing (a) a block schematic diagram of the transmitting terminal, and (b) the signal structure of a speech burst, including the preamble and the first two frames of speech words*

Each frame starts with the 16 symbols (32 bits) of the start of message (SOM) word; the SOM word provides word timing synchronization, enabling the receiving earth station to predict precisely when the first speech word will begin. The frame is then completed by 32 7-bit speech code words. This frame of $32 + 32 \times 7 = 256$ bits is transmitted in 4 ms. Then the next frame begins with its SOM word, followed by its 32 speech code words, and so on. When the speaker stops speaking, and after a short pause, the carrier is switched off. The SOM word, repeated 250 times per second, ensures that there will be no loss of word timing during a speech burst.

At the receiving earth station a group of 32 7-bit received speech words is delivered into a store every 4 ms; a dummy bit is added to each word to form the least significant bit of an 8-bit word, and the speech words are read out of the store at 64 kbit/s, closing the gaps occupied, over the radio link, by the SOMs. It will be necessary to repeat the initial protocols for carrier phase and symbol clock recovery when the start of the next speech burst is detected.

It will be noted that the channel is not open for speech signals until about 2.3 ms after speech is first detected. Thus there will be 'front end clipping'. This is not noticeable with some languages but for others it may cause some loss of intelligibility. If so, clipping can be eliminated by inserting a time delay between the speech detector and the modulator.

More generally, the design of economical SCPC systems turns largely on the means used to assign radio channels to users, on the use of FEC in order to reduce the carrier power required to provide an acceptable quality of service and on the suppression of the carrier when it is not required. These matters are considered further in Section 10.5 Example 4 and Section 12.3.2.

9.4.2 Sequential access to transponders on demand

Satellite communication is uniquely flexible. For example, a temporary link of large information capacity can be set up quickly between a pair of earth stations by arrangement between system operating staff, then the transponder power and bandwidth may be used for another purpose and between a different pair of earth stations as soon as that link is released. However, many of the links which satellites are required to provide on demand are for very small amounts of information; perhaps a few hundred bits for a data-gathering link or even a few bits of housekeeping data for the control of a multiple access system. Slicker ways of handling brief connections of low information capacity are needed.

Time slot allotment and polling

For a star network involving a number of outstations which need from time to time to communicate with a master station, the simplest method is to allot time slots, perhaps one designated second in each minute, during which an outstation may transmit. Polling is a low-cost alternative; the master station automatically addressing an enquiry to each outstation in turn, asking if there is a need to communicate. The polled earth station responds affirmatively or negatively. If the response is negative, the next outstation is polled. If the response is affirmative, communication is set up immediately, polling being resumed when the call is finished.

These methods are very suitable, for example, for collecting environmental data, such as rainfall rates, river levels, etc. from large numbers of isolated observation points. However, the pre-allotted time slot method is very rigid and for both methods the traffic for the whole network is taken in sequence, so that even a negative polling response occupies more than half a second via a geostationary satellite. These methods are not satisfactory for mesh networks, for communication between people in real time, or where very urgent machine-to-machine messages may arise.

The ALOHA principle

The ALOHA system was developed as an experimental facility to provide data links between the various campuses of the University of Hawaii. The same carrier frequency was assigned to all participants. In its original form, any participating station transmitted a carrier burst modulated by a packet of data, preceded by the address of the destination earth station, whenever it was ready to do so. The size of a packet was limited to prevent any one user from denying use of the system to other participants; a long message was broken up and sent as a series of packets of modest size. If a packet was received satisfactorily by the intended recipient, an acknowledgement was sent automatically in reply. If there was interference to the reception of the burst, typically due to collision with a burst simultaneously transmitted from another earth station, or if the acknowledgement itself suffered interference, the packet was sent again after a brief pause, the process being repeated until success was achieved. To avoid repeated collisions of the same two packets of data, the lengths of the pauses before retransmission were randomized.

The basic ALOHA system works well when the channel is lightly loaded, but as the level of traffic rises and collisions become frequent, its effectiveness is reduced. Quaglione [14] notes that, with traffic being initiated randomly, the system does not make use of more than about 18 per cent of the capacity of the channel. Potential throughput can be increased in two ways:

(1) When a basic ALOHA system is heavily loaded, one packet may overlap, for example, with the end of another packet from a different earth station and the start of yet another packet from a third earth station, causing all three attempts to communicate to fail. In slotted-ALOHA, users are not free to start a burst at random instants in time; they must locate their bursts in time slots derived from a system clock. In this way, a smaller proportion of the bursts that are transmitted suffer collisions.
(2) Various slot reservation systems have been devised. For example, a user with several packets of data to send, having used a slot successfully for one burst, may inhibit the use by other earth stations of a designated future slot, thereby keeping it clear for a second burst, and so on. A more elaborate slot reservation system may use memory at a central earth station to register needs for transmission facilities; as slots become free they are allotted to earth stations with registered needs.

Access control networks

A group of earth stations may share a group of telephone or data channels in FDMA, or TDM, any pair of earth stations having the right to seize any free pair of channels or slots on demand. A channel access control system, separate from the channels that are used for traffic, is usually provided in such cases. The Intelsat Spade system [2] used a narrowband TDMA system to enable any participating earth station to announce to all other participants an intention to seize a designated channel, and to pass call setting-up information to the destination earth station. Most systems using shared access to a group of SCPC channels, however, use a simpler system, such as slotted ALOHA, to get access to a register at a control earth station which will assign free channels for use on demand; see, for example, Section 14.3.1 for the methods used in the Inmarsat system.

9.4.3 Time division multiplex

In a simple time division multiplex (TDM) system a continuous bit stream is divided into frames, each containing one word from each of a number of digital channels. Additional bits which do not carry information are included in the bit stream to provide for housekeeping functions such as automatic frame synchronization. At the TDM receiving terminal, a continuous channel can be made up by extracting the word relating to that channel from each frame and assembling them in sequence. Figure 9.14 shows an elementary model TDM system, multiplexing 10 channels with 8-bit words.

TDM systems take various forms and have a wide range of information transmission capacities. Some have quite low capacity; perhaps 24 channels of the 6-bit words of telex channels operating at 100 baud, giving a total information rate (disregarding housekeeping bits) of 2.4 kbit/s. Some, operating at bit rates up to several tens of megabits per second, carry hundreds of telephone channels. Other systems, with a frame structure that can be flexibly reconfigured to meet changing needs, are specially designed to function as transmission elements in a complex digital telecommunication network. Most high bit rate systems carry 24 or 32 telephone channels, or multiples thereof, conforming to ITU standards associated with PCM encoding using the μ or A companding laws to which Section 9.3.1 refers.

TDM systems like these are transmitted over satellite links, modulating a carrier, typically by coherent BPSK or QPSK or by DCPSK, often protected with FEC, and using a level of carrier power scaled to provide an appropriate C/N ratio in the bandwidth they occupy; see Section 10.5 Example 2. The same TDM equipment design is also used for cable and terrestrial radio links. For the satellite application, a system of very large capacity might use all the power or all the bandwidth of a transponder, but most TDM systems share a transponder with other emissions by FDMA.

The ITU standard TDM telephony systems are described briefly below; they are defined in detail in various CCITT texts, a key text being Recommendation

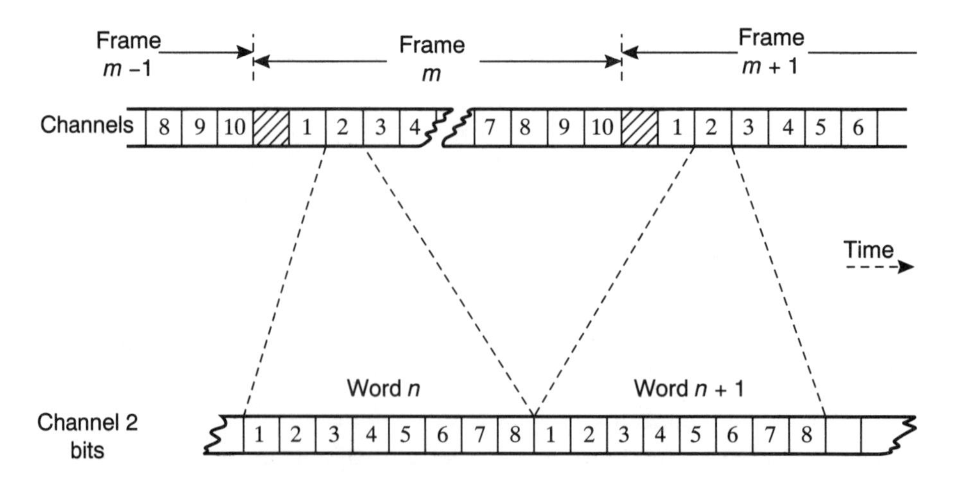

Figure 9.14 *The signal structure of an elementary TDM system carrying 10 channels of 8-bit words*

G.704 [15]. These systems have complex housekeeping subsystems. To facilitate the economic transfer of housekeeping control signals between TDM terminals, it is found convenient to organize frames into 'multiframes' of 12 or 16 consecutive frames. Frame synchronizing bits are transmitted with every frame but some supervisory system bits are transmitted only once per superframe.

These TDM systems are primarily intended to provide 8-bit PCM telephone channels. However, the channels are also used for other services that are commonly carried in the telephone network, such as low-speed data and telex channels on multiple subcarriers or facsimile and medium-speed data via modems. Furthermore, with the appropriate peripheral equipment, each 64 kbit/s channel can be used, for example, for two 32 kbit/s ADPCM channels or one 64 kbit/s data channel.

The ITU primary PCM multiplex operating at 2048 kbit/s

The 2048 kbit/s primary PCM multiplex has a frame length of 256 bits, consisting of 32 8-bit code words, numbered 0 to 31. The zero word is used for various housekeeping functions. Words 1–15 and 17–31 provide 30 channels, identified by the same numbers and encoded with A law companding when used for telephony. Word 16 is available for common channel telephone signalling, but it may be used as an additional telephone channel if not needed for signalling. A total of 8000 frames are transmitted per second, giving $256 \times 8000 = 2048$ kbit/s.

The 256-bit frame is organized in 16-frame multiframes, as illustrated in Figure 9.15. The zero word of the first frame of each multiframe contains the 7-bit frame alignment word (FAW), namely 0 0 1 1 0 1 1. This FAW is repeated in the zero words of alternate frames throughout the multiframe and the first two bits of the zero words that do not carry the FAW provide for multiframe alignment. All the rest of the bits in the various zero words are available for other supervisory purposes.

A TDM receiving terminal seeking to synchronize with an incoming signal examines the polarity of bits in a sequence of words separated by 31×8 bits to

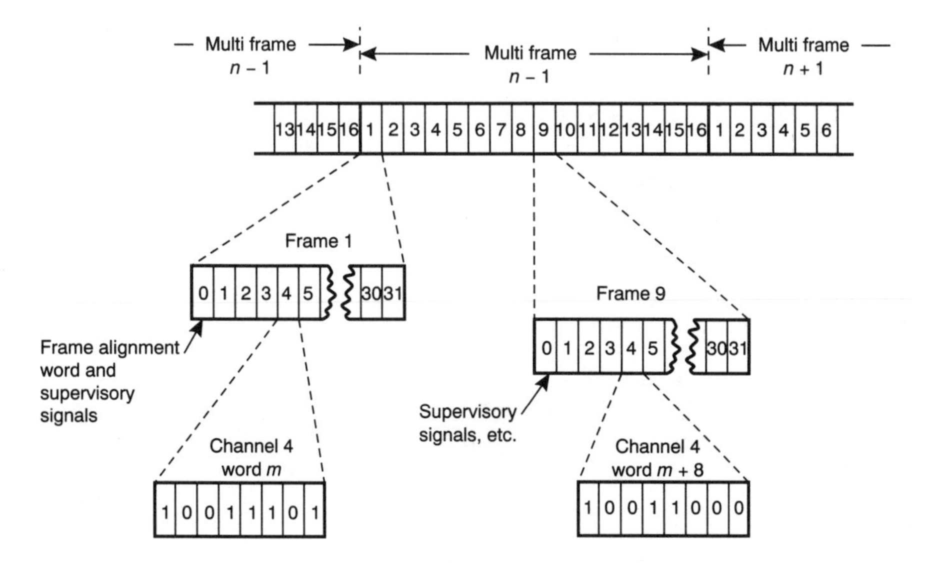

Figure 9.15 *The signal structure of the CCITT primary PCM TDM multiplex operating at 2048 kbit/s*

determine whether they contain the FAW and the multiframe alignment bits. If the FAW is not received correctly and consistently, the terminal is programmed to conclude that frame alignment or multiframe alignment has not been achieved and the terminal then hunts for a new alignment state where the FAW is received correctly.

The ITU primary PCM multiplex operating at 1544 kbit/s

The 1544 kbit/s primary PCM multiplex specification is derived from the Bell T1 system. The frame length is 193 bits, consisting of 24 8-bit code words plus 1 additional bit, the first bit in every frame. The first bit is identified as the zero word; it is designated an F-bit, which indicates that it is used for frame and multiframe timing. The 8-bit code words are numbered 1 to 24; they are encoded using the μ law when used for speech and provide channels 1 to 24 respectively. One of the 8-bit channels may be set aside for common channel telephone signalling; otherwise, the eighth (least significant) bit of channels 6, 12, 18 and 24 may be used ('stolen') for signalling, leaving seven bits for those telephone channels. A total of 8000 frames are transmitted per second, giving $193 \times 8000 = 1544$ kbit/s.

The 193-bit frames are arranged in multiframes of 12 (or 24) consecutive frames. The 12 F-bits of a 12-frame multiframe, or the first 12 of the F-bits of a 24-frame multiframe, are used for the FAW. In other respects the 1544 kbit/s multiplex resembles the 2048 kbit/s multiplex.

Higher-order ITU PCM multiplexes

When these primary PCM TDM systems provide insufficient capacity, four primary multiplexes are combined to make a second-order system. The more complex system so produced requires additional 'housekeeping' functions to be performed, and additional functional bits are needed to serve these functions. In consequence, second-order systems made up of four 1544 or four 2048 kbit/s primary multiplexes have bit rates of 6312 or 8448 kbit/s respectively. In a similar way, a third-order system is produced by combining several second-order multiplexes. And so on.

Interfacing satellite TDM with terrestrial networks

The national digital networks of a number of countries are already synchronized by a single clock source, or virtually so, but there is no global clock. National clocks differ to a small but very significant degree. In addition, even in geostationary satellite systems there is some relative movement between the satellite and its earth stations (see Section 5.3.2), so Doppler effects create additional potential synchronization problems at the interfaces between the satellite system and the terrestrial network. However, there are some interface arrangements which avoid these difficulties.

Sometimes the signals of a TDM satellite link are carried by analogue plant between the earth station and their terrestrial source or their terrestrial destination. Where both backhaul links are analogue, the analogue/digital converters take their bit, word and frame timing from the clocks that control the TDM system.

Where one backhaul link is digital and the other is analogue, there can be a direct digital interface (DDI) between that digital terrestrial system and the TDM system at the associated earth station. The TDM system would take its timing from the terrestial system. The analogue/digital converter at the other earth station would take its timing from the TDM system and thus, indirectly, from the distant terrestrial system.

For other TDM systems both backhaul links are digital, but signals are converted to the analogue state at the receiving earth stations before being reconverted to the digital state, thereby avoiding the synchronization problems that are outlined below. These various arrangements are illustrated in Figure 9.16.

However, where the backhaul links to both earth stations are digital, it is usually technically and economically preferable for the satellite/terrestrial interface to be a direct digital one at both earth stations. This involves timing problems. Doppler effects due to satellite motion relative to the earth stations are

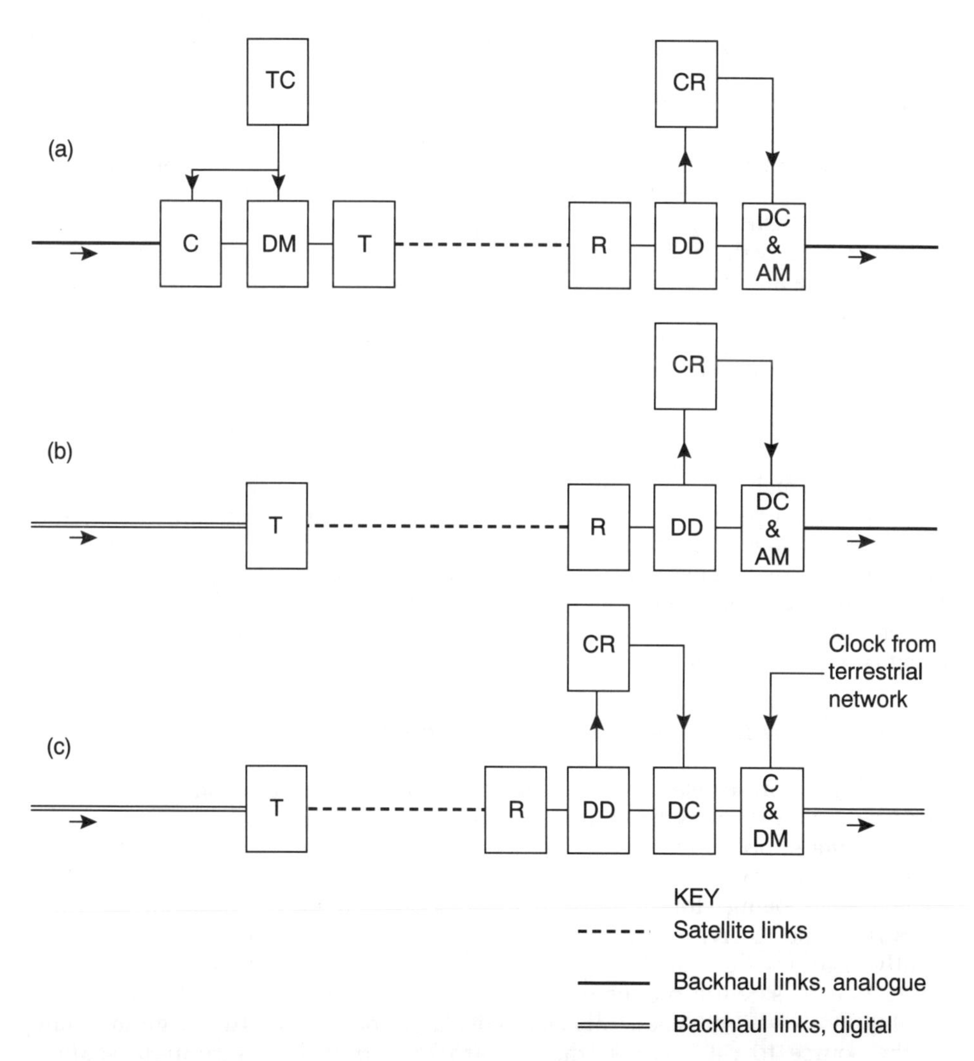

Figure 9.16 *Interfacing TDM satellite links with terrestrial networks: (a) between analogue networks, (b) between an analogue network and a DDI to a digital network, and (c) between digital networks, but with a DDI at one interface only. Key: T, transmitter; R, receiver, C, encoder; DC, decoder; DM, digital multiplexer; DD, digital demultiplexer; TC, TDM clock; CR, clock recovery; AM, analogue multiplexer*

cumulative and significant. Differences between the timings of the various terrestrial digital networks must also be accommodated.

Doppler effect and DDI

With fixed earth stations communicating via a geostationary satellite which is being kept very precisely on its nominal longitudinal station, the distance between an earth station and the satellite varies cyclically during the course of each day over a range of perhaps 30 km. However, some nominally geostationary satellites have an orbital inclination of several degrees and in such cases the daily movements may range up to 500 km. Moreover, if two earth stations using the satellite both have high northern latitudes or both have high southern latitudes, the effect of the daily variations on the length of their mutual satellite link will be additive.

For a day as a whole, these variations will sum to zero. Nevertheless, if both earth stations have DDIs and if it is assumed that the terrestrial networks which the TDM system connects together are synchronous with one another, then for half of each day the satellite link will be delivering bits to the receiving earth station rather faster than they can be accepted by the terrestrial network, and for the rest of the day the delivery will be too slow. For a satellite with precise station-keeping, these alternating surpluses and shortages of bits will involve some tens of thousands of bits for a 2 Mbit/s TDM system, and proportionately more for higher-order TDM systems. For a satellite with substantial orbital inclination, the order of magnitude may be megabits.

The necessary flexibility is provided by an elastic buffer store located after the earth station receiver. Bits are fed into this store at the rate at which the satellite link delivers them and they are clocked out of the store by the terrestrial network to which the receiving earth station is connected. While the length of the propagation path is shortening the number of bits in the store will rise, and while the path length is increasing the number of stored bits will fall. Subject to a number of conditions, and in particular that care is taken to ensure that the store is never completely emptied, the buffer store enables the DDIs to function between terrestrial networks which are synchronous with one another without losing or repeating any bits.

DDI timing with terrestrial networks that are not synchronous

The timing errors caused by the absence of perfect synchronism between the terrestrial networks may be treated by frame slipping or justification.

The frame slipping technique requires the PCM bit stream delivered by the satellite link to flow through an elastic buffer store at the receiving earth station, a few frames being stored before being clocked out in synchronism with the terrestrial link receiving the signal. If the onward terrestrial link is running slightly faster than the clock of the source of the signal (disregarding the Doppler effects due to satellite movement), there will be times when only one frame remains in the store; when this happens, a frame is sent twice, giving time for the source to catch up. If the onward terrestrial link is running slightly slower than the source clock, one frame will occasionally be dumped from the bit stream without being passed on. In both cases, the integrity of the multiframe is preserved. The store used for the frame slipping function can be combined with the much larger store that is required for Doppler correction. With a pair of plesiochronous networks, a 125 μs frame may have to be repeated or dumped once every 72 days.

Justification, also called pulse stuffing, involves adding dummy bits to the onward bit stream at the receiving Earth station or removing dummy bits that had previously been included in the bit stream in case the need should arise. Housekeeping bits which identify dummy bits are included in the bit stream to enable all dummy bits to be eliminated before the signal is passed to its destination, a process called dejustification.

Frame slipping is relatively simple to implement and, within limits, has no noticeable effect on speech; it is, however, objectionable for many data links unless the difference in timing between the terrestrial networks being interconnected is extremely small. Justification leaves the bit stream unimpaired but it is complex to implement and is not compatible with some channel handling techniques, such as digital speech interpolation (DSI).

9.4.4 Demand assignment and digital speech interpolation

In a conventional satellite TDM system, used for the PSTN, channels are pre-assigned in pairs to form permanent two-way circuits, a number of which serve the route between two switching centres, typically located at national gateways or provincial capitals. Such channels are idle for much of the time for the following reasons:

(1) While a telephone call is in progress, the parties do not both speak at the same time and there are times when neither is speaking. Thus, the 'go' and 'return' channels will, on average, be idle for rather more than half of the time that a call is in progress.
(2) Enough circuits are assigned to a route to meet the peak demand. Outside the 'busy hour', the demand is considerably less and at any moment many circuits are not carrying a call.
(3) Even in the busy hour a circuit is not in use all the time.

The flow of telephone calls through a circuit or a group of circuits is measured in erlangs:

$$Ch/T = A \quad \text{(erlangs)} \tag{9.8}$$

where $C =$ the number of calls in T hours, and
$h =$ the average holding time in hours.

With telephone calls arising at random times, the maximum amount of traffic that a circuit can carry is a function of the number of circuits that are available to serve the route. The classic study of the statistics of the flow of telephone calls in automatically switched networks was done by Erlang [16], after whom the unit is named. Erlang's lost-call formula states that the probability p that a new caller will fail to find a free circuit from among N circuits already carrying a total of A erlangs of traffic is given by

$$p = \frac{A^N/N!}{1 + A^1/1! + A^2/2! + \cdots + A^N/N!} \tag{9.9}$$

if the demand for calls is random and all circuits are equally available. Thus, taking a lost call probability of 0.02 as the target grade of service in the busy hour, the traffic capacity per circuit varies with the size of the group of which the circuit is a member as shown in Figure 9.17. It may be argued that the cost of long circuits might justify a less stringent grade of service. Nevertheless, it is clear that small groups of circuits cannot be loaded efficiently.

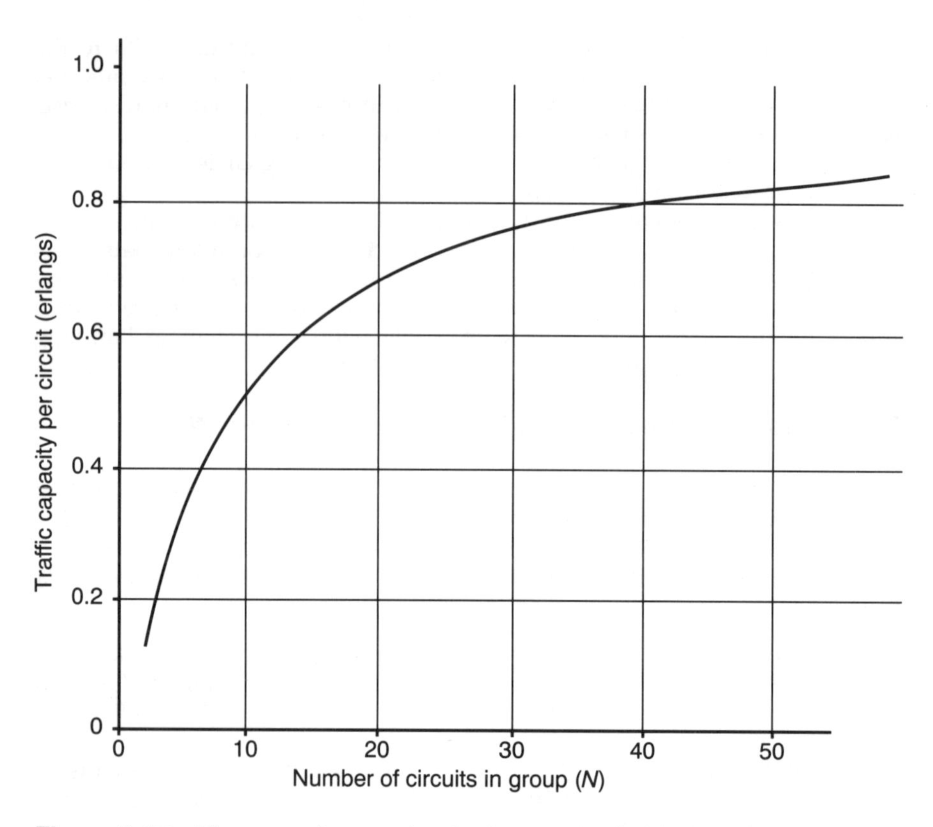

Figure 9.17 *The capacity per circuit of a group of circuits with full availability as a function of the number of circuits in the group for a probability of one lost call in 50*

Substantial incidence of idle channel time is not, of course, peculiar to TDM systems but with other transmission techniques the consequences are mitigated in various ways. For FDM/FM, a main factor determining the bandwidth and power requirement of the emission is the multi-channel loading factor l, which takes foreseen loading levels into account. In addition, many SCPC systems, whether analogue or digital, suppress the carrier when the speaker is silent; if the transponder is power-limited, the number of carriers given access to the transponder can be made greater than the number that the transponder could support if all were continuously present.

Two techniques that enable the efficiency of channel loading in TDM systems to be increased are demand assignment (DA) of channels shared between a group of routes and digital speech interpolation (DSI). DA is also beneficial when used on SCPC channels using bandwidth-limited transponders and it is particularly effective where the busy-hour traffic between two switching centres is small, a few erlangs or less. DSI shows most benefit where the traffic flow is large.

Demand assignment

Consider a group of earth stations, all of which need to communicate with all of the others. Circuits may be pre-assigned for traffic between each pair of earth stations and used solely for that route. Alternatively, a block of circuits may be made available for use on any of the routes, any free circuit being assigned to a call when the demand arises. It is found that the same total amount of traffic can be carried by fewer space segment channels in the DA mode.

A model enables a simplistic comparison to be made between the satellite channel loading efficiency achievable by the pre-assignment and demand assignment of channels. Six earth stations access a satellite using TDM. Earth station A transmits a multiplex emission and earth stations B to F receive it. To complete the model, Earth stations B to F each transmit similar TDM emissions for the five other earth stations to receive. Let it be assumed that the busy-hour traffic flow from each earth station to each other station is 2 erlangs. Thus, the maximum traffic demand on this group of stations would be $2 \times 5 \times 6 = 60$ erlangs.

With pre-assignment of channels, equation (9.9) shows that six circuits would be required from each earth station to every other earth station to carry 2 erlangs of traffic with a lost call probability of 0.02. A total of 180 circuits would be required, each circuit carrying 0.33 erlangs in the busy hour.

With DA, each earth station would be equipped to transmit and receive all the satellite channels used by the group as a whole. Any two free channel frequency assignments could be used for any call that arose. With full availability of all channels obtained in this way, 60 erlangs could be carried by a total of 68 satellite channel pairs, assuming that the busy hour was at the same time on all routes. Additional satellite capacity, perhaps equivalent to one telephone channel per earth station, would be needed for the DA system control. Allowing for control channels, demand assignment would raise the busy-hour traffic load to about 0.81 erlangs per circuit-equivalent. Thus, while the cost of earth station hardware would be increased to provide DA facilities, the space segment cost would be reduced to only 41 per cent of what would be required for pre-assigned circuits. And the potential space segment saving would be greater if the busy hours for the various routes did not coincide.

DA is used in some TDM systems and the principle is widely used where SCPC networks are carried by bandwidth-limited transponders.

Digital speech interpolation

DSI attacks the inefficiency that arises with pairs of pre-assigned channels because they are not used simultaneously in both directions.

With DSI, a block of N TDM satellite channels for each direction of transmission is pre-assigned for telephone calls between two switching centres, and connections for up to M telephone calls at a time have access to those blocks, where M is greater than N. A notional 2-way circuit is established between the switching centres and extended to the calling and called subscribers as soon as the call is initiated, but a (1-way) channel over the satellite portion of the circuit is assigned to the call only while the presence of speech signals in that direction can be detected. At other times the channel is available for a speech burst on any other circuit with access to the DSI group. A fast-acting control channel is required between the earth stations to ensure that the satellite channel assigned for a speech burst at the transmitting earth station is connected very quickly at the receiving earth station to the speaker's correspondent. See [17] and Figure 9.18. In the figure the path of a call from circuit 3 of the West switching centre to circuit 6 of the East switching centre is emphasized, carrying one speech burst in each direction.

In this way, the signals of M switch-to-switch circuits, each with a call in progress, can be transmitted over the satellite link using N TDM channels in each direction, where M is considerably larger than N. The ratio M/N is called the DSI gain.

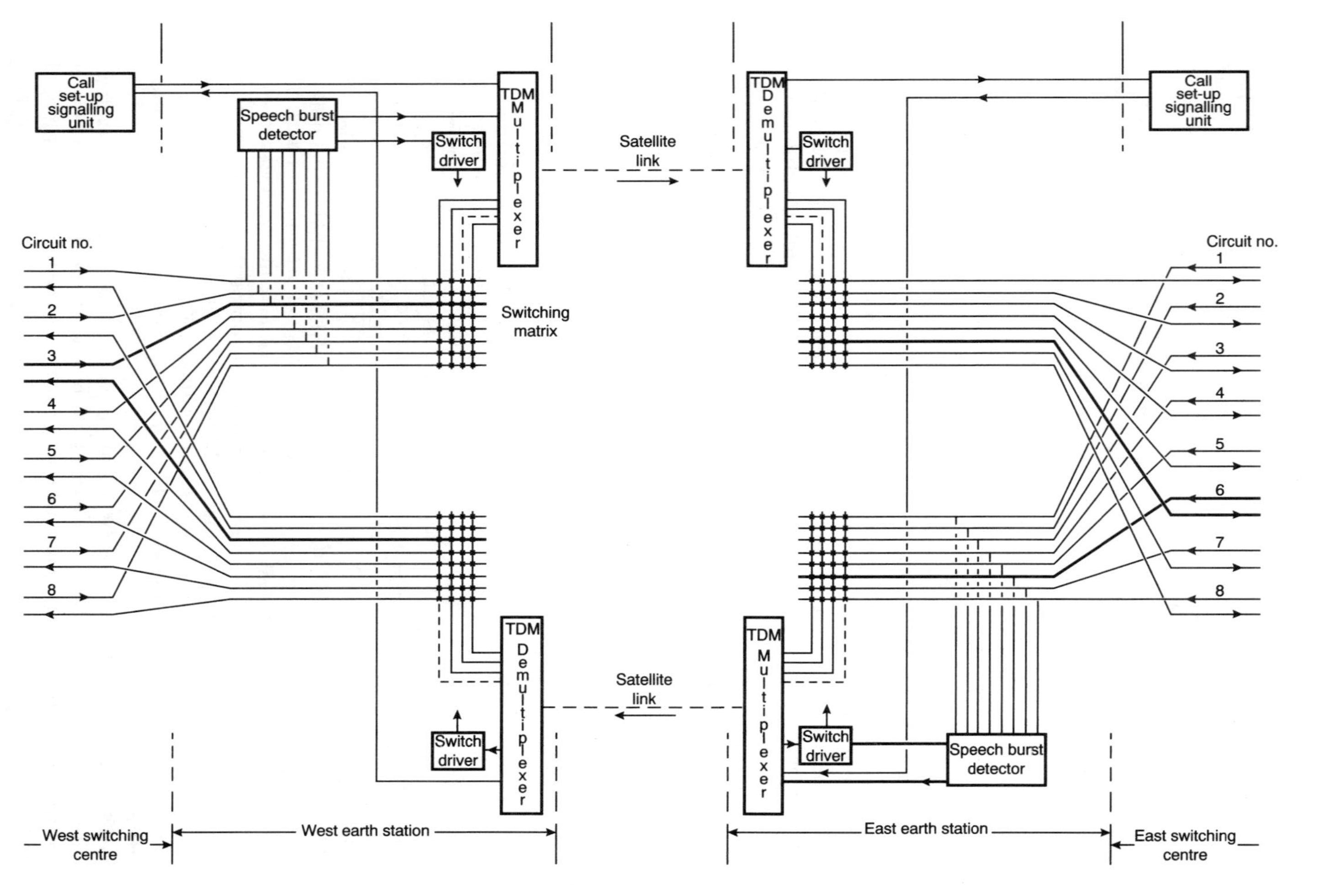

Figure 9.18 *A schematic diagram of a model DSI system carried by a TDM link between earth stations, West and East, showing a matrix of eight circuits and four pairs of channels*

There is no rigid relationship between M and N, but as the number of circuits given access to a given number of satellite channels is increased, the probability rises that a channel will not be available when a speaker begins to speak; the voice will not be heard for the fraction of a second, perhaps longer, before a channel becomes free. This is called freeze-out. In practice, it is usual to limit the number of circuits so that the probability of noticeable freeze-out is small when traffic flow is at its peak. Given this precaution, there is no significant freeze-out at other times.

DSI is not helpful when N is very small, but significant benefit is obtainable without unacceptable freeze-out if $N = 10$. A DSI gain of about 2.2 can be obtained if $N = 100$ channels in each direction, with the possibility of some further increase in the gain for very large routes. Thus DSI can be seen as a form of DCME. When ADPCM and DSI are both applied to a large group of transmission channels, a total circuit multiplication factor around 5 is obtainable. For internationally agreed algorithms for ADPCM/DCME, see CCITT Recommendation G.763 [18].

9.4.5 Time division multiple access

An earth station transmitting a TDM emission seldom requires all of the power or bandwidth of a complete satellite transponder even when that emission is used for the GO legs of circuits with many distant earth stations. Several or many TDM carriers from different earth stations can access a transponder in FDMA, but at a cost in terms of total information capacity because the power amplifier of the transponder must be backed-off by a substantial margin to limit intermodulation noise, etc.

Time division multiple access (TDMA) allows earth stations to transmit their digitally modulated carriers through a single transponder without substantial backoff because the carriers access the transponder, not simultaneously, but in sequence. In consequence the transponder can relay more information, up to the maximum allowed by its bandwidth and power.

In a typical high-capacity TDMA network, each earth station in turn transmits a carrier for a few tens of microseconds, the carrier being modulated at a bit rate high enough to load the information capacity of the transponder fully. The signals sent may be addressed to any or all of the other earth stations participating in the TDMA system. After one or two milliseconds, during which each of the other earth stations will have taken its turn to transmit, the cycle is repeated, each earth station transmitting in turn the information words that have accumulated since its previous transmission, and so on. Earth stations receive all transmissions of interest from other earth stations and extract the signals that are addressed to them.

These brief transmissions are called carrier bursts and the period (one or two milliseconds in this example) during which each earth station has one opportunity to transmit signals to other earth stations is called a frame. As with TDM systems, it may be found convenient to organize frames into groups of frames; for TDMA, the higher-order group is called a superframe.

The earth station equipment that TDMA requires is relatively costly, but it enables more economic use to be made of other costly assets, such as satellite transponders and earth station HPAs. TDMA has other advantages also. In particular many of the network parameters can be implemented in software, providing great flexibility. It may, for example, be easy and quick to reconfigure the facilities available to an earth station, such as opening links between participating earth stations which were not previously connected, changing the number of telephone channels on an established route, or replacing a group of telephone channels with a bit-reduced video channel or a high-speed data channel, without interrupting the traffic flowing through channels which are not to be changed. These changes are relatively cumbersome to implement with conventional FDMA.

For reasons such as these, the use of the TDMA technique may sometimes be economically justified even if the system does not require all of the power or bandwidth of the transponder. In such cases a transponder may be shared with other carriers by means of FDMA, the reduction in total transponder e.i.r.p. through backoff being accepted. Thus, while wideband TDMA systems having a nominal bandwidth of 36 or 72 MHz and requiring the exclusive use of transponders may be the ones most commonly encountered, there are others providing flexible digital facilities which operate in a bandwidth of 2 MHz. Very simple TDMA systems are used to apply the same principle to such limited

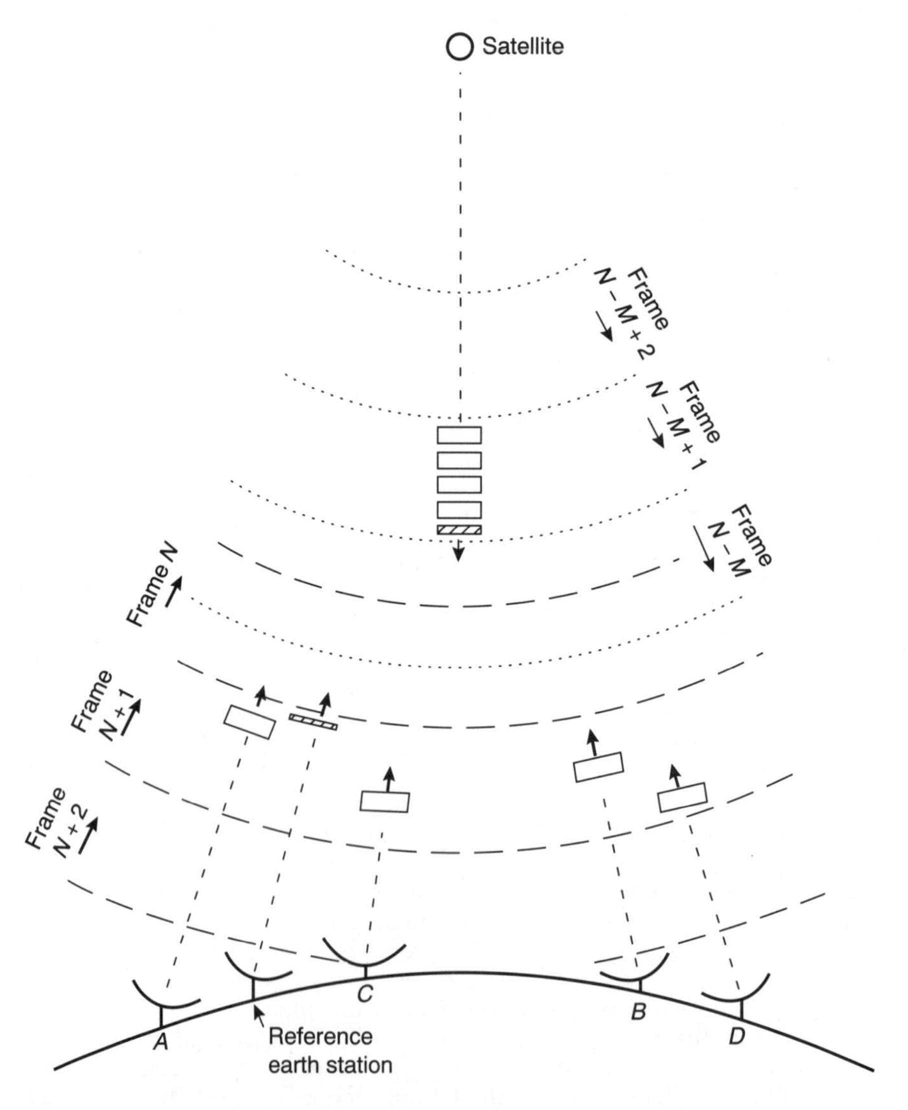

Figure 9.19 *A TDMA system serving earth stations A, B, C and D, showing frame N − M + 1 being downlinked and frame N + 1 being uplinked*

tasks as carrying telex channels and controlling access to a group of SCPC channels.

The description of TDMA systems that follows in this section relates mainly to large-capacity systems and it has been kept rather general, emphasis being given to a description of the functions that are performed rather than the hardware and software that performs them, which can take various forms. There is further information on specific systems in later chapters.

Frame structure

Figure 9.19 shows the carrier bursts from several earth stations which make up TDMA frame $N+1$ on their way to the satellite, and the bursts of frame $N-M+1$ returning to Earth after amplification by the transponder. One of the earth stations, called the reference earth station provides frame reference bursts which establish a timing standard for the system, enabling the other earth stations to keep their bursts accurately within their assigned time slot in the frame.

Burst carriers are usually modulated by coherent or differential QPSK. The modulation of a frame reference burst consists of a preamble, as described in the next paragraph, typically followed by addition words which are used for identifying the station transmitting and for various system control functions. A traffic burst begins with a preamble, similar to that of the frame reference burst, also followed by identification and system control words but followed in turn by one or more data sub-bursts, consisting of a sequence of information words containing the traffic to be communicated. See Figure 9.20.

The TDMA preamble used in a high-capacity system is rather similar in form and function to that of the voice-switched digital SCPC carrier described in Section 9.4.1. It starts with a sequence of around 200 symbols to enable carrier frequency to be acquired, phase ambiguity to be resolved and clock (bit timing) to be recovered. Next comes a 'unique word' (UW) sequence, consisting of perhaps

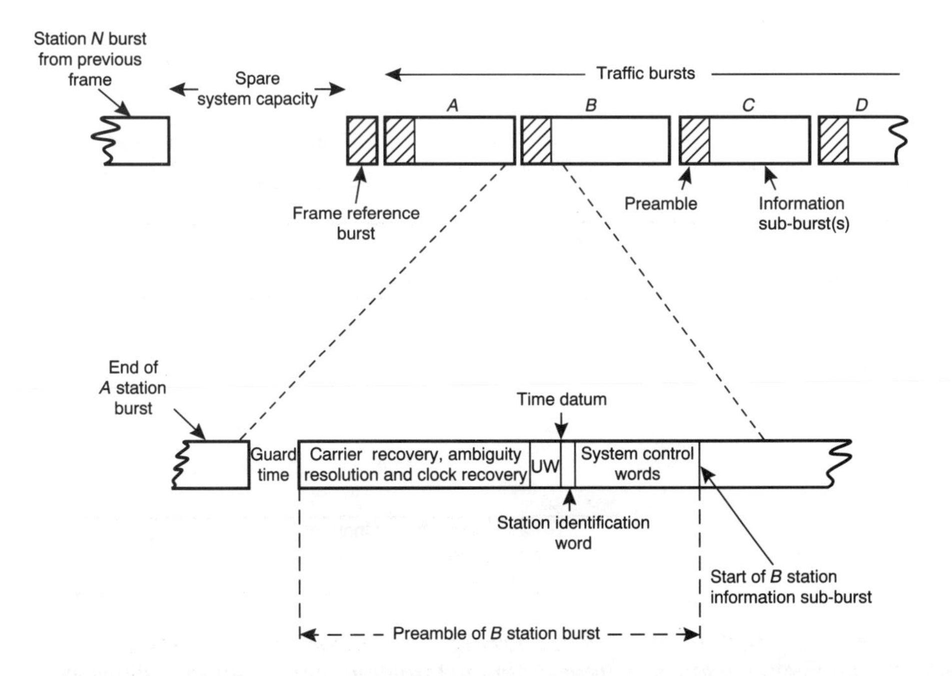

Figure 9.20 *A TDMA burst, showing the structure of a typical preamble*

12 symbols, the final symbol of which provides a precise time datum. This datum provides word-timing synchronism for the identification, system control and information words that follow. For frame reference bursts, the time datum is also used to synchronize the frame as a whole.

The information sub-bursts

The information sub-bursts which follow the preamble of a traffic burst may contain various kinds of signals. Some are PCM telephone multiplex systems, with or without DSI, with or without ADPCM. Other modules may contain an encoded FDM multiplex, a high-speed data channel or a bit-reduced video channel. The whole of the bit stream of one sub-burst may be intended for reception at one other earth station but the various channels making up another sub-burst may be destined for reception at various different earth stations. Many of the features of a typical set of data sub-bursts are illustrated in Figure 9.21.

Frame synchronization

The system master clock which controls the transmission of the frame reference burst is itself driven from a highly stable frequency source. However, the absolute timing of the reference bursts, as perceived by another earth station, will depend on the locations of the two stations, being delayed, for example, if the stations are remote from the subsatellite point. The reference bursts will also drift forward and back in the course of a day by a fraction of a millisecond, due to cyclical changes in the physical lengths of the transmission paths between the earth stations and the satellite, which must to be assumed to be imperfectly

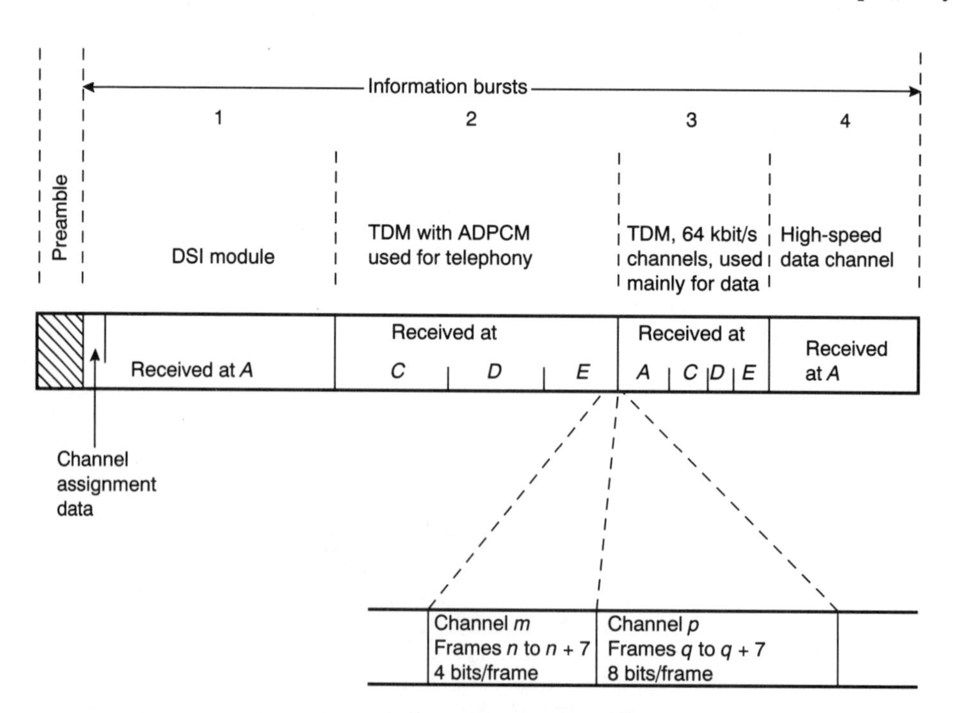

Figure 9.21 *A TDMA traffic burst, showing the structure of typical data sub-bursts. Earth station B is transmitting information sub-bursts to stations A, C, D, E and F, using QPSK and a 2 ms frame period*

geostationary. All earth station clocks, locked to the system master clock, must follow the cyclical drift, so maintaining frame timing. The timing of a traffic burst, relative to the frame reference burst, must be measured after both bursts have been retransmitted by the satellite.

Each earth station in a TDMA network will have been assigned a position in the burst time plan, in an unoccupied time slot of duration $t_1 + t_2$, long enough to accommodate the station's traffic flow, and identified by T, the time elapsing between the UW of the reference burst and the UW of its own prospective burst as observed in the downlink. These symbols are illustrated for earth station B in Figure 9.22. When earth station B is preparing to enter the network, it receives a series of reference bursts and uses the UW to lock its local digital clock to the system master clock. However, before the earth station begins to transmit its normal traffic burst, it must determine the required timing for its uplink signal, relative to reference bursts perceived in the downlink, with sufficient accuracy to avoid interfering with other bursts which are already in operation.

One of the ways that are used to do this depends on precise knowledge, calculated perhaps from the parameters of the orbit and the earth station location, of the distance between the earth station and the satellite at the time when the entry process is about to begin. Given that this distance is d, it will be seen that if station B were to transmit a burst at exactly the same time as it received reference burst n, the B station burst would be retransmitted by the satellite transponder $2d/c$ seconds later than the time of retransmission of reference burst n. Let the number of complete frames transmitted by the transponder during the period $2d/c$ be x. Then the retransmission of this B station burst would have occurred $2d/c - xP_F$ seconds after the retransmission of reference burst $n + x$, where P_F seconds is the frame period. It follows that the correct time for station B to transmit its burst, after the reception of a reference burst, is given by

$$2d/c - xP_F = T \quad \text{(seconds)} \tag{9.10}$$

To lessen the risk of a timing error, earth station B might begin to enter the network by transmitting the preamble of its burst, without the information sub-bursts, at the calculated delay relative to the receipt of a reference burst, plus a small addition δt to provide a safety margin. The shortened bursts will be received at the earth station after retransmission from the satellite starting x frames later. Once station B can receive its own preamble after retransmission by the satellite, it can measure the time difference between its UW and that of the frame reference burst with great precision. Having done so it will make any final timing corrections, omit the safety margin δt and start transmitting the information sub-bursts. Once the complete traffic burst has been established in its proper time slot, its timing must be monitored continuously, in particular to ensure that the cyclical variation of frame reference burst timing is accurately tracked. A brief 'guard time' is scheduled between bursts in case minor mis-timings should occur.

Frame period

The length of the TDMA frame is one of the important parameters of a TDMA system. If it is too short, too much time is spent on preambles and guard time, leaving too little for traffic. If the frame is too long, the cost of storage for data bits at the earth stations may become significant. If it is very long indeed there is an increase in the end-to-end transmission time which may be unacceptable.

For a TDMA network intended primarily for telephone transmission, the obvious choice for the frame period is the same as the analogue-to-digital sampling period, namely 125 μs. However, to take an example, the core of the

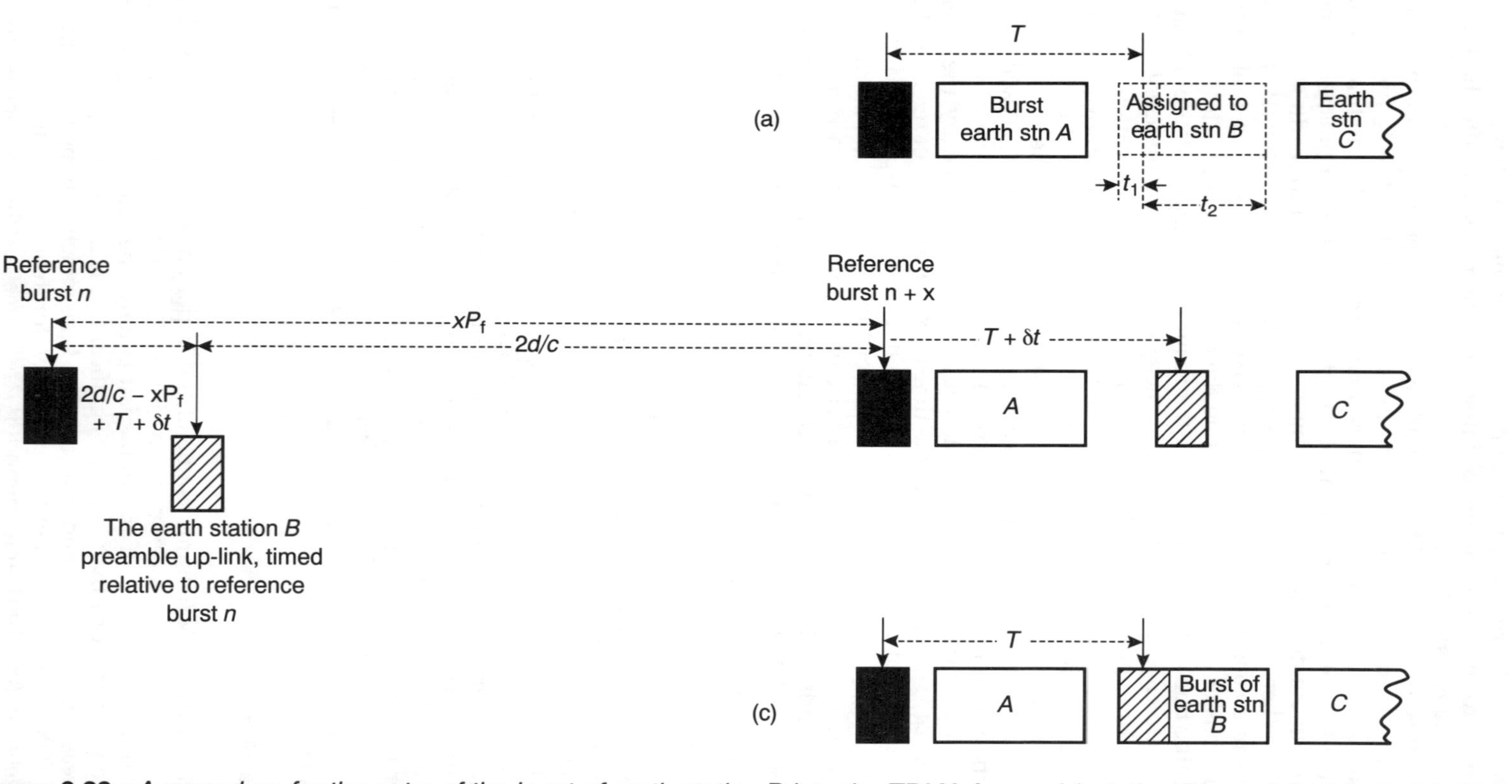

Figure 9.22 *A procedure for the entry of the burst of earth station B into the TDMA frame: (a) station B's assigned time slot; (b) the initial position of the station B preamble, based on computed transmission time; (c) the station B traffic burst in its operating position*

Intelsat TDMA preamble, from the start to the end of the UW, contains 200 symbols and the nominal guard time is 48 symbols. Thus, the preamble and guard time for each burst occupy 4.1 μs of frame time. This would be an unacceptable overhead for a system using a 125 μs frame. A frame period of 2 ms has been chosen by Intelsat as a suitable compromise between usable capacity and data store costs.

Other high-capacity TDMA systems that are used for telephony, and especially those serving smaller geographical areas, have shorter preambles; for these, shorter frames might be found desirable. On the other hand, for a medium-capacity system serving many earth stations, for which the loss of capacity for a given preamble length would be proportionately greater, a longer frame period might be preferred. A much longer frame time may be used for TDMA systems that are not used for telephony, thus avoiding severe constraints on the total transmission time. For example, the Inmarsat TDMA system which is used for telex has a frame period of 1.74 seconds.

The TDMA signal path

Figure 9.23 shows schematically how signals flow through a TDMA network. In brief, signals from their source, typically a telephone switching centre, are received at the earth station and are fed into an appropriate type of terrestrial interface module (transmitting) (TIM(T)). The functions of TIMs are reviewed below. The TIM(T) converts the signals into a form that the TDMA system can accept and passes them to a sub-burst store (transmitting) (SS(T)), where all the information words that are to flow from that source during a TDMA frame period are accumulated. At an instant determined by stored instructions and the time of receipt of a reference burst, the frame and burst timing control (FBTC) unit switches on the burst carrier and sends a preamble from the preamble generator (PG) to the modulator for transmission. The FBTC then calls the information words out of the various SS(T)s in succession, at the high bit rate at which the

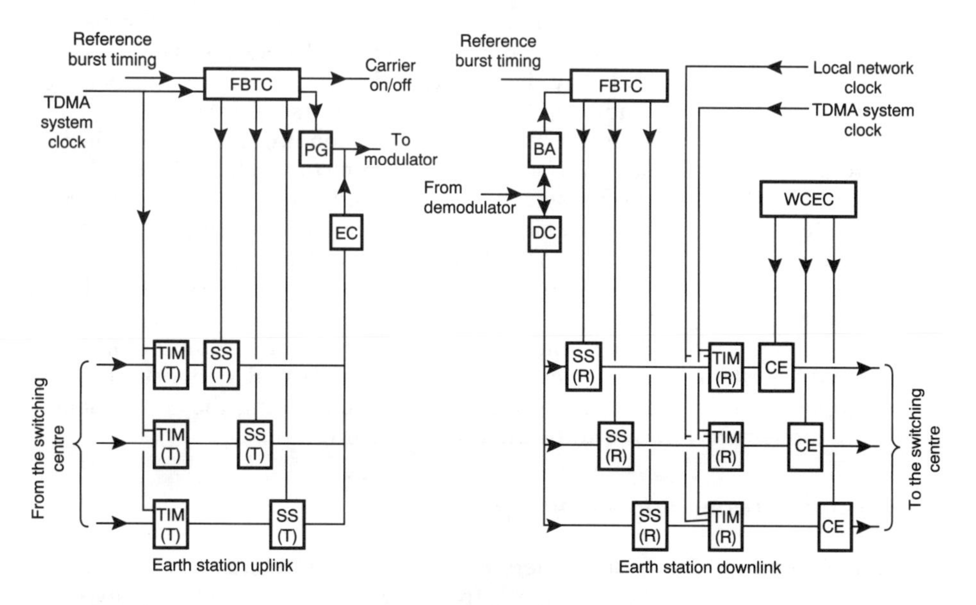

Figure 9.23 *A functional diagram of the path of three data sub-bursts through a TDMA network. For the key to abbreviations, see the text*

TDMA system operates, and they too pass to the modulator for transmission. When all the stored information has been transmitted, the FBTC switches off the burst carrier, and the SS(T)s begin to accumulate information words for the next frame.

At any receiving earth station a burst acquisition (BA) unit locks on to the traffic bursts from all the earth stations in the TDMA network in turn. The FBTC is programmed to identify bursts and sub-bursts that contain information words that are addressed to the earth station in question, and directs sub-burst stores (receiving) (SS(R)) to seize the relevant information words of the relevant sub-bursts. A terrestrial interface module (receiving) (TIM(R)) accepts a continuous bit stream from the SS(R) and converts it, if necessary, into a form that can be passed on to the destination, typically a telephone switching centre. Finally, it may be necessary to use a channel extractor (CE), programmed by a wanted channel extraction control (WCEC) unit, to select channels from a multiplex for onward transmission.

It may be necessary to use FEC to achieve a sufficiently high BER. It will often be necessary to add a pseudo-random bit stream to the data stream for carrier energy dispersal. Encoders (EC) and decoders (DC) for these purposes are located in the system so that the data sub-bursts are encoded but not the preamble.

The TIMs are key components of the system. Where the interface between the terrestrial network and the TDMA system is analogue, the functions of the TIM are relatively straightforward. If an analogue terrestrial network is interfaced at channel level, the conversion of the channels to or from the digital state is clocked from the TDMA system master clock. Likewise, if an analogue terrestrial network is interfaced at FDM multiplex level, a supergroup encoder or decoder or a transmultiplexer can be clocked from the master clock. However, more complication functions, akin to those outlined in connection with TDM links (see Section 9.4.3), arise where direct digital interfaces are used.

If a TDMA network, including all of its earth stations, is operating within a synchronous terrestrial digital network, the master clock at the reference station can be synchronized with the network. The TDMA system master clock, as perceived by other earth stations from reference bursts passed over the system, will not be in perfect synchronism with the terrestrial network clock because movement of the satellite will vary the length of the transmission paths between the satellite and the earth stations. However, in the course of each day the clock errors will sum to zero, assuming that the satellite longitude does not change. Therefore, with direct digital interfaces, it is sufficient for each TIM(T) and each TIM(R) to have an elastic buffer which is big enough to store data that cannot cross the interface when the input side clock is gaining on the output side clock.

However, if the TDMA network is not operating within a synchronous terrestrial digital network, it must be assumed that the interfaces, at entry to the TDMA system and at exit from it, are plesiochronous at best. In such a case, it will be necessary for both TIMs to accommodate differences in the absolute data rates of the TDMA system master clock and the local terrestrial network clock by frame slipping or justification, in addition to accommodating the consequences of satellite movement referred to in the previous paragraph.

9.4.6 Signal processing satellites

With few exceptions the transponders of commercial satellites now in use are designed merely to translate emissions from the uplink band to the downlink band and to amplify them; they are transparent. Some signal degradation also occurs but it is kept acceptably small. Furthermore, while there may be options as to the coverage of a transponder's antennas, the choice, once made, is a semi-permanent one.

More complex transponder systems are coming. Already some transponders in Intelsat satellites that are used for TDMA are connected to their antennas through matrices of switches which are capable of operating within periods that are very short compared with the TDMA frame period. This arrangement, which adds routing flexibility to a TDMA system that is carried by a multi-beam satellite, is called satellite-switched TDMA (SSTDMA). It is reviewed further in Chapter 11.

A number of other possible ways of increasing the value of the facilities that satellites can provide in certain niches in the market are being explored and some are reviewed briefly below.

A regenerating transponder demodulates uplinked digital carriers and regenerates the bit streams, so eliminating the distortions that noise and interference in the uplink have caused. These regenerated signals are then used to modulate downlink carriers that have been generated on board the satellite. See Figure 9.24. Regeneration permits the uplink e.i.r.p. to be reduced considerably below the value that would be needed in the absence of regeneration.

Store-and-forward transponders are already being used with experimental LEO satellites. These demodulate the carriers that are received from earth stations as the satellite passes overhead and regenerate and store the uplinked signals. The signals are called out of the store and retransmitted, either continuously or on demand from the destination earth station, at some later point in the satellite's orbit. Store-and-forward techniques provide a low-cost wide-area message service, although subject to delays of up to several hours, with satellites and earth stations of minimal capital cost.

Akin to SS/TDMA would be the transponder with steerable antenna beams. These beams, typically generated by phased arrays, can sweep or hop rapidly over a wide service area, combining some of the benefits of high satellite antenna gain, namely low uplink transmitter power requirements and high downlink power flux density, with the flexibility of access and interconnection which antennas with big footprints provide. This arrangement may prove to be of particular value to mobile or low-cost fixed earth stations that are thinly scattered over very wide geographical areas. A system of this type is being demonstrated by the NASA ACTS satellite.

An FDMA/TDM transponder would receive a number of digitally modulated carriers in the FDMA mode, and demodulate and regenerate the signals, like the regenerating transponder illustrated in Figure 9.24. The several bit streams would then be assembled into a single TDM that modulates a carrier that has been generated on board, and the modulated carrier is amplified and downlinked.

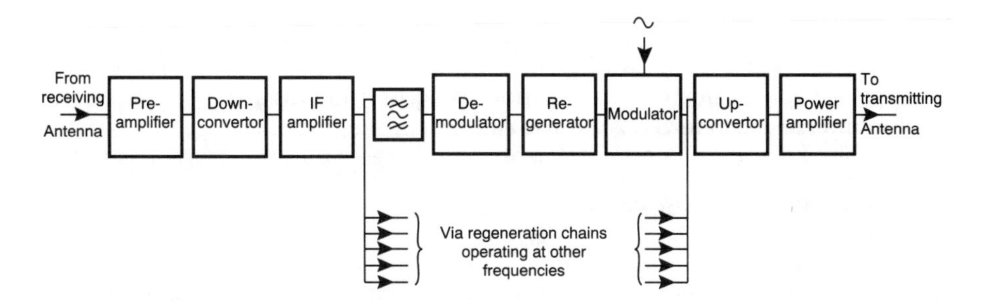

Figure 9.24 *A block schematic diagram of a regenerating transponder*

This technique would offer low earth station capital costs. If the total flow of data were large, the TDM system might need the exclusive use of the power amplifier of the transponder, which could operate efficiently, with little backoff.

The 'telephone exchange in the sky' extends the concept of the FDMA/TDM transponder. Participating earth stations uplink TDM carriers containing channels addressed to various destination earth stations. These multiplexes would be demultiplexed on board, the channels being switched individually to one of several multiplexers that assemble TDM downlinks intended for the various destination earth stations, switching being pre-assigned or demand-assigned. Implemented on a sufficiently large scale, a multi-beam satellite might be used, reducing the carrier power required from both earth stations and transponders. Such a system might combine simple, low-cost, earth station systems with some of the system advantages of TDMA. See, for example, [19].

All of these signal processing satellite concepts increase the complexity of the satellites and may reduce their reliability. They may acquire an important role in mobile satellite systems, especially those using LEOs. There will, no doubt, be niche markets for such systems in the FSS. However, at a time when satellites are being designed for increasingly long working lifetimes in orbit, one of the great commercial strengths of conventional satellite communication is flexibility of use. Transparent transponders impose few constraints on use. When the demand for one application declines, another may develop. Signal processing transponders designed to serve niche markets may be confined to those niches, becoming uneconomic if the required technical parameters change or if the size of the market is found, in time, to be much higher, or much lower, than the capacity that the transponder can provide.

References

1. Sekimoto, T. and Puente, J. G. (1968). A satellite time division multiple access experiment. *IEEE Trans.*, **COM-16**.
2. Edelson, B. I. and Werth, A. M. (1972). Spade system progress and application. *Comsat Technical Review*, **2**.
3. Pascall, S. C. (1991). Digital satellite communications. In *Digital Systems Reference Book* (Holdsworth and Martin, eds.) Chap. 5.13, Butterworth-Heinemann.
4. Spilker, J. J. (1977). *Digital Communications by Satellite*. Prentice-Hall.
5. Wu, W. W. (1984). *Elements of Digital Satellite Communication*. Computer Science Press.
6. Ogawa, A. (1981). Digital satellite communications systems. In *Satellite Communications Technology* (Miya, ed.) Chap. 7, KDD Engineering and Consulting.
7. Pretzel, O. (1992). *Error-Correcting Codes and Finite Fields*. Clarenden Press.
8. Viterbi, A. J. (1967). Error bounds for convolutional codes and an asymptotically optimum decoding algorithm. *IEEE Trans.*, **IT-13**.
9. *Pulse code modulation (PCM) of voice frequencies* (CCITT Recommendation G.711, 1988), CCITT Blue Book, Fascicle III-4. Geneva: ITU
10. *40, 32, 24, 16 kbit/s adaptive differential pulse code modulation (ADPCM)* (CCITT Recommendation G.726, 1990). Geneva: ITU.
11. *5-, 4-, 3-, and 2-bit sample embedded adaptive differential pulse code modulation (ADPCM).* (CCITT Recommendation G.727, 1990). Geneva: ITU.
12. Blankenship, P. E. (1979). A review of narrow-band speech processing techniques. *Proc. IEE Conference No. 180 on Case Studies in Advanced Signal Processing*. London: IEE.
13. Shoham, Y. (1993). High-quality speech coding at 2.4 to 4.0 kbit/s based on time-frequency interpolation. *Proc. IEEE Conference on Acoustics, Speech and Signal Processing*, Minneapolis, MN, USA.
14. Quaglione, G. (1984). Transmission techniques. In *The Intelsat Global Satellite System* (J. Alper and J. N. Pelton, eds.) Chap. 8, American Institute of Aeronautics and Astronautics.

15. *Synchronous frame structures used at* 1544, 6312, 2048, 8488 *and* 44736 kbit/s *hierarchical levels* (CCITT Recommendation G.704, 1995). Geneva: ITU.
16. Erlang, A. K. (1918). Solution of some problems in the theory of probabilities of significance in automatic telephone exchanges. *Post Off. Elec. Eng. J.*, October 1918.
17. Campanella, S. J. (1976). Digital speech interpolation. *Comsat Technical Review*, **6**(1).
18. *Digital circuit multiplication equipment using ADPCM (Recommendation G.726) and digital speech interpolation* (ITU-T Recommendation G.763, 1994). Geneva: ITU.
19. Evans, B. G., Coakley, F. P., El-Amin, M. H. M., Lu, S. C. and Wong, C. W. (1986). Baseband switches and transmultiplexers for use in an on-board processing mobile/business satellite system. *Proc. IEE*, **133**, Part F, No. 4.

10 Carrier power, noise and channel performance

10.1 Losses, gains and C/N requirements

Energy is very costly to provide on board a satellite. However, power can be divided flexibly between different links and the performance of the links can, for most systems and under normal propagation conditions, be predicted precisely from the emission parameters. Thus, it is commercially desirable but also feasible to assign satellite power to emissions very thriftily, ensuring that channel performance objectives are achieved but not wastefully exceeded. Thus the potential information throughput of the satellite as a whole is maximized. The process involves exact prediction of gains, losses, noise and interference levels and C/N requirements for each satellite link and exact implementation of the parameters calculated to be appropriate.

Carrier power

The gains and losses affecting an emission transmitted over a satellite link can be summarized as follows:

$$P^2_{\mathrm{re}} = P^1_{\mathrm{te}} - L_{\mathrm{u}} + G_{\mathrm{sat}} - L_{\mathrm{d}} \quad \text{(dBW)} \tag{10.1}$$

where P^1_{te} is the power of the emission delivered to the transmit port of the antenna of earth station 1 (dBW),
P^2_{re} is the power delivered from the receive port of the antenna of earth station 2 (dBW),
G_{sat} is the gain of the satellite transponder (dB), and
L_{u} and L_{d} are the losses not exceeded in the uplink and the downlink respectively (dB).

In more detail,

$$L_{\mathrm{u}} = L_{\mathrm{bfu}} + L_{\mathrm{mu}} + L_{\mathrm{pu}} + M_{\mathrm{au}} - G^1_{\mathrm{te}} - G^1_{\mathrm{rs}} \quad \text{(dB)} \tag{10.2}$$

$$L_{\mathrm{d}} = L_{\mathrm{bfd}} + L_{\mathrm{md}} + L_{\mathrm{pd}} + M_{\mathrm{ad}} - G^2_{\mathrm{ts}} - G^2_{\mathrm{re}} \quad \text{(dB)} \tag{10.3}$$

where L_{bfu} and L_{bfd} are the free-space basic transmission losses L_{bf} at the uplink and downlink frequencies f_{u} and f_{d} respectively (dB) (see equations (5.3) and (2.8));
L_{mu} and L_{md} are propagation losses additional to L_{bf} in the uplink and the downlink due to absorption, scattering, obstructions, etc. (dB);
L_{pu} and L_{pd} are losses in the uplink and the downlink due to mismatch between the polarization characteristics of the receiving antennas and the waves that they receive (dB);

G^1_{te} and G^2_{re} are the gains of the antennas of earth stations 1 and 2 in the direction of the satellite, in the transmit and receive modes, at f_u and f_d respectively (dBi);
G^1_{rs} and G^2_{ts} are the gains of the satellite antennas, receiving from earth station 1 at f_u and transmitting to earth station 2 at f_d respectively (dBi);
M_{au} and M_{ad} are achievement margins, associated with the uplink and the downlink respectively (dB).

The achievement margins M_{au} and M_{ad} allow for underachievement of the performance that would ideally be available, due to failure to direct the earth station beam accurately on the satellite, sub-ideal performance of demodulators, etc. See also Figure 10.1. These equations apply equally to links with satellites in any orbit but they are developed below primarily for geostationary satellites.

Noise power

Noise within the frequency band necessary for the emission enters a satellite link from many sources. It is convenient to gather these sources into five groups:

(1) noise arising in the earth station transmitter;
(2) satellite receiver system noise, most conveniently reckoned as a system noise temperature T_{ss}, including noise received from Earth by the satellite antenna;
(3) noise arising in the satellite transponder;

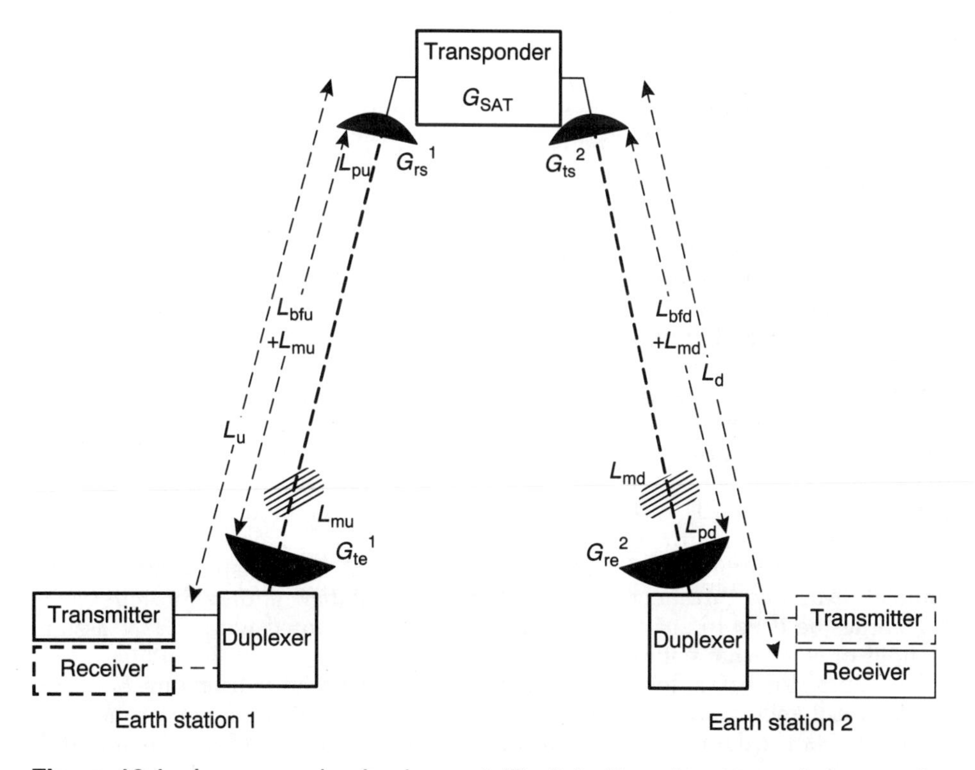

Figure 10.1 *Losses and gains in a satellite link. For a key to symbols, see the text*

(4) atmospheric noise entering the earth station receiver, additional to that occurring when the sky is clear; and
(5) earth station receiver system noise under clear-sky conditions, T_{sE}, referred to the output terminal of the earth station antenna.

These various noise contributions must be summed. This is done after each contribution has been referred to a common point, allowance being made for gains and losses occurring between each noise source and the point of reference. The most convenient reference point is the point of reference of T_{se}, that is, the output terminal of the antenna of the receiving earth station. Summing can be carried out most readily through the relationship

$$(C/N)_T = [(n/c)_1 + (n/c)_2 + (n/c)_3 + \cdots]^{-1} \tag{10.4}$$

where $(C/N)_T$ represents the carrier-to-total-noise ratio at the point of reference and $(c/n)_1$, $(c/n)_2$, etc. represent a succession of C/N ratios for noise from sources 1, 2, etc. It will be noted that these ratios are arithmetical, not logarithmic.

Noise levels are represented in equation (10.4) by the symbol n, indicating a noise power component in the bandwidth necessary for the emission, and this is the convention in general use in this chapter. However, in some circumstances the use of spectral noise density, $n_0 = n/B$ W Hz^{-1}, where B is the pre-demodulator bandwidth, or the noise temperature, $t = n_0/k$ K, where k is Boltzmann's constant, is more convenient.

C/N requirements

Virtually all carriers transmitted by satellite are frequency modulated by an analogue signal or phase shift keyed (or modulated in a closely equivalent way) by a digital signal.

With a conventional FM demodulator there is a performance threshold when the input C/N ratio equals about 10 dB. Below the threshold, the signal/noise ratio S/N falls rapidly as C/N declines. Above the threshold, S/N is determined mainly by the modulation index, although S/N is increased, decibel for decibel, by the amount that C/N exceeds the threshold. See Section 8.1 and Figure 8.2. If the satellite transponder has enough power to enable all of its bandwidth to be used, the pre-demodulator C/N requirement will be taken to be 10 dB, desirably with the addition of a small margin to avoid degradation of performance due to impulsive noise; the modulation index will then be chosen to meet the S/N objective.

If there is power to spare when the transponder bandwidth is fully utilized (that is, the transponder is 'bandwidth limited') it may be feasible to make provision for additional carriers. This is done by assigning some of the surplus power to carriers that are already accommodated in the transponder, enabling channel performance standards to be maintained with a lower modulation index, and assigning the bandwidth set free in that way and the remainder of the surplus power to the additional carriers. If, on the other hand, the transponder is 'power-limited', extended threshold demodulators and a higher modulation index may sometimes be used to enable S/N objectives to be attained with a C/N assignment at or somewhat below 10 dB.

PSK does not have close equivalents to the FM demodulator threshold nor the S/N-enhancing ability of widened FM deviation but the range of the C/N requirements for the many kinds of digital emission is similarly narrow. A BER between 1 in 10^6 and 1 in 10^4 is sufficient for most applications, although a somewhat worse error ratio might be acceptable for systems using powerful FEC. An ideal 2-phase PSK coherent demodulator requires pre-demodulator C/N ratios of 10.5 dB and 8.4 dB respectively to meet these BER objectives; see Section 9.2.1 and Figure 9.7. Higher orders of PSK modulation, 4-phase, 8-phase, and so on, require the addition of 3 dB to the C/N ratio for each doubling of the number of significant phase conditions used. The most commonly used technique is 4-phase PSK. Practical demodulators do not attain this ideal performance; an attainment margin of about 2 dB is usually allowed when estimating the C/N ratio required for a given BER value.

10.2 Polarization mismatch

The electric vector of any wave can be resolved into two linearly polarized components. The major component (peak amplitude $= e_1$) is perpendicular to the direction of propagation of the wave and is aligned with the direction of maximum voltage. The minor component (peak amplitude $= e_2$) is perpendicular to both the direction of propagation and the major component, and is in phase quadrature, leading or lagging, relative to the latter. See Figure 10.2.

In the special case where $e_2 = 0$, the wave is linearly polarized. In the general case where e_2 is equal to neither e_1 nor 0, the wave is said to be elliptically polarized. The sense of the polarization of an elliptically (or circularly) polarized wave is right-hand or left-hand according to the direction of rotation (clockwise or anticlockwise) of the resultant of the two-component vectors when viewed in the

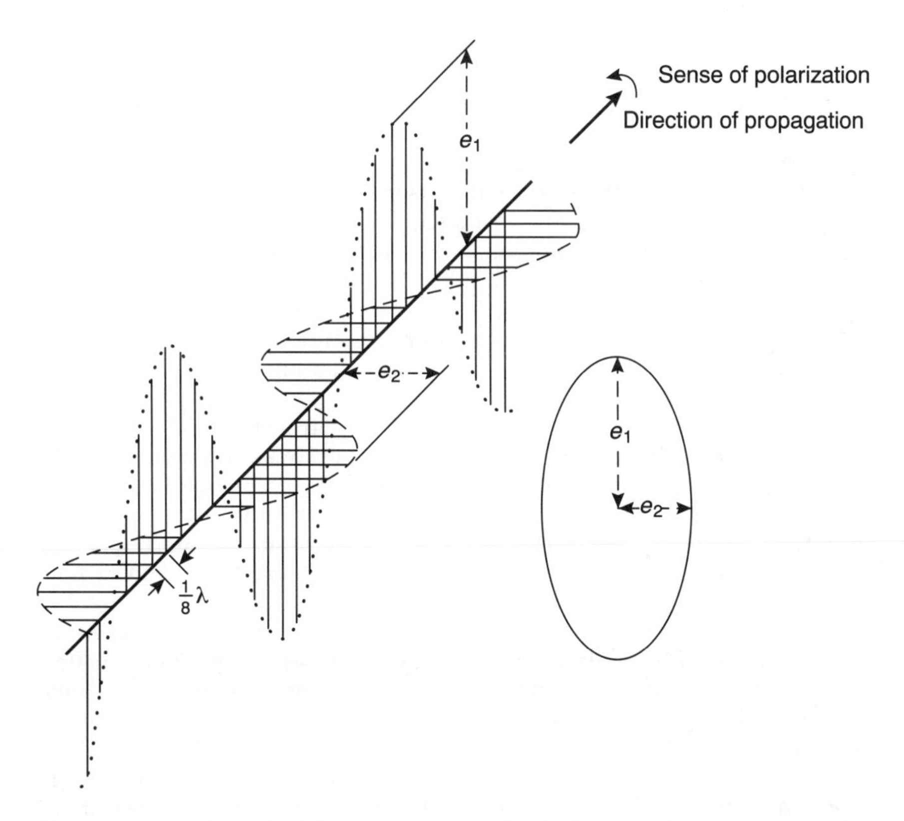

Figure 10.2 *The electric vector of an elliptically polarized wave, resolved into two orthogonal linearly polarized components*

direction of propagation of the wave. In the situation represented in Figure 10.2, the wave is left-hand elliptically polarized. Finally, in the special case where $e_1 = e_2$, the wave is circularly polarized (RHCP or LHCP).

The voltage axial ratio (VAR) of a wave is given by

$$\text{VAR} = e_1/e_2 \tag{10.5}$$

Thus the VAR for an ideal circularly polarized wave is unity and that of an ideally linearly polarized wave is infinite. The logarithmic equivalent of the VAR, called the axial ratio R, is given by

$$R = 20 \log \text{VAR} \quad \text{(dB)} \tag{10.6}$$

To complete a statement of the polarization characteristics of a linearly polarized wave it is necessary to know the angle of the electrical vector relative to some suitable datum. This is called the tilt angle τ. Similarly, for an elliptically polarized wave, τ defines the orientation of the major component e_1. Relative to a satellite, the datum usually chosen is the plane which contains the satellite and the Earth's axis. Relative to an earth station, the most convenient datum is the local vertical.

A transmitting antenna determines the polarization characteristics of the wave that it launches; thus it is appropriate to define the polarization characteristics of the antenna in the same terms, namely axial ratio, sense and tilt angle. Practical antennas are designed to produce either linear or circular polarization, but an elliptically polarized component is usually present along the axis of the antenna. The elliptical component tends to be greater off-axis in the main lobe and it is usually quite large in the side-lobes. For a well-designed circularly polarized antenna, the VAR on-axis is less than 1.05; that is, $R < 0.5$ dB. The corresponding figures for linear polarization would be VAR > 32, so $R > 30$ dB.

If the polarization characteristics of a receiving antenna match an incoming wave in axial ratio and tilt angle and/or sense, as appropriate, the energy in the wave will be received with no mismatch loss (L_{pu} and $L_{pd} = 0$ in equations 10.2 and 10.3). Satellite link antennas are, of course, aligned for optimum match. In practice there are minor mismatches under clear-sky conditions, due for example to imperfect antenna performance, satellite yaw and off-axis operation of antennas, but the mismatch loss should be negligible. The mismatch may become more significant during heavy rain at earth stations but such losses are short-term ones.

However, many satellite antennas are designed to transmit and/or receive both senses of circular polarization or two orthogonal modes of linear polarization simultaneously. This enables the satellite to reuse the spectrum. Some of the associated earth stations may be equipped to operate in only one polarization mode, provided that they are capable of doing so, neither causing interference to, nor suffering interference from, services operated by the satellite in the orthogonal polarization mode. Other earth stations have dual polarization facilities and are capable of operating in either or both radio channels without causing or suffering interference. This ability of earth station antennas to discriminate effectively against satellite emissions of an unwanted polarization is called cross-polar discrimination (XPD). The term cross-polar isolation (XPI) is also sometimes used; the usage of these terms is not clear-cut.

If an emission with perfect LHCP and having power w_1 is received by a dual-polarized antenna, the VAR of which is not unity, the power delivered at the LHCP receive port will be approximately w_1 but power w_3 will be delivered at the RHCP receive port. Then

$$\text{XPD} = 20\log\frac{w_1}{w_3} \quad \text{(dB)} \tag{10.7}$$

$$\approx 20\log\frac{\text{VAR}+1}{\text{VAR}-1} \quad \text{(dB)} \tag{10.8}$$

In practice, however, it is unlikely that the incoming wave will have perfect circular polarization, and this is likely to reduce further the XPD obtained.

10.3 Protection of links from adverse propagation conditions

10.3.1 Protection of links operating above 3 GHz

L_{mu} and L_{md} in equations (10.2) and (10.3) cover many kinds of propagation degradation, ranging from obstruction by buildings, through absorption in the ionosphere and by rain, to wave interference due to multi-path propagation. The various phenomena are considered in some detail in Chapter 5. However, some of these degradations have little impact on systems operating above 3 GHz and the contrary is true of the others. It is convenient to consider the two groups separately.

Systems currently operating above 3 GHz are mainly of the FSS and the BSS. These systems are characterized by earth stations having antennas with gain which is high enough to make multi-path propagation due to reflections from the ground or nearby buildings unlikely when, operating in clear line-of-sight with satellites which are, with rare exceptions, geostationary. They operate at frequencies that are too high for major ionospheric effects and too low for significant absorption by the atmospheric gases.

Faraday rotation

A varying degree of rotation of the plane of polarization arises in the ionosphere (Faraday rotation), making frequency reuse by means of orthogonal linear polarizations impracticable at C band. This problem is avoided by using circular polarization.

Causes of increased propagation loss

The principal propagation degradation for these systems is absorption by rain, with an associated increase of thermal noise entering the link when it is raining at the receiving earth station. Some earth stations may, for a small part of the time, suffer an increase in Earth–satellite losses due to tropospheric and ionospheric scintillation. In addition, depolarization due to rain degrades the performance of systems employing frequency reuse by dual polarizations. Some of the depolarization may be eliminated by adaptive compensation at the earth station where rain is falling. The remainder of the degradations may be treated as an increase in the attenuation, relative to L_{bf}, in the uplink and/or the downlink, and an increase in noise temperature in the downlink relative to T_{ss}. Many of these factors are reviewed in ITU-R Recommendation S.1061 [1].

Absorption due to rain

The severity of the loss due to rain and the associated rise of noise temperature at the receiving earth station depend on the climate at the earth station, the angle of elevation of the satellite as seen from the earth station and the operating frequency. See, for example, Figure 5.3. At C band the effects are negligible for all but a very small percentage of the time except for earth stations in very rainy climates. At Ku band the effects are small at high angles of elevation in temperate climates, although the loss may exceed, say, 10 dB at low angles of elevation for 0.1 per cent of the time, while remaining negligible at any angle for perhaps 80 per cent of the time. However, in very rainy climates the loss in Ku band may reach some tens of decibels even at high angles of elevation, and at Ka band these rain effects are even more severe.

Limited performance objectives

Within limits these various degradations are taken into account in link design by setting performance objectives that are to be attained for most, but not necessarily all, of the time. For example, the BSS 12 GHz Agreement, reviewed in Section 15.2, envisages that a television carrier of stated PFD will be set up at the Earth's surface for 99 per cent of the worst month of a typical year. It is accepted that, for 1 per cent of the time, performance is likely to be degraded to some degree.

Dual performance objectives

Dual performance objectives are used for many systems of the FSS. The long-term objective specifies the fully acceptable channel performance that is to be achieved for 80 or 90 per cent of the worst month of an average year. For a digital system this performance might, for example, be a BER of 1 in 10^6. In addition, a short-term criterion specifies an inferior channel quality (for example, a BER of 1 in 10^4), usable for telephony but considered unsatisfactory, that must be exceeded for all but a small proportion of the time, typically 0.3 per cent of the worst month of an average year. In some cases a third criterion of even worse performance is set for an even smaller proportion of the time. Dual objectives, and typical performance statistics at C band and Ku band, are illustrated in Figure 10.3.

How are these objectives to be met, despite variable rain degradation? In some circumstances no special action is necessary. C band links having parameters that enable them to meet the long-term objective under clear-sky conditions may also meet the short-term objective under adverse conditions, except in very rainy climates. But these relatively tranquil propagation conditions are not found at higher frequencies. There are several strategies for overcoming worse propagation conditions.

Adaptive control of uplink power

Section 15.1 describes the BSS DBS frequency assignment agreements for operation at 12 GHz. There is provision in these agreements for increasing the transmitter power level at the feeder link earth stations, which operate at 15 or 18 GHz, providing adaptive compensation for loss due to rain in the uplink, so that the satellite emission will maintain a constant downlink e.i.r.p. and suffer no deterioration in the uplink interference conditions.

There is no provision for adaptive compensation for downlink rain loss in the BSS agreements, and adaptive compensation is not used in other kinds of satellite communication system for one or other of the following reasons:

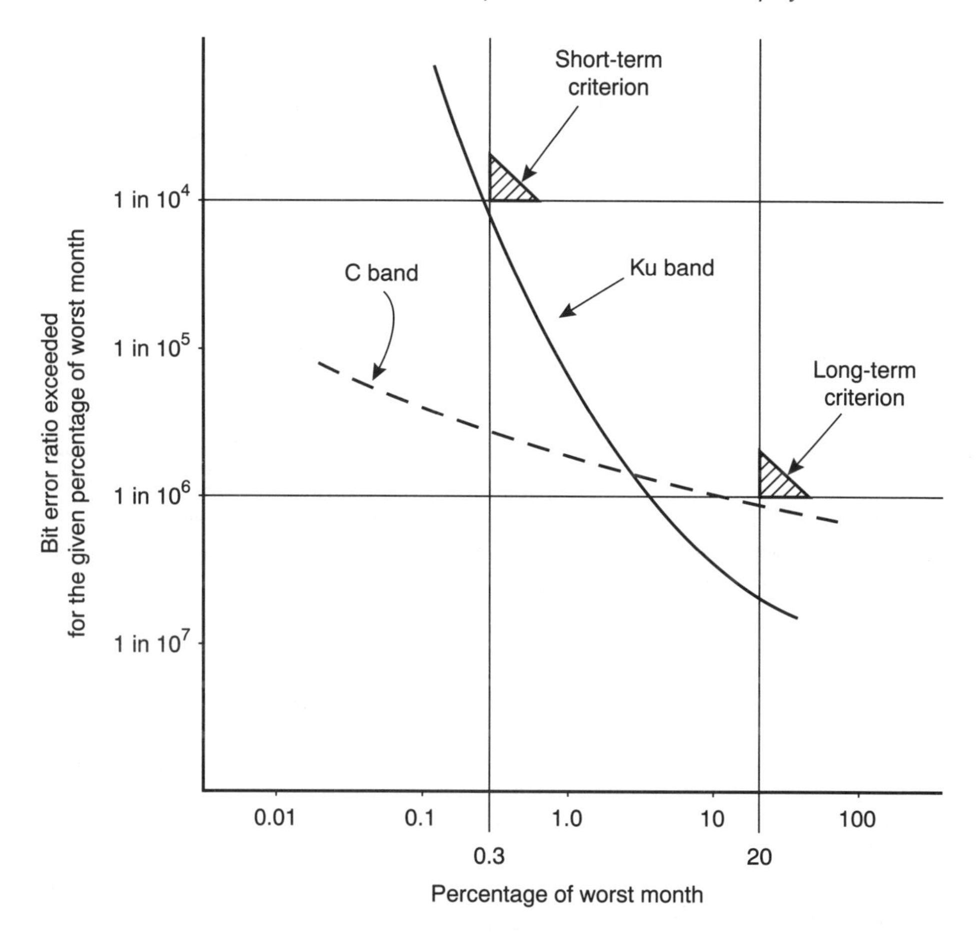

Figure 10.3 *A sketch of the post-demodulator performance of digital satellite links, operating at C band and Ku band, designed to meet dual criteria despite occasional rain attenuation*

(1) Most satellite emissions are received at several or many earth stations spread over a wide area. It may be raining at one or other of the earth stations for much of the time and there is therefore little advantage in adaptive compensation for rain loss. A permanent propagation margin achieves the same result without the complexity of an adaptive system (see below).
(2) Many carriers share satellite transponders with other carriers in FDMA. If the level of one of these carriers is raised to compensate for adverse downlink propagation conditions, the performance of the other carriers may be degraded by signal compression.
(3) An adaptive system may malfunction, a reduction in received level being interpreted as an increase in rain loss although it may be due to another cause, such as an antenna pointing error.

Rain margin

Except perhaps in C band, an FSS system using dual performance objectives may fail to meet the short-term objective if the carrier parameters are chosen to achieve the long-term objective only. A rain margin of a few decibels may be added to the losses assumed in calculating the required earth station transmitter power. This would be designed to compensate for foreseen rain attenuation in the uplink or the downlink and for increased received noise when the rain is in the downlink, enabling the short-term objective to be met. As a result, the long-term objective may be substantially exceeded; see the higher-frequency curve in Figure 10.3.

Earth stations in diversity

When a rain cell is causing a heavy propagation loss for one earth station, another earth station ten or twenty kilometres away may, at the same time, be enjoying considerably better propagation conditions, not being affected by rain cells. Thus one way to escape from the effects of severe rain absorption is to use a pair of earth stations in diversity. There is an analysis of the benefit to be obtained in this way in Section 2.2.4 of the Annex to ITU-R Recommendation PN.618 [2]. However, it is a costly solution in most circumstances and it is little used so far, if at all. However, viable applications may arise in the future; Hodson *et al.* [3], for example, propose an application of site diversity to VSAT earth stations linking together cabled metropolitan area networks (MANs).

Change of modulation parameters

Other techniques have also been devised to enable communication, perhaps at a reduced information rate, to be maintained during periods of severe rain attenuation. For example, the emission may be replaced by another with a smaller information capacity, the uplink having the same e.i.r.p. in a smaller bandwidth or by changing to a more robust modulation technique or an error correcting coding system.

Choice of frequency band

In very rainy climates, for applications in which the short-term performance criterion must be achieved for all but a very small percentage of the time, the rain attenuation may be so severe that it is not practicable to meet the short-term performance criterion in the higher-frequency bands. In such cases the only practical solution may be to use lower frequencies.

10.3.2 Protection of links operating below 3 GHz

Systems operating below 3 GHz, mostly belonging to the MSS but with BSS (sound radio) in prospect, are characterized by earth stations with antennas without enough directivity to prevent the occurrence of multi-path propagation due to reflections from the environment. There may not always be a clear line-of-sight between the satellite and the earth station. The satellites may not be geostationary; some will use circular orbits at medium height or low Earth orbits (LEO). Finally, whereas the effect of rain is negligible, ionospheric degradations may be significant.

Below 1 GHz

Below 1 GHz the effects of Faraday rotation on the plane of linear polarization is severe, and it is very severe in the lowest frequency allocations for satellite communication, around 140 MHz. The use of circular polarization to avoid deep fading is general. For circularly polarized waves, the absorption for most of the time due to passage through the ionosphere would range from a few decibels at 140 MHz for earth stations within the auroral zone down to negligible levels at or above 500 MHz at mid-latitudes. Ionospheric scintillation increases these losses for a small percentage of the time. These sporadic increases in loss might reach 20 dB for 0.1 per cent of the time at 140 MHz but, as before, the effect at or above 500 MHz is small. Finally, there may be very variable but potentially severe losses at all frequencies below 1 GHz due to screening by buildings, hills, etc. However, these propagation losses must be considered in the context of the use of low Earth orbits below 1 GHz. For LEO systems, the distance between a satellite and an earth station will vary considerably from minute to minute due to their relative motion. Consequently, the free-space path loss varies regardless of propagation losses. Similarly, the effects of screening may change rapidly as the ray path moves relative to obstructions. Sometimes there may not be a satellite in sight of an earth station.

Such propagation factors cannot be counteracted realistically for LEO systems by the application of a propagation margin. One solution is the use of a constellation of a sufficiently large number of satellites to ensure that an earth station will have good visibility of at least one satellite at all times. Alternatively, the system must be designed to provide a standard of service appropriate to its role in the market despite the degradations. Accepting that a working link will not be available for all of the time that it might be wanted, the design of the system must ensure that signals will be passed over the link when feasible and without unacceptable inconvenience to the user. It is difficult to operate two-way services, and especially telephony, on this basis, but store-and-forward message systems using FEC and, desirably, ARQ principles can operate satisfactorily.

Transmission loss between 1 and 3 GHz

Between 1 and 3 GHz, for circularly polarized waves, the loss due to the ionosphere is small, no more than a fraction of 1 dB. Some mobile earth station antennas are protected by a radome; the loss due to transmission through a dry radome is very small and it does not exceed a fraction of a decibel when the radome has been made wet by rain. Variations of wave velocity in the ionosphere may have to be taken into account in systems providing radio-determination facilities.

Multi-path propagation and path obstruction due to the environment

Multi-path propagation arises chiefly because the mobile earth station antenna, while transmitting and receiving via the direct path to and from the satellite, does not have enough directivity to reject indirect paths. The phase relationships between rays propagated via the different paths are random and unstable and they interfere when they combine at the receiving antenna, causing fading. If the various rays have comparable amplitudes and if the indirect rays have a substantial coherent component, the fading is likely to be deep. If the differences in the path lengths of strong components is large, inter-symbol interference may occur in high-bit-rate digital emissions. If there is no direct path, it may be feasible to communicate via indirect paths but multi-path propagation may be particularly severe under these conditions.

In the maritime service, the indirect paths arise mainly from reflection from the surface of the sea. Aircraft earth stations are also affected by sea-surface reflections, but in addition aircraft antennas must be located with great care to avoid multi-path propagation due to reflections from wings or the fuselage. Sea-surface reflection is most severe where the angle of elevation of the satellite as seen from the earth station is low. Land mobile earth stations in city centres are particularly susceptible to multi-path propagation arising from reflections from the sheer sides of buildings.

Signal fading due to multi-path propagation can be reduced by using mobile earth station antennas with higher gain. Figure 14.4 shows around 10 dB rejection of rays arriving 12° off the axis of a ship's antenna such as Inmarsat Standard A. Such antennas can limit the sea areas within which multi-path propagation is severe. But the great and growing majority of mobile earth stations have antennas of much less gain and many are virtually omnidirectional.

Obstruction of the direct path between satellite and mobile earth station is not, in general, a problem for ship and aircraft earth stations. The antenna can usually be installed so that there is a clear line of sight towards the sky, without blockage by superstructure; if not, it may be necessary to install two antennas, with automatic selection of the antenna with the better satellite visibility.

However, for the LMSS and the BSS, path obstruction often associated with severe multi-path propagation, raises great problems. The pressures on satellite system operators to solve these problems are made heavier by the large margins of power that the emissions of their terrestrial competitors often provide at centres of demand, providing deep penetration into buildings and the urban canyons that are city streets; satellite system users may expect the same degree of availability. The impact of obstructions on satellite systems can be reduced, especially for non-geostationary systems, by deploying more than the minimum number of satellites that are required to cover the service area, so that two or more satellites are always present to serve a user's earth station from different points in the sky, but this solution brings its own problems, technical and economic.

Summing up, for most mobile earth stations multi-path propagation or screening by natural features and buildings are often the factors that determine the availability of service. Store-and-forward messaging systems can overcome most propagation limitations. Digital systems have been devised for sound broadcasting by satellite which should overcome the multi-path interference problem. It may also be economic to provide a large propagation margin, up to several tens of decibels, for high-value digital services requiring relatively few bits per user, like satellite paging, enabling the satellite emission to penetrate more deeply into buildings. But no more than a small power margin seems likely to be available for satellite hand-held radio telephones; users will have to make their calls from places that can be expected to be in clear view of a satellite.

10.4 Channel performance objectives

ITU performance and interference objectives have been agreed for telephone channels carried by geostationary satellites of the FSS. These objectives relate specifically to channels that form part of international telephone connections and they are harmonized with the objectives applied to channels of corresponding length routed via terrestrial transmission media. These same objectives are often also applied to telephone channels used for domestic connections and to data channels. Corresponding objectives have also been agreed for international video channels routed by FSS satellite.

Channel performance objectives have also been agreed internationally for some applications of the MMSS but so far without interference objectives; see Section 14.3.1. No objectives have been agreed through the ITU yet for AMSS and LMSS channels.

For the BSS frequency assignment plan for television broadcasting at 12 GHz, the achievement of an acceptable relationship between the PFD of the wanted emission within the service area, the noise level and the levels of interference from other planned emissions have been designed into the plan; see Section 15.1. These relationships are not expressed in terms of end-to-end performance because the characteristics of the domestic receiver were not specified in the plan.

Channel performance criteria for the FSS

Substandard performance events in PSTN channels are divided into two catagories. When a channel is required but it is either non-existent or utterly unusable, it is said to be unavailable. At other times its performance is judged in terms of the percentage of a month or a year during which the noise level or the BER is worse than a given value.

Availability

A channel may be unavailable due to equipment failure or because radio propagation is exceptionally bad. Mis-pointing of antennas, loss of frame synchronism in digital systems, solar noise interference when the satellite transits the Sun at equinox and human errors in satellite and earth station operation may all be grouped with equipment failures. Unavailability due to propagation conditions is taken to include abnormal incidents of powerful interference. A channel in an analogue emission may be regarded as unavailable while the emission is being received 10 dB or more below its expected level or while the channel, after demodulation, has a noise level greater than 10^6 pW0. For a digital telephone channel, the limit may be a BER worse than 1 in 10^3. However, a very brief incidence of severe channel degradation having a duration of less than ten consecutive seconds is not treated as unavailability but is included in the statistics of channel noise or BER degradation.

ITU-R Recommendation S.579 [4] recommends that a channel should be engineered and operated so that it is available when scheduled to be available for at least 99.8 per cent of a year. Likewise unavailability due to propagation conditions should not exceed 0.2 per cent of any month. A third criterion, currently under study, would set a limit on total unavailability due to propagation conditions in any year; no figure has been agreed for this yet, but 0.1 per cent is being considered.

Channel performance objectives in the FSS and their sub-allotment

Performance objectives have been agreed for the 99.6 to 100.0 per cent of the scheduled time that a well-engineered PSTN telephony channel is available. It has also been found convenient to subdivide the recommended allowances of channel noise or bit errors into three parts so that three functions, namely system management, frequency coordination with the managers of other satellite systems and frequency coordination with terrestrial radio system operators, may proceed independently. The first and largest part is for degradations arising within the wanted satellite system. The other parts are allowed for interference from other satellite systems and from terrestrial radio systems respectively.

Omitting many details that are not essential to the present issue, the agreed channel performance objectives and the subdivisions of the degradations for FSS channels are shown in Table 10.1 for analogue and digital telephone PSTN channels and analogue video channels operating below 15 GHz.

Table 10.1 *Noise, BER and interference objectives for channels in FSS links operating below 15 GHz, using geostationary satellites*

Source of degradations	*Percentage of worst month*	*Degradation not to be exceeded for more than the stated percentage of the worst month of an average year*		
		Analogue telephony (Note 1)	*PCM telephony (Note 2)*	*Analogue video (Note 3)*
Sum of all sources	20	10000	1 in 10^6	
	1			$S/N \geq 53$ dB
	0.3	50000	1 in 10^4	
	0.1			$S/N \geq 45$ dB
	0.05		1 in 10^3	
	0.01	10^6 pW0		
ITU-R Recommendation		S.353 [5]	S.522 [6]	S.354 [7] S.567 [8]
Interference from other FSS systems				
(a) total	20	2500/2000 (Note 4)	25%/20% (Note 4)	10% of noise allowance
(b) single entry	20	800	6%	4% of noise allowance
ITU-R Recommendation		S.466 [9]	S.523 [10]	S.483 [11]
Interference from all terrestrial FS systems	20	1000	10%	No objectives set
	0.03	50000	1 in 10^4	
	0.005		1 in 10^3	
ITU-R Recommendation		SF.356 [12]	SF.558 [13]	
In-system degradations	20	6500/7000 (Note 4)	65/70% (Note 4)	90% of noise allowance
	0.27	50000	1 in 10^4	
(Note 5)	0.045		1 in 10^3	
(Note 5)	0.01	10^6 pW0		

Note 1. Noise levels given in pW0p except where pW0 is stated.
Note 2. BER degradation objectives are stated in two forms according to the circumstances, either a bit error probability (1 in 10^X) or a percentage of the noise level at the input to the demodulator that causes a BER of 1 in 10^6.
Note 3. S/N as defined in equation (8.16)
Note 4. The first of the two stated values applies if the wanted system does not employ frequency reuse and the second value applies if frequency reuse is employed.
Note 5. Note that these high-level in-system degradation allocations exclude incidents properly covered by unavailability.

Channels of a multiplexed telephony carrier, analogue or digital, are also used for various kinds of non-speech signals by means of modems, including low-speed facsimile and data. The performance objectives shown in Table 10.1 are, in general, adequate for these other signals. However, when a digital channel forms part of an integrated services digital network (ISDN) it may be used as a bearer channel for non-speech signals of higher information rate, such as facsimile and data at 64 kbit/s. Performance objectives which are adequate for digital telephony do not provide acceptable performance for such channels. Accordingly, where ISDN channels are involved, ITU-R Recommendation S.614 [14] recommends that the BER should not exceed

1 in 10^9 bits for more than 10%
nor 1 in 10^6 bits for more than 2%
nor 1 in 10^3 bits for more than 0.03%
} of the worst month in an average year.

ITU-R Recommendation S.735 [15] recommends that interference from other FSS systems should not contribute more than 25 per cent of the total pre-demodulator noise power under clear-sky conditions, the percentage being reduced to 20 per cent for satellite systems with frequency reuse. No special case has been made for the level of interference from terrestrial radio systems; no doubt ITU-R Recommendation SF.558 [13] would be applicable.

All of these performance objectives are expressed in the form of dual or triple standards which, as is explained in Section 10.3.1, is a convenient aid to the planning of satellite links. This method gives fully satisfactory results for telephony. It is not ideal for the ISDN because it takes no account of the way in which bit errors are distributed in time, nor does it provide for the higher performance that is desirable for digital multiplexes with bit rates higher than the 1.5 or 2 Mbit/s primary level. Objectives have been agreed for terrestrial systems providing ISDN channels which directly address these needs; see ITU-T Recommendation G.826 [16]. ITU-R Recommendation S.614 [14] was drawn up to achieve the objectives set for terrestrial systems as well as may be, using the methods traditionally used in the FSS, and an extensive annex to that text explains the thinking behind ITU-T Recommendation G.826. However, ITU-R Recommendation S.1062 [17] recommends that future, and wherever possible, existing satellite links should be engineered to meet the specific objectives of ITU-T Recommendation G.826.

The FSS interference and noise budget

Taking as an example the long-term objectives for an analogue telephone channel in an FSS system that employs frequency reuse, Table 10.1 provides the following breakdown;

Total interference from other FSS systems:	2500 pW0p
Interference from terrestrial FS systems:	1000 pW0p
In-system degradations:	6500 pW0p
Total	10000 pW0p

Where a satellite operates with some form of frequency reuse, the first item is reduced to 2000 pW0p and the third is increased to 7000 pW0p. When considering system design, it is necessary to budget for all of the major elements that make up the in-system degradations and to ensure that the available 6500 or 7000 pW0p is not exceeded.

These major elements are as follows:

(1) uplink and downlink interference due to in-system frequency reuse;
(2) noise generated in the earth station transmitter;
(3) uplink thermal noise n_u, radiated by the Earth or generated in the satellite receiver;
(4) intermodulation noise and AM/PM conversion, n_{im}, arising in the transponder;
(5) downlink thermal noise n_d, radiated by the earth station environment or generated in the earth station receiver;
(6) signal degradations due to group delay distortion.

With good system design and management, it should be possible to limit items (1), (2) and (6) to a total of 1000 pW0p per telephone channel. Thus 5500 or 6000 pW0p remains available to be divided between items (3), (4) and (5). The division between these last three items needs to be optimized according to the operational situation.

Noise budget optimization

The carrier to uplink thermal noise ratio $(C/N)_u$ (item 3) is controlled by the gain of the transponder. Raising the gain increases $(C/N)_u$ and reduces the earth station e.i.r.p. requirement for a given link information capacity, thus reducing earth segment costs. The choice of transponder gain also has a major effect on frequency coordination with other satellite systems, because high transponder gain makes the system more susceptible to uplink interference from other systems and low transponder gain increases the likelihood of causing unacceptible uplink interference in other systems.

The carrier to transponder noise ratio $(C/N)_{im}$ (item 4) is negligible if only one carrier is being amplified in the transponder. $(C/N)_{im}$ can usually be made small by careful frequency assignment planning if there are only two carriers or if the transponder is severely power-limited. However, in a typical FDMA situation, with several or many carriers occupying all or most of the bandwidth of the transponder, $(C/N)_{im}$ is determined by the chosen degree of backoff. Increasing backoff reduces $(C/N)_{im}$ and, within limits, increases the potential information throughput of the transponder, but it demands higher G/T from earth stations.

The carrier to downlink thermal noise $(C/N)_d$ (item 5) is affected indirectly by the choice of transponder gain and backoff. If the sum of $(C/N)_u$ and $(C/N)_{im}$ is increased, then $(C/N)_d$ must be reduced by increasing the downlink carrier level if the sum of the three C/N components is not to exceed the available budget.

Two examples illustrate the process of optimizing transponder gain and backoff.

(1) If the transponder is occupied by a single carrier of high information capacity, $(C/N)_{im}$ is negligible, backoff can be made small and $(C/N)_d$ will be determined by the single carrier saturation e.i.r.p. of the transponder and the G/T of the earth station. The remainder of the 5500/6000 pW0p budget remains available for $(C/N)_u$. This may enable a high value of transponder gain to be used, reducing the earth station e.i.r.p. requirement and therefore the transmitter cost. The foregoing argument is phrased in analogue telephony terms but a corresponding argument applies for a high-power digital system, such as a TDMA system, or a high-performance analogue television channel.
(2) If the transponder is used for multi-carrier FM/FDMA, the degree of backoff will be chosen to maximize the potential information throughput. An input backoff of 10 dB is typical, probably leading to $(C/N)_{im}$ in the range 2000 to 2500 pW0p. With typical modulation parameters, intermodulation noise

of this order would be associated with a value of $(C/N)_{im}$ about 21 dB. This $(C/N)_{im}$ budget will leave around 3500 pW0p to be divided between $(C/N)_u$ and $(C/N)_d$. A relatively low transponder gain setting will probably be chosen, raising Earth station transmitter power requirements, but increasing the potential available throughput of the transponder. As before, a corresponding argument holds for digital carriers in FDMA.

10.5 Power and noise budgets

The basic problem, given the necessary data, is to determine the value of C/N that is required at the input to the demodulator of a receiving Earth station in order that the performance objectives may be achieved. Given this C/N value, the power of the carrier at the input to the transmit antennas at the satellite and the transmitting Earth station can be calculated from equations (10.1) to (10.4). However, system data are not usually presented in the form called for in these equations, and practical considerations sometimes call for minor departures from the objectives.

Thus, the gain of a satellite transponder is usually specified in the form of the PFD at the satellite, Φ_{ts}, of a single carrier, transmitted from an earth station at the −3 dB gain contour of the footprint of the satellite receiving beam, that would drive the transponder to its maximum output (point S in Figure 7.22). Similarly, the potential power output of a transponder is usually expressed as the e.i.r.p. radiated towards the −3 dB gain contour of the footprint of the satellite transmitting beam as a result of that single uplink carrier which saturates the power amplifier. The system noise temperature of receiving systems is often stated in terms of G/T.

Φ_{ts} is related to the earth station e.i.r.p., $P_{te} + G_{te}$, that would produce it by

$$P_{te} + G_{te} = \Phi_{ts} + L_{mu} + M_{au} + 10 \log 4\pi d^2 \quad (\text{dB(W m}^{-2}\text{)}) \tag{10.9}$$

$(C/N)_u$ is related to the transmitting earth station e.i.r.p. by

$$(C/N)_u = P_{te} + G_{te} - L_{bfu} - L_{mu} - M_{au} + (G/T)_s - 10 \log kB \quad (\text{dB}) \tag{10.10}$$

and $(C/N)_d$ is related to the satellite e.i.r.p. by

$$(C/N)_d = P_{ts} + G_{ts} - L_{bfd} - L_{md} - M_{ad} + (G/T)_e - 10 \log kB \quad (\text{dB}) \tag{10.11}$$

where P_{ts} is the power delivered to the satellite transmitting antenna (dBW);
G_{ts} is the gain of the satellite transmitting antenna at the −3 dB contour (dBi);
$(G/T)_e$ is the figure of merit of the earth station antenna/receiver combination (dB K^{-1});
$(G/T)_s$ is the figure of merit of the satellite antenna/receiver combination for an earth station on the −3 dB contour of the antenna footprint (dB K^{-1});
B is the relevant bandwidth (Hz);
k is Boltzmann's constant $= 1.381 \times 10^{-23}$ J K^{-1}; and the other symbols are as already defined.

Clearly, adjustments can be made to these calculations, for example to allow for the differences in the gains of satellite antennas towards the beam edge and towards specified earth stations. However, this would often be inappropriate; a satellite downlink may be received, not at one earth station, but at many in different places in the service area, all of which must have their needs taken into account. In some cases (for example, in a TDMA system) a similar situation arises in the uplink. Thus, a calculation for the least favourably located earth station, typically at the edge of the beam footprint, is often what is required.

The remainder of this chapter consists of a set of examples of power and noise budget calculations. These calculations are a simplified version of the ones that would be performed in a real system. In practice the frequency assignment plan for the carriers amplified in an FDMA transponder would be optimized by computation to minimize the impact of intermodulation noise, taking into account the frequency plans for transponders reusing the same frequency band in the same satellite. Similar consideration might be given to interference to and from the assignments made in other satellite systems. The value of noise due to intermodulation and AM/PM conversion and the interference due to these various sources would then be computed for each individual carrier. Finally an adjustment might be made to the carrier power assignment to compensate for exceptional levels of intermodulation noise. However, the simplifications used in these examples and the use of typical values in some cases enable the basic processes involved to be brought out more clearly.

Example 1. Analogue telephony in FDM/FM/FDMA

A transponder in a geostationary satellite with a bandwidth of 72 MHz, operating in C band, is to be used for FDM/FM telephony in FDMA. In this example, a calculation is made to determine whether sufficient satellite power is available under clear-sky conditions for the transponder bandwidth to be fully occupied with such emissions, except for the 10 per cent of the bandwidth which is reserved for guard bands between emissions. If the transponder is indeed bandwidth limited, what margin of power beyond the FM threshold would there be? It is assumed that the earth stations involved conform to Intelsat Standard A and are near the edge of the footprints of the satellite antennas and close to the extreme range for the GSO ($d \approx 40\,600$ km). Parameters appropriate to links in an international system, such as those of an Intelsat VI Hemi beam, are assumed, including in-system frequency reuse. The basic calculations are set out in the form of a power and noise budget in Table 10.2.

Table 10.2 shows $(C/N)_u = 24.2$ dB and $(C/N)_d = 13.5$ dB. Assuming that $(C/N)_{im} = 21$ dB and applying equation (10.4), it is found that $(C/N)_T = 12.5$ dB. Assuming that interference from all sources has power equal to one-third of the power of the noise, $(C/(N+I))_T = 11.3$ dB. Figure 8.2 shows that this C/N ratio is sufficient to enable conventional FM demodulators (that is, without threshold extension) to operate above the threshold although the margin is small. Thus, under clear-sky conditions, all of the available transponder bandwidth could be assigned to emissions, allowing 10 per cent for guard bands. A review of the effect of rain on the margin and downlink thermal noise at 4 GHz (see Chapter 5) shows that it may be assumed, within the limits of accuracy of calculations of this kind, that demodulators would be operating above threshold for all but a very small percentage of the time, so the short-term channel noise objective could be achieved, except in very rainy climates. Favourably located earth stations, such as those near the centre of the satellite antenna beams, might have sufficient margin even in a rainy climate.

To complete the process, the bandwidth and frequency deviation of the emission which each earth station would need to transmit to provide the necessary number of channels with the appropriate S/N would be calculated, taking more precisely determined interference and intermodulation noise levels into account, using the methods set out in Section 8.2.2. The aggregate uplink e.i.r.p. (item (f)

Table 10.2 *Power and noise budget for FDM/FM/FDMA at C band under clear-sky conditions*

Uplink	
(a) PFD at satellite to saturate transponder (Φ_{ts})	$= -70$ dB(W m^{-2})
(b) input backoff	$= 10$ dB
(c) atmospheric loss (L_{mu})	$= 0.5$ dB
(d) achievement margin (M_{au})	$= 0.5$ dB
(e) wave spreading factor $= 10 \log 4\pi d^2$ ($d \approx 40600$ km); equations (2.8) and (5.5)	$= 163.2$ dB
(f) multi-carrier aggregate earth station e.i.r.p. $=$ a $-$ b $+$ c $+$ d $+$ e; equation (10.9)	$= 84.2$ dBW
(g) uplink free-space basic transmission loss (L_{bfu}), equation (5.3), $f_u = 6.2$ GHz, $d \approx 40600$ km,	$= 200.5$ dB
(h) satellite $G/T\,(G/T)_s$	$= -9$ dB K^{-1}
(i) potential usable bandwidth $B = 0.9 \times 72$	$= 64.8$ MHz
(j) $10 \log kB$($k = 1.38 \times 10^{-23}$; $B = 64.8$ MHz)	$= -150.5$ dB (W K^{-1})
(k) $(C/N)_u =$ f $-$ g $-$ c $-$ d $+$ h $-$ j	$= 24.2$ dB
Downlink	
(l) single carrier saturation beam-edge e.i.r.p.	$= 31$ dBW
(m) output backoff,	≈ 5.6 dB
(n) multi-carrier aggregate satellite e.i.r.p. $= \ell - $ m	$= 25.4$ dBW
(o) downlink free-space basic transmission loss (L_{bfd}), equation (5.3), $f_d = 3.95$ GHz, $d \approx 40600$ km	$= 196.6$ dB
(p) atmospheric loss (L_{md})	$= 0.3$ dB
(q) achievement margin (M_{ad})	$= 0.5$ dB
(r) earth station $G/T\,(G/T)_e$	$= 35$ dB K^{-1}
(s) potential usable bandwidth $B = 0.9 \times 72$ MHz	$= 64.8$ MHz
(t) $10 \log kB$	$= -150.5$ dB (W K^{-1})
(u) $(C/N)_d =$ n $-$ o $-$ p $-$ q $+$ r $-$ t, equation (10.10)	$= 13.5$ dB

in Table 10.2) would be shared between the earth stations in approximate proportion to the bandwidth of their emission; thus, an emission occupying 10 per cent of 64.8 MHz would have an e.i.r.p. of approximately 84.2 – 10 dBW.

For similar service requirements using a Ku band transponder, an approximate indication of the operating situation can be obtained by considering the downlink situation only. L_{bfd} at 11.2 GHz would be 205.6 dB instead of 196.6 dB. The atmospheric loss L_{md} and the achievement margin M_{ad} would be somewhat higher. The G/T for an Intelsat Standard C earth station at 11.2 GHz is 37 dB K^{-1} instead of 35 dB K^{-1}. Thus, disregarding the effects of rain, it is to be expected that the beam edge saturation e.i.r.p. required to enable all the bandwidth to be used while achieving the same threshold margin at the earth station demodulators would have to be about $31 + 9 + 0.5 + 0.5 - 2 \approx 39$ dBW. This value is to be compared with 31 dBW at C band.

However, at Ku band some allowance must usually be made for rain absorption in the downlink and an associated increase in downlink thermal noise. For example, in a temperate climate such as Zone E (see Section 5.2.1 and Table 5.1), rain is predicted at an earth station at a rate of 6 mm h^{-1} for 0.3 per cent of the worst month of an average year. This would lead to additional downlink absorption of about 3.5 dB at 11.2 GHz for earth stations close to the extreme range for the GSO (for example, those seeing the satellite at an angle of elevation of 10°). In addition, there would be a reduction in the effective earth station G/T

of 2.5 dB due to increased thermal noise; see equation (5.17). To compensate for these effects due to rain, it would be necessary to increase the satellite beam edge saturation e.i.r.p. rating by a further 6 dB to 45 dBW.

Example 2. A digital carrier in FDMA at Ku band

A digital emission with an information rate of 2048 kbit/s is to be transmitted using coherent QPSK, with other digital emissions, in FDMA via a geostationary satellite using a Ku band transponder, 72 MHz wide, having Intelsat VI characteristics. The earth stations meet Intelsat Standard C, they both see the satellite at an angle of elevation of 30° and they are located in rain climate zone E. This example calculates the power that the earth station transmitter must deliver to its antenna if the BER is not to exceed 1 in 10^6 under clear-sky conditions nor 1 in 10^4 for more than 0.3 per cent of the worst month. The use of FEC is considered.

Using the methods of Example 1, a preliminary iteration of the problem shows that there will not be sufficient power available in the transponder to permit the whole of the bandwidth to be occupied by emissions of this kind if the stated performance objectives are to be achieved. Thus the transponder would be power-limited. There would, however, be enough power to enable about 70 per cent of the bandwidth to be occupied, that is, 45.4 MHz plus 10 per cent for guard bands. Without FEC, QPSK at 2048 kbit/s requires a bandwidth of $\pm(0.6 \times 0.5 \times 2.048) = 1.23$ MHz. Thus, one carrier would require about 2.7 per cent of the available power. The result of this first iteration is used in Table 10.3 as the basis for a second iteration, examining the performance of a single carrier.

Table 10.3 shows $(C/N)_u = 25.4$ dB. $(C/N)_d = 17.2$ dB under clear-sky conditions and 15.2 dB for 0.3 per cent of the worst month of an average year. Using the same assumptions as to $(C/N)_{im}$ and interference levels such as were used in Example 1, $(C/(N+I))_T = 14.0$ and 12.6 dB for clear-sky and rainy conditions respectively. Figure 9.7 shows that a QPSK emission would be received with a BER of about 2.5 in 10^7 and 1 in 10^5 for clear-sky and rainy conditions respectively, comfortably exceeding the performance objective. And Table 10.3, item (h), gives the required earth station transmitter power as 2 dBW.

A third iteration, reviewing backoff and uplink power level, might show how a small increase in the load carried by the transponder might be made without causing performance to fail to meet the objectives.

The addition of, for example, a $\frac{3}{4}$-rate FEC system to the emissions carried by this transponder would provide a coding gain of one or two decibels, enabling the whole of the bandwidth of the transponder to be occupied by emissions which would achieve the performance objectives. However, the addition of FEC would increase the bandwidth required for each emission. It seems that FEC is not likely to increase the potential flow of information through the transponder in this particular situation. However, FEC would be advantageous in other circumstances, for example if the transponder power density were smaller or the earth station G/T were lower.

Example 3. A high-capacity TDMA system at C band

In this example a TDMA system operating at a bit rate of 120 Mbit/s with coherent QPSK has exclusive access to a C band transponder in a geostationary satellite with the same characteristics as an Intelsat VI Hemi beam transponder. The earth stations meet Intelsat Standard A. In this example a calculation is made to determine whether the performance objectives set out in Table 10.1 for PCM telephony would be met. The clear-sky power budget is in Table 10.4.

Table 10.3 *QPSK emissions in FDMA at Ku band without FEC*

Uplink	
(a) PFD at satellite to saturate transponder (Φ_{ts})	$= -75$ dB(W m^{-2})
(b) input backoff (initial working assumption)	$= 10$ dB
(c) PFD at satellite for one 2 Mbit/s emission $= a - b + 10 \log 0.027$	$= -100.7$ dB(W m^{-2})
(d) atmospheric loss (L_{mu}), clear-sky	$= 1$ dB
(e) achievement margin (M_{au})	$= 1$ dB
(f) wave spreading factor $= 10 \log 4\pi d^2$, equation (2.8), $\delta = 30°$, $d \approx 38570$ km	$= 162.7$ dB
(g) earth station e.i.r.p., equation (10.9) $= c + d + e + f$	$= 64$ dBW
(h) earth station transmitter power $= g - G_{te}$	$= 2$ dBW
(i) uplink free-space basic transmission loss (L_{bfu}), equation (5.3), $f_u = 14.25$ GHz, $d \approx 38570$ km	$= 207.3$ dB
(j) satellite $G/T (G/T)_s$	$= 3$ dB K^{-1}
(k) bandwidth of emission (B)	$= 1.23$ MHz
(l) $10 \log kB$	$= -167.7$ dB (W K^{-1})
(m) $(C/N)_u = g - i - d - e + j - l$	$= 25.4$ dB
Downlink	
(n) single carrier saturation beam-edge e.i.r.p.	$= 42$ dBW
(o) output backoff (initial working assumption)	$= 5.6$ dB
(p) e.i.r.p. for one 2 Mbit/s emission $= n - o + 10 \log 0.027$	$= 17.7$ dBW
(q) downlink free-space basic transmission loss (L_{bfd}), equation (5.3), $f_d = 11.2$ GHz, $d \approx 38570$ km	$= 205.2$ dB
(r) atmospheric loss (L_{md}), 11.2 GHz, clear-sky	$= 1$ dB
(s) atmospheric loss (L_{md}), 0.3% of worst month	$= 2$ dB
(t) achievement margin (M_{ad})	$= 2$ dB
(u) earth station G/T	$= 37$ dB K^{-1}
(v) earth station G/T, adjusted for 0.3% of worst month	$= 36$ dB K^{-1}
(w) $(C/N)_d$, clear-sky, equation (10.10) $= p - q - r - t + u - \ell$	$= 17.2$ dB
(x) $(C/N)_d$, 0.3% of worst month, equation (10.10) $= p - q - s - t + v - \ell$	$= 15.2$ dB

Table 10.4 shows that $(C/N)_u = 23.7$ dB and $(C/N)_d = 15.7$ dB. With only one carrier present in the transponder and a modest input backoff, $(C/N)_{im}$ would be negligible. Using the same assumption as to interference as in the earlier examples, $(C/(N+I))_T = 13.8$ dB for beam edge earth stations. Figure 9.7 shows that, for QPSK, this would be sufficient to provide a BER of about 4 in 10^7. This would be ample to meet the short-term performance objective, but only just sufficient for the long-term objective for beam edge earth stations; indeed, beam edge earth stations with very rainy climates might fail to achieve the long-term objective.

This rather marginal situation would probably be considered insufficiently robust. There are two evident ways of providing more margin while using the same satellite and earth station parameters. The bit rate of the TDMA system might be reduced, enabling the pre-demodulator bandwidth to be reduced with no corresponding reduction in the downlink e.i.r.p. Alternatively a TDMA system with very robust frame and burst synchronization protocols and FEC coding to protect the information sub-bursts might be specified.

Table 10.4 *High-capacity TDMA at C band, clear-sky*

Uplink	
(a) PFD at satellite to saturate transponder (Φ_{ts})	$= -78$ dB(W m^{-2})
(b) input backoff	$= 2$ dB
(c) atmospheric loss (L_{mu}), clear-sky	$= 0.5$ dB
(d) achievement margin	$= 0.5$ dB
(e) wave spreading factor $= 10 \log 4\pi d^2$ ($d \approx 40600$ km), equations (2.8) and (5.5)	$= 163.2$ dB
(f) earth station e.i.r.p. $= \text{a} - \text{b} + \text{c} + \text{d} + \text{e}$	$= 84.2$ dBW
(g) earth station antenna gain at 6.2 GHz	$= 59.0$ dBi
(h) power into earth station antenna $= \text{f} - \text{g}$	$= 25.2$ dBW
(i) uplink free-space basic transmission loss (L_{bfu}), equation (5.3), $f_u = 6.2$ GHz, $d \approx 40600$ km	$= 200.5$ dB
(j) satellite $G/T\,(G/T)_s$	$= -9$ dB K^{-1}
(k) required bandwidth (bit rate $\times$ 0.5 $\times$ 1.2)(B)	$= 72$ MHz
(l) 10 log kB	$= -150$ dB(W K^{-1})
(m) $(C/N)_u = \text{f} - \text{i} - \text{c} - \text{d} + \text{j} - \text{l}$	$= 23.7$ dB
Downlink	
(n) single carrier saturation beam-edge e.i.r.p	$= 31$ dBW
(o) output backoff	$= 0.2$ dB
(p) downlink free-space basic transmission loss (L_{bfd}), equation (5.3), $f_d = 3.95$ GHz, $d \approx 40600$ km,	$= 196.6$ dB
(q) atmospheric loss	$= 0.5$ dB
(r) achievement margin	$= 3$ dB
(s) earth station G/T	$= 35$ dB K^{-1}
(t) $(C/N)_d = \text{n} - \text{o} - \text{p} - \text{q} - \text{r} + \text{s} - \text{l}$	$= 15.7$ dB

Example 4. A domestic digital SCPC network at C band

In this example a domestic network serving earth stations with relatively small antennas (5 metres in diameter) in a thinly populated region operates through a C band transponder with a bandwidth of 36 MHz. The satellite is geostationary. The single carrier saturation beam edge e.i.r.p. = 35 dBW. Digital SCPC carriers, QPSK modulated at 64 kbit/s plus $\frac{1}{2}$-rate FEC provide channels used for demand-assigned telephone circuits, data channels or control signals for the demand assignment telephony system. The FEC system has a coding gain of 2.0 dB. The channel performance objective is as shown for PCM telephony in Table 10.1. The climate in the service area is temperate and it may be assumed that, if the long-term objective is achieved, the short-term objectives will be achieved also.

It is assumed that the carriers used for telephony are voice-switched, being suppressed when the subscriber is silent, but the data channel carriers are continuously radiated. When traffic is at its daily peak, 75 per cent of the available channels are used for telephony and 25 per cent are used for data services or for system control purposes. It may be assumed that each pair of telephony channels is in use for a call for 80 per cent of the time at the daily peak of traffic and, on average, each carrier is switched on by the presence of the user's voice for 45 per cent of the time that a call is in progress. The purpose of the calculation is to determine whether the capacity of the network would be limited by the power or the bandwidth of the transponder. If the transponder should be power-limited, how many channels could be operated at the daily peak of traffic? The clear-sky power budget is in Table 10.5.

Table 10.5 *A domestic SCPC network at C band*

Uplink	
(a) PFD at satellite to saturate transponder (Φ_{ts})	$= -70$ dBW m^{-2}
(b) input backoff	$= 10$ dB
(c) atmospheric loss (L_{mu})	$= 0.5$ dB
(d) achievement margin (M_{au})	$= 1$ dB
(e) wave spreading factor $= 10 \log 4\pi d^2$ ($d \approx 38570$ km, $\delta = 30°$), equations (2.8) and (5.5)	$= 162.7$ dB
(f) multi-carrier aggregate earth station e.i.r.p. $=$ a $-$ b $+$ c $+$ d $+$ e; equation (10.9)	$= 84.2$ dBW
(g) uplink free-space basic transmission loss (L_{bfu}), equation (5.3), $f_u = 6.2$ GHz, $d \approx 38570$ km,	$= 200$ dB
(h) satellite $G/T\,(G/T)_s$	$= -5$ dB K^{-1}
(i) potential usable bandwidth $B = 0.9 \times 36$	$= 32.4$ MHz
(j) $10 \log kB$ ($B = 32.4$ MHz)	$= -153.5$ dB(W K^{-1})
(k) $(C/N)_u =$ f $-$ g $-$ c $-$ d $+$ h $-$ j	$= 31.2$ dB
Downlink	
(l) single carrier saturation beam-edge e.i.r.p.	$= 35$ dBW
(m) output backoff	$= 5.6$ dB
(n) multi-carrier aggregate satellite e.i.r.p. $=$ l $-$ m	$= 29.4$ dBW
(o) downlink free-space basic transmission loss (L_{bfd}), equation (5.3), $f_d = 3.95$ GHz, $d \approx 38570$ km,	$= 196.1$ dB
(p) atmospheric loss (L_{md})	$= 0.3$ dB
(q) achievement margin (M_{ad})	$= 2.5$ dB
(r) earth station $G/T\,(G/T)_e$	$= 22$ dB K^{-1}
(s) $(C/N)_d =$ n $-$ o $-$ p $-$ q $+$ r $-$ j,	$= 5.8$ dB

Table 10.5 shows that $(C/N)_u$ and $(C/N)_d$ would be 31.2 and 5.8 dB respectively if the available bandwidth of the transponder (that is 36 MHz, the actual bandwidth, minus 10 per cent for guard bands) were filled with 422 active carriers, all continuously present, each with a bandwidth of 76.8 kHz. Assuming that $(C/N)_{im} = 21$ dB and that interference is at the same relative level as in earlier examples, then $(C/N)_T = 5.7$ dB and $(C/(N+I))_T = 4.4$ dB.

Figure 9.7 shows that QPSK emissions with an FEC coding gain of 2.0 dB require $(C/(N+I))_T$ to reach 8.5 dB if the BER is to equal 1 in 10^6. Thus, the transponder could transmit only 164 carriers simultaneously. However, three-quarters of the channels would be used for telephony, and these would be radiated for only 36 per cent (45% of 80%) of the time, even when the traffic is at its daily peak. Taking this into account, it is found that the transponder would indeed be power-limited, but it would be capable of supporting 79 unidirectional data or system control channels and 118 bidirectional telephone circuits, together occupying 75 per cent of the bandwidth of the transponder.

Example 5. An analogue video channel in a C band transponder

An analogue video channel (picture standards I/PAL, 625/50), without an accompanying sound channel, is assigned all of the power and bandwidth of a C band transponder, with 41 MHz bandwidth centred on 6402.5/4177.5 MHz, having the characteristics of an Intelsat VI global beam transponder. The satellite is geostationary. Earth stations to Intelsat Standard A access the transponder, seeing the satellite at an angle of elevation of 5°. How powerful an FM emission must be delivered to the antenna at the transmitting earth station to provide the maximum S/N ratio at the receiving earth station without over-deviation, and what would that S/N ratio be?

Table 10.6 shows that $(C/N)_u$ and $(C/N)_d$ would be 23.9 and 15.8 dB respectively under clear-sky conditions if the full bandwidth of the transponder were used. With no intermodulation noise, $(C/N)_T$ would be 15.2 dB. With the same assumptions as to interference as have been used in the other examples, $(C/(N+I))_T = 13.9$ dB. This provides an ample threshold margin for conventional FM demodulators, and it may be assumed that sufficient margin would remain under rainy conditions except for earth stations in very rainy climates.

Table 10.6 *Analogue video channel in C band transponder*

Uplink	
(a) PFD at satellite to saturate transponder (Φ_{ts})	$= -77$ dBW m^{-2}
(b) input backoff	$= 0$ dB
(c) atmospheric loss (L_{mu})	$= 0.5$ dB
(d) achievement margin (M_{au})	$= 0.5$ dB
(e) wave spreading factor $= 10 \log 4\pi d^2$ ($d \approx 41130$ km), equations (2.8) and (5.5)	$= 163.3$ dB
(f) earth station e.i.r.p. = a − b + c + d + e	$= 87.3$ dBW
(g) earth station antenna gain	$= 59$ dBi
(h) power into earth station antenna = f − g	$= 28.3$ dBW
(i) uplink free-space basic transmission loss (L_{bfu}), equation (5.3), $f_u = 6402.5$ MHz, $d \approx 41130$ km	$= 200.9$ dB
(j) satellite G/T	$= -14$ dB K^{-1}
(l) $10 \log kB$ ($k = 1.38^{-23}$, $B = 41$ MHz)	$= -152.5$ dB (W K^{-1})
(m) $(C/N)_u$ = f − i − c − d + j − l	$= 23.9$ dB
Downlink	
(n) single carrier saturation beam-edge e.i.r.p.	$= 26.5$ dBW
(o) output backoff	$= 0$
(p) downlink free-space basic transmission loss (L_{bfd}), equation (5.3), $f_d = 4177.5$ MHz, $d \approx 41130$ km	$= 197.2$ dB
(q) atmospheric loss (L_{md}), clear-sky	$= 0.5$ dB
(r) achievement margin (M_{ad})	$= 0.5$ dB
(s) earth station G/T	$= 35$ dB K^{-1}
(t) $(C/N)_d$ = n − o − p − q − r + s − l	$= 15.8$ dB

The bandwidth of a Standard I video signal is 5.5 MHz. From equation (8.3), it can be seen that the peak frequency deviation F_p would be 15 MHz if the highest modulating frequency f_u were 5.5 MHz and the Carson's rule bandwidth were 41 MHz. Thus F_{tt} the test tone peak-to-peak deviation would be $2 \times 15 = 30$ MHz if the test tone were located at the emphasis curve crossover frequency. Thus, from equation (8.18), the post-demodulator S/N ratio in accordance with the convention of equation (8.16) is 64.6 dB.

The result shows a channel performance well in excess of the value recommended for international channels used for the distribution of video signals that are to be broadcast, namely 53 dB; see Table 10.1. Fully acceptable performance could be obtained using a receiving earth station with a substantially lower G/T, although if G/T were lower than about 33 dB K^{-1} it would be necessary to reduce the deviation and the bandwidth in order to maintain an adequate threshold margin. It may also be noted that the required earth station transmitter power is large (approaching 700 W, see item (h) of Table 10.6) for an Intelsat Standard A station and would become inconveniently large if the gain of the antenna were much less. Alternatively, and this is a common practice, the transponder could be used in FDMA for two video emissions. This reduction of bandwidth and power causes a substantial reduction in S/N, some of which is made good by means of over-deviation without causing an unacceptable degree of degradation of the subjective quality of the picture.

References

1. *Utilization of fade countermeasures, strategies and techniques in the fixed-satellite service* (ITU-R Recommendation S.1061, 1994), ITU-R Recommendations 1994 S Series. Geneva: ITU.
2. *Propagation data and prediction methods required for the design of Earth–space telecommunications systems* (ITU-R Recommendation PN.618-3, 1994), ITU-R Recommendations 1994 PN Series. Geneva: ITU.
3. Hodson, K., Spracklen, C. T. and Heron, R. (1992). VSAT networks and wide area diversity. *Proc. Military Microwaves Conference*, Brighton, UK.
4. *Availability objectives for a hypothetical reference circuit and a hypothetical reference digital path when used for telephony using pulse-code modulation, or as part of an Integrated Services Digital Network hypothetical reference connection, in the fixed satellite service* (ITU-R Recommendation S.579-3, 1994), ITU-R Recommendations 1994 S Series. Geneva: ITU.
5. *Allowable noise power in the hypothetical reference circuit for frequency-division multiplex telephony in the fixed-satellite service* (ITU-R Recommendation S.353-8, 1994), ITU-R Recommendations 1994 S Series. Geneva: ITU.
6. *Allowable bit error ratios at the output of the hypothetical reference digital path for systems in the fixed-satellite service using pulse-code modulation for telephony* (ITU-R Recommendation S.522-5, 1994), ITU-R Recommendations 1994 S Series. Geneva: ITU.
7. *Video bandwidth and permissible noise level in the hypothetical reference circuit for the fixed-satellite service* (ITU-R Recommendation S.354-2, 1994), ITU-R Recommendations 1994 S Series. Geneva: ITU.
8. *Transmission performance of television circuits designed for use in international connections* (CCIR Recommendation 567-3, 1990), Recommendations of the CCIR 1990, Volume XII. Geneva: ITU.
9. *Maximum permissible level of interference in a telephone channel of a geostationary-satellite network in the fixed-satellite service employing frequency modulation with frequency-division multiplex, caused by other networks of this service* (ITU-R Recommendation S.466-6, 1994), ITU-R Recommendations 1994 S Series. Geneva: ITU.
10. *Maximum permissible levels of interference in a geostationary-satellite network in the fixed-satellite service using 8-bit PCM encoded telephony, caused by other networks of this service* (ITU-R Recommendation S.523-4, 1994), ITU-R Recommendations 1994 S Series. Geneva: ITU.
11. *Maximum permissible level of interference in a television channel of a geostationary satellite network in the fixed-satellite service employing frequency modulation, caused by other networks of this service* (ITU-R Recommendation S.483-2, 1994), ITU-R Recommendations 1994 S Series. Geneva: ITU.
12. *Maximum allowable values of interference from line-of-sight radio-relay systems in a telephone channel of a system in the fixed-satellite service employing frequency modulation, when the same frequency bands are shared by both systems* (ITU-R Recommendation SF.356-4, 1994), ITU-R Recommendations 1994 SF Series. Geneva: ITU.
13. *Maximum allowable values of interference from terrestrial radio links to systems in the fixed-satellite service employing 8-bit PCM encoded telephony and sharing the same frequency bands* (ITU-R Recommendation SF.558-2, 1994), ITU-R Recommendations 1994 SF Series. Geneva: ITU.
14. *Allowable error performance for a hypothetical reference digital path in the fixed-satellite service operating below 15 GHz when forming part of an international connection in an Integrated Services Digital Network* (ITU-R Recommendation S.614-3, 1994), ITU-R Recommendations 1994 S Series. Geneva: ITU.

15. *Maximum permissible levels of interference in a geostationary-satellite network for an HRDP when forming part of the ISDN in the fixed-satellite service caused by other networks of this service below 15 GHz* (ITU-R Recommendation S.735-1, 1994), ITU-R Recommendations 1994 S Series. Geneva: ITU.
16. *Error performance parameters and objectives for international, constant bit rate paths at or above the primary rate* (ITU-T Recommendation G.826, 1993). Geneva: ITU.
17. *Allowable error performance for a hypothetical reference digital path operating at or above the primary rate* (ITU-R Recommendation S.1062, 1994), ITU-R Recommendations 1994 S Series. Geneva: ITU.

Part 4
Applications for Satellite Communications

11 Satellite systems serving the international PSTN

11.1 Introduction

Satellite systems serving fixed earth stations take many forms and it is one of the strengths of the FSS as a telecommunications medium that the same satellite can provide the space segment for radically different applications. Nevertheless, in considering technical aspects of those applications, and especially those of the earth segment, it is convenient to divide networks into three groups, as follows.

When commercial satellite communication began, it was used mainly for international links between national PSTNs. The earth stations required very large antennas. International PSTN links are still an important market for satellite communication and in most cases they still involve earth station antennas around 10 metres in diameter or larger. These big-antenna applications differ in major ways from other FSS applications and it is convenient to label them 'international networks', although they also carry some domestic traffic (that is, traffic that is wholly within the boundaries of one country). These international network applications are the subject of the present chapter.

There is another distinct category of FSS applications of growing importance which use earth stations with small antennas; the stations are called very small aperture terminals (VSATs). The term VSAT can conveniently be applied to earth stations having antennas not more than about 3 metres in diameter. Their use is considered in Chapter 13.

Between these big-antenna and small-antenna applications, there lies a large field for earth stations of intermediate size. Such systems are used largely, though not exclusively, for domestic PSTN and private networks and they are considered in Chapter 12.

Satellite systems for international links

Section 1.3 recalls how two organizations, Intelsat and Intersputnik, were set up in 1964 and 1971 respectively to provide space segment facilities with world-wide coverage for international links. Intelsat has become a very large system, providing a Global Network for the PSTN as well as extensive space segment facilities for other purposes. Intersputnik has remained relatively small. Other space segment providers, set up to serve more limited geographical regions, also provide a space segment for international links. Eutelsat is a good example of a regional system. The use of Intelsat and Eutelsat for international links is described in this chapter.

The special features of the Intelsat Global Network arise from the need to relay a very large number of telephone channels between many earth stations spread over a very large service area through the least number of satellites. In this way the earth segment costs of countries using relatively few channels distributed between many routes can be minimized. However, to achieve this objective, earth stations are required to have a high and uniform figure of merit (G/T) and Intelsat satellites must employ an exceptionally high degree of frequency reuse.

Frequency allocations

Wide frequency bands are allocated to the FSS between 3.4 and 31 GHz. However, almost all of these bands are shared with other radio services and some are reserved for specified purposes, such as the FSS frequency allotment plan or feeder links for the BSS frequency assignment plan. The allocations that are currently in general use for FSS systems that serve the PSTN are as follows:

$$\text{In C band} \begin{cases} 5925\text{–}6425\,\text{MHz for uplinks,} \\ 3700\text{–}4200\,\text{MHz for downlinks.} \end{cases}$$

$$\text{In Ku band} \begin{cases} 14.0\text{–}14.5\,\text{MHz for uplinks,} \\ 10.95\text{–}11.2 \text{ and } 11.45 - 11.7\,\text{GHz for downlinks.} \end{cases}$$

In the Intelsat VI satellite series, the C band spectrum extends down to 5850 and 3625 MHz. Spectrum in Ka band is also available for these international systems, around 19 and 29 GHz, but its use presents major problems for earth stations located where substantial amounts of rain fall, and it remains unused for the present.

11.2 Intelsat and its space segment

11.2.1 The system

For a general description of the Intelsat system and its background, see [1]. Intelsat has 24 satellites in operation at the time of writing, all geostationary. For the Global Network, Intelsat divides the world into three regions: the Atlantic Ocean, the Indian Ocean and the Pacific Ocean regions. Earth stations in these regions can be served by satellites located in the GSO within a few degrees of 330° E, 60° E and 177° E longitude respectively. The three regions together include virtually all land areas between latitudes 70° N and 70° S. A fourth group of satellites is planned at about 90° E to expand Intelsat's coverage of the Asia-Pacific region. Other Intelsat satellites elsewhere in the GSO provide transponders for leasing and are used outside the Global Network.

The level of demand for the Intelsat Global Network is so great that it cannot be met by a single satellite in even the least heavily loaded region. To ensure high connectivity (that is, to allow many earth stations to communicate with many other earth stations via a single satellite) and thus to keep acceptably small the earth segment cost for the smaller users of the system, one satellite in each region is designated the primary satellite. In principle, every earth station in a region has access to the primary satellite and uses it for traffic streams which are not very large. Serving the Global Network in each region there is also a second, and perhaps a third, satellite. These latter are called the 'major path' satellites. Earth station owners with a large total traffic flow install additional antennas and route some of their traffic through these other satellites.

The current Intelsat satellites

In 1996 the last of eight satellites of the Intelsat V series is approaching the end of its useful life after 15 years in orbit. The five satellites of the Intelsat VA series that were carrying the heavier traffic loads of the Global Network in the late 1980s are now being used for the lighter loads, mainly leased transponder services. The five satellites of the Intelsat VI series, with considerably more channel capacity than those of the earlier series, have taken over the main Global Network functions. The launching of satellites of the Intelsat VII and VIIA series has begun, and spacecraft of two new series, Intelsat VIII and Intelsat VIIIA, are currently being procured. All of these seven series of satellites, Intelsat V, VA, VI, VII, VIIA, VIII and VIIIA, operate in both C band and Ku band. A single satellite designated Intelsat K, operating in Ku Band only, was launched in 1992. Intelsat also leases capacity in satellites owned by other space segment operators.

The Intelsat VI platform is spin-stabilized with a despun payload. The others are all three-axis stabilized and there are close similarities between the Intelsat VII and Intelsat VIII generations.

The design of the communications payloads of each generation of Intelsat satellites up to and including Intelsat VI has been driven primarily by the need to keep pace with the ever-growing demands on the Atlantic Ocean region primary satellite. The footprints of satellite beams optimized for this role, taking into account some limited options for footprint adjustments which can be telecommanded in orbit, have matched the requirements of the other regions well enough, enabling satellites of standard designs to be used. This has provided cost savings in procurement and flexibility in operation. However, the scale of the system as a whole, and of each region individually, is now so great that the purchase of satellites with designs optimized for each individual region is becoming economically justified. For example, Intelsat VII and Intelsat VIII were optimized for the Pacific Ocean region and both have more powerful C band transponders than Intelsat VI.

Thus, the Intelsat space segment comprises several different kinds of satellite. They show differences of physical structure and payload configuration, but the transmission characteristics of the different satellite types have much in common. The description in Section 11.2.2 of the Intelsat VI communications payload is broadly applicable to the other satellite series also. For an outline description of the Intelsat VII satellites, see [2].

11.2.2 Intelsat VI satellite transmission characteristics

Attitude stabilization of the Intelsat VI satellites is provided by spinning the cylindrical body about the pitch axis, the communications payload including the antennas being housed on a despun platform. There are 48 transponders in total, 38 operating in C band and 10 operating in Ku band. The solar panels are designed to provide over 2 kW of primary power at end-of-life. The mass of the satellite at the start of life in the GSO was 2235 kg, including about 450 kg of fuel for thrusters. See Figure 11.1.

The 6/4 GHz antenna beams

Intelsat VI has three sets of 6/4 GHz antenna beams, called the Global, Hemi and Zone beams. The footprints of the Hemi and Zone beams, for a satellite located at longitude 335.5° E, are shown in Figure 11.2.

The Global beams do not, of course, have global coverage but they cover as much of the Earth's surface as can be seen from a geostationary satellite. The beams are generated by 6 and 4 GHz conical horns, with frequency reuse by means of orthogonal circular polarizations. The beamwidth between −3 dB gain contours is about 17°, providing a gain of about 17 dBi at the edge of the Earth's disc as seen from the satellite.

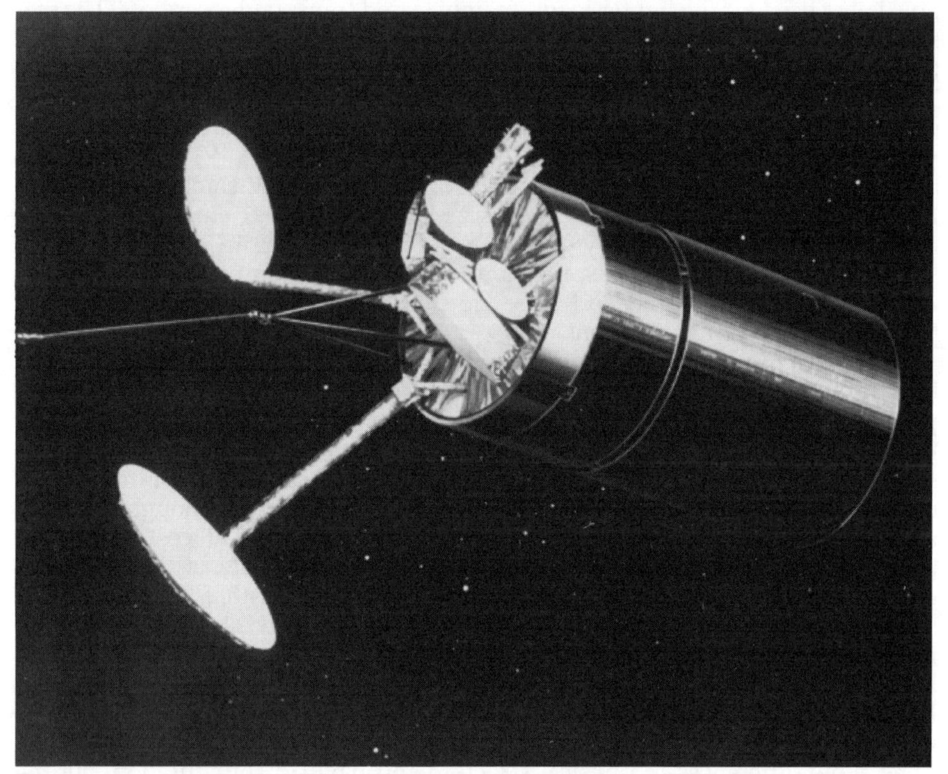

Figure 11.1 *An Intelsat VI satellite. (Reproduced by permission of Intelsat.)*

There are two 6 GHz Hemi beams and two matching 4 GHz Hemi beams, shaped to cover the areas on the Earth, continental in extent, within which most of the Global Network earth stations are located. For example, from the Atlantic Ocean region orbital locations, the footprint of the East Hemi beam covers as much of Europe and Africa as is visible from those satellite locations. And the West Hemi beam similarly covers almost all of the visible parts of North and South America.

The two 4 GHz Hemi beams are a conflation of high-gain beams generated by an array of 143 feed horns in the focal plane of a circular offset-fed paraboloidal reflector, 3.2 metres in diameter. Figure 11.3 shows the feed horn array. A feed matrix serving the right-hand circular polarization ports of these horns enables the phase and level of the input to each individual horn to be adjusted to achieve three objectives. Interference between the East and West Hemi beams must be made acceptably low at the location of any earth station foreseen to need to use one of the beams, so that spaced-beam frequency reuse is feasible. The second objective is to produce a PFD distribution within each Hemi beam footprint that is relatively uniform at the Earth's surface. Thirdly, pure right-hand circular polarization is required, to enable the Hemi beams and the Zone beams, which have common frequency occupation, overlapping footprints, but orthogonal polarizations, to reuse the spectrum.

A similar array of horns feeding a reflector 2 metres in diameter generates the 6 GHz Hemi beams; the same three objectives guide the final adjustment of the feed matrix, the required polarization in this case being left-hand circular.

Four Zone beams, two within each Hemi beam, are formed at 6 and 4 GHz by the same circular reflectors and subsets of the same arrays of feed horns

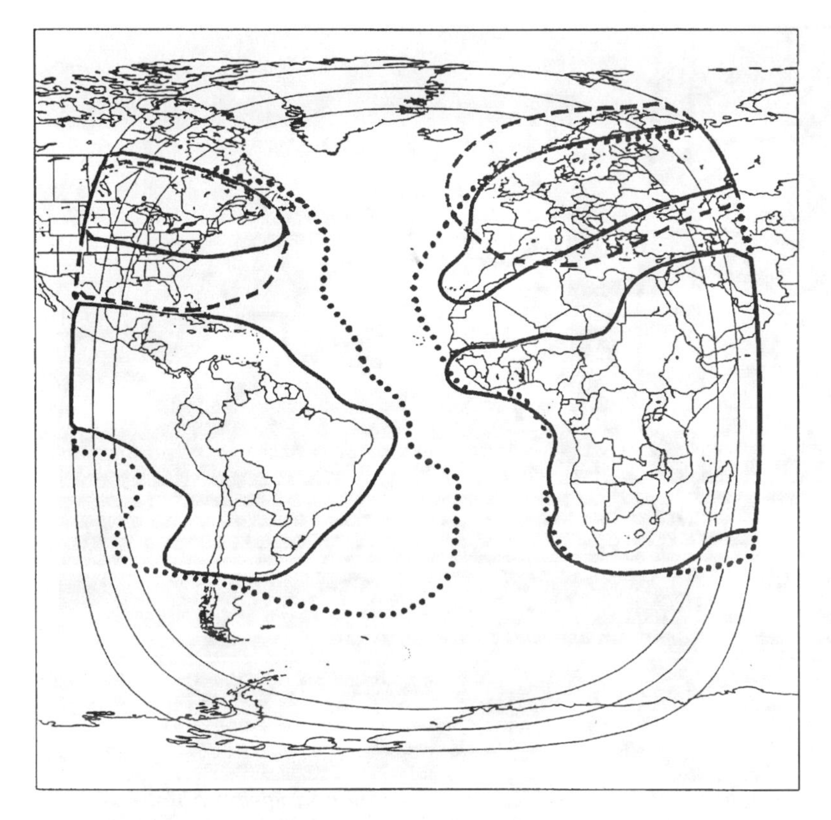

Figure 11.2 *The Intelsat VI Hemi, Zone and Spot beam footprints from a satellite stationed at 335.5° E longitude. (Reproduced by permission of Intelsat.)*

but using different feed matrices. These matrices serve feed ports which provide polarization orthogonal to that provided for the Hemi beams. The footprints of the Zone beams cover areas on Earth which contain most of the earth stations with the heaviest total traffic demands. For example, in the Atlantic Ocean region, the North-East Zone footprint covers western Europe and the South-East Zone footprint covers a large area of Africa south of the Sahara; the North-Western Zone footprint covers earth stations in the United States and Canada and the South-Western Zone footprint covers most of South America and part of Central America. See Figure 11.2. These four Zone beams have sufficient mutual angular separation at the satellite to permit fourfold spaced-beam frequency reuse, despite a common polarization. Thus, the Zone beam feed matrices are adjusted to allow frequency reuse between the Zone beams, sufficiently uniform power flux density within each footprint, and uniformly pure circular polarization, orthogonal to that of the Hemi beams.

There are substantial differences between the optimum Zone footprint locations for the three regions. Consequently, three different zone feed matrices are provided for each feed horn array. The appropriate matrix is selected in orbit by telecommand if the satellite should be moved from one ocean region to another in the course of its in-orbit lifetime.

Figure 11.3 *The Intelsat VI feed horn array that generates the 4 GHz Hemi and Zone beams. (Reproduced by permission of Intelsat.)*

The Intelsat VI satellites occupy a nominal 2 × 575 MHz of spectrum in C band, but an actual usable bandwidth of 2 × 2026 MHz is derived from it by means of dual-polarization and spaced-beam frequency reuse.

The 14/11 GHz antenna beams

On Intelsat VI, offset-fed paraboloidal reflectors having diameters of the order of 1 metre provide Ku band beams, fully steerable by telecommand, with a half-power beamwidth of about 1° for the East Spot beam and about 2° for the West Spot beam. Spaced-beam frequency reuse can be employed, provided that the beams have sufficiently well-separated pointing directions. These antennas both transmit and receive, using a single linear polarization mode. The footprints of the Spot beams, deployed as for an orbital longitude of 335.5° E, are shown in Figure 11.2.

The transponder frequency plan

The nominal frequency bands used by Intelsat VI satellites are 5850–6425 and 14 000–14 500 MHz for uplinks and 3625–4200, 10 950–11 200 and 11 450–11 700 MHz for downlinks. This spectrum is divided into nine channels

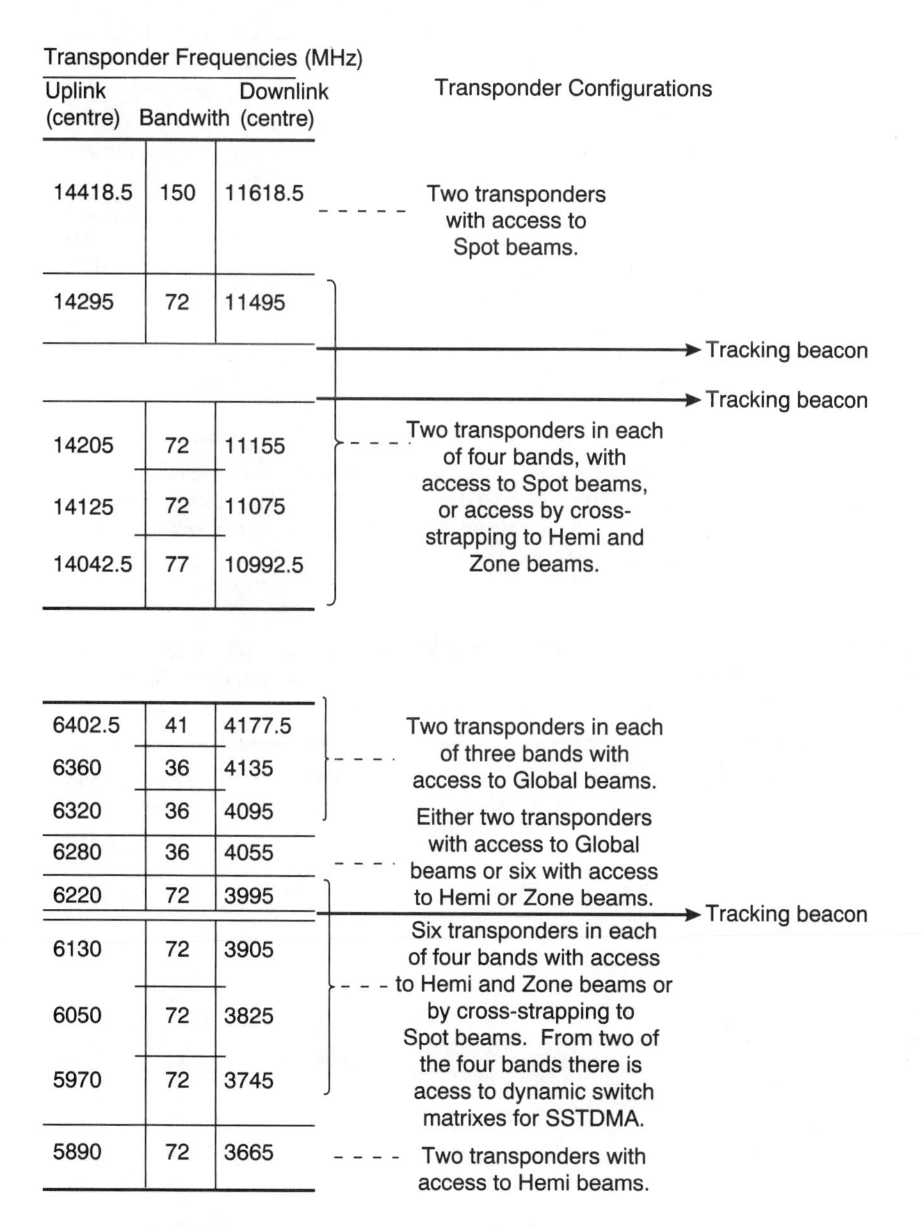

Figure 11.4 *The Intelsat VI transponder frequency plan*

for C band transponders and five for Ku band transponders, as shown in Figure 11.4. There are narrow guard bands at the edges of the frequency allocations and between transponders. The guard bands centred on 3950, 11 198 and 11 452 MHz are occupied by downlink carriers which carry telemetry signals to TTC&C stations on Earth and serve as tracking beacons for all earth station antennas.

The 6/4 GHz global beam transponders

Intelsat VI has six transponders operating as three pairs with global coverage, using orthogonal circular polarizations to provide frequency reuse. Reception is between 6300 and 6425 MHz and transmission is between 4075 and 4200 MHz. The bandwidth of these transponders is either 36 or 41 MHz. A fourth pair of transponders can operate in a 36 MHz wide frequency slot at 6260–6300/4035–4075 MHz, also with global coverage. (Alternatively this frequency slot can be used for a set of six transponders with Hemi or Zone beams.)

There are wideband GaAs FET low-noise preamplifiers covering 6260–6425 MHz, one for each polarization state, having a noise figure of about 3 dB. They are followed by down-converters which change the frequency of the uplinked emissions to their prospective downlink frequency in the 4 GHz band. An input multiplexer then separates the received signals into wideband paths 36 or 41 MHz wide. Each wideband path is routed via an attenuator, adjustable by telecommand, to a transistor driver amplifier and a TWT power amplifier. After amplification the wideband paths are filtered to eliminate out-of-band intermodulation products. The paths are then recombined by output multiplexers and pass to the feed port of the global transmit horn which is orthogonal to the polarization in which the signals had been uplinked. The block schematic diagram in Figure 7.5 shows this arrangement. The basic parameters of these transponders are set out in Table 11.1.

Table 11.1 *The basic parameters of Intelsat VI transponders (A range of values is quoted in some cases, the parameter in any particular case depending on the bandwidth, the attenuator setting or the antenna gain)*

Beam	*Bandwidth* (MHz)	*G/T* ($dB\,K^{-1}$)	*Uplink single carrier saturation PFD* ($dB(W\,m^{-2})$)	*Beam-edge downlink single carrier saturation e.i.r.p.* (dBW)
C band				
Global	36/41	−14	-74 ± 4	26.5/23.5
Hemi	72/77	−9.4	-72 ± 5	31
Hemi	36	−9.4	-74 ± 4	28
Zone	72	−2/−7.5	-72 ± 5	31
Zone	36	−2/−7.5	-74 ± 4	28
Ku Band				
Spot	72/150	+1/+6	-80 ± 7	41/44

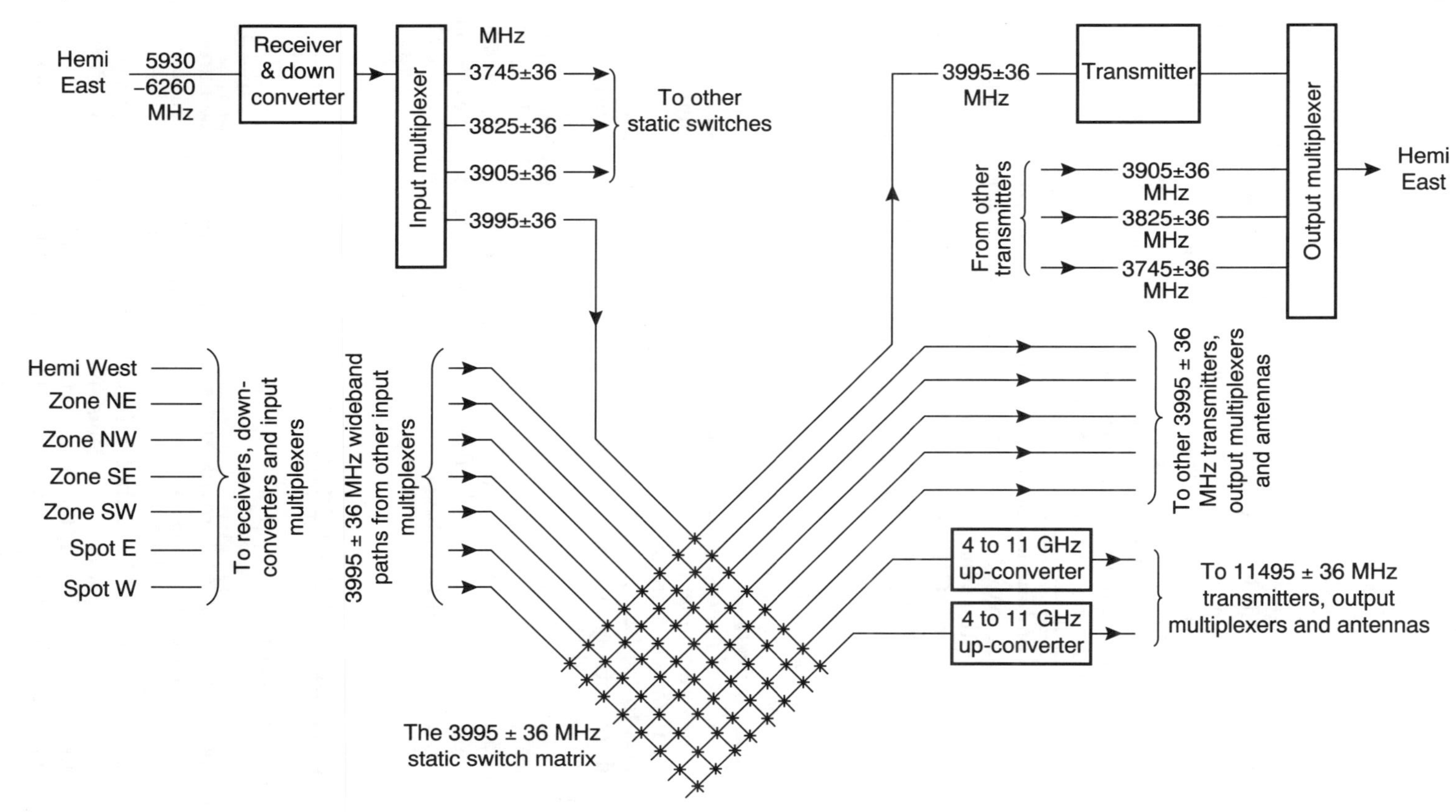

Figure 11.5 *A simplified block schematic diagram of the wideband paths (bandwidth = 72 MHz) through Intelsat VI, served by Hemi, Zone and Spot beams*

The narrow beam transponders, 6/4 and 14/11 GHz

Intelsat VI has four Zone beams and two Hemi beams with uplinks at 5930–6260 MHz and downlinks at 3705–4035 MHz, plus two Spot beams with uplinks at 14.0–14.355 GHz and downlinks between 10.95 and 11.535 GHz. Eight streams of signals from earth stations, one from each up-beam, enter the narrow-beam transponder system. See Figure 11.5. Each stream is fed to a wideband receiver. After amplification and down-conversion to the 4 GHz band, input multiplexers extract four wideband paths, 72 MHz wide, from each of the eight main streams. Advantage was taken of an opportunity to widen some of the paths to 77 MHz, but this point is disregarded here. Thus, at this point in the system there is a set of eight wideband paths in the band 3995 $\pm$ 36 MHz and three more sets of eight paths centred on 3905, 3825 and 3745 MHz respectively. These 32 paths are passed, via flexibility fields, to 32 transmitters, which incorporate 4/11 GHz up-converters where necessary. The outputs from the 32 transmitters are then filtered and combined in fours into eight main streams by output multiplexers, one for each transmit beam.

The four flexibility fields each consist of an 8 $\times$ 8 matrix of switches which permit any desired combination of up-beams and down-beams to be selected by telecommand for each frequency band; these are called the static switches. It will be noted that these options include coupling 6 GHz uplinks to 11 GHz downlinks and 14 GHz uplinks to 4 GHz downlinks, an option called cross-strapping. In addition, for two of the frequency bands there are matrices of high-speed switches which enable selected Hemi and Zone beams and the associated receivers and transmitters to be operated in the SSTDMA mode. These are called the dynamic switches; they are programmable to reconfigure the up-beam and down-beam combinations used for a TDMA system as many times as required in the course of the 2 ms TDMA frame; see Section 11.4.5.

The 6 GHz receivers are similar to those used for the global beams. The 14 GHz LNAs have a noise figure around 5 dB. The 4 GHz transmitters have solid state power amplifiers but TWTs are used in the 11 GHz transmitters. Basic transponder parameters are in Table 11.1.

In addition to the 32 narrow-beam transponders which operate in a flexible mode and the six global transponders, Intelsat VI has three further sets of transponders. Firstly, reference was made above to the frequency slot 6260–6300/4035–4075 MHz which can be used at will for a fourth pair of global beam transponders or a set of six transponders with Hemi and Zone beams, the bandwidth being 36 MHz in either case. Secondly, a pair of C band transponders, operating at 5850–5930/3625–3705 MHz, have access to the Hemi beams only. And thirdly, a pair of Ku band transponders, 150 MHz wide, centred on 14 418.5 and 11 618.5 MHz, have access to the Spot beams only.

11.3 Eutelsat and its space segment

11.3.1 The system

Eutelsat was set up in 1977 to provide space segment facilities for the European area, much as Intelsat provides facilities on a world-wide scale. Initially the services provided were limited to PSTN channels by TDMA and temporary video links coordinated by the European Broadcasting Union (EBU) and serving European terrestrial broadcasting systems. Nowadays many transponders are used for direct-to-home (DTH) television broadcasting and others are operated in FDMA providing Intermediate-rate Digital Carriers (IDC) for public networks and Satellite Multi-Services (SMS) for private networks. Eutelsat also provides Euteltracs, offering communication and radiodetermination facilities for vehicles on land; see Section 14.5.2.

In addition to the management of Eutelsat satellites, Eutelsat has responsibility for sales of international telecommunication facilities in certain European national satellites.

Eutelsat started operation using the Orbital Test Satellite (OTS), a developmental satellite launched in 1978 and made available by the European Space Agency (ESA). Five satellites, originally called the European Communication Satellites (ECS) but renamed Eutelsat I, were ordered in 1979 and four were successfully launched between 1983 and 1988. Three are still in operation in 1996, although with diminished capabilities. For a description of Eutelsat I, see below. Of five Eutelsat II second-generation satellites, four were launched successfully between 1990 and 1992. In each series one satellite was lost in launching. The transmission characteristics of Eutelsat II are outlined in Section 11.3.2. All of these satellites are located in the GSO between 1° E and 26° E.

A sixth satellite to the Eutelsat II basic design but with a substantially modified payload has been procured and will be used for DTH television broadcasting; it is to be known as Hotbird 1. Two more satellites, optimized for DTH and broadly similar to Hotbird 1, are also being procured, the first for launch in 1996; they will be called Hotbird 2 and 3. There are also plans for a third-generation of Eutelsat satellites intended primarily for telecommunication services.

Eutelsat I

Eutelsat I is a three-axis-stabilized satellite with an on-station mass of about 750 kg. The frequency plan for all but the first flight model provides for 14 transponders, each with a bandwidth of 72 MHz, operating in frequency reuse pairs in Ku band with orthogonal linear polarizations. TWT power amplifiers are used, rated at 20 watts single carrier saturated output. However, the power available from the solar array is not sufficient for more than ten of the TWTs to be operated at the same time. The Eurobeams, transmit and receive, cover a wide area from the Azores to the Caspian Sea and three transmit-only spot beams provide higher downlink e.i.r.p. in the more important parts of the service area. Additional beams, transmit and receive, covering continental Europe are used for SMS. The uplink G/T at beam edge is $-3\,\mathrm{dB\,K^{-1}}$ for the Eurobeam and $-1\,\mathrm{dB\,K^{-1}}$ for the SMS beam. The beam edge downlink single carrier saturated e.i.r.p. values for the various beams are in the range 37 to 41 dBW.

11.3.2 Eutelsat II satellite transmission characteristics

The design of Eutelsat II evolved directly from that of Eutelsat I. Capacity per satellite has been increased. A changing market has led to additional facilities, optimized for SMS. Shaped antenna beams reflect more accurately the geographical distribution of demand.

There are 16 transponders, each rated at 50 watts single carrier saturated output. The frequency plan is shown in Figure 11.6. Seven transponders have 72 MHz bandwidth, like Eutelsat I, but nine are 36 MHz wide, providing higher downlink spectral power density and thus making the transponders better suited for downlinks to earth stations with small antennas. Six transponders can operate downlinks in the band 11.45–11.7 GHz or alternatively in the band 12.5–12.75 GHz; this increases the proportion of the total operating bandwidth available for downlinking in the latter band from the 17 per cent for Eutelsat I to 48 per cent for Eutelsat II. This is important because 12.5–12.75 GHz is not

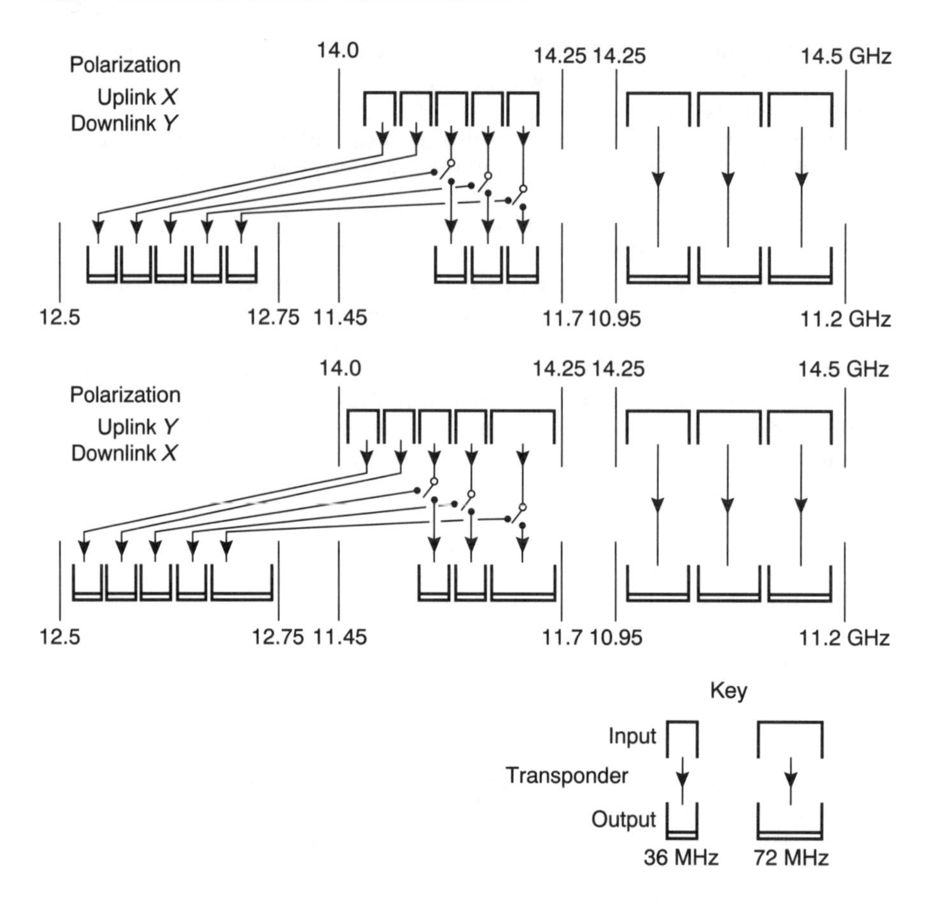

Figure 11.6 *The Eutelsat II transponder frequency plan*

generally used for terrestrial radio services in Europe, greatly easing frequency coordination for earth stations. To the extent feasible, the frequency bands of the 36 MHz transponders are interleaved with those of the orthogonal polarization, greatly increasing XPI for wideband frequency modulated emissions.

The two sets of shaped beams make adequate provision for the outer reaches of the service area but concentrate gain in the geographical areas where demand is more intense. As a result, the uplink G/T in the central areas of western Europe has been raised to $+2\,\mathrm{dB\,K^{-1}}$. The single carrier saturated downlink e.i.r.p. for much of Europe exceeds 44 dBW per transponder, and it exceeds 50 dBW at the centre.

11.4 The earth segment

11.4.1 Earth station sites

For most satellite applications the location of the earth station is determined, to a large degree, by the location of the user. However, there can be a much wider choice of location for earth stations providing international links for the PSTN, and important issues arise to guide that choice.

Some of the basic requirements for the site of an international earth station are like those for any other major industrial installation. The site must not be liable to flooding. The subsoil should be strong enough to support heavy structures without unduly costly foundations. Easy access by road, easy access to power, water, drainage and a socially acceptable place for staff to live are all important. In their absence the cost of building and running the station will be substantially increased. The area of land required will be quite small if the site is to be equipped with one antenna only, but many hectares may be needed for a multi-antenna station. There may be aesthetic objections to the intrusion of highly visible engineering plant into a rural landscape.

In addition to these general factors, there are others which arise from the station's telecommunication function. Earth stations should be built where interference to or from terrestrial radio stations will not raise insuperable problems. The angle of elevation, measured at the Earth station, of the satellite which is to be accessed, must not be too low. It may be important to consider the microclimate of an otherwise satisfactory location. In addition, the cost of the backhaul link to the gateway to the terrestrial network should be taken into account.

Terrestrial interference

Interference to and from radio relay systems is a common problem for satellite systems like Intelsat and Eutelsat which serve the international network. The problem arises because the frequency bands that these systems use are also in very widespread use for terrestrial radio relay systems, whereas many of the FSS systems discussed in Chapters 12 and 13 use frequency bands that are allocated, locally or more widely, exclusively for satellite systems. These earth stations transmit a particularly high e.i.r.p., making them very liable to cause interference to terrestrial receiving stations. Moreover, their antennas, with high gain and low noise in the receiving mode, make the earth stations particularly susceptible to interference from terrestrial transmitting stations. Reduction to an acceptable level of interference between satellite and radio relay systems can be achieved much more readily if the potential problem is taken into account in choosing station sites. Interference from spurious emissions radiated by other powerful microwave transmitters, and especially radars, can also be very troublesome at earth stations.

To avoid such problems completely, an earth station would have to be located in a region, some hundreds of kilometres in extent, free from other radio stations which are potential sources or victims of interference; see Section 6.4.1. Such regions are becoming rare. However, provided there are no terrestrial stations likely to be sources or victims of interference within a few tens of kilometres of the earth station, there are several strategies for choosing a specific site which limit the severity of the interference problems that arise at greater distances. For example:

(1) Ranges of hills or mountains might screen the earth station site from more distant radio stations.
(2) A cautious view should be taken of the presence of open water near a potential earth station site. Tropospheric ducts tend to form over warm seas and large lakes, and these ducts are capable of propagating strong interference from radio stations beyond the further shore even though the over-water path may be several hundreds of kilometres long.
(3) Radio propagation losses from known existing sites and potential new sites for radio relay stations and radars within a radius of several hundreds of kilometres of a potential earth station site should be reviewed individually. A relatively small change in the location of the stations may make a large difference to the interference potential.

(4) No new earth station site should be adopted without careful measurements of interference levels, sufficiently prolonged to ensure that any liability to abnormal radio propagation conditions can be identified and evaluated.
(5) If no site sufficiently remote from potential interference or screened by natural features can be found, it may be possible to take advantage of man-made physical screening. Large buildings near the earth station may provide screening, although equally they may reflect interference into the antenna. An earth station antenna, even a large one, constructed on the floor of a large, dry, disused quarry, may enjoy a considerable degree of screening from other radio stations.
(6) The possibility arises that rain falling through the main lobe of an earth station antenna will scatter energy transmitted from the earth station, causing interference to a distant radio relay receiver using the same frequency. The interference may well be significant if the gain of the terrestrial station antenna is high in the direction towards the location where scattering is occurring. Interference from a radio relay transmitter may enter the satellite network by the same mechanism. The risk is greater if the earth station is to operate at a low angle of elevation. In such cases, in areas subject to heavy rain, it is desirable to locate the earth station so that the main lobe of its antenna radiation pattern has no common volume within the troposphere with the main lobe of any terrestrial station sharing a frequency band with it.

These precautions having been taken to the extent possible, interference at some level will probably still arise. It may be feasible to eliminate a single strong interference entry by the use of an adaptive interference canceller (see, for example, ITU-R Recommendation S.734 [3]) but, at the present state of the technology, this treatment cannot be extended to wideband interference entries from several distant sources. Other interference entries must be identified, evaluated and found tolerable or be made tolerable, for example by the means summarized in Section 6.4.1.

A particular problem may arise with the terrestrial link, called the backhaul link, used to connect the earth station to the gateway to the PSTN. Ideally, this link, or the portion of it close to the earth station, would be by cable. In practice a radio relay system is often preferred for reasons of cost. If the radio relay system uses a frequency band that is also to be used by the earth station, it is very difficult to keep interference between the two radio systems down to an acceptable level. If the earth station has more than one antenna, it is probably quite impracticable for the terrestrial and satellite services to share a frequency band. The usual practice is to use a backhaul radio relay system operating in a frequency band which is not expected to be used for satellite links at any foreseeable future time.

Constraints imposed by satellite location

If a C band earth station is to be given access to an Intelsat satellite it must see the satellite it is to operate with at an angle of elevation not less than 5°. A larger angle of elevation is considered desirable for a Ku band earth station; consequently Intelsat considers the grant of access to the system on a case-by-case basis for earth stations having angles of elevation below 10°. Satellites of at least one Intelsat region can be seen at a sufficiently high elevation from some part of the territory of almost every country. Sites must be chosen with care, however, if access is required from the same earth station site to satellites of more than one region.

Microclimate

There may be considerable differences in the climate at locations only a few tens of kilometres apart. At Ku band the antenna gain requirement may be increased significantly if a site is chosen which is subject to particularly heavy rain. Snowfall also varies greatly from place to place. An exposed site may have strong winds, which demand particularly robust antennas and powerful antenna drive motors. Heavy snow lying in the main reflector reduces antenna gain and adds considerable load to the structure, while snow melting on the feed window causes loss, transmitting and receiving, and raises the receiver noise temperature. Heating systems can melt the snow on the main reflector and a hot air blast will keep the feed window clear, but the amount of power that is required for this may increase operating costs considerably. Consideration given to microclimates in choosing a site can reduce earth station capital and operating costs considerably.

11.4.2 Earth station standardization

Governments license earth stations to operate within their jurisdiction and they may impose constraints on the technical characteristics of the equipment. These constraints will usually include those made mandatory by or recommended by the ITU to ensure that the GSO and the radio spectrum are used efficiently; see Section 6.5.

Intelsat [4] and Eutelsat also impose technical constraints on the earth stations that are granted access to their space segments and compliance with these constraints has to be demonstrated. High standards are enforced for two main reasons. Maximization of the number of channels of acceptable quality that can be transmitted through each satellite demands that earth stations should attain or exceed certain key technical parameters. It is particularly important that standards of G/T should be achieved, because multi-channel satellite emissions are often received at several different earth stations and the emission parameters may have to be determined to suit the worst-performing earth station. Also both consortia are major users of the GSO and have a direct interest in its efficient use and therefore in the minimization of interference between satellites adjacent in orbit. The principal mandatory requirements are mentioned in the review of earth station equipment which follows.

11.4.3 Earth station antennas

Intelsat has three sets of requirements for earth stations having access to its satellites for the Global Network. Standards A and B relate to stations for C band access and Standard C is for Ku band. Eutelsat's operations in the international network are centred mainly on a high capacity TDMA system, for which a high-performance antenna is necessary. However a range of IDC options using TDM/PSK has recently been introduced, primarily for international links, enabling a smaller antenna type to be standardized by Eutelsat also.

Intelsat Standard A antennas

Prior to 1986, Intelsat required Standard A antennas to have a figure of merit (G/T) not less than 40.7 $dB\,K^{-1}$ at 4 GHz and many earth stations meeting this standard are still in service. In 1986 the G/T required for access to Intelsat VI satellites was reduced to

$$G/T \geq 35.0 + 20 \log \tfrac{1}{4} F_c \quad (\text{dB K}^{-1}) \tag{11.1}$$

where F_c is the frequency in GHz. This requirement applies to any frequency between 3.7 and 4.2 GHz and for the angle of elevation at which the antenna is to operate. Earth station owners are invited to consider whether climatic conditions at the earth station are likely to justify the specification of higher performance in the receiving mode in order to make sure that channel performance objectives recommended by the ITU (see Section 10.4 and Table 10.1) will be achieved. Typical antenna gain values and primary reflector diameters for Intelsat Standard A and for the other standards used for international links are given in Table 11.2. The table shows that a typical post-1986 Standard A antenna has a primary reflector about 17 metres in diameter.

Table 11.2 *Typical basic parameters of earth station antennas for international networks*

System and standard	*Typical antenna*				
			Gain		
	Mid-band G/T (dB K^{-1})	*Typical* T_s *(note 1)* (K)	*Receive* G_{re} (dBi)	*Transmit* G_{te} (dBi)	*Diameter of primary reflector* (m)
Intelsat A	35	100	55	59	17
(pre-1986)	(40.7)	(100)	(60.7)	(64.5)	(32.5)
Intelsat B	31.7	110	52	56	12
Intelsat C	37	300	61.8	64	13
(pre-1986)	(39)	(300)	(64)	(66)	(16.5)
Eutelsat/TDMA	39	300	64	66	16.5
Eutelsat/IDC	34	300	58.5	60.5	9

Note 1. Arbitrary additions have been made to T_s to cover the effect of rain, appropriate for earth stations in a temperate climate.

For earth stations built in very rainy locations, it might be wise to give careful consideration to the possible need to design for a clear-sky G/T in excess of the mandatory value, in order that channel performance objectives should be attained. There may also be other reasons for installing an antenna having a higher gain than is required to meet Standard A, thus:

(1) The 120 Mbit/s TDMA emission uses all of the available power and bandwidth of an Intelsat VI Hemi or Zone beam transponder and, as is suggested in Example 3 in Section 11.5, link performance may be marginal for earth stations with $G/T = 35$ dB K^{-1}.
(2) Similarly, performance with $G/T = 35$ dB K^{-1} may be marginal when an analogue video signal is being transmitted by FM using over-deviation in half the bandwidth of a 36 MHz global transponder, as is usual.
(3) An antenna having a transmitting gain greater than the 59 dBi associated with $G/T = 35$ dB K^{-1} may also be advantageous, regardless of rain climate,

especially at an earth station which transmits several wideband carriers. A large degree of backoff is required when several carriers are amplified in a common HPA, and higher antenna gain may avoid the use of amplifier tubes of exceptionally high output power rating, the cost of which is disproportionately high.

Intelsat Standard B antennas

Intelsat Standard B defines an earth station of lower performance than Standard A. The figure of merit to be achieved is given by

$$G/T \geq 31.7 + 20 \log \tfrac{1}{4} F_c \quad (\text{dB K}^{-1}) \tag{11.2}$$

where, as before, F_c (GHz) is any frequency between 3.7 and 4.2 GHz. Table 11.2 shows that an antenna conforming to this standard requires a primary reflector having a diameter of about 12 metres and this would be significantly less costly than a Standard A antenna. However, other things being equal, the access of earth stations with performance inferior to Standard A requires more bandwidth and power from the satellite for a given information flow. To limit this impact on satellite capacity, constraints are applied to the use that is made of Standard B earth stations. In particular, types of emission are limited to SCPC and FDM/FM systems which are made up of companded channels (CFDM). One consequence of this is that the use of Standard B may not economically attractive for traffic loads exceeding a few dozen telephone channels.

Intelsat Standard C antennas

Significant degradation of propagation conditions by rain is much more likely at 11 GHz than at 4 GHz and Intelsat does not leave the provision of adequate margin against the effects of rain to the discretion of the earth station owner. A Standard C antenna requiring access to an Intelsat VI satellite must be shown to meet dual performance criteria. The following standard has been required since 1986 at the operational angle of elevation under clear-sky conditions to meet the first criterion:

$$G/T \geq 37.0 + 20 \log(F_{Ku}/11.2) \quad (\text{dB K}^{-1}) \tag{11.3}$$

where F_{Ku} is the operating frequency in GHz. The frequency range to be covered may be 10.95–11.2 GHz or 11.45–11.7 GHz, or both of these bands if both are to be used. Before 1986 the requirement was 2 dB higher.

The second Intelsat criterion is intended to ensure the maintenance of channel performance objectives, based on the ITU recommendations. Intelsat provides enough rain margin to enable short-term performance objectives to be achieved despite downlink propagation conditions about 12 dB worse than clear-sky values. If, at the earth station and for the intended antenna pointing direction, the rainfall statistics indicate that the degradation would exceed that value for more than around 0.02 per cent of an average year, the excess is to be eliminated by a compensating increase in the earth station G/T specification. It is recognized that a severe rain climate may make it necessary to use earth stations in space diversity in order to achieve the requirements of Standard C. Table 11.2 shows that the current Intelsat C antenna is likely to need a primary reflector at least 13 metres in diameter.

Eutelsat antennas

Antennas with access to the Eutelsat 120 Mbit/s TDMA system require $G/T = 39\,\text{dB}\,\text{K}^{-1}$, equal to the pre-1986 Intelsat C standard and necessitating a primary reflector diameter around 16.5 metres, depending to some extent on climate and the angle of elevation of the satellite.

International IDC facilities are also available via these big antennas, but a new antenna standard, I-1, has recently been defined specifically for earth stations that use IDC and which do not need the high capacity and connectivity that TDMA makes available. Standard I-1 requires $G/T = 34\,\text{dB}\,\text{K}^{-1}$. Depending on climate, etc. this performance is provided by an antenna 8 or 9 metres in diameter.

Polarization

Intelsat Standard A earth station antennas are required to be operable for transmission and reception in both senses of circular polarization simultaneously, although transmitters and receivers for both polarizations need not be provided unless both modes are to be used. It is required that the VAR in the transmit mode within a square box $\pm 0.02°$ in azimuth and elevation relative to the main beam centre does not exceed 1.06. It is recommended that the axial ratio be no worse in the receive mode. An axial ratio of 1.09 is accepted for antennas built before 1977.

Intelsat Standard A earth stations are not required to be equipped for adaptive compensation of depolarization by rain, but Intelsat urges that antennas should be designed so as to facilitate equipping for compensation of differential phase shift in both the uplink and the downlink at a later stage if it should be found necessary.

Intelsat VI satellites do not use dual polarization at 14/11 GHz but its use is already planned for subsequent satellite generations. Accordingly, Standard C earth stations requiring access to Intelsat VI are required to be equipped for the appropriate sense of linear polarization, adjustable to within 1° of the satellite signal's polarization. In the transmit mode a discrimination of 30 dB against the orthogonal polarization anywhere within a cone angle defined by the pointing error is required, and it is recommended that this discrimination be exceeded for reception also.

All Eutelsat satellites operate frequency reuse with orthogonal linear polarizations and earth stations must be equipped to operate in this mode.

Antenna steerability and tracking capability

It is required that an Intelsat Standard A or Standard C antenna can be steered to point at any geostationary orbit location planned for use by Intelsat for the ocean region in which the antenna is to operate. It is recommended that antennas should be capable of being directed at any point in the GSO visible from the earth station. In this context, 'visibility' is taken to mean that the satellite would have an angle of elevation at the Earth of not less than 5° at a C band antenna and of not less than 10° at a Ku band antenna.

Standard A and C antennas must have both manual tracking and automatic tracking facilities to ensure accurate alignment of the main beam with the satellite. It is Intelsat's policy to maintain Intelsat VI satellites within 0.05° of their nominal station, east–west and north–south, although much larger north–south excursions may arise for reasons outlined in Section 2.2.3. The satellites transmit beacons at various frequencies in the vicinity of 3950 MHz and at 11 198 and 11 452 MHz to facilitate tracking by earth stations.

The precision with which the satellite must be tracked by these high-gain antennas is not specified by Intelsat, but equations (7.5) and (7.6) provide an indication of what must be achieved if small performance margins are to prove adequate and foreseen carrier to interference ratios are to be achieved. The uplink half-power beamwidths θ_0 of post-1986 Standard A and Standard C antennas are about 0.19° and 0.12° respectively. Thus, if a performance margin of 0.5 dB were assigned for antenna mis-pointing, the tracking errors would have to be limited to about 0.04° and 0.025° respectively.

Antenna structures

The axisymmetrical Cassegrain configuration has been found capable of providing a good balance of the required characteristics, namely high main-lobe gain, low waveguide loss, wide bandwidth, high polarization purity, good physical stability, acceptably low radiation density outside the main beam and acceptably low cost. It is the universal choice for the larger earth station antennas. With few exceptions these antennas are fully steerable, that is, they can be made to point in any direction above the horizontal plane. Usually the antenna assembly turns about a horizontal axis (the elevation axis) which can itself be rotated about a vertical axis (the azimuth axis); this is known as an azimuth–elevation (Az–El) mounting. Most of the massive C band antennas required to meet Intelsat Standard A before 1986 are of the kingpost design, although some have the wheel-and-track azimuth mechanism.

A kingpost antenna mount consists typically of a massive vertical shaft within a tower, its lower end, probably at ground level, resting on a thrust bearing and supported near the top by bearing surfaces that transmit sideways thrusts from the rotating structure to the supporting tower. This shaft forms the azimuth axis. A large yoke at the top of the shaft, perhaps 10 to 15 metres above the ground, supports trunnions which carry the elevation axis to which the antenna assembly proper is attached. The azimuth drive motor located at the top of the tower turns the shaft and the elevation drive motor on the yoke steers the antenna assembly in the vertical plane, typically through a large wheel-and-pinion gear. See Figure 11.7. A much lighter structure is adequate for smaller antennas using the same basic geometry; see Figure 11.8.

The azimuth bearing of a typical wheel-and-track mount consists of a circular railway on which bogies run, the bogies being driven to provide steering in azimuth. The elevation axis, carrying the antenna, is mounted on a gantry which is carried by the bogies. See Figure 11.9.

The primary reflector for a typical large antenna is made of aluminium alloy panels. The panels are stretch-formed over moulds accurately shaped to the specified profiles, then assembled on a backing structure consisting mainly of a central hub and radial ribs, also made of aluminium alloy. The reflector surface must conform precisely to the specified profile and must retain that profile despite wind, temperature changes, changes in the relative direction of application of gravitational forces arising from antenna movement in elevation and the passage of time. There is significant loss of gain unless the r.m.s. profile error is less than 1 mm for a Standard A antenna or 0.3 mm for a Standard C antenna. Corresponding requirements apply to the profile of the subreflector and maintenance of the correct relative location of the primary and secondary reflectors. Heating elements for de-icing are attached to the back of the primary reflector where necessary.

Many of the larger antennas have an equipment cabin behind the pole of the primary reflector. The rear end of the feed assembly projects into this cabin and the LNAs are located there. For an antenna that is not expected to carry much traffic, this cabin or another mounted elsewhere on the moving structure might

Figure 11.7 *Antenna number 3 at BT's Goonhilly earth station. This is an axisymmetrical Cassegrain antenna with a kingpost Az–El mount, the diameter of the primary reflector being 30 metres. (BT Corporate Pictures; a BT photograph.)*

accommodate the HPA also. These cabins must be accessible for maintenance of equipment at any time when the antenna is directed towards any satellite with which it is to operate. To facilitate access, the reflector is often mounted well forward of the elevation axis and counterbalanced with ballast weights. Electrical connections to this moving cabin are made, using coaxial cables, power cabling and flexible waveguides, via a cable turning assembly.

Figure 11.8 *Antenna number 7 at BT's Goonhilly earth station. This is an axisymmetrical Cassegrain antenna with an Az–El mount, the diameter of the primary reflector being 13 metres. (BT Corporate pictures; a BT photograph.)*

Figure 11.9 *A group of antennas at BT's Madley earth station. These are axisymmetrical Cassegrain antennas with wheel-and-track Az–El mounts, the diameter of the primary reflectors being 32 metres. (BT Corporate Pictures; a BT photograph.)*

Alternatively, a system of passive reflectors, plane and concave, may be used to present the received wave, focused by the subreflector, to a feed assembly at a fixed location, typically near ground level and conversely for transmitted signals; see Figure 11.10. In the receiving mode, a plane reflector *AA*, located at the point of intersection of the azimuth and elevation axes, deflects along the elevation axis the beam focused by the subreflector, so that the beam reaches a concave reflector *BB*. *BB* directs the collimated beam downwards to a second concave reflector *CC*, located near the ground. *CC* diverts the beam back to the azimuth axis, focusing it so that, after reflection at a plane reflector *DD*, it enters the feed horn, located in the building upon which the antenna assembly is mounted. This so-called beam waveguide system has very low feeder losses for both directions of transmission, it facilitates the accommodation of multiple HPAs and it avoids a need for access to a movable cabin which may be awkward to provide. It is particularly suitable for wheel-and-track antennas with heavy traffic commitments.

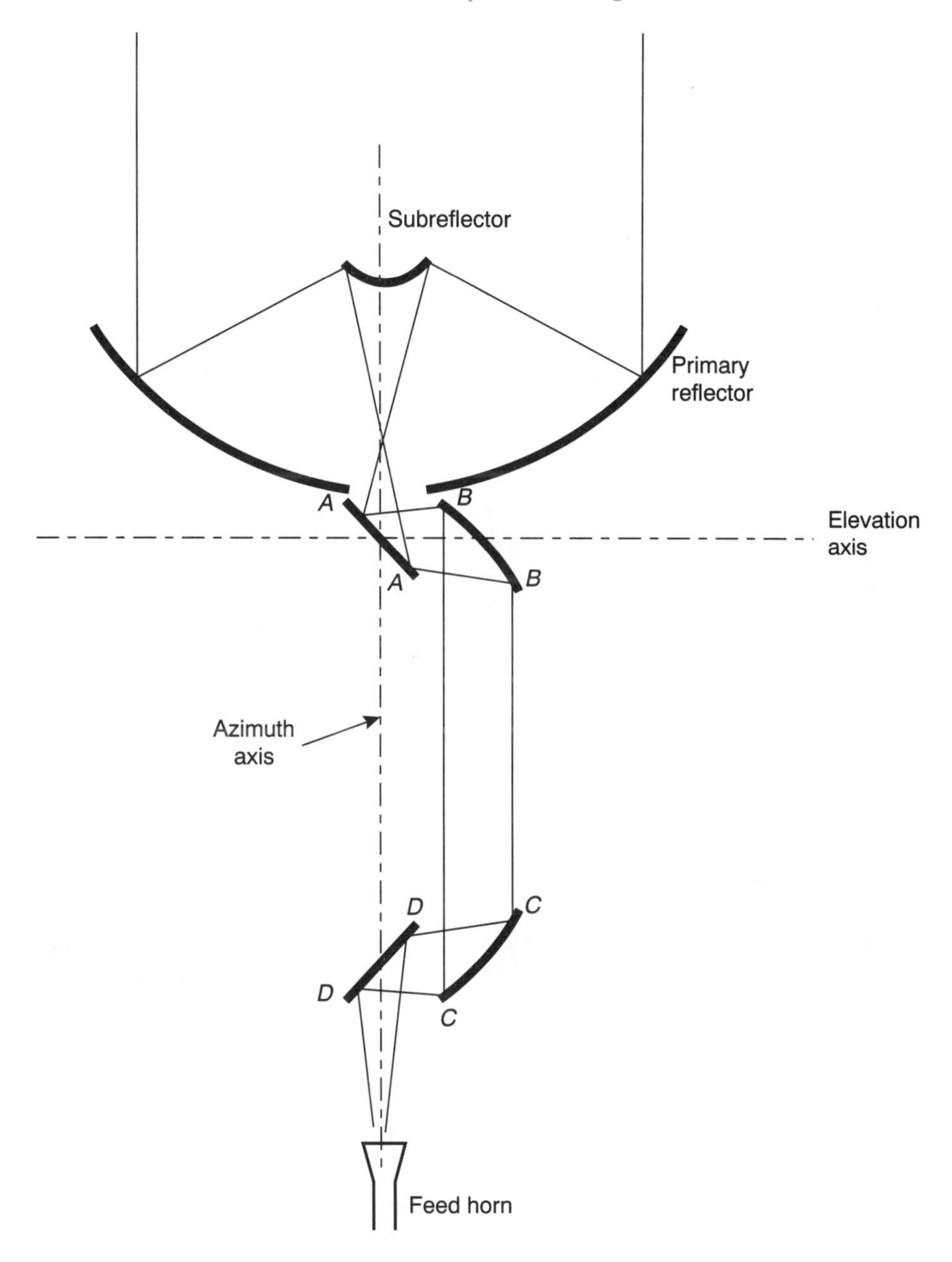

Figure 11.10 *A beam waveguide feed system*

11.4.4 Transmission systems in use

In the early years of Intelsat operations, almost all circuits for the PSTN were analogue, transmitted using FDM/FM/FDMA, although the Spade SCPC system carried a few circuits in digital form. In the 1980s there was growing use of digital transmission by means of TDMA/PSK or TDM/PSK. This development was driven by the global trend towards digital transmission and switching in the PSTN itself and also by the higher efficiency which digital transmission allows in the utilization of satellite transponders. By 1992–93 digital transmission had become the dominant satellite technique. By the year 2000 the use of analogue transmission for telephony may have been phased out completely but at the time of writing it is still important.

Intelsat VI and Eutelsat II are used for international channels of the PSTN by the transmission techniques shown in Table 11.3. In addition, both satellite systems carry sound and video signals, mostly in analogue form using FM, which are used to provide contributions to broadcasting programmes. There is a brief discussion of these techniques below, except for TDMA and SCPC, which are considered in Sections 11.4.5 and 12.3.2 respectively.

Table 11.3 *Transmission techniques and earth station standards for international PSTN services*

Earth station standard	*FMDA*				*120* Mbit/s
	FDM/FM	*CFDM/FM*	*SCPC*	*TDM/PSK*	*TDMA*
Intelsat A	X	X	X	X	X
Intelsat B		X	X		
Intelsat C	X			X	X
Eutelsat/TDMA				X	X
Eutelsat/IDC				X	

FDM/FM/FDMA

The structure of a typical FDM/FM baseband is reviewed in Section 8.2.2 and illustrated in Figure 8.9. The channels to be used for traffic are assembled into the smallest possible number of 12-channel FDM groups, arranged in the baseband upwards in frequency from 12 kHz. Where, as is usual, the transmitting earth station has links with several or many receiving earth stations within the coverage area of the down-beam from the satellite, all of the outgoing channels may be combined into the same baseband. In this mode, each distant earth station receives the carrier and extracts from the baseband the channels which carry its traffic.

Intelsat has a modular system for assigning spectrum for emissions in FDMA transponders, bandwidth being assigned in units of 1.25 MHz, 2.5 MHz and multiples of 2.5 MHz up to 25 MHz. If required, the full bandwidth of a transponder would be assigned to a single carrier without backoff. The test tone deviation and carrier power level required to provide the recommended transmission quality in all its channels are determined for each carrier. A second iteration of the process may then be used to optimize the parameters for all the emissions to be amplified by the transponder, having regard to the distribution of intermodulation noise and interference. See Section 8.2.2 and Example 1 in Section 10.5, in which the parameters of Standard A earth stations and an Intelsat VI Hemi beam are assumed.

CFDM/FM/FDMA

The standard CCITT compandor is described in Section 8.2.1 in the context of SCPC systems, but it is also used in some circumstances for channels in FDM.

CFDM uses less carrier power and/or bandwidth compared with an FDM emission of the same channel capacity. Thus if the demand for channels to earth stations located in a down-beam exceeds the available satellite e.i.r.p. using FDM, CFDM may enable more channels to be operated. Also, since CFDM reduces the uplink e.i.r.p. requirement per channel, conformity by Standard B earth stations with the limits on off-beam radiation contained in ITU-R Recommendation 524 (see Section 6.5 item 5) may be facilitated.

TDM/FDMA

Much of the digital PSTN traffic carried by Intelsat in the mid-1990s is transmitted by Intermediate Data Rate (IDR) TDM systems in transponders operating in the FDMA mode. The use of the same technique under the name IDC is growing fast in Eutelsat. A range of standard aggregate bitrates between 64 kbit/s and 45 Mbit/s is available via Intelsat. Eutelsat provides for 2 and 8 Mbit/s. FEC is used and carriers are modulated by QPSK.

Intelsat aims in general for a clear-sky BER of 1 in 10^7 for IDR channels, in line with ITU-R Recommendation S.614 (see Section 10.4). However, Hemmati *et al.* [5] discuss improvement in system error performance by the application of concatenated error correction coding, and it is foreseen that this could lead to the achievement of the objectives set out in ITU-R Recommendation S.1062.

TV/FM

Like most satellite operators, both Intelsat and Eutelsat lease transponders for the distribution of broadcasting programmes, sound and especially television, for reception and broadcasting at small-antenna earth stations located at terrestrial broadcasting stations and cable system head-ends. These satellite emissions are increasingly being treated as DTH satellite broadcasting; see Section 15.2. But Intelsat and Eutelsat and the associated big-antenna earth stations supply another service to broadcasters. This is the provision of audio and especially video links between earth stations for brief periods, typically half an hour or less at a time, often at short notice, for the transmission of material from news-generating events to the locations where it will be incorporated into programmes and broadcast. Satellite news gathering (SNG), which is broadly equivalent to this service but using mobile earth stations at the news source, is discussed in Section 14.5.1.

In most Intelsat VI satellites the global beam transponder at the top end of C band is reserved for these temporary transmissions. Sometimes the whole of the bandwidth and power of the transponder is used for a single channel; at other times the transponder provides two channels of lower quality in FDMA. Similarly, the EBU has a permanent lease on a transponder in one of the Eutelsat II satellites and manages its use on behalf of the European broadcasting agencies. Often, and especially when sporting and major news events are being reported on, several or many earth stations will receive a single emission simultaneously.

The video channel performance that is obtained depends on the way in which these facilities are used. As will be seen from Section 8.2.3 and Example 5 in Section 10.5, if the whole power and bandwidth of an Intelsat VI global beam transponder is used for a single video channel, the video signal to noise ratio received at a Standard A earth station will exceed the standard recommended by the ITU by a substantial margin. The same is true of the EBU facility via Eutelsat. However, most Intelsat video transmissions operate with over-deviation in only half of the transponder's bandwidth and power.

11.4.5 Time division multiple access systems

TDMA systems have been operated in Intelsat and Eutelsat since 1985 and 1986 respectively. The two consortia use equipment designed to the same basic specification. It is described in Intelsat documentation; see, for example, [6, 7]. See also [8, 9]. The system configuration departs from the principles described for high-capacity TDMA systems in Section 9.4.5 in only one basic respect; the multi-beam Intelsat VI requires a more complex frame synchronization system.

TDMA is particularly useful where an earth station requires a substantial number of digital channels with high connectivity. Thus, Bedford *et al.* [10] consider that TDMA's main field of economic application lies with earth stations requiring at least one hundred 64 kbit/s circuits divided between at least 10 destinations. SCPC or multi-destination TDM/PSK systems are more economic for earth stations needing fewer circuits in total and single destination TDM/PSK carriers are often preferred for the heavier routes of stations with large numbers of circuits.

Intelsat is currently operating TDMA in C band transponders in two Atlantic Ocean region satellites and one Indian Ocean Region satellite. One of the Atlantic satellites has fixed beam operation (that is, without on board transponder switching) in two transponders; the other two satellites operate SSTDMA. Eutelsat is operating TDMA in four transponders.

An outline description follows of the current TDMA system as used in Intelsat and Eutelsat, with emphasis given to the ways in which it differs from the basic principles outlined in Section 9.4.5. However, much of the TDMA terminal equipment in use at earth stations is approaching the end of its economic life. It is bulky, it is becoming costly to maintain and it would be costly to replace with equipment of the same kind. Intelsat has sponsored the development of a new generation of terminal equipment to a system specification that is simplified in some details although operationally compatible with the original specification. See [10], and also [11, 12]. Using current technology, this new equipment is expected to be more compact, less costly, more reliable and easier to operate.

Radio-frequency characteristics

The system operates at 120.832 Mbit/s and uses coherent QPSK. Thus the symbol rate is 60.416 megasymbols per second and a pre-demodulator noise bandwidth of 72 MHz is sufficient. FEC using a $\frac{7}{8}$-rate Bose–Chauduri–Hocquenghem block code is applied to data sub-bursts to improve performance where necessary. It should not, however, be essential to use FEC for sub-bursts which are to be received only at earth stations which are close to the centre of the down-beam footprint or which have the higher value of G/T required of Standard A earth stations before 1986. A pseudo-random binary sequence is added to the modulating signal of all bursts, starting at the end of the unique word, to disperse strong lines in the emission spectrum that are due to repetitive sequences in the bit pattern when the multiplex is lightly loaded, and thus to ensure that the mandatory limits on the downlink PFD are not exceeded.

Frame structure

The system uses a 2 ms frame which contains traffic bursts from all participating earth stations and reference bursts from main and 'hot standby' reference stations. Superframes made up of 16 frames structure the housekeeping functions of the system. Among these automatic functions is provision for changing the burst time plan without interrupting the operation of the system. When it is necessary to change the lengths of bursts or their location within the frame, the details of the new burst time plan can be stored at all terminals in advance. The new plan is implemented automatically throughout the network in the first frame of a designated superframe. An 0.8 μs guard period between bursts is used to prevent accidental overlapping of bursts due to small errors of timing.

Traffic bursts can carry telephone channels in DSI sub-bursts or 'digital non-interpolated' (DNI) sub-bursts. The duration of each sub-burst equals $n \times 64$ symbols $= n \times 1.059\,\mu s$, where n is the number of 64 kbit/s channels actually required for transmission. That is, burst lengths are tailored to meet current commitments, with no spare channels reserved for growth. DCME techniques such as ADPCM as well as DSI are applied to increase the traffic flow per 64 kbit/s bearer channel.

Frame synchronization

The method of burst timing and synchronization described in Section 9.4.5 and illustrated in Figure 9.28 requires each earth station to receive the reference burst and its own traffic burst after retransmission by the satellite. Thus, earth stations ensure that their own bursts maintain their assigned time slot within the frame. This closed-loop method is used in most high-capacity TDMA systems, including Eutelsat's. It could be used with Intelsat VI with a transponder that has uplink and downlink footprints which cover the same areas. However, it cannot in general be used with a multi-beam satellite because the bursts which an earth station uplinks to the satellite are usually downlinked in a beam that does not cover the transmitting earth station.

For example, consider two Intelsat VI coverage areas, designated by footprints X and Y (see Figure 11.11) linked both ways by transponders. Earth station X_A located in the X footprint could receive a reference burst transmitted from a reference station located in the Y footprint but it cannot receive a traffic burst carrying information to earth stations Y_A, Y_B and Y_C, which earth station X_A had itself transmitted.

For use with multi-beam transponders in Intelsat VI, there is a reference station and a hot standby reference station in both footprints X and Y with their timing clocks precisely synchronized and stable to within 1 part in 10^{11}. Each earth station sending traffic bursts is kept in its assigned position in the frame by means of control signals originating at the reference station in the distant footprint. The frame synchronization process, simplified by the omission of reference to the functioning of the hot standby reference stations and with some further simplification, is as follows:

(1) The reference stations are continuously updating their knowledge of the precise location of the satellite. This is done by measuring the time that elapses before a signal that has been transmitted through the satellite is returned by three earth stations at precisely known locations.
(2) One of the reference stations is the master reference station for the system; let it be the reference station in the X footprint. Reference station X transmits a series of reference bursts which reference station Y receives. Reference station X computes the time delay that must elapse between the reception of an X reference burst at reference station Y and the transmission of the first of a series of reference bursts by reference station Y that will reach the satellite in a predetermined time relationship with the arrival there of a subsequent X reference burst. This time delay instruction is passed to reference station Y via a control and delay channel (CDC) in the X reference bursts, and it is updated from time to time as the satellite moves. Reference station Y

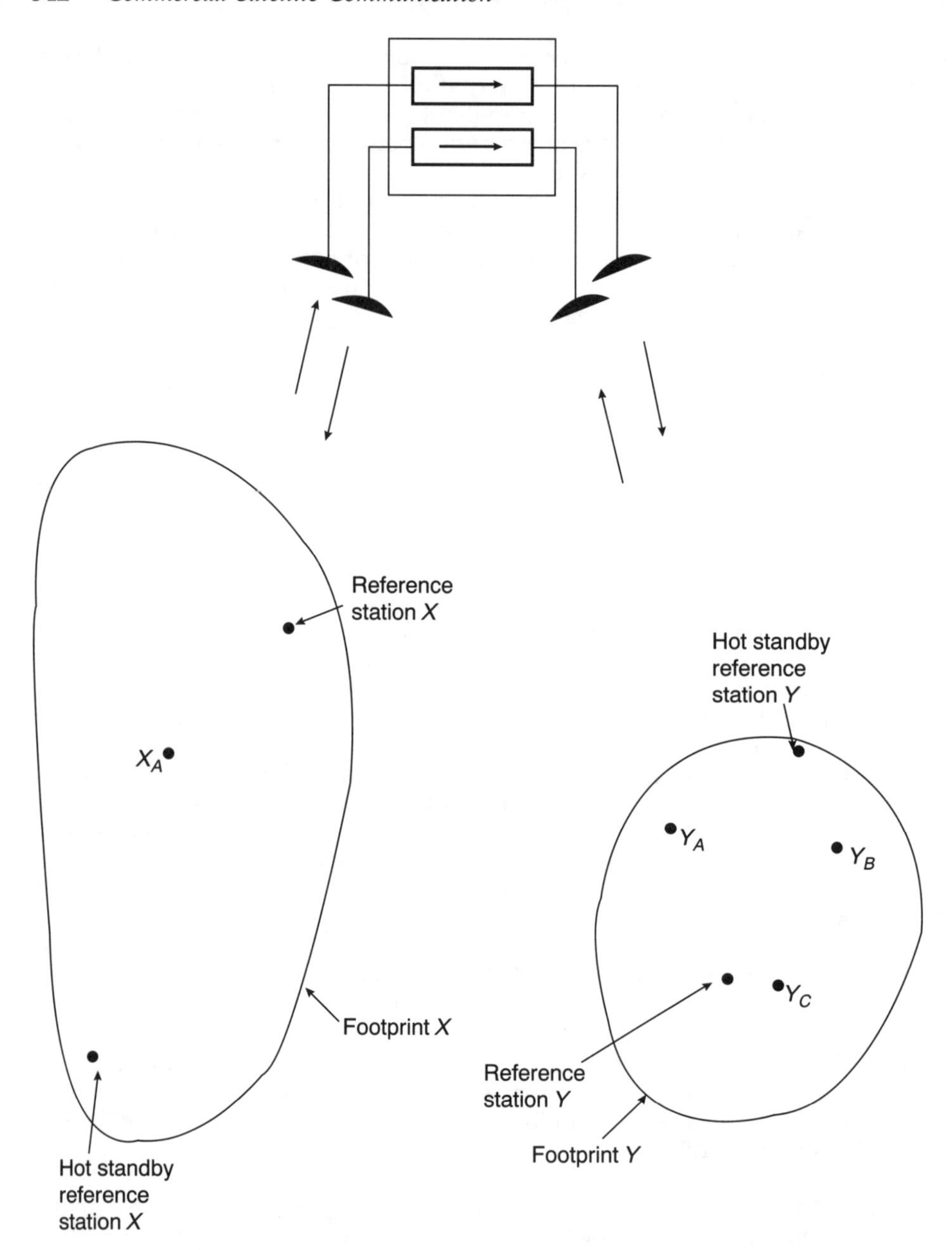

Figure 11.11 *Reference stations serving an Intelsat TDMA system linking earth stations in the footprints of beams X and Y*

transmits reference bursts accordingly. In normal operation the X and Y reference bursts are maintained in their proper timing relationship under this open loop control.

(3) Let it be assumed that earth station X_A is to enter the TDMA system to pass traffic to earth stations in the Y footprint. The position in the frame assigned for traffic bursts from earth station X_A is T seconds later than the X reference burst, both burst times being defined by the end of the unique word in the relevant preamble. Reference station Y computes the time delay D_A, relative

to the instant of reception of a Y reference burst at earth station X_A, at which that station's traffic burst should be transmitted if the station is to take up its proper position in the frame. The value of D_A, with a small addition δD_A to minimize the risk that station X_A will interfere with the end of the traffic burst of another earth station in footprint X, is transmitted to station X_A via a CDC in the Y reference bursts.

(4) Earth station X_A receives the Y reference burst, extracts the time delay message from the CDC and starts transmitting short bursts, consisting of the preamble only, at a time $D_A + \delta D_A$ after receiving the Y reference bursts. Reference station Y measures the position of these short bursts in the frame relative to X reference bursts and revises the value of D_A if necessary. When the timing of the short burst is correct, the Y reference station advises the X_A earth station accordingly. The quantity δD_A is subtracted from the timing delay, the traffic sub-bursts are added to the X_A earth station's bursts, and traffic can pass.

(5) Once earth station X_A's bursts are established with correct timing relative to the X reference burst, reference station Y maintains closed loop control on that timing via the CDC by varying D_A as the satellite moves.

(6) Reference station Y controls the timing of the bursts of the other earth stations in beam footprint X in the same way, and reference station X does likewise for the bursts from footprint Y.

Common timing for TDMA systems and transponder hopping

If a satellite is carrying TDMA signals in more than one transponder, the frame and superframe timings of all the systems can be synchronized. This has three consequences.

(1) If an earth station participates in TDMA in more than one transponder of a satellite, the basic timing functions for each of those TDMA systems can be performed by a single control unit, producing significant savings in capital cost.

(2) The timing within the frame of each earth station's transmit bursts can be planned so that the bursts to be send via different transponders do not overlap. Thus, a single TDMA terminal can be used at the earth station for the bursts sent via any transponder of the same satellite. Furthermore, if the various transponders use the same polarization (that is, if the uplink beams are all Hemi or all Zone beams) it is possible to use the same transmitting chain and a wideband HPA at the earth station to access all of the TDMA transponders. To implement the facility, the frequency of the local oscillator of the second up-converter is changed from burst to burst to provide the final radio-frequency required for each transponder. This cost-saving technique is called 'transponder hopping'. Similarly there can be transponder hopping in the receiving mode.

(3) The satellite-switched TDMA technique (SSTDMA) becomes available for use with satellites equipped with a dynamic transponder switch matrix.

On-board TDMA switching (SSTDMA)

Since the frame and superframe timing of all the TDMA systems that a satellite carries can be synchronized, the dynamic switch matrices, with which two of the wideband paths are equipped, can be used to reconfigure, in the course of each TDMA frame, the combinations of up-beams, down-beams and transponders which constitute a TDMA path through the satellite. This may make economic the use of TDMA on wideband paths that would be too lightly loaded to justify a full-time link.

Figure 11.12 shows how a selection can be made from the Hemi and Zone beam receivers to enable the signals from selected up-beams and down-beams to be connected to the dynamic switch matrix. The matrix takes signals from one receiver, and then from another as required, and passes them to one or other of several transmitters, in accordance with a programme stored on board. The switch timing is driven from an on board clock, corrected as necessary from a reference station, which also supplies frame timing synchronization. A revision of the switching programme can be uplinked from the reference station when required.

It may be noted that it is feasible to include in the SSTDMA frame a period when each uplink beam is connected through a transponder to a downlink

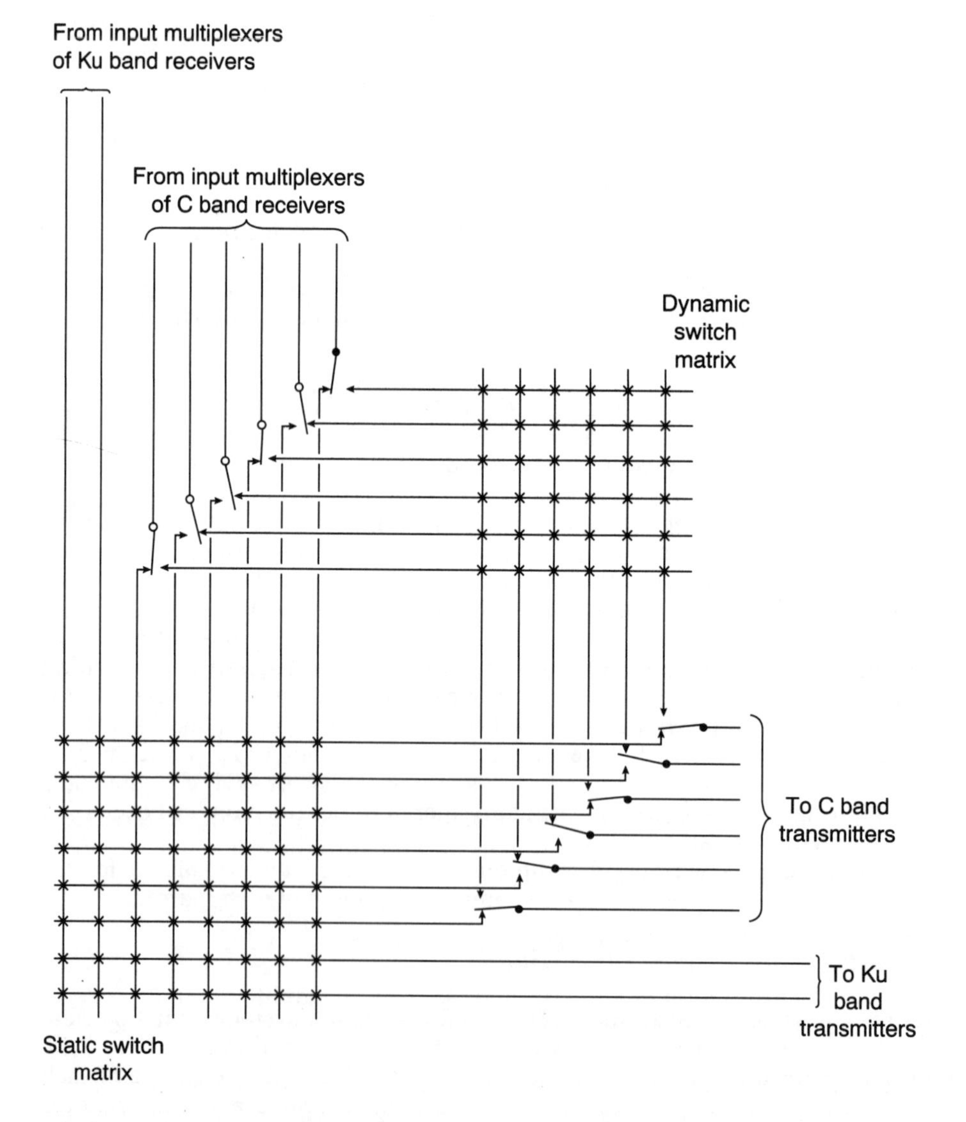

Figure 11.12 *An Intelsat VI dynamic TDMA switch matrix*

beam serving the same footprint. A reference station located in each footprint could transmit a reference burst during that period. If this were done, earth stations located in that footprint could synchronize their traffic bursts in the way described in Section 9.4.5, thus avoiding the need for the more complex open loop synchronization process described above.

11.4.6 Transmitting and receiving chains

The earth station block schematic

The block schematic diagram of an earth station (see Figures 7.1 and 7.3) and the accompanying description in Section 7.1.1 illustrate, in a general way, how the internal equipment of an earth station providing links for the PSTN is configured. However, the equipping of earth stations varies a lot, depending on the scale of its traffic load and on whether TDMA is used. Much of the system is conventional radio relay practice, but what follows relates to aspects more specific to earth stations.

The terrestrial/satellite interface

Signals in narrowband channels to be linked with various, often many, distant national PSTNs are connected between the earth station and its PSTN gateway by radio relay or cable as time division or frequency division multiplexes, often of large capacity. It is usually necessary to convert these aggregates to and from the form in which they are to be transmitted over satellite links. This will typically involve:

(1) Breaking down high-capacity FDM systems containing channels intended for several or many destinations into groups or supergroup units containing channels destined for a single distant earth station. These single-destination multiplexes are then combined into baseband assemblies that can be transmitted over the satellite system using a single FDM carrier. Alternatively the channels to be transmitted over the satellite system using a single carrier may be digitally encoded, singly or as FDM assemblies by means of a transmultiplexer, and transmitted in the TDM or TDMA mode.
(2) The rearrangement of digital multiplexes by processes corresponding to those indicated in (1).
(3) Passing single telephone and low- or medium-speed data channels to SCPC modulators, often with signal activity detectors to activate the carrier when required and sometimes with encoders to convert analogue signals to the digital form.

These signal handling processes will be repeated in the reverse order to repackage signals received from distant earth stations into the form in which they are to be passed on to the local PSTN gateway.

Modulators and up-chain IF amplifiers

The design of earth station modulators and up-chain IF amplifier systems follows radio relay practice closely, but the following specific points should be noted:

(1) The carrier frequency of a wideband emission may be subject to a mandatory tolerance. An automatic frequency control system will probably be used for SCPC emissions, tying the frequency of all carriers to a system pilot carrier.

(2) The level of carriers, as reflected in the e.i.r.p. radiated to the satellite, and the test tone deviation of FM carriers must be accurately set to, and maintained at, the level assigned by the system manager.
(3) Double frequency changing is used to provide adequate control of the bandwidth occupied by the emission and to assist in the suppression of spurious emissions arising in the up-chain.
(4) It may be required that intermediate frequency filters meet stringent standards of suppression of signal products outside the assigned bandwidth. In-band, the variation of loss and group delay across the bandwidth of the signal must be closely controlled. In addition, earth stations may be required to apply group delay pre-equalization to uplinked emissions to compensate for unequalized group delay arising in the satellite transponders and multiplexers.
(5) If TDMA is to be used, frequency hopping may be necessary, requiring step control of the frequency of the local oscillator of the second up-converter, and the first down-converter.

High-power amplifiers

The number and e.i.r.p. of the carriers that an earth station transmits depend a great deal on the traffic load, on the communication technologies that are used, on the G/T of the receiving earth station and on the gain of the satellite transponder and its antennas.

For example, an Intelsat Standard A earth station with a heavy traffic load might transmit to other Standard A stations:

(1) A TDMA emission hopping between two Hemi transponders, its e.i.r.p. about 84 dBW, using RHCP. Disregarding losses between the HPA and the antenna feed, this would require a carrier power of about 250 W.
(2) Several TDM or FDM emissions accessing various transponders in the FDMA mode, the e.i.r.p. being in the range 70 ± 8 dBW according to the number of channels provided by the multiplex, using LHCP. On the same basis as in (1), the HPA output power per carrier required would range from a few watts upwards towards 100 W.
(3) On occasion, an analogue video channel, using FM with over-deviation in half of the bandwidth of a global beam transponder, the emission having an e.i.r.p. of about 77 dBW, using RHCP. The carrier power required, disregarding losses, would be of the order of 60 W.
(4) Several dozens of SCPC carriers having an e.i.r.p. of about 60 dBW per carrier, using LHCP. These carriers would be assembled into an FDMA aggregate at low level using a simple lossy combining network. Subsequent power amplifiers would amplify the aggregate, raising each carrier to a level of about 1 W.

The arrangement of HPAs and combiners shown in Figure 11.13 might be used for these emissions. It will be noted that the power ratings proposed in the figure for an HPA amplifying more than one emission allows for a large degree of backoff. An input backoff of 15 dB is not unusual to reduce intermodulation products to an acceptable level. Linearization of the HPA by predistortion may be used to reduce further the radiation of spurious emissions towards the satellite.

On the other hand, a Standard A earth station with a similarly heavy traffic load which did not use TDMA would probably require two or three additional FDM or TDM carriers. A lightly loaded Standard A station might be able to use

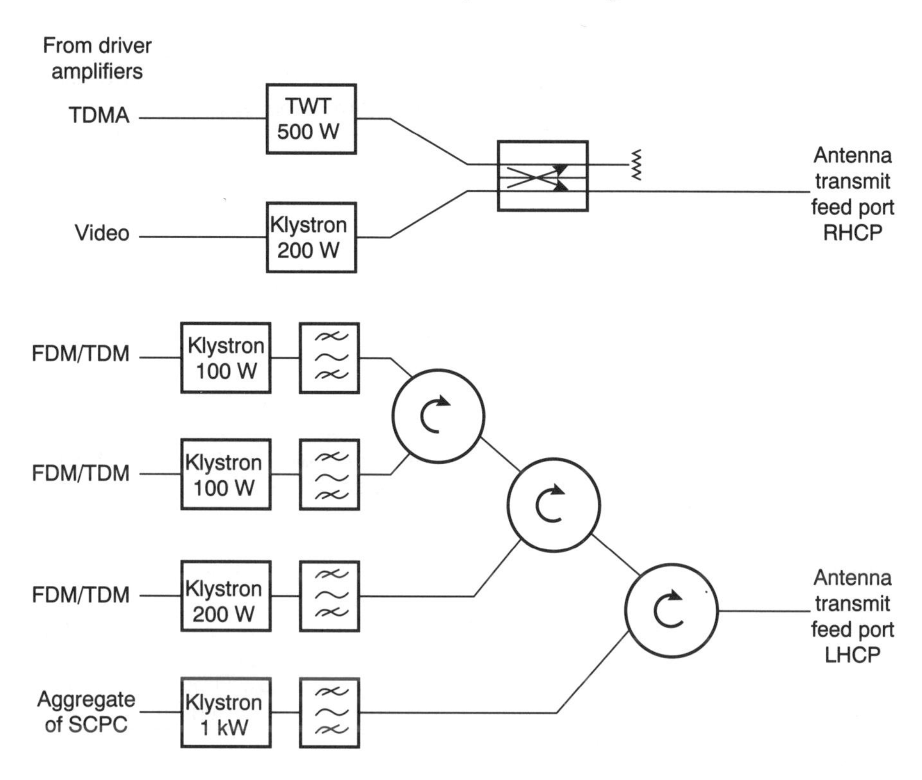

Figure 11.13 *A scheme of HPAs and combiners that might be used for a heavily loaded Standard A earth station*

TDMA for all PSTN channels that were not transmitted by SCPC, or a single TWT HPA for all its emissions.

Constraints on downlink PFD and carrier energy dispersal

In almost all frequency bands, FSS satellites must limit the spectral PFD of downlinks, as measured at the Earth's surface, to avoid unacceptable interference to terrestrial radio systems. These terrestrial systems are typically radio relay links operating in the FSS downlink frequency band; see 'interference mode 3' in Section 6.4.1.

Table 10.2 shows a typical power budget for an Intelsat VI Hemi beam transponder operating in the FDMA mode. An Intelsat Standard A earth station transmitter delivering to the antenna an emission having a spectral power density of +8 dBW in a bandwidth of 4 kHz would cause the satellite transponder to radiate an emission having a maximum spectral PFD of $-152\,\text{dB}(\text{W}\,\text{m}^{-2}\,(4\text{kHz})^{-1})$. Under worst-case conditions, this PFD would be the maximum permitted by the ITU *Radio Regulations* at 4 GHz. The transmitter power required at a Standard A earth station for an SCPC emission or for a low-capacity FDM or TDM emission is usually below these levels. Thus these emissions would conform to the constraint even if the carrier had no modulation at all.

For multiplexed FM/analogue emissions with a wider bandwidth and consequently a higher carrier power, the spectral power density in the vicinity of the carrier will usually be dispersed, to some degree, by modulation due to the FDM pilot tones which are continuously present and perhaps by digital services, transmitted by modem and which may also be present all the time; see ITU-R Recommendation SF.675 [13]. Similarly the spectrum of TDM emissions will be dispersed considerably by the repetitive pattern of modulation that is present even when there are no traffic signals present. Nevertheless, emissions occupying larger bandwidths will often need artificial carrier energy dispersal if their downlink PFD is to remain within the regulatory constraints, even under the favourable circumstances arising from the use of Standard A earth stations. The need for artificial carrier energy dispersal may arise even for narrowband carriers if the antenna of the receiving earth station has low G/T.

Constraints on uplink off-beam radiation density

As will be seen from Section 6.5, item 5, limits are recommended by ITU-R for the spectral e.i.r.p. density radiated by earth stations off-beam, in directions within 3° of the GSO, to ensure that the GSO can be utilized efficiently. These constraints are also requirements of the satellite operating organizations. Simplifying a little from Figure 6.8, the limits are as follows:

(1) At 6 GHz, for all emissions except SCPC with voice-activated carriers,

$$\text{e.i.r.p.}_{\text{max}} = 35 - 25\log\phi \quad (\text{dB(W}\,(4\,\text{kHz})^{-1})) \tag{11.4}$$

for values of ϕ, the off-beam angle, between 2.5° and 48°. (The constraint is 3 dB more stringent between 2.5° and 7° for antennas installed since 1988.)

(2) At 6 GHz, for voice-activated SCPC/FM emissions and at 14 GHz, for all emissions,

$$\text{e.i.r.p.}_{\text{max}} = 42 - 25\log\phi \quad (\text{dB(W}\,(40\,\text{kHz})^{-1})) \tag{11.5}$$

where ϕ is limited as before. (The constraint is 3 dB less stringent for SCPC/PSK emissions at 6 GHz. For 14 GHz emissions, the constraint is 3 dB more stringent between 2.5° and 7°.)

Many studies have shown that the envelope of the peak gains of the side-lobes of axisymmetrical Cassegrain antennas can be represented fairly well by an expression in the form

$$G_{\text{sl}} = E - 25\log\phi \quad (\text{dBi}) \tag{11.6}$$

where E is a constant. See also equation (6.2). For most big antennas, including many old ones, the substitution of 32 for E covers all but a small proportion of the side-lobe peaks and for many recently installed antennas $E = 29$ would be appropriate. Section 6.5, item 7, refers to the ITU-R recommendation that $E = 29$ be adopted as a design objective for big new antennas.

Thus, for example, an emission which is delivered to the transmit port of an Intelsat A antenna with a spectral power density of +3 dBW per 4 kHz band is not likely to set up off-beam radiation levels in excess of equation (11.4). A modern antenna with well-suppressed side-lobes might well have a safety margin of 3 dB. For SCPC emissions, the constraint given by equation (11.5) specifies a sampling bandwidth of 40 kHz, providing additional margin.

It may be necessary to use artificial carrier energy dispersal to avoid exceeding the downlink spectral PFD density limit if the transmitting antenna input has a spectral power density exceeding +8 dB(W (4kHz)$^{-1}$). A small increase in the degree of carrier energy dispersal would ensure that the uplink constraint was also met. However, this results from the specific link parameters of the example chosen; it is not a general rule. Reducing the gain of any of the antennas, satellite or earth station, and reducing the gain of the transponder all tend to lead to an increase in off-beam radiation at the transmitting earth station.

Receiver input stages

High-performance, highly linear, GaAs FET LNAs are usually employed at earth stations providing international connections. Some C band GaAs FETs are thermoelectrically cooled; the remainder, probably including all at Ku band, operate at ambient temperature.

The LNA is followed by an RF amplifier. This is a wideband amplifier, normally covering the whole of the operating bandwidth of the satellite, typically 500 MHz. Its presence ensures that the C/N ratio of the received signals will not be significantly degraded in the IF amplifiers, despite the loss in the link from the LNA and in a lossy splitter. Very high linearity is required of this amplifier to avoid degradation of the C/N ratio of received emissions due to intermodulation between them.

Down-chains

The wideband RF amplifier is followed by a splitter, which provides an input for each down-chain. For an earth station providing international connections there may be a considerable number of these down-chains, one for each wideband emission that is to be received and one for a group of SCPC emissions. A heavily loaded Intelsat Standard A station which does not use TDMA may receive several dozens of FDM or TDM aggregates. Frequency modulated emissions are usually demodulated conventionally but extended threshold demodulators may be used, for example, when receiving narrowband video emissions.

Station control and monitoring

To ensure the reliability of services, the safety of equipment and the integrity of the satellite system as a whole, earth stations have comprehensive monitoring and control subsystems. These are designed to ensure that all malfunctioning of equipment is identified quickly and logged, remedied automatically wherever possible and drawn to the attention of staff for action where automatic remedial action is not feasible. Where earth stations are not continuously staffed, the control equipment must ensure that a fault which may be detrimental to the services of other earth stations leads to the prompt switching off of the unserviceable equipment.

References

1. Alper, J. and Pelton, J. N. (eds.) (1984). *The Intelsat Global Satellite System*, Progress in Astronautics and Aeronautics Vol. 93. The American Institute of Aeronautics and Astronautics.
2. Thompson, P. T. and Silk, R. (1990). Intelsat VII: another step in the development of global communications. *Journal of the British Interplanetary Society*, **43**(8).
3. *The application of interference cancellers in the fixed-satellite service* (ITU-R Recommendation S.734, 1994), ITU-R Recommendations 1994 S Series. Geneva: ITU.
4. Intelsat's standards of performance for earth stations requiring access to its satellites are contained in the Intelsat Earth Station Standards (IESS) series of documents.

5. Hemmati, F., Miller, S. and Schweikert, R. (1995). Improved Intelsat IDR and IBS services. *Proc. Tenth International Conference on Digital Satellite Communications*, Brighton. London: IEE.
6. Basic concept of the TDMA/DSI system. Intelsat document ESS-TDMA-1-4 H/11/80, 22 October 1980.
7. The Intelsat TDMA/DSI system. Intelsat document ESS-TDMA-1-5 H/11/80, 22 October 1980.
8. Pontano, B. A., Campanella, S. J. and Dicks, J. L. (1985). The Intelsat TDMA/DSI system. *Comsat Technical Review*, **15**(2B).
9. Lewis, J. R. and Elliot, D. T. (1985). Time-division multiple-access systems for satellite communications. *British Telecommunications Engineering*, **4**.
10. Bedford, R., Chaudhry, K. and Smith, S. (1995). Improvement and application of the Intelsat (SS) TDMA system. *Proc. Tenth International Conference on Digital Satellite Communications*, Brighton. London: IEE.
11. Mittal, S., D'Ambrosio, A., Mazzola, E. and Eagling, T. (1995). Next generation TDMA terminal for Intelsat system users. *Proc. Tenth International Conference on Digital Satellite Communications*, Brighton. London: IEE.
12. Lunsford, J., Thorne, R., Lindstrom, R. *et al.* (1995). Intelsat second-generation TDMA terminal. *Proc. Tenth International Conference on Digital Satellite Communications*, Brighton. London: IEE.
13. *Calculation of the maximum power density (averaged over 4 kHz) of an angle-modulated carrier* (ITU-R Recommendation SF.675-3, 1994), ITU-R Recommendations 1994 SF Series. Geneva: ITU.

12 Domestic public networks and private systems

12.1 Applications and basic network characteristics

Many FSS earth stations do not require high connectivity with many distant earth stations in a very large network; they differ in this respect, for example, from the earth stations of the Intelsat Global Network. For earth stations with limited needs, the means for achieving a very large per-satellite capacity which are applied for the Intelsat Global Network and in particular the requirement for a high earth station G/T, may be irrelevant and pointlessly costly. If the design of such stations were economically optimized, giving due weight to the variation with earth station G/T of the cost of space segment facilities, it would usually be found that antennas smaller than those described in Chapter 11 would be appropriate. An aperture diameter in the range 3 to 10 metres would be typical.

Typical applications

Many countries use satellites for links in the domestic PSTN. In very large countries satellite communication may be the most economical medium for long-distance trunk links. This is especially true where the terrain, being broken up by sea, mountains or desert, makes terrestrial telecommunication facilities particularly costly to provide or where a low population density prevents the achievement of the economies of scale of modern terrestrial systems. In very sparsely populated regions, satellites may be used to provide service directly to isolated settlements and even individual homesteads. Satellites are also sometimes used to provide a temporary high-grade digital overlay network to augment an unsatisfactory terrestrial network while a general programme of modernization is carried out.

Major industrial organizations with establishments at a number of different locations have found it economic to set up private satellite networks. Often the earth stations are on the user's own premises, but satellite service providers, including PTOs, also build earth stations in metropolitan areas to provide access to private space segment facilities for the local business community. Some of these multi-user stations have several or many antennas, giving access to a number of satellites; they are called 'teleports'. These private networks are often tailored to serve the particular needs of the user.

A very important use of satellite communication is for the transmission of standard definition television signals. This application takes several forms, such as:

(1) The distribution of broadcasting programmes to terrestrial broadcasting stations and the head-ends of cable broadcasting networks, typically via receive-only earth stations at the destination locations. When such signals are transmitted in a conventional analogue form, they may also be intercepted by members of the public using their own antennas; this kind of 'direct to home' (DTH) satellite broadcasting is considered further in Section 15.2.

(2) The distribution of television signals to a limited number of receive-only earth stations for display to closed groups of viewers for training purposes, commercial presentations, video conferencing, etc.
(3) Links, usually temporary, carrying contributions from the scene of news, cultural or sporting events to a switching point where this material can be incorporated into broadcasting programmes. Mobile earth stations are often used at the source of the signal, in which case this practice is called satellite news gathering (SNG); see Section 14.5.1.

A fourth important application for antennas of medium size is for the earth stations used as hub stations for networks of very small aperture terminals (VSATs). These are considered in Chapter 13, together with the VSATs themselves.

The space segment options

In a country where there is a large market for domestic satellite communications there may be an economic case for setting up a satellite system wholly or mainly for that country's use. A substantial number of such national satellite systems have been set up already, virtually all of them geostationary.

However, a single FSS satellite provides a great deal of capacity, its working lifetime in orbit may be 12 to 15 years, initial costs are high and running costs are substantial. Many countries do not need to use enough space segment capacity to justify setting up a system for their own use and some needs are temporary. For such cases there are two other options. If the domestic need is long term, a satellite system might be set up as a joint venture with other countries or with the intention of leasing some of its capacity for use by other countries. Alternatively, whole transponders or a specified bandwidth within a transponder can be leased from one of the international satellite consortia or from a domestic satellite under foreign ownership. The latter may be feasible because the footprints of the antenna beams of domestic satellites often extend beyond the boundaries of the country licensing the system.

Most satellites to be used for domestic networks have their frequency assignments coordinated with those of other satellites nearby in orbit in accordance with the procedures summarized in Section 6.2. However, domestic systems may also be set up under the ITU FSS Allotment Plan in frequency allocations set aside for that purpose. The administrative procedures used in the implementation of the plan are outlined in Section 6.3.2 and some technical aspects of the plan are reviewed in Section 12.4.

Most modern satellites designed to provide domestic networks have a transponder frequency plan like the one shown in Figure 12.1a. Typically a bandwidth of 500 MHz is divided between twelve transponders. Allowing for guard bands, the bandwidth per transponder available for use is about 36 MHz. A further set of 12 transponders operating in the orthogonal polarization often doubles the capacity of the satellite by frequency reuse. Where a dual-polarization satellite is to be used mainly for wideband FM/analogue carriers, typically FM/television, the frequency plan for the two sets of transponders may be staggered, as shown in Figure 12.1b. This staggered arrangement reduces slightly the usable bandwidth per transponder but it ensures that much potential cross-polar interference falls into the guard bands between the transponders having the orthogonal polarization, eliminating many XPI problems.

C band is strongly preferred and widely used where rainfall is heavy. The bands used for systems which are not part of the FSS allotment plan are 5925–6425 MHz for uplinks and 3700–4200 MHz for downlinks, the same as are used for international networks. Where Ku band radio

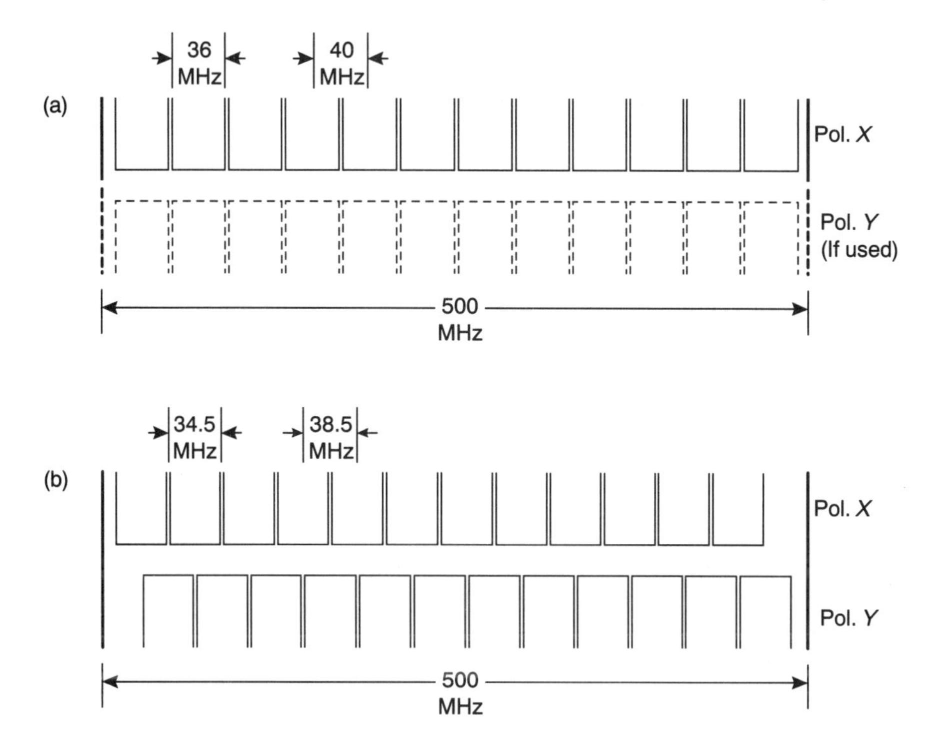

Figure 12.1 *Typical domestic satellite transponder frequency plans*

propagation is satisfactory, the bands which have least terrestrial use in the service area tend to be favoured for domestic satellites. Thus, satellites serving Europe are likely to include 12.5–12.75 GHz for downlinks, although 10.95–11.2 GHz and 11.45–11.7 GHz are also used. Satellites serving North America use 11.7–12.2 GHz for downlinks. The universal Ku band uplink choice is 14.0–14.5 GHz. Where the demand for domestic satellite services is strong, satellites may carry transponders for both C band and Ku band. The power rating of typical domestic satellite C band transponders is 10 to 20 W single carrier saturated output. For Ku band transponders, the typical range is 20 to 50 W.

The size of the footprint, and therefore the gain, of satellite antennas depends, of course, on the area which the satellite is designed to serve. The basic market will often lie in one country and the minimum footprint will be large, or small, accordingly. If the home country is very large, it may be found desirable to subdivide the service area between two beams. On the other hand, if the home country is small, and particularly if the demand from that market is not expected to require the use of the whole of the satellite's capacity, the minimum footprint may be deliberately extended beyond the home country's territory to open the use of the satellite to foreign markets. Beams with a half-power beamwidth of 2° to 3° are typical.

12.2 Earth station antennas

12.2.1 The trade-offs

The diameters of earth station antennas used for domestic networks are typically between 3 and 10 metres. The use of these smaller antennas, compared with the 10 to 17 metre antennas typically required for international PSTN links, has important effects on the design and economics of the station. The more important of these consequences can be summarized as follows and some trade-offs are considered in more detail below.

(1) The capital and running costs of the antenna structure become smaller, sometimes much smaller. This factor may have a dominant impact on the economic feasibility of systems, above all where a limited amount of information is to be passed between a lot of earth stations.
(2) The area of land required for an earth station and the impact of the earth station on the environment are both reduced. This may make it less difficult to obtain permission to construct an earth station in or near an urban area, thereby reducing the cost of backhaul links. It may be feasible to install the antenna at or on a down-town office building, eliminating backhaul links altogether. For the smallest antennas, with low wind loading, the cost of structural modifications of the building to enable the antenna to be installed may be minimal.
(3) The on-axis gain of the antenna in the transmitting mode is reduced. Consequently the transmitter will have to supply more power to produce the required uplink e.i.r.p. This may lead to a need for the gain of the antenna side-lobes to be particularly low.
(4) Since the on-axis gain of the antenna in the receive mode is also reduced, the G/T ratio will become smaller, increasing the e.i.r.p. required from the satellite for given communication facilities and probably increasing space segment costs.
(5) The width of the main lobe of the antenna, transmitting and receiving, is increased. This will make less critical the tracking of the satellite by the earth station antenna. If the satellite is kept close to its nominal orbital station in both the N–S and E–W planes, and the diameter of the earth station antenna is sufficiently small, there may be no need for active tracking, significantly reducing cost.
(6) The increases in uplink and downlink emission power level and in antenna beamwidth increase the likelihood of interference to other networks, satellite and terrestrial, making frequency coordination more difficult to achieve.
(7) Antenna and antenna mount configurations that are not practicable for very big antennas become feasible. These configurations may be technically better or substantially cheaper than a fully steerable axisymmetrical Cassegrain antenna with an Az–El mount.

Aperture versus G_0 and G/T

The approximate on-axis gain G_0 of an antenna can be calculated from equations (7.1) to (7.4). This has been done for antennas with primary reflectors having diameters in the range 3 to 10 metres and for frequencies from 3.9 to 14.5 GHz; see Figure 12.2. For an axisymmetrical Cassegrain antenna, with the subreflector obstructing part of the aperture, the aperture efficiency η tends to decline as the size of the primary reflector is reduced. This decline occurs because the size of the subreflector cannot be scaled down proportionately without reducing the efficiency of illumination of the primary reflector. Here it has been assumed that η changes from 65 per cent when the diameter is 10 metres to 60 per cent when the diameter is 3 metres. This assumption is probably rather optimistic for small axisymmetrical antennas and rather pessimistic for configurations in which there is no blocking of the main aperture by a subreflector or a feed assembly.

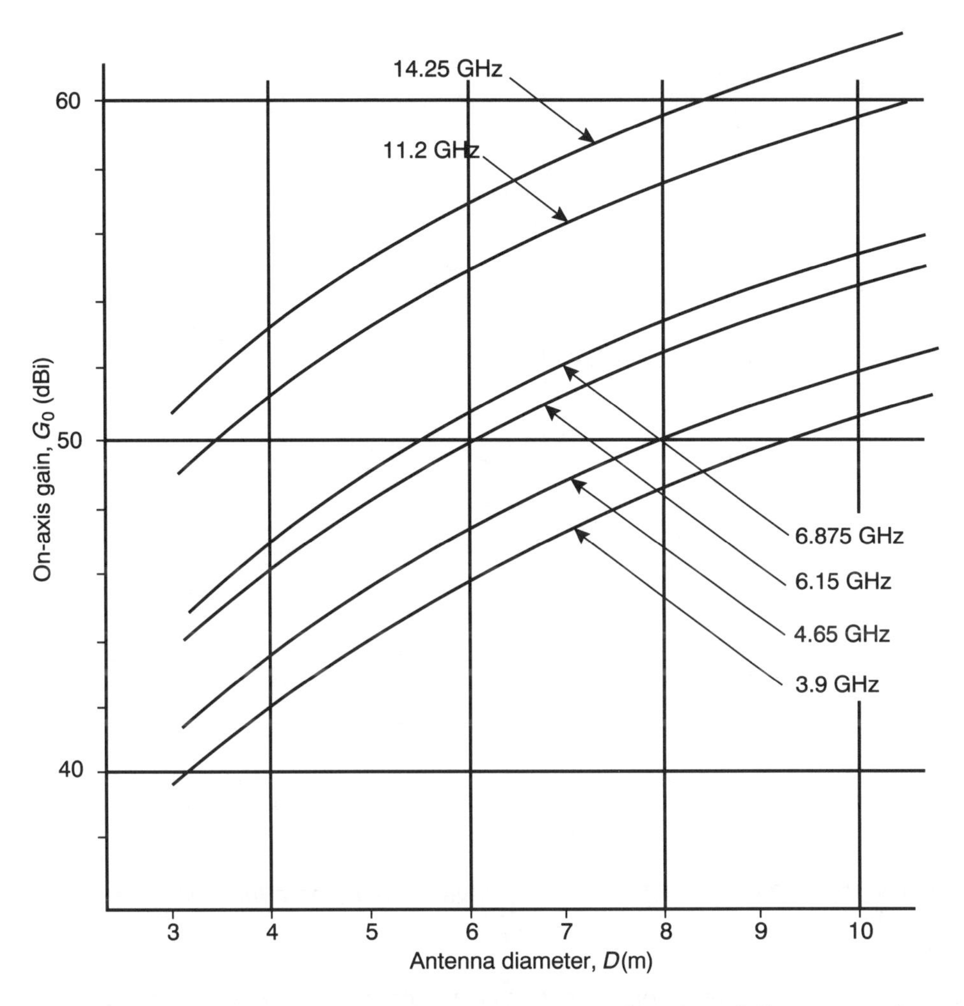

Figure 12.2 *The on-axis gain of an antenna as a function of diameter and frequency*

A reduction in the antenna gain, other parameters remaining unchanged, will demand a corresponding increase in earth station transmitter power. This may have little impact on cost if a low-power solid state HPA has to be replaced by a more highly rated solid state amplifier. The cost impact will, however, be quite significant if it becomes necessary to use a klystron in place of a solid state amplifier or if a klystron rated at several kilowatts output must take the place of a low-power klystron.

Wideband, highly linear, GaAs FET low-noise amplifiers operating at the ambient temperature are currently available for use in the first stages of earth station receivers having noise temperatures around 70 and 130 K at 4 and 11/12 GHz respectively. Allowing for other sources of noise (see Section 7.4), it may be expected that the clear-sky system noise temperature T_s will rise from 100 to 150 K at 4 GHz and from 200 to 250 K at 11/12 GHz as the diameter of the antenna falls from 10 to 3 metres. Values of G/T calculated on these assumptions are shown in Figure 12.3.

An increase in the downlink power required for an emission, made necessary by a reduction in earth station G/T, may be expected to raise space segment

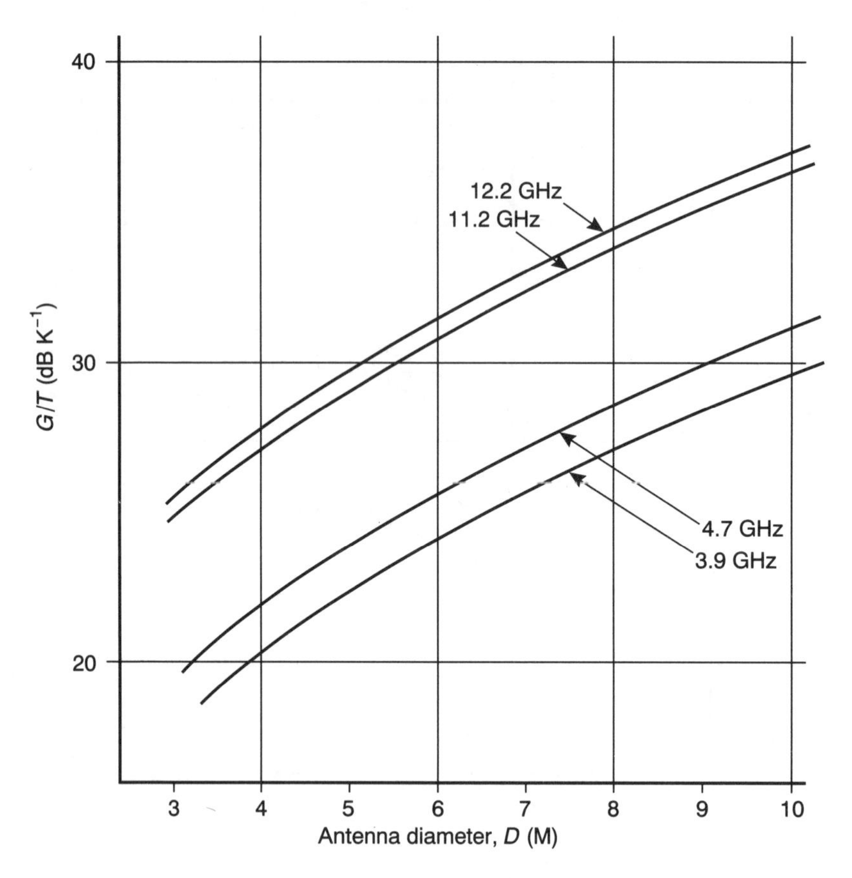

Figure 12.3 *The figure of merit (G/T) of an antenna as a function of diameter and frequency*

costs in the same proportion. This may be a matter for concern if the bandwidth occupied by the carrier is large and the number of earth stations receiving the carrier is small. The increase may, on the other hand, be thought trivial if the bandwidth is small. Furthermore, increased space segment costs may be a sound economic trade-off for a point to multi-point network if the number of earth stations involved is large.

However, whether the cost impact of using smaller earth station antennas is large or small, it will clearly be important for the emission parameters to be optimized to minimize costs. Analogue signals with frequency modulated carriers will tend to have wider deviation. Extended threshold demodulators may have to be used. But digital modulation predominates in these systems, and various forms of signal processing to reduce the bit rate, and FEC coding to reduce the downlink e.i.r.p. requirement further are widely used.

Aperture versus tracking requirements

Equipping an earth station antenna to track its satellite adds to the capital cost of the antenna, and the relative cost of active tracking becomes quite significant if the basic cost of the antenna is small. However, the bigger beamwidth of a smaller antenna makes the system more tolerant of satellite motion. Tracking may even become unnecessary if the orbit is controlled sufficiently precisely.

The maximum daily angular excursion of a geostationary satellite from its nominal position, as seen from an earth station, is given approximately by the square root of the sum of the squares of its E–W and N–S excursions from the nominal position. The ITU *Radio Regulations* require the E–W station-keeping error to be less than $\pm 0.1°$ (see Section 6.5, item 3) but there is no explicit mandatory limit on the N–S excursion, which is caused by inclination of the satellite's orbital plane. A substantial mass of thruster fuel is required, in the course of the lifetime of a satellite in orbit, to neutralize perturbations of the angle of inclination of the orbit. This fuel requirement increases considerably the total mass of a satellite at the start of its life, and thus raises the cost of launching. As the working lifetimes of satellites increase, satellite owners have become more inclined to underprovide thruster fuel. Many satellites are operated with significant orbital inclination towards the end of their lives, causing a N–S motion, seen from an earth station, of several degrees.

Figure 12.4 relates the on-axis gain G_0 of an antenna to the off-axis angle ϕ at which the reduction of gain $G_0 - G_\phi$ is equal to 3 dB, that is, where $\phi = \frac{1}{2}\theta_0$. In calculating θ_0, it has been assumed that the value of k in equation (7.5) varies from 65° when $\lambda/D = 0.0025$ ($G_0 \approx 60$ dBi) to 70° when $\lambda/D = 0.025$($G_0 \approx 40$ dBi). Figure 12.4 also shows the values of ϕ for which the reduction of gain is equal to 1.0, 0.5 or 0.2 dB; see equation (7.6).

Take as an example an antenna without tracking facilities and with an on-axis gain of 50 dBi. Figure 12.2 shows that this might be a 3 metre antenna operating at 13 GHz or a 6 metre antenna operating at 6 GHz. Figure 12.4 shows that this antenna would suffer a maximum gain loss of 0.2 dB due to the pointing error if the N–S and E–W excursions of the satellite were not greater than $\pm 0.05°$. This would rise to about 0.6 dB if the E–W excursion remained at $\pm 0.05°$ but the N–S excursion rose to $\pm 0.1°$.

Losses of this order would probably be found acceptable in exchange for the elimination of tracking. If a large number of earth stations were served by a single satellite carrier, the cost saving due to the elimination of tracking might well justify a considerably larger mis-pointing loss. However, it will be seen that there would be a loss of up to 1.0 dB for this same antenna if the N–S excursion approached $\pm 0.2°$. Moreover, many earth stations have bigger antennas and many satellites have a bigger N–S movement. In such circumstances the losses arising from a fixed pointing angle would probably be too large to accept. However, given a system that, having a broad main beam, was relatively tolerant of a residual pointing error, a relatively simple programme tracking mechanism would often be sufficient to reduce the mis-pointing loss to an acceptable value.

Aperture versus interference potential

Networks of the FSS are subject to international regulatory constraints in the form of downlink spectral PFD limits and uplink spectral e.i.r.p. limits outside the beam of the earth station antenna. See Section 6.4 and Section 6.5 item 5. The uplink constraint is globally applicable and the downlink constraint applies almost everywhere. These constraints apply to earth stations with big antennas (see Section 11.4.6) as well as the smaller ones used for domestic and private networks but they are more onerous for systems using smaller antennas.

(a) The downlink constraint

The limit on downlink power is most stringent where the angle of incidence at the Earth's surface is less than 5°. In this worst case, the power density fed to the satellite transmitting beam must be limited so that the PFD does not exceed

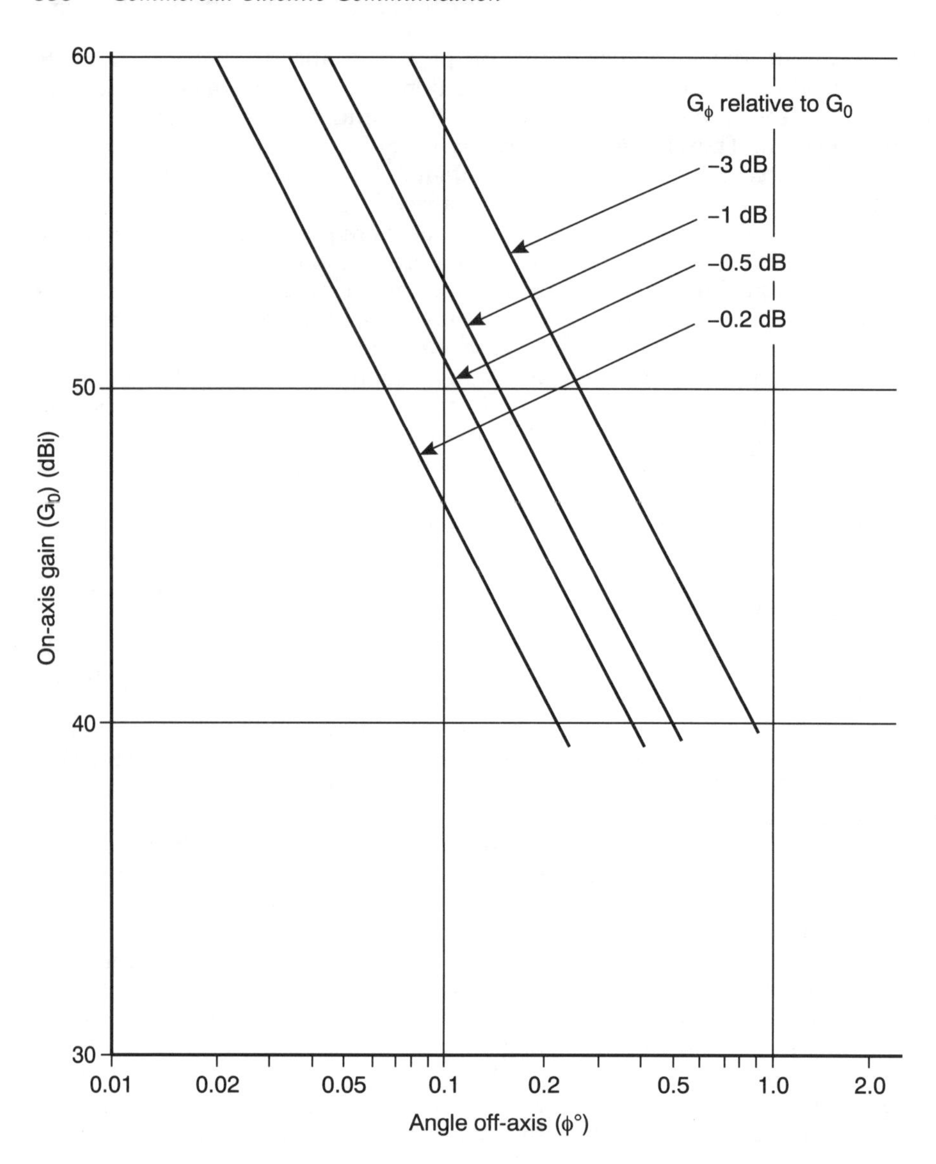

Figure 12.4 *The angle off-axis (ϕ°) at which the gain G_ϕ of the main lobe of an antenna is 0.2, 0.5, 1.0 and 3.0 dB less than the on-axis gain G_0*

$-152\,\mathrm{dB(W\,m^{-2})}$ in any 4 kHz band at 4 GHz, nor $-150\,\mathrm{dB(W\,m^{-2}\,(4\,kHz)^{-1})}$ at 11 GHz nor $-148\,\mathrm{dB(W\,m^{-2}\,(4\,kHz)^{-1})}$ at 12.5 GHz. Equation (5.5) shows that this is likely to limit the e.i.r.p. of that beam towards an earth station to about 11, 13 and 15 dBW in any 4 kHz band at 4, 11 and 12.5 GHz respectively.

Rearranging equation (10.11) and disregarding minor factors, it is found that, with a clear sky,

$$(C/N)_d \approx w_{ts} - L_{bfd} + (G/T)_e - 10\log kB \quad \text{(dB)} \tag{12.1}$$

where $(C/N)_d$ is the downlink C/N ratio (dB),
w_{ts} is the satellite e.i.r.p. (dBW),

L_{bfd} is the free-space basic transmission loss at the downlink frequency (dB),
$(G/T)_e$ is the G/T of the receiving earth station (dB K^{-1}),
k is Boltzmann's constant ($= 1.38 \times 10^{-23}$ J K^{-1})
B is the bandwidth (Hz).

Evaluating the impact of the downlink constraint on system design can be further simplified by disregarding interference and assuming that downlink noise is half the total noise at the receiving earth station. Then equations (12.1) and (5.3) can be used to relate the worst-case maximum permissible value of w_{ts} to the pre-demodulator C/N level in a 4 kHz band. It is found that the following approximate maximum C/N ratios are obtainable without exceeding the worst-case downlink constraints:

$$C/N = (G/T)_e - 4\,\text{dB} \quad \text{at } 4\,\text{GHz} \qquad (12.2a)$$

$$C/N = (G/T)_e - 11\,\text{dB} \quad \text{at } 11\,\text{GHz} \qquad (12.2b)$$

$$C/N = (G/T)_e - 10\,\text{dB} \quad \text{at } 12.5\,\text{GHz} \qquad (12.2c)$$

Three-metre antennas with figures of merit given by Figure 12.3 could provide pre-demodulator C/N ratios around 15 dB in each of these frequency bands with downlink emissions that did not break the downlink constraint, provided that the energy of the emission was spread evenly across the band occupied by the emission. This would provide adequate threshold margin for FM/analogue emissions and a BER that is sufficient for most purposes for PSK/digital emissions, with a rain margin of a few decibels.

The assumption of an even spread of spectral energy, in the context of a 4 kHz sampling bandwidth, is clearly optimistic. Carrier energy dispersal techniques in current use fall short of this standard, especially for FM emissions. On the other hand, the beam geometry assumed above is the worst case. If the footprint of the satellite beam was such as to ensure that the Earth's surface was not illuminated at high gain at low angles of elevation, the e.i.r.p. density towards earth stations would be less constrained. Furthermore, substantial additional margin could be obtained by using low-threshold demodulators for FM emissions and FEC for digital ones.

(b) The uplink constraint

The level of carrier energy dispersal that must be applied to emissions from smaller antennas in order to meet the downlink constraint contributes also to the reduction of the off-axis spectral radiation density from the antenna. See also Section 11.4.6. But the off-axis radiation level is affected by many factors in addition to carrier energy dispersal. High satellite antenna gain and high transponder gain tend to reduce uplink e.i.r.p. requirements and therefore off-beam radiation. Low on-axis antenna gain at the earth stations have the reverse effect. The most important factor of all may be the gain of the side-lobes of the transmitting earth station antenna itself. Perhaps the only valid generalization is that the uplink constraint becomes more difficult to meet as the on-axis gain of the earth station antennas is reduced.

There are cost penalties when most of the factors listed above are applied to reduce uplink off-beam radiation, and some penalties are severe. One method that involves a relatively low penalty might be to increase the degree of carrier energy dispersal; this can be effective for the larger earth station antennas in the domestic system range. However, where additional carrier energy dispersal does not provide sufficient reduction, the improvement of antenna side-lobe suppression in the directions of concern, mainly those within 3° of the GSO, may be the least costly option. Well-designed axisymmetrical Cassegrain antennas have side-lobe levels falling below the envelope given by equation (6.5). Other antenna configurations have substantially better side-lobe suppression; they are considered briefly in Section 12.2.3.

(c) Control of interference by frequency coordination

The means used for limiting interference between one satellite with its associated earth stations and another using the same frequency bands are outlined in Section 6.2. The use of earth stations with smaller antennas increases both susceptibility to interference and liability to cause it; this may make it more difficult to achieve success in frequency coordination. However, the need for high downlink power densities for reception at earth stations with low G/T ratios often causes transponders to be power-limited. This may provide opportunities for manipulating frequency plans to minimize interference, carrier frequencies used in one satellite system being interleaved with those of other systems likely to cause or suffer interference.

The means used for limiting interference between earth stations and nearby terrestrial stations sharing a frequency band are outlined in Section 6.4.1. Where the earth stations serve sparsely populated areas, interference of this kind is not usually a problem. Elsewhere, in areas where there is substantial use of terrestrial radio, the presence of domestic and private satellite systems raises rather intractable interference problems. For example, the use of distance and screening to isolate terrestrial stations from earth stations is difficult to implement. In such cases there are great advantages in avoiding this interference problem altogether by using frequency bands for the satellite service which are not used locally for terrestrial systems.

12.2.2 Earth station antenna G/T standards

Intelsat and Eutelsat have both established standards of earth station antenna G/T for use with their leased transponders. There are, however, no global standards and earth station operators are, in general, free to optimize this parameter.

Intelsat standards

Intelsat has three C band standards and three Ku band standards for domestic and private systems, calling for the following values of G/T:

At C band

Standard F1 $G/T = 22.7 + 20\log \tfrac{1}{4}F_C \quad \text{dB K}^{-1}$

Standard F2 $G/T = 27 + 20\log \tfrac{1}{4}F_C \quad \text{dB K}^{-1}$

Standard F3 $G/T = 29 + 20\log \tfrac{1}{4}F_C \quad \text{dB K}^{-1}$

At Ku band

Standard E1 $G/T = 25 + 20\log \tfrac{1}{11}F_{Ku} \quad \text{dB K}^{-1}$

Standard E2 $G/T = 29 + 20\log \tfrac{1}{11}F_{Ku} \quad \text{dB K}^{-1}$

Standard E3 $G/T = 34 + 20\log \tfrac{1}{11}F_{Ku} \quad \text{dB K}^{-1}$

where F_C and F_{Ku} represent the frequency of operation in C band and Ku band respectively, in GHz. In addition, Standard D1, available for the Vista service which provides for small groups of SCPC channels serving the PSTN, has the same G/T requirement as Standard F1.

Eutelsat standards

Eutelsat has established three standards for Ku band antennas used for IDC. Standard I-1 calls for $G/T = 34\,\mathrm{dB\,K^{-1}}$; reference has been made to this standard in Section 11.4.3. Standards I-2 and I-3 require values of G/T of 29 and $26\,\mathrm{dB\,K^{-1}}$ respectively.

Three standards have also been provided for SMS. Standards 1, 2 and 3 call for values of G/T of 30, 27 and $23\,\mathrm{dB\,K^{-1}}$ respectively. Standard 3 is typically implemented with an antenna 2.4 metres in diameter and it can be considered as a VSAT standard. However, Eutelsat does not limit access for SMS service to earth stations having these G/T values unless the user requires inter-operability with the earth stations of other users.

12.2.3 Antenna configurations

Big earth station antennas, with rare exceptions, have axisymmetrical Cassegrain geometry and Az–El mounts. Massive antenna structures can be supported best in this way, the antenna can be steered to any satellite in sight of the earth station and can in principle be driven to track any satellite movement. Antennas at the upper end of the size range for domestic and private networks are also of this type. However, many earth stations do not need unlimited steerability and axisymmetrical Cassegrain antennas in the middle and lower parts of the size range are likely to have unacceptably high levels of off-beam gain. Other antenna geometries are necessary, and other mount designs are feasible, for the smaller antennas.

Antenna geometry for low-gain side-lobes

Some improvement in the side-lobe suppression of a conventional axisymmetrical Cassegrain antenna can be obtained by placing material that absorbs radio-frequency energy round the edge of the subreflector, so reducing spill-over radiation. Where this does not provide sufficient improvement, it is usually necessary to adopt one of the offset antenna configurations described in Section 7.3.3 to ensure that ITU-R Recommendation S.524 is met; see Section 6.5, item 5.

Off-set front-fed antennas have no subreflector and the feed horn is not directed towards the GSO; thus the worst causes of spill-over radiation are eliminated. However, it may be difficult to design a low-loss waveguide connection between the antenna feed and the transmitter without loss of antenna steerability. This configuration is widely used for small receive-only antennas which have a down-converter located at the feed, the output of which is connected to the receiver input by a flexible coaxial cable.

Off-set Cassegrain or off-set Gregorian geometries are used at Earth stations that transmit. They have subreflectors but it can be arranged that all significant subreflector over-spill radiation is directed more than 3° to the north or the south of the GSO, thereby escaping the constraint imposed by ITU-R Recommendation S.524. Figure 12.5 shows an off-set Gregorian antenna. The antenna in Figure 14.13 has off-set Cassegrain geometry.

Figure 12.5 *An off-set Gregorian antenna (left) and an axisymmetrical Cassegrain antenna (right). (BT Corporate Pictures: a BT photograph.)*

Antenna mounts

Like the Az–El mount, a 'polar' or 'equatorial' mount also allows movement of the antenna about two axes. Polar mounts follow the basic design of mounts for astronomical telescopes and the same terminology is used for the principal features. The lower axis, called the 'hour-angle' axis, is made parallel with the Earth's polar axis; in other words it is perpendicular to the equatorial plane. The upper axis, on which the antenna structure is mounted, is made orthogonal to the hour-angle axis; it is called the 'declination axis'.

For an earth station located on the equator, the hour-angle axis of a polar mount is horizontal, aligned north–south. The antenna beam scans round the GSO from the east horizon to the west horizon when the hour-angle axis is driven and the declination axis holds the antenna at a constant angle. Movement about the declination axis may be needed to track the north–south movement of a satellite with significant orbital inclination but the range of this movement should not exceed a few degrees. By adding a small angular bias to the orientation of the hour-angle axis, the simplicity of this tracking system can be extended, to some considerable extent, to earth stations that are somewhat north or south of the equator.

Polar mounts are sometimes used for the smaller steerable antennas instead of Az–El mounts, particularly in low latitudes. However, in cases where the antenna is very small and light and where there is no requirement for it to be directed, first at one satellite and then quickly at another, the mount may be simplified still further. The antenna is secured at three points of attachment to a gantry, to concrete plinths or most commonly to a wall facing in a suitable direction. The antenna is free to swivel about one of these fixed points, and is attached to the others by legs of the length necessary to direct the antenna beam towards the satellite. If satellite tracking is needed, the length of one or both of these legs can be made remotely adjustable by incorporating electrically or hydraulically actuated jacks in them.

12.3 Transmission systems

12.3.1 Domestic trunk networks for the PSTN

A typical domestic satellite network providing trunk circuits for the PSTN would have several, perhaps many, earth stations at key provincial locations in the terrestrial PSTN. Each earth station would transmit one multiplexed carrier and receive carriers from several, perhaps all, of the other stations. Depending on the scale of the network, these carriers would, in aggregate, probably occupy one or several transponders operating in FDMA. The choice between analogue and digital transmission for the satellite links would be determined by the needs of the PSTN.

Choice between C band and Ku band would depend on rain climate, on what space segment facilities are available for the service area and on their price. The earth station antennas used are likely to be at the upper end of the size range, probably 8 to 10 metres diameter for C band and 7 or 8 metres for Ku band, although smaller antennas might be used if the satellite antenna gain within the service area were particularly high. Baseband structures would be similar to those described for FDM, CFDM and TDM using FDMA in Section 11.4.4. Approximate emission parameters may be calculated using the methods of Examples 1 and 2 in Section 10.5.

In addition to multiplexed carriers for the PSTN, a network of this kind is often used to distribute sound radio and television programme signals from a central point to terrestrial broadcasting stations nation-wide.

12.3.2 Thin-line public network systems using SCPC

To provide domestic PSTN services for a territory with sparse population and little terrestrial infrastructure, a satellite network consisting of SCPC channels in FDMA is often chosen. Transmission may be analogue or digital, and the channel equipment described in Sections 8.2.1 and 9.4.1 is typical. There are two basic network configuration options when small earth station antennas are used, the hubbed star and the mesh.

The hubbed star configuration

For a service area which needs a large number of earth stations, each having a very small telephone traffic flow, it would be essential to minimize the cost of each earth station; antennas about 3 metres in diameter would often be chosen. However, these small antennas require high downlink e.i.r.p., leading to high space segment costs for each telephone call made. A high-power downlink also involves a correspondingly high uplink e.i.r.p., a potentially costly transmitter and an uplink radiation density problem that is difficult to solve. To minimize costs and solve some of the technical problems, a star network configuration may be adopted, with a hub earth station having a considerably bigger antenna, perhaps 8 or 9 metres in diameter, serving the 3 metres outstations.

In such networks, most telephone calls are likely to be between an outstation and the more densely populated areas which are served by the terrestrial PSTN. The hub station would have direct access to the terrestrial PSTN and calls of this kind would involve a high-power downlink in the outbound direction only, that is, from the hub to the outstation. When a call is required from one outstation to another, it would be relayed at the hub station, the long transmission delay associated with a two-hop path via a geostationary satellite being tolerated. Relaying at the hub station eliminates the need for outstation transmitters powerful enough to communicate directly with other outstations. This is a technique much used in VSAT networks; see Figure 13.1a.

The mesh network

Where the traffic per earth station is greater, the cost of bigger earth station antennas may be economically justifiable. Under these conditions a mesh network might be viable, any pair of earth stations communicating directly without an intermediate hub station; see Figure 13.1b.

Optimization of space segment utilization

The cost of a thin-line access link to the PSTN will be very high compared with typical short cable or terrestrial radio access, so careful system optimization is essential. In carrying out this optimization it will be necessary to make assumptions about the space segment; a good initial assumption would be that there will, sooner or later, be economic use for all the transmission capacity that the satellite can provide. Thus, a basic objective will be to ensure that

$$w_n/b_n \approx W_T/B_T \qquad (12.3)$$

where w_n and W_T are the downlink multi-carrier e.i.r.p. used by the network and available from the transponder respectively, and
b_n and B_T are the bandwidth occupied by the network and the bandwidth of the transponder respectively.

The optimization process operates at two levels: channel optimization and network optimization. See Example 4 in Section 10.5.

First, the channel and modulation parameters are chosen to give acceptable performance for least satellite occupation. Whether 'least satellite occupation' should be interpreted as minimum bandwidth or minimum downlink e.i.r.p. will depend on the parameters of the transponder and of the earth station. Optimum parameters may have to be determined iteratively, taking the results of the network optimization process into account and with earth station G/T as a variable. For a PSTN using analogue transmission, analogue FM satellite links would probably be chosen and the factors to be studied would include limitation of channel bandwidth, the use of emphasis and companding and the subjective signal to noise ratio. For a digital PSTN, an economical speech coding system such as ADPCM would often be employed for satellite telephony, although with appropriate provision being made for the performance needs of low-speed data and facsimile signals that might be sent by modem. Separate provision might have to be made for medium-speed data channels. For all digital emissions, a choice must be made between the modulation system options, effectively between 2-phase and 4-phase PSK, and between the various FEC options.

The network optimization process will apply to all emissions except those used for uninterrupted data transmissions. Voice switching, to suppress the carrier while a telephone user is silent or no call is in progress, saves a large proportion of the downlink power that would otherwise be used. The potential benefits of demand assignment of channels, explained in Section 9.4.4 mainly in terms of shared TDM systems, are equally available in an SCPC system involving many earth stations. One or other of these options may be vitally important in optimizing a network, depending upon whether the transponder is power limited or bandwidth limited respectively.

Demand assignment of telephone channels adds two more requirements to the system, namely instantaneous flexibility in the carrier frequency of all emissions transmitted and received and a control network to enable that flexibility to be put to use. Frequency synthesizers provide the flexibility. A typical control network would involve an ALOHA system, enabling any earth station to pass brief control messages to any other station, directly or via a hub station, in order to seize free go and return channels from a pool of free channels, to be used for a call and then to release them back into the pool.

Automatic frequency control

Doppler effect on the frequency of carriers transmitted via a geostationary satellite, due to the diurnal motion of the satellite relative to fixed earth stations, is negligible when the inclination of the satellite's orbit is small. However, the frequency shift may reach some tens of kilohertz if the inclination is allowed to rise to several degrees. This shift may be aggravated by local oscillator drifts at earth stations and in the satellite. Carrier frequency errors of this order are not important for most emissions used in the FSS, which occupy wide bandwidths. However, the bandwidth occupied by typical SCPC emissions is in the range 20 to 80 kHz. If the transponder carrying these narrowband emissions is not severely power limited, an effective automatic frequency control (AFC) system must be used to maintain the proper spacing between adjacent channels.

AFC may be implemented at each earth station by comparing the carrier frequency of traffic emissions after retransmission by the satellite with a pilot carrier which has been transmitted by one designated earth station. The local oscillator frequency of an up-converter can be varied to correct any relative frequency error.

12.3.3 Video signal distribution

Satellite links are used in the FSS for video signals in various ways, mostly in point to multi-point, quasi-broadcast, modes or for temporary facilities. The medium is particularly attractive for point to multi-point connections. This is because the cost of the space segment facility is the same regardless of the number of earth stations receiving the signal, a feature for which long-distance terrestrial alternatives have no equivalent. Different applications have different picture quality requirement and the size of the earth station antennas required ranges from 3 metres at Ku band to 10 metres at C band.

Broadcasting signal distribution

Video signals and the associated sound programme signals, made up into complete television programmes, may be distributed by satellite from a programme centre to terrestrial broadcasting stations and to cable broadcasting network head ends. The signal is usually transmitted in analogue form, frequency modulating the carrier, with emphasis and carrier energy dispersal as described in Section 8.2.3. The emission usually occupies the whole of the bandwidth of a transponder 36 MHz wide. An S/N ratio substantially better than 40 dB, defined as in equation (8.16), will usually be required for rebroadcasting.

Satellite distribution of digital video signals, for terrestrial broadcasting with standard definition (625/50 or 525/60 picture standards) after conversion to analogue form, may be used in some instances. Digital signal processing will be necessary to reduce the bit rate, to limit the power and bandwidth required for the satellite links. An information bit rate of 8 Mbit/s, plus FEC, can be taken as current practice but the required bit rate is being reduced by the introduction of more advanced codecs. Substantial changes in this area can be expected in the mid-term as digital transmission is introduced for the final link to the domestic receiver.

Conference video links

It may be convenient to hold a conference or a presentation simultaneously at two or more venues, with audio and video links between them to enable all participants to share as fully as possible in the occasion. Such conferences may extend over a few hours or a few days but they are essentially temporary. Satellite communication is attractive for such purposes since it is feasible to rent the transmission medium economically for the brief period for which it is required; this is not usually possible for a wideband system via terrestrial links.

Video conference signals have a limited menu of picture formats. The most important format is the head of a speaker or the heads of a small panel of speakers; there is little or no movement in most of the area of such a picture. Another standard format shows a diagram, a photograph or the image of an object that the speaker is using to illustrate his presentation; such a picture may need fine detail to be reproduced, but there will be no movement at all to transmit for a period of many seconds. The third commonly occurring format is a distant view of an audience, for display at other venues, which will contain some movement but does not need to be reproduced in fine detail. In each case, substantial bit rate reduction is feasible with specialized equipment, reducing the bit rate required for transmission to about 2 Mbit/s.

12.3.4 Private corporate links and networks

Many commercial organizations use private satellite links to provide communications facilities which they cannot obtain, or cannot obtain at an acceptable price, from the PSTN. For example, Eutelsat's SMS channels are used for point to multi-point commercial information, news and market information broadcasts to subscribers. High-speed digital emissions are operated, continuously, by schedule or on demand, for computer file transfer and for the transmission by high-speed facsimile of made-up newspaper text from the editorial office to distant presses for printing. Private video channels may be used as an aid to discussion. Companies with major offices at several locations may find it cheaper to use the telephone channels of conventional TDM systems, transmitted by satellite, to interconnect branch telephone exchanges at their various establishments, instead of leasing channels or making calls through the PSTN. Some of these options are particularly attractive where the PSTN is unreliable, the public service operating agency has outdated tariff policies or the terrestrial network is ill-suited to digital transmission. Most characteristic of the large private satellite communication networks is the use of very flexible TDMA systems, perhaps having a bit rate between 2 and 10 Mbit/s, capable of meeting a wide range of communications needs as they arise.

12.4 The FSS Allotment Plan

12.4.1 Introduction

The objective of the ITU FSS frequency allotment agreement of 1988 is to ensure that every country in the world will retain into the indefinite future the opportunity to use and register interference-free frequencies for a domestic satellite system serving its territory, despite growing congestion in the GSO. The administrative provisions are outlined in Section 6.3.2 and some technical aspects are considered briefly below.

The frequency bands reserved for planned systems are as follows:

In C band 6725–7025 MHz for uplinks,
4500–4800 MHz for downlinks.

In Ku band 12.75–13.25 GHz for uplinks,
10.7–10.95 and 11.2–11.45 GHz for downlinks.

These bands are also allocated for terrestrial radio services.

Unlike other ITU frequency plans such as the satellite broadcasting plans reviewed in Sections 6.3.1 and 15.1, it was accepted as a basic requirement that an FSS plan must be flexible. There was to be as little constraint on the ways in which an allotment could be used as was compatible with providing allotments of adequate scope. This flexibility has been provided through a two-stage process. The plan defines model domestic satellite systems in ways that place upper limits on the interference that they may cause to other networks and lower limits on their susceptibility to interference. No specific assumptions were made about the frequency assignments that will be used or the kinds of emissions that the satellites and earth stations will transmit or receive. On that basis an approximate location has been computed for each country's satellite, such that unacceptable interference should not arise between satellite systems.

In the second-stage, which is initiated when a country decides to set its system up, the parameters proposed for its antennas and emissions, still expressed in general terms, are declared. This enables the ITU to determine whether the proposals are compliant with the model on which the plan is based and any necessary small adjustments can be made to take into account the situation then arising in the GSO in the vicinity of the planned satellite location.

12.4.2 The network model used in the plan

The essence of the model is as follows:

(1) *Satellite antenna characteristics.* In 1988 each country defined its required service area by means of a set of geographical 'test points', all within its own territory. The satellite antenna beams assumed in planning were circular or elliptical in cross-section, with a footprint for which the −3 dB contour is the best fit for the test points. If this rule were applied rigorously, a small country might find itself called upon to use an impracticably large satellite antenna; to avoid this, the smallest −3 dB beamwidth assumed was 1.6° at 6/4 GHz and 0.8° at 13/11 GHz. Fairly stringent assumptions were made for the rate of roll-off of the gain of the main lobe and the side-lobe envelope. It was assumed that deviation from the nominal beam pointing direction would be limited to 0.1° and that the rotation of non-circular beams would be limited to ±1.0°.
(2) *Satellite receiver noise temperature.* 1000 K at 6 GHz and 1500 K at 13 GHz.
(3) *Earth station antenna.* Antenna diameter = 7 and 3 metres respectively at C band and Ku band. Aperture efficiency = 70 per cent. Envelope of the peak gain of side-lobes = $(32 - 25 \log \phi)$ dBi, where $\phi°$ is the off-axis angle.
(4) *Earth station receiver noise temperature.* 140 K at 5 GHz and 200 K at 11 GHz.
(5) *Polarization.* No assumptions were made about polarization. Dual polarization may be used and no polarization discrimination was assumed in calculating interference levels.
(6) *Interference levels.* The objective is an aggregate C/I ratio of 26 dB or higher under free-space propagation conditions.
(7) *Performance of links.* The objective is a carrier-to-total-noise ratio of 16 dB under rain fading conditions, the rain attenuation margin being limited to 8 dB.
(8) *Emission parameters.* No assumptions were made about modulation characteristics and parameters and none was made about power levels apart from those implicit in the assumptions about the performance of links and the interference objectives.

12.4.3 Demonstration of compliance

When a country is ready to start implementing a domestic system in its orbit/spectrum allotment, the ITU is informed of the technical details of the proposed network, and in particular:

(1) the preferred orbital location for the satellite;
(2) the characteristics of the earth station and satellite antennas;
(3) the average spectral power density of emissions, uplink and downlink.

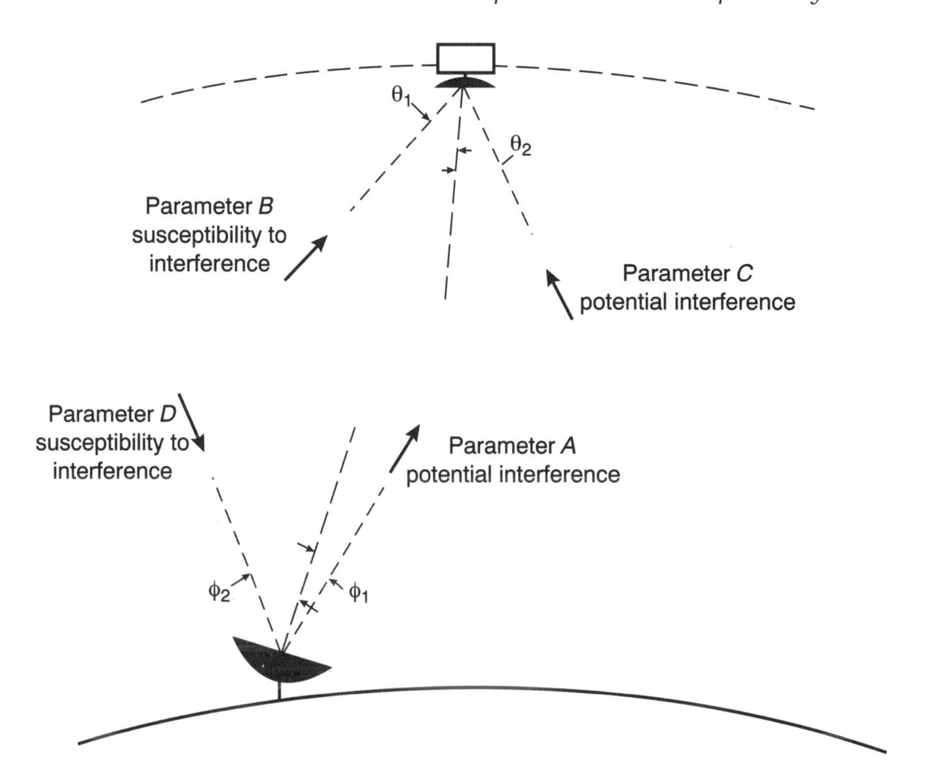

Figure 12.6 *The four generalized parameters, A, B, C and D, used by the ITU to determine whether a satellite system planned for implementation under the FSS frequency allotment plan conforms with the model used when the plan was drawn up*

The ITU examines this information, testing it for conformity to the plan, in particular by the use of 'generalized parameters' to characterize the risk of interference that the emissions and antennas used will cause or suffer. See Figure 12.6. There are four of these generalized parameters, *A*, *B*, *C* and *D*, having the following significance.

A is a measure of the power of the strongest emission of an earth station, radiated off the axis of its antenna, expressed as a function of the off-axis angle.
B is an indirect measure of the susceptibility of the network to interference entering at the satellite antenna, expressed as a function of off-axis angle.
C is a measure of the power of the strongest emission from the satellite, expressed as a function of the antenna off-axis angle.
D is an indirect measure of the susceptibility of the network to interference entering at the earth station off the antenna axis, expressed as a function of off-axis angle.

If the proposal is found, by this examination, to be in conformity with the plan, the proposing administration is asked to confirm that the concept of 'macrosegmentation' (see below) is to be applied to the arrangement of carriers across the frequency bands which the network uses. This provision is intended to reduce interference between satellite systems. If the reply is affirmative, the processes of the allotment plan are complete and the administration notifies frequency assignments to the ITU for registration in the usual way, after any necessary coordination with terrestrial stations in neighbouring countries (see Section 6.4). If macrosegmentation is not to be applied, it may be necessary to coordinate the frequency assignments with those of other satellite networks which are set up under the allotment plan.

Macrosegmentation in this context involves locating high-density carriers in the upper 60 per cent of each allocated frequency band and low-density carriers in the lower 40 per cent. High-density carriers are defined for this purpose as emissions for which the peak spectral density power, averaged over 4 kHz, is more than 5 dB greater than the average for the emission as a whole.

13 VSAT networks in the FSS

13.1 Very small aperture terminals

VSAT networks

The satellite networks discussed in Chapter 12 have earth stations with antennas at least 3 metres in diameter. This limit is arbitrary. Large numbers of earth stations in the FSS have smaller antennas, but it is convenient to treat them as a separate category; they are known as very small aperture terminals (VSATs). VSAT systems are in a state of rapid evolution and there is an extensive literature. Comprehensive accounts of VSAT systems are to be found, for example, in an ITU handbook [1] and Everett [2].

Almost all VSATs in current use function as the outstations of star networks served by a central hub station. The outstations have satellite links with the hub station only; if an outstation needs to communicate with another outstation, the hub station connects the two star links in tandem; see Figure 13.1a. The hubbed star configuration is used for the technical and economic reasons given in Section 12.3.1, which apply with even greater force to VSAT networks. But the hubbed star is also functionally convenient for most VSAT applications, since the outstations usually need to communicate, not with one another, but with a single central point, typically a source of information, a collecting point for information or an administrative office with direct communication links with the hub station. Mesh networks (see Figure 13.1b), which would enable a VSAT to communicate with another VSAT without the involvement of a hub station, are preferable for some applications but major technical problems arise and they are little used at present.

Hubbed star networks are made up of unidirectional or bidirectional links. Most unidirectional networks provide point to multi-point communication facilities from a central point via the hub station to outstations controlled by that central point or to other parties that buy the information, entertainment programmes, or whatever other material the central source provides. In bidirectional star networks, usually called 'interactive networks', there is two-way communication between the hub and each outstation. Most interactive networks provide a low-capacity communication network between the various locations at which a single organization operates. Hubbed star networks are often large, some involving hundreds or thousands of outstations and some having nation-wide or international service areas.

VSAT antennas are typically 1.2 or 2.4 metres in diameter although some are as small as 0.6 metres or as large as 3.0 metres. With such small antennas and with links of relatively small information capacity, the cost of the hardware and the cost of installing it is low. This provides savings that more than compensate for high space segment charges if the number of outstations is big enough, enabling VSAT networks to compete with terrestrial communications media for a number of niche markets. In addition to this commercial competitiveness, VSAT networks are often preferred because they avoid the unreliability and delays in circuit provision of many terrestrial networks. Most VSATs are located at the

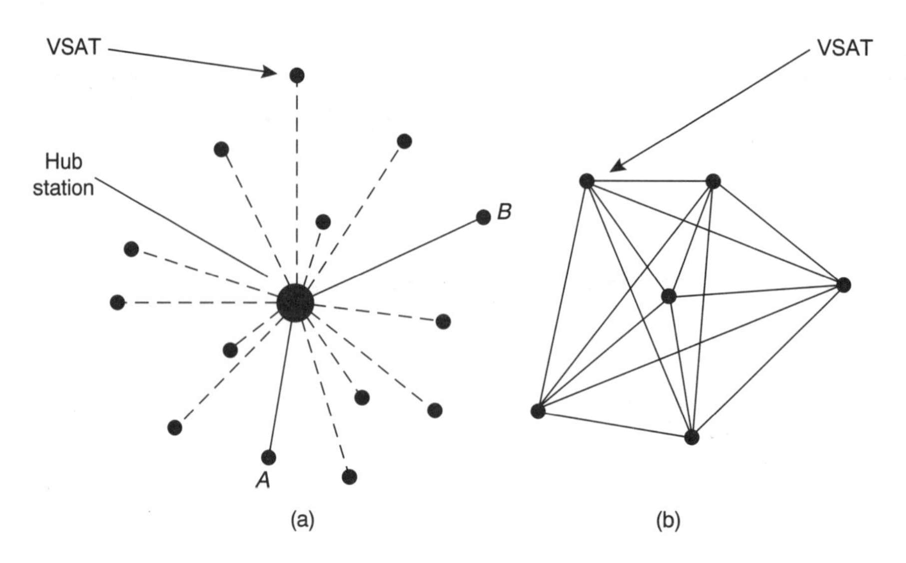

Figure 13.1 *Links in VSAT networks: (a) star and (b) mesh*

user's premises, the larger antennas often being mounted on the roof of an office building and the smaller ones being attached to an outside wall.

It is necessary for both technical and economic reasons for the hub station to have an antenna that is much larger than an outstation antenna, usually 6 to 9 metres in diameter. Many hub stations are operated, not by the organizations using the networks, but by a service provider, perhaps a teleport operator. One hub station antenna may serve several separate VSAT networks, all using the same satellite.

Frequency allocations used for VSAT networks

VSAT systems may use the same FSS satellites as are used for the international and domestic systems described in Chapters 11 and 12 and are free to use the same frequency allocations. However, many outstations are located in city centres, mounted high off the ground on tall buildings Consequently frequency coordination with the terrestrial radio relay stations that share most of this spectrum in most countries is particularly difficult to achieve. Furthermore, with a low G/T ratio for the earth station, the limit on downlink power imposed to protect radio relay receivers (see Section 6.4.1) is a major constraint on VSAT system design. These two factors may influence the choice of operating frequencies for VSAT networks.

Satellites set up under the FSS Allotment Plan are limited to the frequency bands listed in Section 12.4.1. For other satellites, C band frequency assignments are mostly between 5925 and 6425 MHz for uplinks and between 3700 and 4200 MHz for downlinks. For Ku band, 14.0–14.5 GHz is available for uplinks and 10.95–11.2 GHz, 11.45–11.7 GHz, 11.7–12.2 GHz and 12.5–12.75 GHz are available for downlinks, depending on the country in which the outstation is located. All of these bands are shared, to some extent, with terrestrial services but the bands least affected and therefore greatly to be preferred for VSATs are 14.0–14.5 GHz and 11.7–12.2 GHz in the Americas and 14.0–14.25 GHz and 12.5–12.75 GHz elsewhere.

There is another pair of frequency bands, allocated world-wide for the FSS and free from sharing with terrestrial services in many countries, namely 29.5–30.0 GHz (uplinks) and 19.7–20.2 GHz (downlinks). Congestion of FSS frequency allocations at lower frequencies may lead to the use of these bands for VSAT networks using geostationary satellites within a few years.

Finally, a WRC in 1995 allocated 28.7–29.1 GHz (uplinks) and 18.9–19.3 GHz (downlinks) for LEO satellites of the FSS, having use for VSAT systems in mind. The constraint on non-geostationary satellite systems using frequency bands allocated to the FSS which is applied in general by paragraph 2613 of the ITU *Radio Regulations* (see Section 6.2.1) no longer applies to these bands.

The development of the VSAT medium

VSAT networks were coming into use in the USA around 1984. Growth of the medium was stimulated by difficulties in leasing private channels in the terrestrial network that followed the breakup of the American Telephone and Telegraph Company, by the high rentals being charged for channels in the terrestrial network and by a liberalization of the radio station licensing policy of the Federal Communications Commission. By 1987 the number of VSATs in service, excluding those used for receiving television, was approaching 30 000, mostly operating in Ku band although with substantial numbers in C band. Most of these early VSATs were in unidirectional star networks, although interactive networks were coming into use.

The VSAT population in the USA continues to grow, particularly for interactive networks. The use of VSATs also started early in several other geographically extensive countries, most notably in Canada, Australia and India. Furthermore, Intelsat made Intelnet facilities available for VSATs world-wide, subject to the approval of the national administrations concerned.

VSATs in Europe

Substantial numbers of unidirectional star networks were set up in Europe in the second half of the 1980s for limited-access television and data broadcasts but few interactive networks were in service by the end of that decade. It may be that the distribution of population and of commercial activity in Europe differs significantly from that of parts of the world where the use of VSATs grew rapidly. Perhaps potential VSAT users found the facilities provided by the European terrestrial networks relatively more acceptable. But other factors undoubtedly arose.

Perhaps the most important of these other factors was the structure of the telecommunications service industry in Europe. In the 1980s, with few exceptions, the European PTOs were publicly owned and had national monopolies of the supply of telecommunications equipment and services to users. The PTOs were the national signatories of the international space segment consortia, in particular Intelsat and Eutelsat, and as such could control access by other parties to the satellite facilities that were then available. The PTOs also had close organizational ties with the ministries that determined policy in the telecommunications field and controlled the use of radio frequencies. Most PTOs had little enthusiasm for providing facilities by satellite when broadly equivalent facilities could be provided by their existing terrestrial networks.

Secondly, the market for VSAT services in Europe is, to a considerable degree, not national but international, and the thirty or forty administrations that licensed the use of radio in that international market place had no coordinated policy towards the use of VSATs. That situation placed obstacles in the way of a would-be wide area VSAT user.

Nevertheless, it was apparent that VSAT networks had characteristics of potential value to business. Some specialized telecommunications facilities could be set up more quickly and more cheaply by VSAT than by terrestrial media. VSAT networks share the ability of all satellite networks to bridge areas of inadequate terrestrial infrastructure and the barriers arising from national boundaries. Moreover, these networks could in principle be set up entirely independently of a PTO monopoly, giving governments, if they wished, a means of providing a measure of competition for the PTO without seriously disrupting what remained a basically monopolistic system.

In 1984 the British government had transferred British Telecom, then the sole UK PTO, to the private sector and had licensed Mercury Communications Limited as a national competitor. In 1988–89 seven other companies were licensed to operate unidirectional VSAT networks, initially serving outstations in the United Kingdom, but subsequently extended to outstations elsewhere in Europe, subject to the approval of the foreign administrations concerned.

Within the European Common Market, soon to become the European Union, the Commission of the European Communities (CEC) sought, as a long-term objective, to liberalize the supply of telecommunications equipment and services and to break down national barriers within the European Union. A CEC Green Paper [3] published in 1987 proposed, as an interim measure, a number of changes of the practices then current, including the following;

(1) The regulatory and operational functions of telecommunications administrations should be separated.
(2) The PTO might continue to operate a number of basic services, and especially the PSTN, as a monopoly where this was considered necessary to maintain an efficient infrastructure but provision of other services should be liberalized.
(3) There should be no constraints on the supply and operation of receive-only earth stations, apart from type approval.

The document favoured liberalization of the supply and operation of transmit-and-receive earth stations where this would not do serious damage to the viability of the PTO but recognized that licensing would need to be considered on a case-by-case basis.

The Green Paper was followed by Directives 88/301/EEC and 90/388/EEC, confirming a liberal regime for earth stations. A second Green Paper [4] sought, among other objectives, free access to space segment capacity and measures to facilitate the provision of Europe-wide satellite services, such as the mutual recognition of licensing and type approval procedures and cooperation in frequency coordination.

To facilitate mutual acceptance of type approval procedures, standards for Ku band VSAT outstations and their network control systems have been drawn up by the European Telecommunications Standards Institute (ETSI) [5]. These standards form, in effect, a common type approval specification for the countries that have approved them. The standards relate in large part to the aspects of equipment design that determine fitness for use, such as the antenna's resistance to wind, protection from lightning, electrical safety, local oscillator stability, potential accuracy of antenna pointing and polarization alignment and immunity from electromagnetic interference. Where major aspects of the performance of the VSAT in the systems environment are included, requirements and recommendations are well aligned with those of the ITU. The signal handling equipment and such basic radio parameters as antenna size and LNA noise temperature are left to be determined by the system designer. Standards for C band VSATs are in preparation.

A further important step to facilitate the establishment of international networks was the agreement in 1993 by the radio regulatory authorities of

Holland, France, Germany and the United Kingdom to a Memorandum of Understanding, whereby an operating licence issued by one state will be accepted by the others.

These changes, sought by the CEC, have not been implemented vigorously by all countries of the EU so far, but a good deal of progress has been made. By the mid-1990s, progress towards these objectives and the new traffic demands that had arisen from the major political and economic changes in eastern Europe since 1990 had led to substantial growth in the number of VSAT networks. While Europe is not yet a user of VSATs on the same scale as the USA, over 8000 interactive VSATs and nearly 40 000 receive-only VSATs were in use by mid-1995 and growth in numbers is accelerating.

13.2 VSATs in star networks

13.2.1 Applications for VSAT star networks

Unidirectional network applications

Typical uses for VSATs in unidirectional star networks are illustrated by the following examples:

(1) The distribution of television signals or multiple sound programmes, not to the public in general but to cable networks, hotels, schools and defined groups of recipients. There is no clear technical distinction between VSATs used in this way and the earth stations with larger antennas considered in Section 12.3.3 which serve the same purpose where higher signal quality is required or the available downlink power is less. Digital transmission is usual, typically between 2 and 8 Mbit/s per channel for video signals and perhaps 2 Mbit/s for several sound programmes, plus FEC. An earth station used only to receive television signals is called a television receive-only (TVRO) station.

(2) The distribution of digital signals containing data, typically in the form of news material, market reports and fascimile to subscribers, including for example newspaper and broadcasting news editors and share and commodity dealers. The information rate would typically be in the range 64 to 300 kbit/s. This data stream might be subdivided by TDM to enable selected material to be addressed to specific recipients. Alternatively the data aggregate might be used to update a database at each outstation, the users accessing the database as required. As before, there are close similarities with systems using larger antennas; see Section 12.3.4.

Interactive network applications

An interactive VSAT network might find application in almost any situation where virtually instantaneous digital communication of high reliability but limited information content has substantial value. Whether an application is economically viable will depend, to a large extent, on the degree to which the same needs can be met by the terrestrial media and at what cost. Applications that have been found of potential value fall into two broad categories:

(1) Networks in which a large number of users, perhaps numbering thousands, need access to a central location to supply or obtain information or to deliver or respond to instructions. Usually there is a relatively large aggregate flow

of data outbound, from the hub to outstation, and messages are brief in the inbound direction, usually requiring the transfer of no more than a few hundreds of information bits per contact. There are, however, applications where the reverse occurs, information flowing inbound from the outstations, with the central office doing little more than collect the data and control the traffic flow.

(2) Private communication networks, typically a scaled-down version of the corporate satellite networks discussed in Section 12.3.4.

13.2.2 Hubbed star VSAT systems

A typical interactive VSAT star network resembles a local area network (LAN). A number, often a large number, of data terminals and computers, staffed or automatic, are enabled to cooperate by means of communication links. But whereas the work stations of a typical LAN are a few metres or a few kilometres apart, those of a VSAT network may be thousands of kilometres apart. The data handling equipment and its software, which may be complex, is outside the scope of this book (but see, for example, [2]).

An interactive star network typically involves the following elements:

(1) A hub earth station with an antenna diameter of 6 to 9 metres, located at the central node of the network. It may be at the central office of the network user but it is often located elsewhere, in good communications contact with the central office but operated by a service provider.
(2) Outstations – there may be tens or hundreds of them, sometimes thousands, all participating in the same network. Ku band VSATs with antennas 1.2 metres in diameter and C band VSATs with 2.4 metres antennas can be taken as illustrative of current practice. Such antennas would have G/T values of about 16 and 17 dB K^{-1} respectively. Their half-power beamwidths θ_0 would be about 2° at C band and 1.3° at Ku band and the gain would be 1 dB below G_0 at about 0.6° and 0.4° off-axis respectively. Satellites with low orbital inclination are needed to avoid a requirement for satellite tracking at the earth station.
(3) An outbound emission, transmitted from the hub station and received at all of the outstations. The form that this emission takes will depend on the nature of the network. It will typically consist of a BPSK or QPSK emission, continuously radiated, carrying 64 to 300 kbit/s of information, usually multiplexed by TDM and protected by FEC. Sometimes spread spectrum modulation is used, to provide carrier energy dispersal.
(4) Inbound emissions from any of the outstations and received at the hub station. The nature of the user's business determines the form that these emissions take. For example, in some networks a carrier, radiated for a fraction of a second and only when necessary, modulated by BPSK or direct sequence spread spectrum modulation, carries a packet of information bits that form the outstation's contribution to each transaction. In other networks each outstation has continuous access to, or can demand access to, a time slot in a TDMA system. The information rate during these burst transmissions seldom exceeds 64 kbit/s. In a lightly loaded network, all of the outstations may be assigned the same carrier frequency, but if there are many outstations with a substantial aggregate amount of data to transmit, the stations may be divided between several groups, each group having its own assigned frequency.
(5) If inbound emissions use PSK modulation but have no permanently assigned time slots in a TDMA system, it may be necessary to use some mechanism to

prevent outstations interfering with one another. If spread spectrum modulation is used, the CDMA properties of the modulation system should prevent interference, although it may be necessary to prevent too many outstations from transmitting on the same frequency at the same time. In lightly loaded networks an ALOHA system might be adequate and economical or the hub station may poll the outstations periodically, inviting them to transmit in a time-limited slot if they have need to do so. With heavier traffic an ALOHA system might be used to enable an outstation to reserve a time slot in a TDMA system operating on a different frequency assignment.

(6) Finally a network supervisory system is required to maintain network integrity. See Section 13.2.3.

A undirectional star network is similar but, having no inbound emissions, no network integrity system is necessary.

Constraints on the outbound links

Consider a 300 kbit/s signal, protected by $\frac{1}{2}$-rate FEC, modulating a carrier with BPSK and occupying a bandwidth of 720 kHz. Let it be assumed that a VSAT sees the satellite at an angle of elevation of 30° and that half of the noise at the input to the demodulator arises in the downlink under clear-sky conditions. Assume also that a rain margin of 1 dB is required at C band in a region subject to heavy rain or 4 dB at Ku band in a temperate region to ensure that the target BER of 1×10^6 is achieved for almost all of the time. Then applying equation (10.11) to get an approximate estimate, the downlink e.i.r.p. required by a C band earth station with an antenna 2.4 metres in diameter will be about 20 dBW. The corresponding e.i.r.p. for a 1.2 metres Ku band antenna will be 34 dBW.

Applying the argument summarized in equations (12.1) and (12.2), it will be seen that a 2.4 metres C band antenna will not be able to obtain a C/N ratio better than about 12 dB without breaking the ITU downlink spectral PFD constraint. The corresponding figure for a 1.2 metres Ku band antenna is about 7 dB. Furthermore, while equations (12.1) and (12.2) are based on worst-case assumptions about beam geometry, the assumptions about carrier energy dispersal and radio propagation conditions are optimistic.

As with the larger earth station antennas considered in Chapter 12, a substantial easing of the downlink constraint can be obtained with more advantageous beam geometry, allowing higher C/N ratios to be obtained. In favourable locations, particularly in North America, the secondary status of the terrestrial radio services which share the band 11.7–12.2 GHz with the FSS may enable the VSAT network to avoid the downlink constraint completely. Link performance could be further improved for analogue FM emissions by means of low-threshold demodulators.

Nevertheless, if account is taken of the non-uniformity of the spectrum of the outbound emission that will be unavoidable in practice and of the probable need for a rain margin at Ku band, it may be concluded that it will not always be possible to obtain the required link performance standard when serving earth stations with such low G/T ratios. Spread spectrum modulation is sometimes used to avoid this problem.

Limitation of interference at the outstation is an important aspect of minimizing the downlink power. Guidance has been given on standards of antenna side-lobe suppression; this is noted below in connection with the inbound link. ITU-R has also recommended a standard of antenna cross-polar isolation. ITU-R Recommendation S.727 [6] calls for the discrimination of the antenna and its feed against waves orthogonal to the wanted polarization to be not less than

25 dB within the −0.3 dB contour of the main beam and not less than 20 dB elsewhere. ETS 300 157 [5a] sets a rather more stringent recommendation for VSATs used for data distribution; the standard recommended for TVRO VSATs by ETS 300 158 [5b] is similar to that of ITU-R. Both authorities warn that some systems may demand a tighter specification.

Like the larger antennas considered in Section 12.2.1, the 6 to 9 metres antennas at hub earth stations will not usually be constrained by the limit on off-axis radiation density, given an antenna designed to achieve good suppression of side-lobes. However, there may be exceptions; with a combination of adverse factors, spread spectrum modulation might have to be used to comply, not with the downlink, but with the uplink, constraint. These adverse factors include low outstation G/T, low satellite G/T and low transponder gain.

Constraints on the inbound links of an interactive network

With a relatively big antenna at the hub station and a correspondingly high G/T ratio, the downlink constraint on inbound links should raise no problems.

However, with a relatively powerful transmitter at the VSAT to compensate for the low on-axis gain of the antenna, the uplink constraint may be a matter for concern. For Ku band VSATs, ITU-R Recommendation S.728 [7] recommends maximum levels of e.i.r.p. within any band 40 kHz wide at an angle $\phi°$ from the axis of the antenna in any direction within 3° of the GSO. The limit takes the form

$$e = f(\phi) - 10 \log N \quad \text{(dBW)} \tag{13.1}$$

where $f(\phi)$ is 6 dB less than Curve E in Figure 6.8 and N is the maximum number of VSATs that are expected to transmit simultaneously in the same 40 kHz band. The factor $(10 \log N)$ has been included to take account of systems using CDMA on the inbound link. Pending the agreement of a recommendation for C band VSATs, ITU-R Recommendation S.524 applies; see Section 6.5, item 5.

ITU-R has recommended a design objective for the side-lobe levels of earth station antennas; see Section 6.5, item 7, and equation (6.5). This recommendation applies to antennas installed after 1995 with a diameter larger than 50λ that is, for diameters exceeding 3.75 metres at C band and 1.25 metres at Ku band. Thus, many VSAT antennas fall outside this recommendation. Agreement has yet to be reached on global standards for smaller antennas. However, standards ETS 300 157/158 [5a, 5b] recommend that equation (6.5) should be achieved for off-axis angles between 2.8° and 7°, with a somewhat relaxed specification outside 7°.

The transmitter power required at an outstation depends on many factors. Two key objective are to keep the transmitter power low enough to permit a solid state HPA to be used and to keep the off-axis e.i.r.p. within the uplink constraint.

As for unidirectional VSAT networks, there is concern to ensure that outstation radio equipment is capable of operating efficiently in a crowded radio spectrum, not causing unacceptable interference to other radio systems and resisting interference from other systems to the extent possible. So, in addition to recommending off-axis e.i.r.p. density limits (see above), ITU-R Recommendation S.728 [7] also recommends, for Ku band VSATs, that the off-axis e.i.r.p. of components orthogonal to the wanted polarization for values ϕ of between 2.5° and 7° should not exceed

$$23 - 25 \log \phi \quad \text{(dBW in any 40 kHz band)} \tag{13.2}$$

nor should the e.i.r.p. exceed 2 dBW for values of ϕ between 7° and 9.2°.

Finally, ITU-R Recommendation S.726 [8] lays down limits for spurious emissions (that is, energy that is radiated outside the bandwidth occupied by the intended emissions, serving no useful purpose). Within the band, allocated for the FSS, in which the VSAT transmits, the on-axis e.i.r.p. should not exceed 4 dBW in any 100 kHz band. Much more stringent limits apply to spurious emissions falling into other frequency bands.

13.2.3 Interactive network integrity

All advanced communication systems have monitoring and alarm facilities that enable their satisfactory operation to be verified and draw attention to failures. Some monitoring facilities automatically restore satisfactory operation in the event of failure by replacing a failed unit with a serviceable spare one. However, the particular nature of interactive VSAT networks exposes them additionally to a different and serious failure mode.

Outstations, if staffed at all, are not staffed by personnel with telecommunications skills. If, owing to mis-operation of the terminal or a control system failure, an outstation transmits an inbound emission in a way that the system design does not allow for, the functioning of the whole network might be interrupted. With the network and also, perhaps, the faulty outstation, disabled in this way, it might be impossible to communicate quickly with the outstation to ask for the inbound emission to be switched off. Furthermore, if a VSAT antenna suffered storm damage, its emissions could cause serious interference to systems using the orthogonal polarization of the same satellite or to the systems of another satellite if the emissions could not be inhibited. For a network serving a large number of outstations, such events might not be rare. Means for controlling the network to prevent these malfunctions must be incorporated in the design of the network. See ITU-R Recommendation S.729 [9].

The form which the control subsystem takes must depend to a considerable extent on the nature of the communication subsystem but the minimum set of facilities required by the ETSI standard [5d, 5e] may be taken as an example. This control subsystem has three main elements. There is a centralized control and monitoring function at the hub station, a monitoring and control unit at each outstation and a processor to processor communication system to pass supervisory messages over the network between the hub and the outstation, each outstation being individually addressable.

The ETSI outstation monitoring and control unit

Monitoring facilities are built into the outstation equipment to enable frequent, automatic checks to be made to verify that the outstation's communications equipment is functioning correctly. This outstation monitoring equipment also verifies that the network supervisory system, including remote control of the outstation transmitter from the hub station, is also functioning correctly. The following checks are made:

(1) that the processors at the outstation are functioning correctly;
(2) that the outstation receiver can lock on to the carrier received from the hub station, and demodulate and decode signals from the centralized control and monitoring equipment;
(3) that the frequency generation subsystem of the outstation transmitter is functioning correctly;

(4) that supervisory signals received at the outstation from the centralized control equipment contain the outstation's own identity code and identify the channel by means of which the hub station will send control signals to enable or disable transmission from the outstation;
(5) that the signal in that control channel enables transmission to take place; and
(6) that, while the outstation is transmitting, a signal is being received from the centralized control equipment confirming that the outcome of checks (7) and (8) which are being made at the hub station (see below) is satisfactory: these signals 'validate' the outstation emission.

The ETSI hub station centralized control and monitoring function

The centralized control equipment at the hub station can identify the outstation which is the source of the emission it is examining at any instant and
(7) verify that the carrier level lies within acceptable limits;
(8) determine whether the carrier frequency is within tolerance;
(9) display the results of the tests for the information of the network control staff at the hub station.

The hub station staff have the means to enter a signal which disables transmissions from any outstation or group of outstations via the control channel referred to in checks (5) and (6). The requirement for staff to intervene in a basically automatic process should ensure that a VSAT is not locked out of the network without a good reason.

The logic of control in the ETSI system

(a) A satisfactory result for checks (2), (4) and (6) may be taken as confirmation that the outstation's antenna is pointing with sufficient accuracy at the right satellite.
(b) The results from checks (1) to (4) reveal the operational status of the outstation; this status result is passed periodically to the hub station while the outstation is transmitting.
(c) A satisfactory result for checks (1) to (5) permits the outstation to begin transmitting.
(d) If any check produces an unsatisfactory result, the outstation transmission is suppressed by the local monitoring and control unit or by a disabling signal from the hub station. The outstation transmitter does not then recommence transmitting until checks (1) to (4) are all once more giving satisfactory results and the hub station staff has supplied the enable signal.

13.3 VSAT developments in prospect

Refinement of VSAT system design to enable future networks to meet users' requirements more effectively and at less cost are going forward by many routes but two more fundamental development objectives are also being pursued. If the growth of systems of the FSS in general and VSAT systems in particular is not to be impeded, other frequency bands, probably in Ka band, may have to be brought into use. Secondly there is interest in VSAT networks which are capable of carrying much more information, preferably in mesh networks.

Use of Ka band

The frequency bands 29.5–30.0 GHz (uplinks) and 19.7–20.2 GHz (downlinks) are available for VSAT networks using geostationary satellites and without downlink PFD constraints. The bands 28.7–29.1 GHz (uplinks) and 18.9–19.3 GHz (downlinks) are available for VSAT networks using LEO satellites. Prototype Ka band VSAT hardware has been produced in several countries and subjected to extended trials with experimental satellites such as Olympus, Italsat, Kopernikus and ACTS. Good results are being reported; see, for example, [10].

The chief remaining problem area for Ka band FSS systems lies with space–Earth propagation. The transmission loss due to rain is much higher at Ka band than at Ku band. It may be supposed that Ka band will not be usable with geostationary satellites in climates where there is heavy rain or protracted moderate rain. Elsewhere, large rain margins might provide a solution in favourable locations. Various other methods for dealing with rain loss are for consideration, including:

(1) use of earth stations in diversity;
(2) a reduction of the rate of information flow in the link while heavy rain is falling, an improvement in performance being obtained by a reduction of the pre-demodulator bandwidth or by the introduction of more powerful FEC;
(3) use of adaptive uplink power control to maintain performance during heavy rain in the uplink;
(4) use of LEO satellites in constellations containing a sufficiently large number of satellites to ensure that there will always be at least one of them in sight of an earth station, despite environmental obstructions (the smaller path loss relative to that for a geostationary satellite might compensate for a larger rain margin as well as lower antenna gains);
(5) use for kinds of traffic that can tolerate unavailability of communication for a significant percentage of the time.

These methods may not be broadly applicable to typical current VSAT networks, but there will be niches in the market where they could be used. Some of the innovations mentioned below, as solutions to the need for higher information capacity and more readily practicable mesh networks, might also enable larger rain margins to be provided at acceptable cost.

High VSAT network capacity and mesh connectivity

Many of the limitations on VSAT networks, including those operating in Ka band, would be eliminated if the gain of the antennas of geostationary satellites were made high enough or if LEO satellites were used. But high-gain satellite antennas have small footprints and LEO space segments seem likely to be costly. The needs of some VSAT users could be met within a small footprint but the larger problem is to combine the advantages of high-gain beams and large coverage areas.

The possible solutions to this problem include the use of satellites with multiple high-gain beams (or perhaps beams, generated by phased arrays, that hop or scan over the service area) and on-board signal processing, enabling signals received at the satellite via one beam to be retransmitted to Earth via whichever of the available transmitting beams serves the intended receiving earth station. Various possible ways of implementing this are reviewed briefly in Section 9.4.6. It is, however, unclear whether systems operating on such a manner could serve economically the VSAT medium that has developed as a low-cost option for small information streams. Such systems seem more likely to find application for major traffic loads, such as those considered in Sections 12.3.1 and 12.3.4.

References

1. *VSAT systems and earth stations.* Supplement No. 3 to the *Handbook on Satellite Communications* 1994. Geneva: ITU.
2. Everett, J. (ed.) (1992). *VSATs: Very Small Aperture Terminals.* Peter Peregrinus.
3. *Towards a dynamic European economy* (Green paper on the development of the common market for telecommunications services and equipment, Document COM(87)290, 1987). Brussels: Commission of the European Communities.
4. *Towards Europe-wide systems and services* (Green paper on a common approach in the field of satellite communications in the European Community, Document COM(90)490, 1990). Brussels: Commission of the European Communities.
5. ETSI has produced the following standards for VSATs operating in the 14/12/11 GHz bands:
   (a) ETS 300 157 – *Satellite Earth Stations: Receive-only VSATs used for data distribution* (August 1992).
   (b) ETS 300 158 – *Satellite Earth Stations Television Receive-only (TVRO) satellite earth stations* (August 1992).
   (c) ETS 300 159 – *Satellite Earth Stations: Transmit/receive Very Small Aperture Terminals (VSATs) used for data communications operating in the Fixed Satellite Service (FSS) 11/12/14 GHz frequency bands* (December 1992).
   (d) ETS 300 160 – *Satellite Earth Stations: Control and monitoring functions at a VSAT* (August 1992).
   (e) ETS 300 161 – *Satellite Earth Stations: Centralised control and monitoring functions for VSAT networks* (November 1992).
6. *Cross-polarization isolation from very small aperture terminals (VSATs)* (ITU-R Recommendation S.727, 1994), ITU-R Recommendations 1994 S Series. Geneva: ITU.
7. *Maximum permissible level of off-axis e.i.r.p. density from very small aperture terminals (VSATs)* (ITU-R Recommendation S.728, 1994), ITU-R Recommendations 1994 S Series. Geneva: ITU.
8. *Maximum permissible level of spurious emissions from very small aperture terminals (VSAT)* (ITU-R Recommendation S.726-1, 1994), ITU-R Recommendations 1994 S Series. Geneva: ITU.
9. *Control and monitoring function of very small aperture terminals (VSATs)* (ITU-R Recommendation S.729, 1994), ITU-R Recommendations 1994 S Series. Geneva: ITU.
10. Ananasso, F. (1995). 20/30 GHz satellite communications: the technology is mature. *Proc. Tenth International Conference on Digital Satellite Communications*, Brighton, UK. London: IEE.

14 Satellite communication for mobile earth stations

14.1 The satellite–mobile services

Commercial communication by satellite with ships began in 1976. Service for road vehicles has been operating at a modest level since the late 1980s, but a stage of rapid development began in the mid-1990s and service is being extended to include hand-portable mobile terminals in the near future. Service to aircraft is growing fast in the 1990s from a beginning in 1992. So far all the satellites used are geostationary.

With few exceptions these systems connect a calling user at a mobile earth station (MES) to an earth station at a fixed location on land, almost invariably on demand, typically for a few minutes or a few seconds, and the land earth station extends the connection via the PSTN to the called user, or vice versa. When the call has been completed, the channel through the satellite network is made free for another call, usually involving a different mobile earth station and often a different land earth station. Mobile-to-mobile calls are usually made by two hops via a land earth station.

In the accepted terminology, an earth station at a fixed location on land providing a gateway between the satellite network and the PSTN is called a shore earth station if it serves ships, an aeronautical earth station if it serves aircraft and a base earth station if it serves land-based mobile stations. Such distinctions are, however, not always clear-cut. If a satellite system offers service to more than one kind of mobile station, the associated land earth stations will probably do the same. This trend is growing, since most frequency bands recently allocated for the mobile–satellite services may be used for any kind of mobile station. To avoid confusion the term land earth station (LES) is used in this chapter for all three applications. The mobile earth stations are called ship earth stations (SES), aircraft earth stations (AES) or land mobile earth stations (LMES) as appropriate.

All links between satellites and mobile stations are part of the mobile–satellite service (MSS) but subcategories are also used. Thus the maritime mobile–satellite service (MMSS) covers links with an SES, the aeronautical mobile–satellite service (AMSS) covers links with an AES and the land mobile–satellite service (LMSS) covers links with an LMES. The links between a satellite and an LES are feeder links; they are technically similar to uplinks and downlinks in the FSS, they are formally a part of the FSS, and they share frequency bands with fixed satellite systems.

The internationally agreed definition of a mobile earth station [1] covers stations that are operated only when the station is halted at an unspecified place, as well as those that can be operated when in motion. This provision has little effect where ships or aircraft are concerned but it leads to two quite different kinds of LMES, those that can be operated on the move and transportable stations that can only be operated at rest.

Frequency allocations for the mobile–satellite services

The following combination of characteristics distinguishes the MSS from other satellite services:

(1) Almost all mobile applications involve very small traffic flows per mobile station, so low MES capital costs are essential for commercial viability.
(2) If an MES antenna has substantial gain, it must track the satellite to compensate for the movement of the vehicle as well as for any apparent motion of the satellite. This is feasible for an SES on a large ship but virtually impossible on other kinds of vehicle. Consequently, most MES antennas have low or negligible gain.
(3) The primary power supply for an MES, and therefore the available uplink transmitter power, is usually quite limited, and especially for hand-portable terminals. The need to avoid a health hazard to the user and to bystanders may place a further constraint on transmitter power levels.

These characteristics make high transmission loss between MESs and satellites a severe problem. Equation (5.3) shows that the transmission loss rises with the square of the operating frequency and the square of the path length. Accordingly, it is desirable for MSS systems to operate at relatively low frequencies, although not low enough to encounter the very high levels of galactic and man-made noise which arise at VHF. And the reduction of path length, relative to the GSO, that LEO satellites provide emerges as a significant advantage for systems using non-directional earth station antennas.

The frequency allocations in current commercial use are in L band at 1525–1559 MHz (downlinks) and 1626.5–1660.5 MHz (uplinks). Some allocations differentiate between the MMSS, the AMSS and the LMSS and they are somewhat complex in detail, but the general situation is shown in Figure 14.1. Except for some for the LMSS, the allocations all have primary status and there are no mandatory sharing constraints such as the downlink PFD limits that constrain the FSS in most of its allocations. There are allocations for other services within these frequency bands in some countries but most of these sharing allocations are of secondary status.

In addition to the L band allocations, there is an LMSS allocation at 14.0–14.5 GHz, secondary to the FSS allocation. It is used, for example, for SNG. Some administrations permit the use of frequency assignments in the 5925–6425 MHz band for the same purpose. Finally there are MSS allocations at 29.5–30 GHz (uplinks) and 19.7–20.2 GHz (downlinks). Commercial operation in Ka band would involve new technical problems; these are being studied, in particular under the NASA ACTS project. These allocations may provide a feasible opportunity for growth if the allocations below 3 GHz prove to be insufficient to meet the demand for commercial systems.

The new allocations

The L band allocations are becoming heavily loaded with systems using geostationary satellites. In addition, new MSS systems using non-geostationary orbits are in early prospect, and geostationary satellites and satellites in other orbits cannot share spectrum efficiently. Thus, other allocations are needed, including some reserved for systems using non-geostationary orbits.

New frequency bands below 3 GHz were allocated for satellite–mobile services at a WARC in 1992 and a WRC in 1995. However, it was very difficult to reach agreement on new allocations in this particularly crowded part of the spectrum. Consequently some of the new allocations are not globally applicable and some have secondary status. A third group do not come into effect for a number of

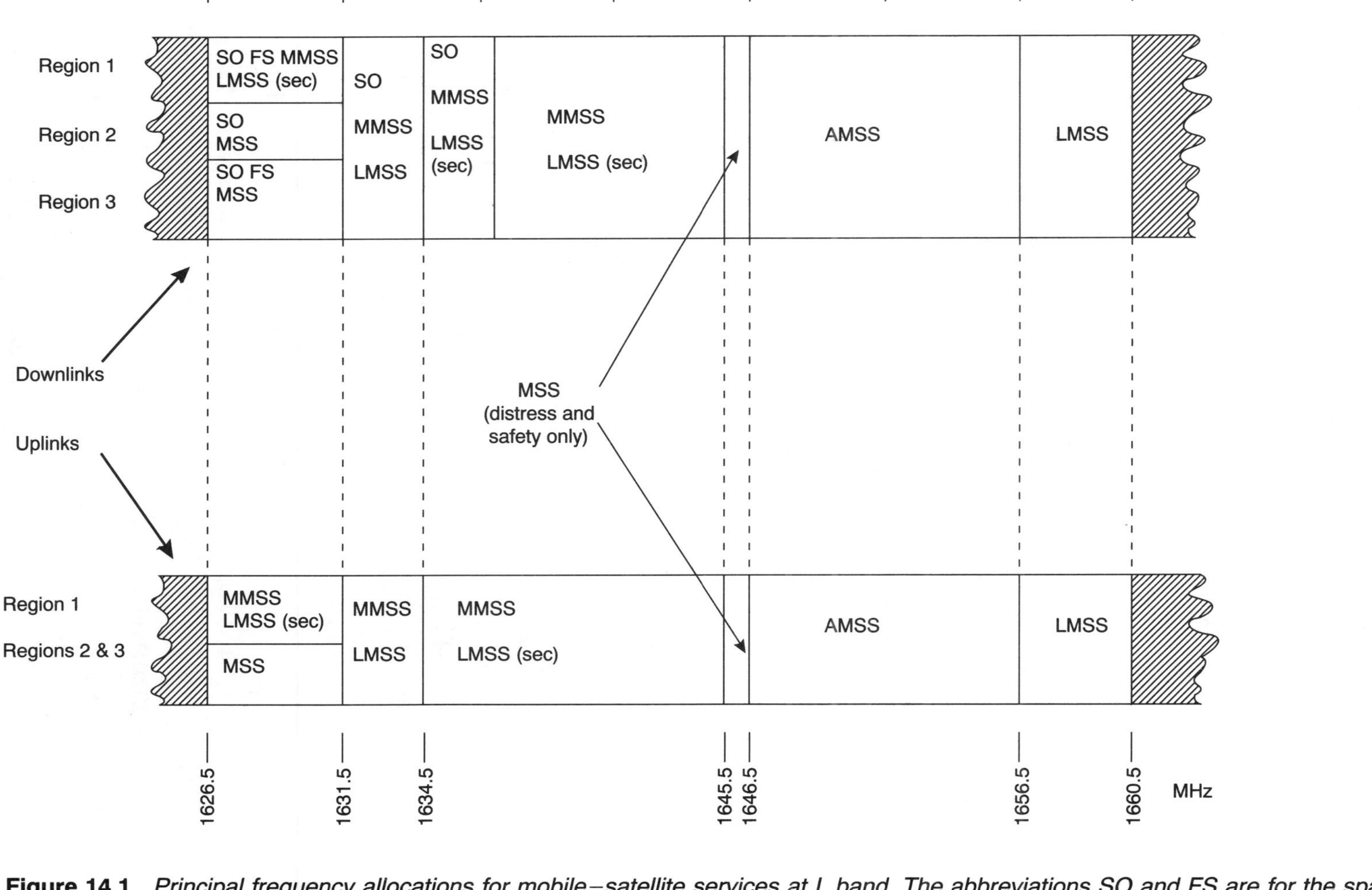

Figure 14.1 *Principal frequency allocations for mobile–satellite services at L band. The abbreviations SO and FS are for the space operation service and the (terrestrial) fixed service respectively*

years, to give time for stations already operating in these bands to be accommodated elsewhere. With minor simplifications, these allocations are shown in Table 14.1. Virtually all of these new allocations were made to the MSS, not to specific service subcategories.

Table 14.1. *New MSS frequency allocations made in 1992 and 1995*

Global allocations (MHz)		*Region 2 only* (MHz)	
Uplinks	*Downlinks*	*Uplinks*	*Downlinks*
148–150.05	137–138	455–456	459–460
399.9–400.05	400.15–401		
406–406.1	400.15–401		
312–315	387–390		
1610–1626.5	2483.5–2500	1675–1710	1492 1525
1980–2010*	2170–2200*	1930–1970	2120–2160
2670–2690*	2500–2520*	1970–1980*	2160–2170*

*These allocations take effect in 2000 or 2005.

The new allocations below 1 GHz are very narrow and geostationary satellites are specifically excluded from most of them; these are sometimes called the 'little LEO' bands. Decisions will be taken soon as to which, if any, of the new bands above 1 GHz should be used for systems using geostationary satellites, the rest being available for 'big LEO' systems and satellites in other non-geostationary orbits.

Feeder links

The feeder links between LESs and satellites of the various mobile–satellite services operate in frequency bands allocated to the FSS but paragraph 2613 of the ITU *Radio Regulations* (see Section 6.2.1) places heavy constraints on the use of FSS allocations by non-geostationary satellites. However, WRC-95 made several bands available for feeder links to non-geostationary satellites without those constraints; these bands are 5150–5250 MHz and 15.45–15.65 GHz (uplinks) and 15.4–15.7 GHz (downlinks). A second downlink band at 5150–5216 MHz is planned, to take effect in the year 2010 and meanwhile there is a temporary arrangement whereby frequency assignments can be made between 5091 and 5150 MHz.

Space–Earth radio propagation

There are brief notes on the main propagation phenomena that affect L band radio links with mobile earth stations in Chapter 5 and Section 10.3.2. A detailed treatment of the subject is to be found, for the MMSS, AMSS and LMSS respectively, in ITU-R Recommendations PN.680 [2], PN.682 [3] and PN.681 [4].

14.2 The geostationary space segment

14.2.1 The space segment operators

Two of the satellites in the NASA ATS series, launched in 1966 and 1967 respectively, were equipped for operation at frequencies around 140 MHz. These facilities were used successfully for experimental communication with ships and aircraft.

In the late 1960s, when the NASA Apollo programme had reached the stage of manned spaceflight, it was decided that terrestrial communication facilities were not sufficiently reliable for some of the links between the launching site at Cape Canaveral and the ships tracking spacecraft in mid-Atlantic in the immediate post-launch phase. Accordingly, temporary ship–shore circuits were set up by satellite using Intelsat C band transponders.

In 1971 frequency bands around 1.6 GHz were allocated for satellite communication with ships and aircraft. In 1976 Comsat had three Marisat satellites launched into the GSO and stationed at 15° W, 176.5° E and 73° E longitude. These satellites had a dual role; they provided space segment facilities that were leased to the US Navy for communication with naval ships and they provided transponders for Comsat itself to operate for communication with merchant ships, virtually world-wide.

The merchant ship element of the Marisat payload (see [5]) used the newly allocated L band frequencies for its MMSS links and C band for feeder links. Land earth stations at Southbury, Connecticut, at Santa Paula, California and at Fucino in Italy gave access to the satellites, linking ships at sea through the PSTN with subscribers ashore for telephone, telex, facsimile and data calls. The service was welcomed by merchant shipping and by 1982 around one thousand ships were equipped to use it.

However, the governments of many countries were not content for control of communication with their ships to rest with a foreign commercial corporation. In 1976, under the aegis of the Intergovernmental Maritime Consultative Organization, now called the International Maritime Organization (IMO), an agreement was drawn up for the establishment of an international organization to provide satellite services for ships.

This new organization, the International Maritime Satellite Organization (Inmarsat), with functions and a constitution broadly similar, within its field of activity, to those of Intelsat, was established in 1979 and started operational service in 1982. It is funded and controlled by PTOs authorized by governments to provide telecommunications services to mobile stations. Originally limited to maritime services, its role has since been extended to cover all mobile–satellite services. By 1995 over eighty nations had joined the organization. Inmarsat provides and manages the space segment, the national operating bodies own and operate LESs that provide gateways with the PSTN and the owners of ships, aircraft, vehicles, etc., wishing to use the system provide the MESs.

ICO Global Communications, a company affiliated with Inmarsat, was set up in 1995 to own and operate the space segment of a MEO satellite system providing service to radiotelephones, vehicle-borne and hand-portable. In February 1996 discussions began, developing proposals for developments in the structure of Inmarsat. The organization is likely to remain an intergovernmental organization, but with wider participation and a management and financial structure more closely resembling a commercial undertaking.

Several other organizations are setting up regional or national geostationary satellite systems operating MSS facilities at L band, alone or in combination with FSS facilities at C or Ku band. Volna and Marafon (Russia), MSAT (a joint project of the USA and Canada), Aussat (Australia) and Solidaridad (Mexico) are likely to be sharing the L band allocations with Inmarsat soon, mainly serving mobile stations on land. However, the Inmarsat system can be seen as the exemplar of commercial mobile–satellite system technology in the mid-1990s. The Inmarsat space segment and the associated LESs are described below. The facilities that geostationary satellites provide for ships, aircraft and land-based users, and the earth segment equipment involved are considered in Sections 14.3 to 14.5.2. The chapter closes with a necessarily preliminary look at the very different, non-geostationary, systems that are emerging to serve the LMSS towards the end of the century.

14.2.2 The initial Inmarsat satellite network

When Inmarsat opened service in 1982, taking over the role of the Marisat merchant shipping network, it leased for its space segment various space segment facilities that were already available or were becoming available. The L band transponders of the three Marisat satellites were leased from Comsat. Intelsat had arranged for L band transponders to be added to five Intelsat V satellites that were under construction while Inmarsat was coming into being; these satellites were launched between 1982 and 1984 and three of the 'maritime communication subsystems' were leased to Inmarsat. Two Marecs satellites, which had been designed and constructed as a development project for ESA, were launched in 1981 and 1984 respectively, becoming the third element of the INMARSAT space segment. Fortunately, the operating frequencies of all three elements were similar.

Together these leased space segment facilities enabled the new system to meet the growing needs of shipping for reliable communications facilities throughout the 1980s. By 1990 one or more LESs had been built in 25 countries to enable the PTOs to meet the needs of the public ashore and their shipping industries. However, by that time most of the satellites providing the space segment were approaching the end of their lives in orbit and could no longer serve adequately a demand that was growing fast. It was time for a new space segment, and Inmarsat bought and launched the Inmarsat 2 series of satellites.

14.2.3 The Inmarsat 2 satellite generation

Inmarsat contracted in 1985 with a consortium led by British Aerospace for the supply of four Inmarsat 2 spacecraft. All were successfully launched between 1990 and 1992. Stationed in the geostationary orbit at 64.5° E, 179.5° E, 55.5° W and 15.5° W longitude, they serve the Indian Ocean region, the Pacific Ocean region, the Atlantic Ocean West region and the Atlantic Ocean East region respectively. Together they provide virtually world-wide coverage between 70° N and 70° S latitude; see Figure 14.2. The first generation space segment has since been taken out of use.

The Inmarsat 2 spacecraft is described by Berlin [6]; see also Figure 4.6. The three-axis stabilized Eurostar bus design was used, similar to the bus used for Marecs and ECS. The solar arrays are designed to provide more than 1000 W of power after 10 years in orbit. The central objectives of the communications payload design are to deliver the maximum uniform aggregate multi-carrier downlink PFD to mobiles in all parts of the global service area and to provide flexibility for optimizing operation with a variety of kinds of mobile earth station. To simplify the account which follows, mobile earth stations are assumed to be maritime.

The forward link transponder (shore-to-ship)

For the forward links, a global coverage seven-element array of cup dipoles receives circularly polarized emissions from LESs in the band 6425–6443 MHz. After amplification ($G/T = -14\,\mathrm{dB\,K^{-1}}$) and frequency changing to 1530–1548 MHz, the signals are band-limited and passed to a remotely controllable variable attenuator. The main function of this attenuator is to compensate for any drift that may arise in the gain of the power amplifier stage. There

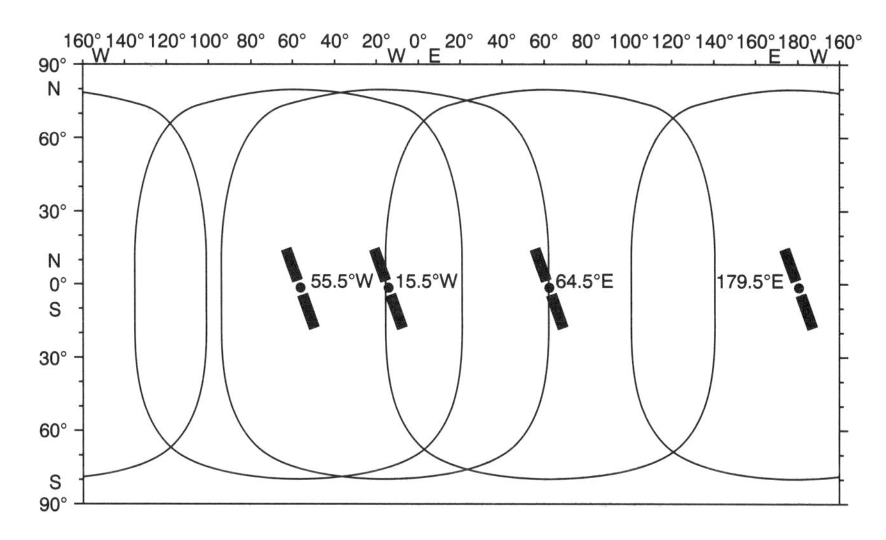

Figure 14.2 *The coverage areas of the Inmarsat 2 satellites, out to an angle of elevation at the Earth's surface of zero degrees*

follows an automatic level control stage to limit non-linear operation of the power amplifier under overdrive conditions. The HPA stage follows, linearized by predistortion, consisting of four travelling wave tubes in parallel. There is a total of six TWTs, including two spares. After a bandpass filter to eliminate out-of-band distortion products, the output of the HPA passes to the circularly polarized transmitting antenna.

The L band transmitting antenna is an array of 43 cup dipoles. This array generates a circularly polarized global beam. The beam is shaped to provide additional gain towards the edge of the Earth's disc to compensate for the increase in space–Earth path loss that arises when the satellite, as seen from an earth station, is at a low angle of elevation. The backed-off multi-carrier aggregate e.i.r.p. is 39 dBW.

The return link transponder (ship-to-shore)

For the return links, circularly polarized emissions from ships are received at the satellite in the band 1626.5–1649.5 MHz. The receiving antenna is a global coverage nine-element cup dipole array. After amplification ($G/T = -12.5\,\text{dB}\,\text{K}^{-1}$) the wideband signal aggregate is translated to an intermediate frequency of about 50 MHz. An input multiplexer made up of highly selective surface acoustic wave (SAW) filters splits the band into four frequency subchannels (see Figure 14.3). The levels of the signals in these four subchannels are separately adjustable by remotely controllable attenuators, enabling the effective gain of the transponder for each of these four subchannels to be optimized for various kinds of MES. The four subchannels are then recombined, translated to the band 3600–3623 MHz and the level of the aggregate is automatically controlled. A TWT HPA follows, linearized by predistortion. Out-of-band distortion products are removed by a bandpass filter, and the downlink signal is radiated with circular polarization and an e.i.r.p. of 24 dBW from a seven-element cup dipole antenna with global coverage.

One of the sub-bands into which the return-link transponder is split is 1626.5–1631.5 MHz. There is no forward-link frequency band corresponding with this sub-band, a consequence of an anomaly in the international frequency

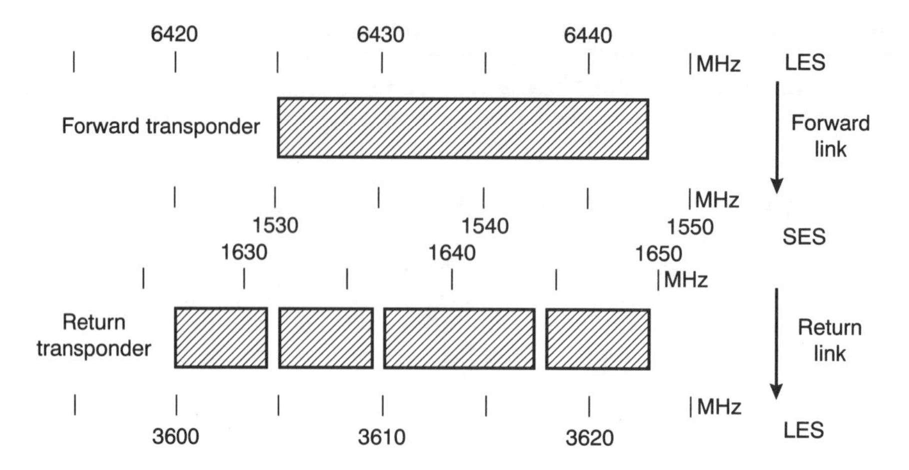

Figure 14.3 *The frequency plan of the Inmarsat 2 transponder*

allocation table at the time when the satellites were built. This sub-band is not used for two-way traffic. It might, however, be used for a wideband data channel to shore from a ship, or an off-shore platform.

An SCPC telephone emission for reception at an SES conforming to Inmarsat Standard A (see Section 14.3.1) requires an e.i.r.p. from the satellite of 15.6 dBW. Thus the greatest number of carriers simultaneously sustainable by the 39 dBW available from the forward transponder is approximately 200. With a frequency separation between carriers equal to 50 kHz, 200 emissions would require a total bandwidth of 10 MHz. Thus, with a total bandwidth of 18 MHz, the forward transponder is power-limited. However, since a single transponder serves the whole band from 1530 to 1548 MHz, the available power can be distributed as required within the downlink frequency band.

Feeder link earth stations

Inmarsat 2 needs to transmit approximately 200 SCPC emissions to the LESs but the aggregate feeder downlink e.i.r.p. is a modest 24 dBW. Consequently the LESs are required to have high performance, comparable with those used for the Intelsat Global Network.

A minimum G/T of 32 dB K^{-1} at 3.6 GHz was specified by Inmarsat for the antennas of LESs having access to the first generation of satellites. This requires an antenna about 10 metres in diameter. (For subsequent satellite generations the minimum LES G/T currently required is 30.7 dB K^{-1}.) However, the uplink e.i.r.p. required per analogue telephony channel addressed to a Standard A mobile terminal was 65 dBW. The gain of a 10 metre antenna at 6.45 GHz is about 54 dBi, indicating a transmitter output power requirement of about 12 W per channel. Thus, if an LES has to be capable of transmitting a substantial number of channels simultaneously, the power output rating required of the HPA would be very high. Thus there might be advantage in installing a larger antenna, perhaps 12 or 14 metres in diameter, in order to limit the required single-carrier saturation rating of the power amplifier to perhaps 3 kW.

Automatic frequency control

Carrier frequency errors, substantial in comparison with the 50 kHz separation between carriers, could arise in the system. To some extent these errors arise from oscillator drift at LESs and SESs, although the specified tolerances are stringent. The drift of local oscillator frequencies on board the satellite is larger. Doppler shifts due to movement of the satellite relative to earth stations are also significant. The situation has since been made more critical by the introduction of new SES standards using smaller carrier separations. It is essential that relative frequency errors be reduced to negligible proportions, in order to optimize the use of spectrum and to allow transmitter and receiver tuning at the SES to be made completely automatic.

Separate automatic frequency control (AFC) loops are used to counteract the major sources of frequency error in the forward and return directions. In order to provide for AFC (and also system monitoring) an LES is equipped to transmit to and receive from the satellite at L band as well as C band. These facilities can be obtained using a separate L band antenna or by equipping the main C band antenna with a dual purpose feed.

To correct errors arising in the forward direction, a nominated LES transmits a pilot carrier at f_u, an accurately known frequency in the feeder uplink band, bypassing the up-converter that it uses for traffic emissions. The frequency changer at the satellite subtracts a nominal f_{CL} from the frequency of the pilot carrier, as received, and the satellite retransmits it at L band. All LESs receive the pilot carrier after retransmission by the satellite, at frequency $f_u - f_{CL} + \delta f$, where δf is the sum of all the frequency errors arising in the earth station to earth station link. The LESs subtract δf from the accurately known local oscillator frequency of the up-converters used for their traffic emissions. This corrects the frequency, as perceived by SESs, of all carriers in the forward direction. If the satellite has substantial orbital inclination, it may be necessary to use several frequency control loops per satellite, each loop serving LESs in a limited latitude range. To compensate for frequency errors arising in the return direction, each LES transmits a pilot carrier at L band, receives it at C band and adjusts the local oscillator frequency of its C band down-convertor accordingly.

Frequency reuse with Inmarsat 2

All of the Inmarsat 2 satellite antennas have global beams and operate in a single-polarization mode. Thus there is no opportunity for frequency reuse within any one satellite.

Furthemore, whereas in the FSS it is always possible for geostationary satellites which are not close together in orbit to use the same frequency bands without mutual interference, this is not necessarily true of the MSS. The feeder link earth station antennas have high gain and a few degrees of orbital separation between satellites is sufficient to prevent significant interference between feeder links accessing different satellites. But the directional discrimination of the MES antennas is very limited at best, and is tending to get less as mobile antennas get smaller. Moreover, the footprints of satellite antenna beams overlap extensively. Consequently, interference between links from different MESs accessing the same frequency channel via different satellites is often severe and must usually be avoided.

14.2.4 The Inmarsat 3 satellite generation

Since the Inmarsat 2 generation of satellites was specified, maritime traffic has grown fast and new opportunities, but also new problems, have emerged for Inmarsat, thus:

(1) The demand for service to LMESs is expected to rise rapidly, the LMSS soon becoming the dominant commercial use of the MSS. There is also a trend towards the use of simpler, cheaper, maritime earth stations with lower antenna gain. Initially these applications might be satisfied with a quite limited communications capability but ultimately a comprehensive range of telecommunications facilities is likely to be demanded, increasing the need for space segment facilities.
(2) The International Civil Aviation Organization (ICAO) has recently decided that there should be substantial use of satellite systems for Air Traffic Control (ATC) communication. There is interest in using the same basic system to provide communication in flight between aircrews and airline operational management. Furthermore, the availability of ground–air radiotelephone links over large land areas, enabling airline passengers to get access to the PSTN while in flight, has stimulated interest in similar facilities by satellite when the aircraft is beyond the reach of terrestrial radio systems.
(3) ICAO also supports the use of a satellite navigation system such as the Global Positioning System (GPS). However, in the short term at least, there is a need for a satellite communication channel, operating in the frequency band used for satellite navigation (that is, about 1575 MHz), for the rapid dissemination to aircraft of information on the integrity of the navigation system.
(4) Early deployment is expected of a number of new national and regional MSS systems using geostationary satellites and the frequency allocations which Inmarsat is already using. These systems will compete with Inmarsat for traffic and it will be necessary for Inmarsat to coordinate its use of the spectrum and the GSO with the new systems. It must also be expected that new systems using non-geostationary satellites will compete for some part of Inmarsat's present and potential future LMSS market.

Competition adds an element of unpredictability to Inmarsat's market forecasts. Nevertheless the decade ahead seems likely to be one of strongly rising demand combined with new technical challenges. To meet this prospect, Inmarsat is procuring a third generation of satellites. A contract for Inmarsat 3 spacecraft was placed in 1991 with Martin Marietta, Matra Marconi Space being the payload subcontractor. Five spacecraft were ordered and the first entered service on 11 May 1996. The orbits will be geostationary, at approximately the same locations as are now used for Inmarsat 2; if the first four launches are all successful, the fifth spacecraft is intended for a longitude of 34° W.

The new satellites will differ from the Inmarsat 2 generation in various ways, the most important of which are as follows:

(1) At L band the transponders will cover all of the contiguous bands allocated for mobile–satellite services, that is, 1525–1559 MHz for forward links and 1626.5–1660.5 MHz for return links; see Figure 14.1. The feeder links are in the same part of the spectrum as those of Inmarsat 2 but the frequency bands are, of course, wider to match the bandwidth available at L band.
(2) There will be global coverage beams at L band, but in addition, up to six spot beams, transmitting and receiving, can be generated by complex feed arrays serving reflectors over two metres in diameter. The beam footprints are located so as to cover the areas on Earth where the most intense traffic flows originate.
(3) SAW filters will be used to divide the bandwidth of the single pair of transponders into about a dozen radio subchannels, each a few megahertz wide. There will be means for connecting each subchannel either to global beam antennas or to one of the spot beams. In order to reduce the complexities

of the C band/L band couplings that this flexible arrangement permits, the bandwidth available for feeder links is being increased through frequency reuse by means of orthogonal circular polarizations.
(4) The directivity of the spot beam antennas will enable a limited amount of frequency reuse to be employed within each satellite.
(5) There will be a separate transponder for the satellite navigation integrity channel, transmitting at 1575 MHz.
(6) The solar arrays will supply about 3 kW of primary power.

It is intended that mobile earth stations with low antenna gain should be assigned frequencies in radio channels served by spot beams, whereas those with relatively high gain antennas will usually be served through the global beams.

With more primary power available and extensive use of high gain antennas, the effective multi-carrier aggregate L band e.i.r.p. of Inmarsat 3 is expected to reach 48 dBW, that is, 9 dB more than Inmarsat 2. It will be feasible to assign selected beam/channel combinations to adjacent Inmarsat satellites so that the spectrum can be reused without mutual interference. Furthermore, the spot beams will provide Inmarsat with a powerful new means for coordinating its use of the spectrum with satellites of other systems with regional coverage operating in the same frequency bands.

14.3 Satellite service to ships

14.3.1 The Inmarsat Standard A system

Inmarsat has defined several sets of standards for the radio equipment of mobile earth stations and with each standard a system is also specified for the control of access on demand for channels through the system. Some of these standards were defined for maritime use and these are described in Section 14.3, although all of them are also applied, with minor changes where necessary, to aeronautical or land mobile applications.

When Inmarsat began operating it took over the Marisat specification for an SES, making only the changes that were necessary to allow operation in the framework of the new international system. SESs conforming to the Marisat specification became Inmarsat Standard A terminals and are still in widespread use. The prime function of a Standard A SES is to connect an authorized ship via the Inmarsat space segment and an LES to the PSTN for a telephone or telex call. Most other standard telecommunications facilities of medium or low information rate are also available.

Ships' above-decks equipment

The SES antenna is required to have a G/T not less than $-4\,\text{dB}\,\text{K}^{-1}$. Paraboloidal reflector antennas are used, protected by a radome, mounted high on the ship's superstructure in order that the direct line-of-sight view of the satellite from the antenna shall not be obstructed. In early designs the diameter of the main reflector was typically 1.2 metres. More recently, improved LNAs having become available, smaller main reflectors down to 0.9 metres diameter are being used, giving on-axis gains around 20 dBi. Larger antennas, up to about 2.2 metres in diameter, are sometimes used on ships, typically large cruise ships, that may require to transmit several SCPC channels simultaneously.

Compared with other mobile earth station antennas, the Standard A antenna has substantial directivity, although measurements show that there is considerable variation in side-lobe gain between samples; see, for example, in CCIR Report 922 [7]. To provide an internationally agreed basis for estimating interference levels, a reference radiation pattern has been defined as a function of main reflector diameter for these antennas, with a margin allowed for variations from case to case; see ITU-R Recommendation M.694 [8]. Figure 14.4 shows the pattern, calculated for a reflector one metre in diameter.

With such directivity it is necessary for the antenna to track the satellite; a tracking error of 2° could cause a significant loss of performance margin. The mean pointing direction for the antenna is usually computed from information on the ship's heading and position supplied by the master compass and an automatic navigation system. The dozen or so manufacturers that supply Standard A SESs employ various antenna mounting, stabilizing and tracking methods to compensate for the roll, pitch and yaw of the ship. A step-track system may be added to correct for minor stabilization errors and the apparent motion of the satellite, due mainly to orbital inclination. Gyroscopic stiffness may be given to the attitude of the antenna by a pair of flywheels, mounted orthogonally. Typically a three-axis mount is used, because a two-axis mount on a rolling ship would not maintain accurate tracking when the satellite was aligned with one of the axes as, for example, near the zenith of an Az–El mounted antenna.

Much of the tracking equipment is necessarily located, along with the high-power amplifier and the input stages of the receiver, in the radome with

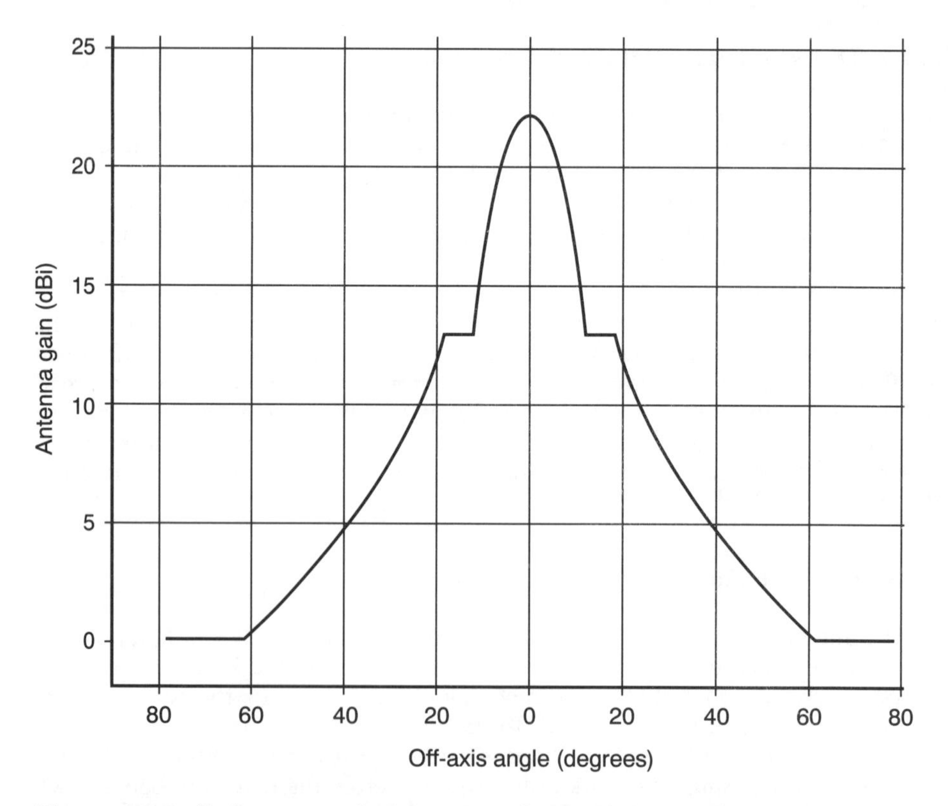

Figure 14.4 *Reference radiation pattern for an Inmarsat Standard A ship earth station antenna, assumed to be 1 metre in diameter. (Based on ITU-R Recommendation M.694 [8].)*

the antenna. The rest of the SES is housed below decks, with a user-friendly control panel.

Service requirements

Ships need several kinds of telecommunication service, presented in various ways. These services include telex and telephony and signals such as facsimile and data communication that can be passed over a telephone channel using modems. Some services are required to be deliverable, person to person or machine to machine, in duplex or in conference form. Alternatively they may be transmitted unidirectionally to one specified recipient, by limited broadcast to a predetermined list of recipients or by broadcast to all ships. But the basic requirements are for user to user duplex telephony and unidirectional telex, and the description that follows concentrates on the main features of providing them.

However, Inmarsat is a commercial system. Its automatic call set-up and control software must maintain the system's commercial integrity, limiting access to the system, except for distress and safety calls, to ships authorized to use it and enabling charges to be brought to account. A relatively complex network control system is required to supply these needs at least running cost and in a way that makes least demands on the attention of ships' crews. The following account of the control network goes into some detail because, while it applies specifically to Inmarsat A standard SESs, it is typical of the control systems of other Inmarsat standard networks.

The control network

Each Inmarsat satellite is served by a number of LESs which act as gateways between ships and the PSTN. Their basic radio characteristics are outlined in Section 14.2.3. One of the LESs is equipped with a suite of control equipment which forms the network coordination station (NCS) for all the Standard A SESs registered as users of that satellite. The main function of the NCS is to coordinate the use made of a pool of telephony and telex channels that are available on demand to all users of the network, by means of signals passed over control channels. There are three main elements in this control network, all of which function automatically.

(1) The NCS transmits a binary PSK carrier with a bit rate of 1200 bit/s which all LESs and SESs receive continuously. (Exceptionally, an SES suspends reception of this carrier while a telex call is in progress.) It is called the common signalling carrier (CSC).
(2) All LESs transmit one or more TDM emissions continuously, also binary PSK and also with a bit rate of 1200 bit/s. The frame period is 290 milliseconds and each frame contains 348 bits; see Figure 14.5. The frame starts with a preamble containing a unique word of 20 bits, used for frame synchronization. The next 63 bits (including FEC) of each frame are available for sending information from the LES to the NCS. This information includes, for example, advice to the NCS that a subscriber ashore wishes to call a ship.
(3) An SES transmits a 'request message' on one of a number of designated carrier frequencies when a call is to be originated from the ship. A request message consists of a carrier burst of 35 milliseconds duration, modulated with binary PSK at a rate of 4800 bit/s. The preamble consists of words for carrier phase and bit timing recovery and a unique word for burst synchronization. The remainder of the burst contains 39 information bits protected by FEC, identifying the calling ship, the LES that is to handle the shore end

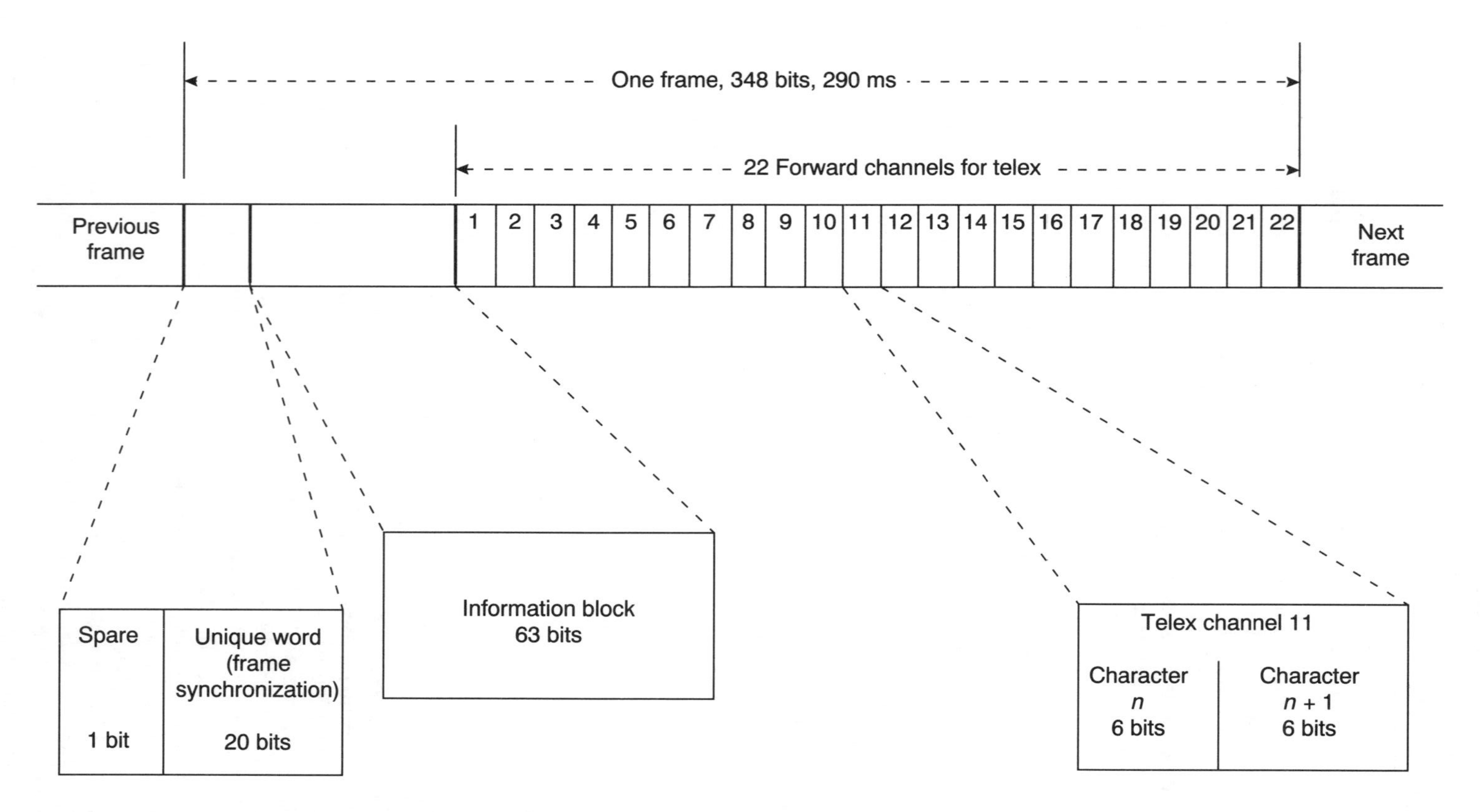

Figure 14.5 *The frame structure of the TDM signal, transmitted by an LES in an Inmarsat A network*

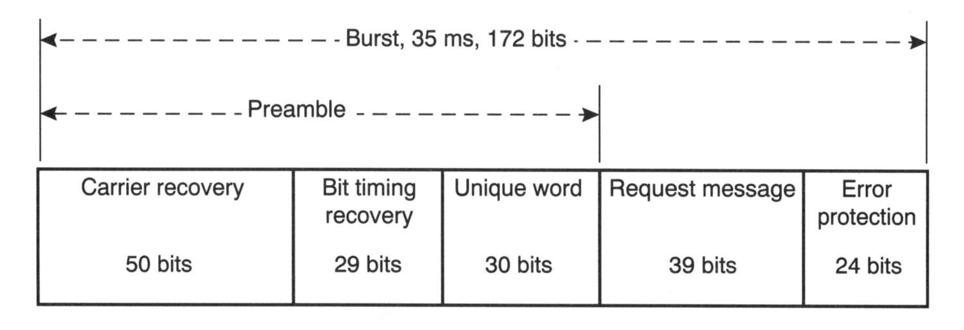

Figure 14.6 *The burst format of a 'request' message transmitted by an Inmarsat A ship earth station*

of the call, the nature of the facility required, etc. See Figure 14.6. All LESs monitor all of the frequencies designated for request messages. If a request message is satisfactorily received at the appropriate LES, the LES sends an acknowledgement to the NCS by way of the information block of its TDM emission and the NCS passes on the acknowledgement to the calling ship through the CSC. If a request message is lost because, for example, another ship transmitted a request message simultaneously at the same frequency, no acknowledgement is sent. The calling ship then repeats the message after a delay of randomized duration until an acknowledgement is received. Thus, the request channel operates in the ALOHA mode.

The setting-up procedure is similar if the call originates on land. From this point onwards, the control of a call depends on whether a telephone channel or a telex channel is required.

The transmission of a telephone call

The NCS has a register of all the available SCPC telephony channels and information on their operational state. When the NCS has been informed that a telephone call is required, it identifies a pair of free channels and sends an assignment message via the CSC listing the SES, the LES involved and the channels that have been assigned. The SES and the LES establish communication using the assigned channels and the call is set up, end-to-end, using in-band signalling (that is, using tones that are transmitted over the channels that have been assigned for communication). When the call has been completed, the ship and the LES cease transmitting the carriers and the LES informs the NCS that the channel is available for another call. The LES records the data necessary for billing the customer and for the internal accounting of the system.

The system uses analogue transmission for the telephony channels. The baseband is limited to 300–3000 Hz. The carrier is frequency modulated, the peak deviation corresponding to a nominal baseband level of 0 dBm0 being 12 kHz. The mean speech level objective is −14 dBm0, producing a frequency deviation of 3.8 kHz r.m.s. and 2:1 syllabic companding is used. The compandor is switched out of circuit when data or facsimile is to be transmitted with a modem. Pre-emphasis is not used. ITU-R Recommendation M.547 [9] recommends that the subjectively equivalent speech signal to psophometrically weighted noise level for a speech signal higher than −20 dBm0 at the transmitting earth station should be not less than 26 dB at the receiving earth station under non-fading conditions when the angle of elevation of the satellite, as seen from the ship, exceeds 10°. The Inmarsat system is designed to achieve this standard.

The transmission of a telex call

As with a telephone call, whether a telex call is initiated from the ship or from ashore, the NCS and the concerned LES and SES are informed, directly or via the control network. The LES assigns one of the telex channels in its TDM system for the forward leg of the call and informs the NCS. The NCS sends an assignment message to the SES via the CSC, indicating the channel that has been assigned by the LES. Corresponding with each TDM system that a LES transmits, it is also equipped to receive a TDMA emission which can carry telex channels.

Having received the assignment message from the NCS, the SES retunes the receiver that had been used to receive the CSC to the frequency of the TDM signal from the LES. It selects the assigned telex channel to receive the forward leg of the call. The SES also accesses the corresponding TDMA channel in order to transmit the return leg.

The frame structure of the TDM carrier is shown in Figure 14.5. In each frame 264 bits are available for forward legs of telex channels, time-divided into 22 channels of 12 bits per frame. Thus, with a frame period of 290 milliseconds, 41.4 information bits are available per channel per second.

The frame period of the TDMA system is 1.74 seconds. The frame structure is shown in Figure 14.7. There is provision for carrier bursts from 22 SESs in each frame, each burst conveying a telex return leg corresponding to one of the forward legs carried in the TDM system. A SES transmits a carrier burst, 37.7 milliseconds in duration, modulated with binary PSK, the bit rate being 4800 bit/s. Each burst contains 109 bits of preamble and 72 information bits. Thus there is an information rate of $72/1.74 = 41.4$ bit/s in the return leg also. With any necessary buffering, this provides for one telex channel operating at 50 bauds, using CCITT International Telegraph Alphabet No. 2. The telex call is set up by in-band signalling. The connection is taken down when the call has been completed, and the LES records the details of the call for billing purposes, etc.

A very simple burst timing system is used for this TDMA system. In every sixth frame of the associated TDM system, the unique word in the preamble is replaced by its complement. The arrival of this event at an SES signals the start of the frame of the TDMA system. This system gives very imprecise timing because the interval that elapses between the emission of a complemented unique word at the satellite and its arrival at the ship varies with the angle of elevation of the satellite as seen at the ship; see equation (2.8). So, for example, the entry into the TDMA channel of a burst from a ship at the edge of the satellite's global coverage area would be 39 milliseconds later than that of a ship immediately below the satellite. However, the long guard time between consecutive bursts in this system prevents overlapping of bursts.

Ship station transmitters

A Standard A station may transmit request message bursts, telephone SCPC carriers and telex TDMA bursts but never more than one emission at any time. The uplink e.i.r.p. used for any of these emissions is 36 dBW, calling for a transmitter output power rating between about 17 W and about 30 W for antennas having diameters between 1.2 metres and 0.9 metres. Solid state power amplifiers are used.

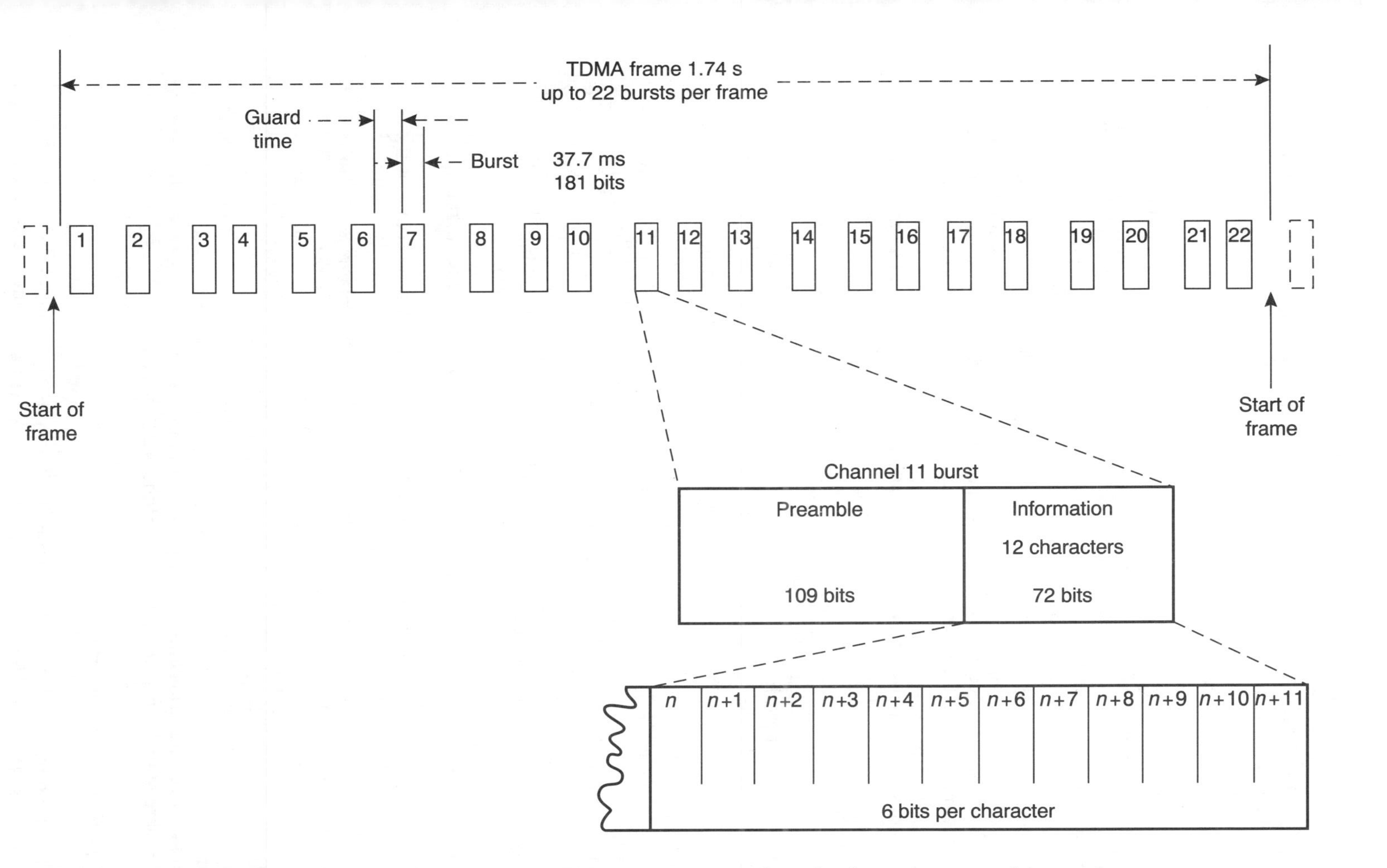

Figure 14.7 *The frame and burst structure of the TDMA system, used for telex in an Inmarsat A network*

14.3.2 The Inmarsat Standard C system

Ships with SESs meeting the Inmarsat A and B standards (see Sections 14.3.1 and 14.3.3) have access to two-way telephone and telex facilities, facsimile and low-speed data. However, the ship's equipment is big, heavy and rather costly. For a small ship that sends few messages and can get by without telephony, a simpler terminal of lower performance would be adequate and might be more economical. Standard C has been defined for this market. Operational use of the system started in 1990.

The Standard C SES has an antenna of very low gain, around 2 dBi. Typically a short helix antenna is used. The gain varies little over the whole upper hemisphere and down to 10° below the horizontal when the ship is on an even keel. The G/T requirement is $-23\,\text{dB}\,\text{K}^{-1}$ at an angle of elevation of $+5°$ and not less than $-24.5\,\text{dB}\,\text{K}^{-1}$ at any angle above $-10°$. Thus, the antenna does not need to track the satellite; communication is maintained without tracking even when the ship is rolling $\pm 10°$ at the edge of the coverage area. The terminals are inexpensive. The antenna under its radome is only about 30 to 40 cm in diameter and height, the below-decks equipment is small and only a few kilograms in weight and the power consumption is less than 50 W. It can thus be installed on virtually any sea-going ship.

Service is limited to the store-and-forward transmission of telex or data messages up to 32 kilobytes long from a ship to a subscriber ashore or vice versa or from ship to ship. The operation of the system is described in some detail by Warry and Hawkes [10]. Basically, however, the system functions like a Standard A network in its telex transmission mode except that, since it is a store-and-forward system, messages can be sent in sequence instead of being interleaved in TDM and TDMA. Thus, a message originating ashore is divided into packets, each protected by half-rate FEC, and transmitted in consecutive frames of the PSK carrier of the LES. A message originating at the ship is transmitted in packets, also protected by half-rate FEC, to the LES on a PSK carrier at a frequency assigned by the LES. When all the packets of a message have been sent, the receiving earth station arranges for the repetition of any packets that have not been satisfactorily received before the connection is broken.

14.3.3 The Inmarsat Standard B and Standard M systems

Two new kinds of SES came into service in 1993. The main objective of the Standard B system design was to align the downlink power density requirement of the SES with the capabilities of Inmarsat 3 satellites, thereby enabling the full potential capacity of such satellites to be deployed. It is foreseen that the Standard B terminal will eventually replace the Standard A terminal. The Standard M system is intended to occupy the gap in capabilities between Standards A (or B) and C, providing for telephony but with a simple antenna. In each case the opportunity has been taken to provide for new user options and the introduction of satellites with spot beam antennas. Both of the new systems use digital transmission for all emissions but in the structure of their network control system they are similar in principle to the Standard A system.

The Standard B system

The Standard B SES has an antenna similar to that of Standard A and G/T must be not less than $-4\,\text{dB}\,\text{K}^{-1}$. Any radio carrier transmitted by an SES, for whatever purpose, has a nominal rate of 24 kbit/s.

For both directions of transmission, a 16 kbit/s ADPCM signal with signalling bits added is used for telephony. The signal is protected by $\frac{3}{4}$-rate FEC and transmitted in the SCPC mode using OQPSK. The telephone channel is capable of supporting data or facsimile, transmitted by modem, at rates not exceeding 2.4 kbit/s. A 9.6 kbit/s data channel or CCITT Group 3 facsimile can replace the telephony channel, in which case $\frac{1}{2}$-rate FEC is used. The carrier spacing module is 20 kHz.

For 50 baud telex, the LES transmits the forward channel in a TDM system and the SES transmits the return channel in a TDMA burst. Half-rate FEC and binary PSK are used in both directions. The TDM carrier and the TDMA burst operate at 6 kbit/s and 24 kbit/s respectively. Separate but similar provision is made for 300 bit/s data.

The Standard M system

The Standard M SES is required to have a G/T not less than $-10\,\mathrm{dB\,K^{-1}}$. This figure of merit is associated with an antenna gain of about 14 dBi, typically realized by means of a short vertical array of radiating elements, protected by a small radome. The polar diagram is symmetrical about the vertical axis and element phasing may be switched, according to the location of the ship, to provide a broad lobe at the approximate angle of elevation of the satellite. Active tracking is not necessary.

All radio emissions transmitted by an SES are at a nominal rate of 8 kbit/s. For telephony, speech is encoded at 4.12 kbit/s using the improved multi-band excitation algorithm. With FEC, the required transmission rate rises to 6.4 kbit/s which is transmitted in the SCPC mode using OQPSK. Alternatively, the speech channel can be replaced by FEC-protected data or a facsimile signal with an information rate of up to 2.4 kbit/s. The carrier spacing module is 10 kHz.

14.3.4 EPIRBs and the Inmarsat Standard E terminal

The saving of human life in danger at sea was the historical stimulus for the first important commercial use of radio, in the first few years of the twentieth century and the use of radio for distress and safety purposes remains of the greatest importance. In the absence of satellite communication facilities, ships in distress use terrestrial radio at one of the international distress frequencies, namely 500 kHz (using morse telegraphy), 2182 kHz (using speech) or 156.8 MHz (using speech) to call for help. These means are reliable for ranges of a few hundred kilometres at 500 and 2182 kHz and fifty kilometres at 156.8 MHz. In coastal waters a distress message may be heard by a coast radio station; the alarm would then be raised and a search for the ship in distress can be mounted. But in mid-ocean the initial alert depends on the interception of the distress message by another ship that is sufficiently close to receive it; this is seriously insecure.

The final stages of the search for a ship in distress is often aided by a radar transponder or an Emergency Position-Indicating Radio Beacon (EPIRB) carried by the ship. (The term 'radar transponder' is used here to mean a radio receiver/transmitter combination which is stimulated by a received radar pulse to transmit a signal, typically carrying information on the identity etc. of the vessel carrying the transponder.) An EPIRB does not need the stimulus of an interrogating radar signal; it transmits automatically in an emergency situation. An EPIRB signal may be picked up by a searching aircraft. However, an EPIRB signal that can be received ashore via a satellite, indicating the position of a ship needing help, overcomes the range limitations of terrestrial emergency radio systems.

Two kinds of satellite EPIRB are coming into use, operating at 406 and 1646 MHz respectively. Both are small, lightweight transmitters, battery powered, that can be carried in survival craft and may be mounted on rafts that will float off the deck of a ship if it should sink.

The 406 MHz EPIRBs are designed to operate in conjunction with Cospas navigation satellites and Sarsat meteorological satellites which are in low, highly inclined, circular orbits. The main radio parameters of the beacons have been internationally agreed; see ITU-R Recommendation M.633 [11]. In brief the beacon, when activated in an emergency, transmits a carrier burst, 0.5 seconds long, every 50 seconds. The carrier is phase modulated with a digital message which identifies the ship and the nature of the emergency; it may also state the location of the ship. The signal, received at a satellite as it passes close to the beacon, is relayed to an earth station which raises the alarm. In the absence of data in the message indicating the location of the ship, an approximate location can be calculated from the pattern of Doppler frequency shift observed in the carrier received from the EPIRB as the satellite passes over it. To be effective, this method requires precise EPIRB carrier frequency stability, and a stringent short-term stability requirement of 2 parts in 10^9 is specified. Other EPIRBs of this type, carried by aircraft, operate at 121.5 and 243 MHz.

The 1646 MHz EPIRB was developed for use with geostationary satellites. ITU-R Recommendation M.632 [12] gives the agreed parameters. In brief, in an emergency a series of transmissions are made, each of several minutes' duration, having an e.i.r.p. of about 0 dBW. These transmissions are modulated with an emergency message which, repeated throughout the transmission, indicates the ship's identity, the nature of the emergency, etc. As before, the signal is relayed by the satellite to an earth station which sets the search and rescue system into action. The satellite, being geostationary, and the ship have no substantial relative motion, so Doppler shift of the carrier frequency cannot be used for determining the ship's position. It is therefore essential for the transmitted message to include the ship's position. This would usually be ensured by a feed of up-to-date position information from the ship's navigation system to the EPIRB right up to the point in time when the emergency begins. There is, of course, one major limitation of this type of EPIRB from which the 406 MHz EPIRB does not suffer, namely the lack of cover at high latitudes.

The Inmarsat Standard E terminal is a 1.6 GHz EPIRB. Access to the Inmarsat system is uncontrolled but collisions between the transmissions from two ships are very improbable, for two reasons. Firstly, a sub-band within the 1645.5–1646.5 MHz 'Distress and Safety' band is reserved for EPIRB signals and the narrowband frequencies assigned to EPIRBs are spread over this sub-band. Secondly, the duty cycle of the transmissions from any one ship is only 1 per cent.

14.3.5 The maritime satellite market

By 1993 about 16 000 Standard A terminals had been installed on ships and new installations were running at a steady rate of about 1500 per annum. For Standard C the corresponding figures were 6000 and 2500. Installations of Standard B and Standard M terminals were just beginning. The space segment's share of the revenue from this maritime traffic was about US$229m in that year, up from about US$210m in 1992 [13]. Clearly maritime satellite communication are serving an important market.

It is not surprising that this should be so. Maritime transport is an industry of great economic importance. Long-distance terrestrial maritime telecommunication, depending mainly on HF links, is limited in capacity, labour intensive in operation, subject to delay, geographically constrained and out of date in the service options it provides. Inmarsat provides the traditional services efficiently, with new options such as closed-group broadcasts and with ample capacity. New narrowband services such as facsimile and data, which have been in growing use ashore for many years, have become available via satellite at economic rates also. As a result, many administrative functions can be transferred from ship to shipping company headquarters on land. There can be direct access from a ship to information sources, such as databases, on land. Satellite communication provides much better protection against the loss of life due to disasters at sea.

A special facility introduced with the Inmarsat Standard C terminal but available as an enhancement on Standard A terminals is the Extended Group Call (EGC) system. EGC enables a message from an LES to be addressed to all equipped ships in a specified sea area. A system called Safetynet is being set up by Inmarsat and the IMO which will enable maritime and meteorological agencies ashore to broadcast warnings of navigational and meteorological hazards to all equipped ships that might be affected. Another new factor which makes satellite communication more attractive to shipping arises from imminent changes in the International Convention for the Safety of Life at Sea (SOLAS) [14].

The SOLAS Convention

SOLAS, last fully revised in 1974 but amended in 1978, 1981 and 1983, includes the following requirements for emergency communication facilities:

(1) Passenger ships of any size and cargo ships of not less than 1600 tons gross must have radio equipment for morse telegraphy at the international radiotelegraphy distress frequency at 500 kHz, crewed by a wireless operator or, in some cases, two operators.
(2) Cargo ships between 300 and 1600 tons are to have radiotelephone equipment operating at the international distress frequency for radiotelephony at 2182 kHz. At least one crew member (two if the ship exceeds 500 tons) must be qualified radiotelephone operators. Alternatively the ship may be equipped and crewed for morse at 500 kHz.
(3) In addition to the foregoing, cargo ships above 300 tons and all passenger ships are to have VHF radiotelephone equipment covering 156.8 MHz.

However, the cost of meeting these requirements, and especially that of the 500 kHz equipment and its specialized crew, is significant and the protection they provide is flawed. Now that better means using satellite communication have become available, the IMO has developed new concepts for safety of life at sea in the form of the Global Maritime Distress and Safety System (GMDSS) [15]. The introduction of this system, in as far as it affects radio system requirements, was approved in 1988 in the form of amendments [16] to the 1974 SOLAS Convention. Implementation began in 1992 and completion of implementation is planned by 1 February 1999.

Under the revised convention there will no longer be a requirement for radiotelegraphy at 500 kHz. However, requirements have been considerably expanded at other frequencies, the scale of provision for emergency communications now depending on the sea areas in which the ship operates. For example, any passenger ship or a cargo ship of more than 300 tons gross which goes beyond the range of coast stations operating at 2.1 MHz but not into polar waters is to be provided with:

(1) VHF equipment equipped to send and receive distress alarms using digital selective calling (DSC) and radiotelephony between 156.525 and 156.8 MHz;
(2) radar transponders operating in the 9 GHz band;
(3) a receiver operating at 518 kHz or in the HF frequency range for the reception of Navtex transmissions if the ship sails in any area where international Navtex service is provided (Navtex is an international automatic direct-printing service distributing urgent navigational or meteorological information), although an Inmarsat SES equipped for EGC/Safetynet is an acceptable alternative; and
(4) either (a) an Inmarsat SES capable of providing telephony and various data and emergency services, plus radiotelephone systems operating at various frequencies near 2 MHz, plus a satellite EPIRB, or (b) two independent multi-purpose MF/HF radio installations, plus means of raising distress alerts via an EPIRB or the Inmarsat system.

All ships sailing in deep water will be required to be equipped with satellite EPIRBs after 1999, and there will be a considerable simplification of their other radio installations if an Inmarsat SES is included. The same is true, to some degree, for ships that remain in coastal waters, although their terrestrial radio alternative will be less complex.

14.4 Satellite communication for aircraft

The requirements

For aircraft flying over land areas where the terrestrial infrastructure is well developed, air-ground VHF radiotelephone links operating around 125 MHz are reliable. This is the established medium for air traffic control (ATC) communication, supplemented in some countries by data signals transmitted from air to ground in response to interrogation by secondary surveillance radar (SSR) Mode S systems. However, over the oceans and the less developed land areas of the world, beyond the reach of land-based VHF stations, the medium for ATC communication is HF radio, typically using morse telegraphy. This is unsatisfactory to both the ATC authorities and the airlines because it is unreliable. It is also costly and slow because of the need to route messages between the air traffic controller and the pilot via telegraphists at both ends of the link.

Air–ground VHF links are also used for airline in-flight business communication, although the number of channels available is insufficient in some areas. In some parts of the world, access to the PSTN for telephone calls is available for airline passengers, using frequencies around 800 MHz or in the bands 1670–1675 MHz and 1800–1805 MHz. However, beyond the range of land-based VHF and UHF stations, there is no practical means of serving either of these needs by terrestrial radio.

The emergence of aeronautical satellite communication

There were demonstrations of the feasibility of satellite links with aircraft very early in the history of satellite communications, using the Syncom 3 satellite in the early 1960s and the ATS-I and ATS-III satellites in the late 1960s. These tests were made using frequencies around 150 MHz. There was some interest at that time in setting up a satellite system for aircraft operating in this part of the spectrum and a number of aircraft that were then under construction were equipped with VHF antennas for satellite systems. However, the frequency bands that were allocated for the AMSS in 1971 were at L band, and interest in an operational system faded.

ICAO had set up the Astra Panel in 1968 to consider in depth the use of satellites for ATC communication. The report of the panel in 1971, favourable to satellites, led to collaboration between Canada, the US Federal Aviation Administration and ESA in developing proposals for a satellite system, called Aerosat, dedicated to aeronautical communication. This scheme was abandoned late in the 1970s, due in part to a financial crisis in civil air transport caused by a leap in the price of fuel. Another attempt to set up a dedicated aeronautical satellite system was launched by Aeronautical Radio Inc. (Arinc) in the mid-1980s; see [17]. This project, AvSat, also failed. However experimental work continued in Europe and North America; particular mention should be made of the Prosat and Prodat programmes coordinated by ESA.

ICAO studies of possible solutions to the communication and navigation problems of civil aviation continued. A special committee on Future Air Navigation Systems (FANS) started work in 1983 with a remit which covered communications and surveillance as well as navigation. The final report of FANS to the ICAO Council in 1988 declared that the only viable solution to the shortcomings of the existing system involved the use of satellite-based communications, navigation and surveillance systems. However, the report also stressed that terrestrial systems would remain important, at least for the near future, where they were available.

A major problem for schemes like Aerosat and AvSat for providing a new satellite system dedicated to civil aviation had been financial. The initial cost of the space segment would be high, yet use in the early years would be small because retrofitting all existing aircraft with AESs would take a considerable number of years. The Inmarsat system is not ideal for aeronautical use, in particular because some scheduled flights go far north, beyond the coverage of geostationary satellites. Nevertheless Inmarsat could supply readily what civil aviation needed most, namely a proven space segment in place, technically suitable in most respects, economically viable, with no start-up costs for the aviation industry and without requiring long-term commitments. Civil aviation has begun to make use of this option.

The first trials of telephone service for passengers via Inmarsat took place in 1988 and regular small-scale use followed, involving a gradually increasing number of aircraft. The successful field trials, starting in 1991, of the use of Inmarsat data calls between airliners and air traffic controllers for Automatic Dependent Surveillance (ADS) have been particularly significant. (ADS involves gathering data on the location and operational status of an aircraft into an on-board data store, then transmitting the data automatically to the controller on demand or routinely at whatever interval is appropriate.) The incorporation of ADS by satellite into ATC procedures has begun.

Other aeronautical uses of the Inmarsat space segment are also developing. For example, some airline companies are using the data service for administrative traffic, and, as mentioned in Section 14.2.4, the Inmarsat 3 satellites will be equipped with a special transponder, transmitting at 1575 MHz, to distribute information on the integrity of satellite navigation systems to civil aircraft. By the end of 1995, AESs had been commissioned on 647 airliners and 311 corporate jets [18] and the numbers are rising fast.

For the opening phases of its aeronautical service Inmarsat has defined main parameters for Standard L, H and I AESs; see below. These are preliminary definitions, defining radio-frequency characteristics but with simplified provisions for channel control. It is intended that enhanced system specifications will be prepared later, when the needs of aircraft can be more firmly determined. At this initial stage an NCS will not be used, although access control arrangements will be otherwise similar to those used for the maritime field. The LESs which serve aircraft, singly or in consortia providing global coverage, will operate their own, essentially independent, networks serving the airlines with which they have operating agreements. As with the maritime systems discussed in Section 14.3, SCPC channels are used for telephony and a combination of TDM and TDMA emissions for data traffic. For aircraft that do not need telephony or duplex data facilities, an aeronautical version of the Inmarsat Standard C earth station provides economical store-and-forward data facilities.

The Inmarsat Standard L AES

A Standard L AES has a low-gain antenna, nominally 0 dBi, typically taking the form of a single element blade antenna mounted on top of the fuselage. The G/T requirement is $-26\,\mathrm{dB\,K^{-1}}$. Service is limited to packet data, normally at an

Figure 14.8 *A phased array AES antenna (Inmarsat Standard H) being installed on a Gulfstream IV aircraft. (Reproduced by permission of Ball Aerospace & Technologies Corp.)*

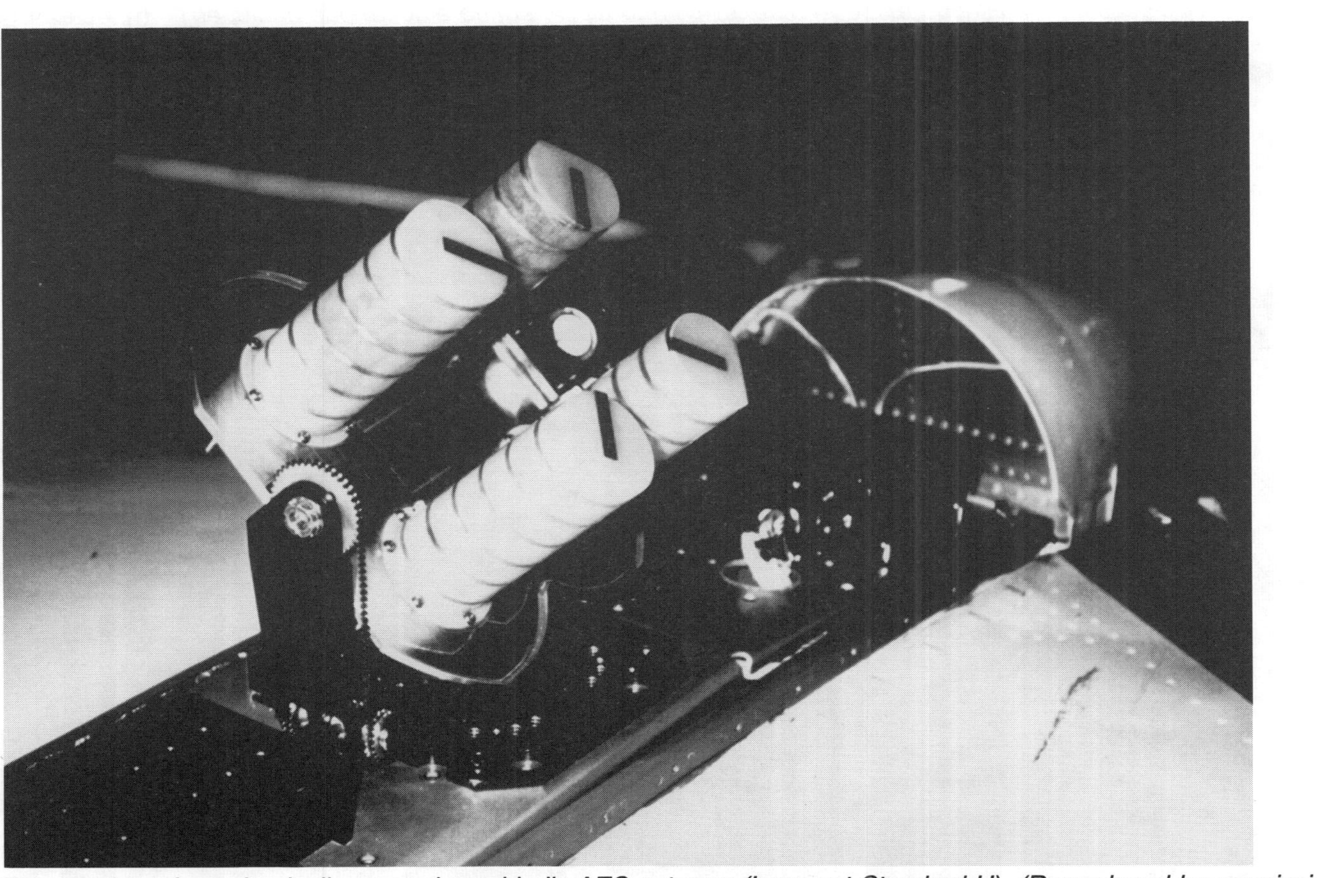

Figure 14.9 *A mechanically steered quad helix AES antenna (Inmarsat Standard H). (Reproduced by permission of Racal Avionics.)*

Figure 14.10 *The fairing and radome on the tail fin of a Gulfstream IV aircraft housing a mechanically steered AES antenna. (Reproduced by permission of Racal Avionics.)*

information rate of 300 bit/s plus $\frac{1}{2}$-rate FEC, although operation at a higher rate may be made available if transmission conditions allow it.

The Inmarsat Standard H AES

A high-gain antenna is required for a Standard H AES, with gain not less than 12 dBi. Such an antenna has appreciable directivity and active satellite tracking is necessary. The G/T requirement is $-13\,\text{dB}\,\text{K}^{-1}$.

Antennas take one of two forms. More usually, a phased array is mounted on top of the fuselage, under a blister radome designed to cause minimal aerodynamic drag. The phase of the feeds to the elements of the array is varied to steer the beam to the direction computed for the satellite. Better performance is obtained at low angles of elevation if two arrays are used, one on each side of the fuselage. The Ball Avionics array design, with particularly low drag, takes the form of a rectangular plate, 0.41 metres high, 0.81 metres long and 9.5 millimetres thick, the profile of the plate being curved to conform with the curvature of the fuselage; see [19] and Figure 14.8.

Alternatively, an antenna under a radome is mechanically steered to track the satellite. For example, the Racal Avionics antenna is a quad helix array, typically housed in a fairing at the top of the tail fin of the aircraft. See Figures 14.9 and 14.10.

A Standard H AES is equipped for packet data facilities. However, it may also be equipped to transmit and receive up to six SCPC carriers, with OQPSK modulation, for telephony, encoded at 9.6 kbit/s plus framing overheads which, with $\frac{1}{2}$-rate FEC, gives a transmission rate of 21 kbit/s per channel.

The Inmarsat Standard I AES

Inmarsat Standard I is came into service in 1996, having a similar relationship, relative to Standard H, as Standard M has to Standard B in the maritime field. With an antenna gain requirement of 6 dBi and a codec for telephony requiring an information rate of 4.8 kbit/s, the new standard addresses economically the needs of short-haul and medium-haul commercial aircraft and light corporate jets.

14.5 Satellite communication for mobile stations on land

14.5.1 The GSO and transportable earth stations

Inmarsat systems operating in L band

Earth stations to Inmarsat Standards A, B, C and M are available in forms that are readily transportable on land. Thus, conventional Inmarsat Standard A or B terminals may be mounted on a light road vehicle, with the antenna under a radome to protect it from mechanical damage. If the station is not to be used while the vehicle is in motion, the antenna needs no elaborate mountings to track the satellite and to compensate for movements of the vehicle. Usually, these terminals also dispense with a radome. For even easier transportation, using an antenna with a reflector that demounts into petals, the whole terminal can be packed for transport in a suitcase.

A complete land-transportable Inmarsat Standard M battery powered terminal can be fitted, ready for use, into a case no bigger than a briefcase. The antenna is built into the lid of the case. Unlike the maritime design, in which the antenna has a horizontally symmetrical polar diagram which is adjustable to match the angle of elevation of the satellite, the briefcase lid antenna has a main lobe that is rather narrow in the horizontal plane but broad in the vertical plane, and the user

Figure 14.11 *An Inmarsat Standard M briefcase LMES. (Reproduced by permission of GEC-Marconi Communications.)*

rotates the briefcase to steer the lobe to the azimuth where the signal received from the satellite is strongest; see Figure 14.11. Alternatively, for semi-static use, a more conventional design separates the antenna from the indoors unit; see Figure 14.12.

Figure 14.12 *An Inmarsat Standard M LMES for semi-static use, showing the antenna unit and (on page 412) the indoors unit. (Reproduced by permission of GEC-Marconi Communications.)*

Figure 14.12 *(continued)*

Where a store-and-forward data service is sufficient, the land mobile form of the Inmarsat Standard C terminal can be even more readily portable than its Standard M equivalent.

The main parameters specified for land mobile Standard A and B terminals are the same as for the maritime versions. For Standard M terminals operating in the more benign environment ashore, the requirements for G/T and uplink e.i.r.p. are relaxed by 2 dB relative to the maritime version. The facilities available to any of these terminals on land are the same as for the maritime version.

Wideband links: satellite news gathering

Transportable earth stations of the LMSS, operated when stationary in Ku band and C band, are used to provide temporary video and sound programme connections from places where news, sporting, cultural or entertainment events are occurring to locations where the signals can be incorporated into broadcasting programmes. High-quality audio signals intended for sound broadcasting may be transmitted in the same way. The transportable earth station uplinks the programme signals to an FSS satellite, to be downlinked at a permanent FSS earth station. Uplink frequencies in the bands 14.0–14.5 GHz or 5.925–6.425 GHz are usual. Two-way audio control channels often accompany the programme channels. This practice is electronic news gathering (ENG) using satellite communication; it is called satellite news gathering (SNG).

A typical SNG earth station is truck- or trailer-mounted, with as large an antenna as can be housed on the vehicle in a state that can be quickly set up for use. A generator capable of supplying several kilowatts of primary power may be carried for use where a mains power supply is not available. Typical antennas are

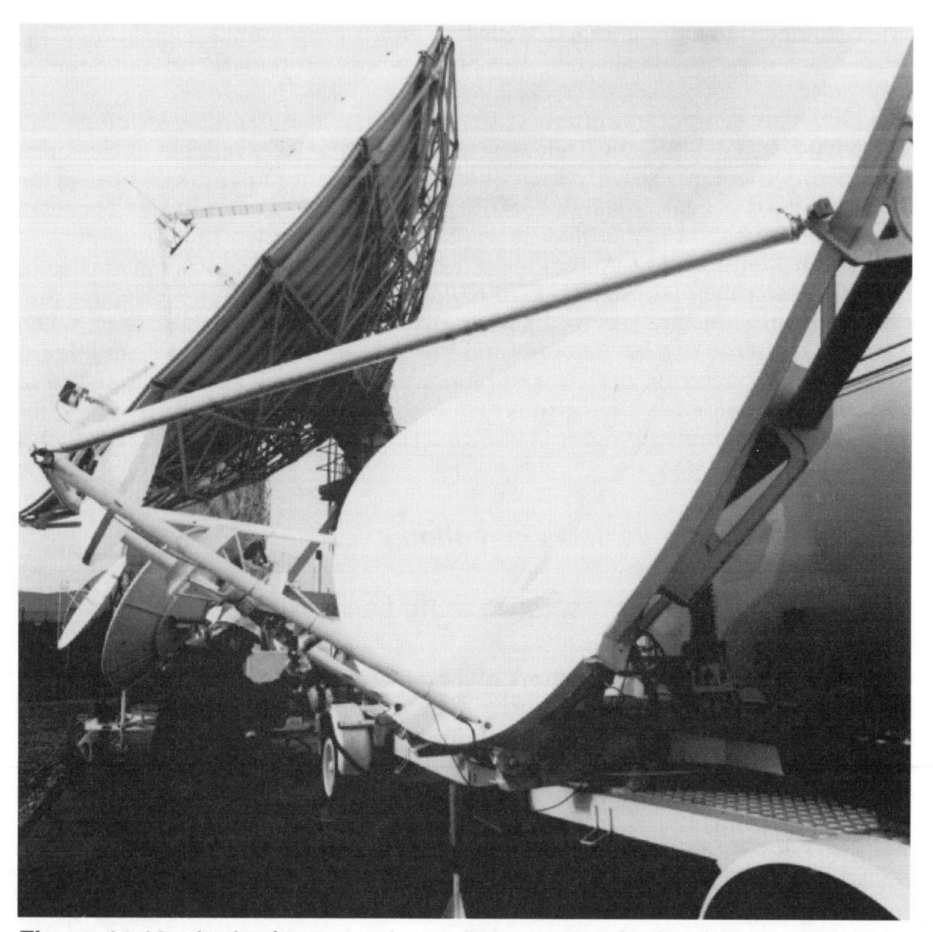

Figure 14.13 *In the foreground, a trailer-mounted SNG antenna, operating at Ku band, with an elliptical primary reflector and offset Cassegrain geometry. (BT Corporate Pictures; a BT photograph.)*

circular, around 2 metres in diameter. Where more gain is necessary, an elliptical main reflector may be used, its long axis being aligned with the road when en route but with the GSO when operating; see Figure 14.13.

Good antenna side-lobe suppression is essential, since ITU-R Recommendation S.524 (see Section 6.5, item 5) applies to all earth stations transmitting to geostationary satellites in FSS allocations which are not specifically covered by other regulatory constraints on side-lobe radiation. A transmitter output power of several hundred watts is usually necessary, mainly determined by the frequency band used, the gain of the satellite receiving antenna, the degree of bandwidth compression applied and the constraint on off-axis spectral e.i.r.p. density.

Both analogue transmission using FM and bit-reduced digital transmission using PSK modulation are employed for video SNG and in both cases efficient carrier energy dispersal is essential. For digital video channels, a bit rate of 8 Mbit/s plus FEC is typical. A spare satellite transponder, not relaying other emissions, is made available for SNG whenever possible. It is desirable for all accompanying narrow band channels such as the programme sound and control channels to be combined with the video programme signals in TDM to avoid the need for backoff in the SNG power amplifier and the satellite transponder.

SNG links are temporary, often set up in a hurry in unfamiliar locations and using whatever satellites and transponders happen to be available at the time. Consequently there is a risk of severe interference arising to emissions using the same satellite or a nearby satellite. So that the source of such interference can be identified quickly and notified, it has been recommended that an automatic transmission identification system should be applied to all SNG transmissions; see ITU-R Recommendation SNG.1070 [20].

SNG transmissions are received simultaneously at many earth stations when the subject matter is of wide interest, and this makes it particularly desirable for equipment and operating practices to be uniform. To some extent a measure of uniformity is provided by the requirements of the space segment providers but see also ITU-R Recommendations SNG.722 [21], SNG.1007 [22], SNG.771 [23] and SNG.770 [24].

14.5.2 The GSO and vehicles in motion

Omnitracs and Euteltracs, operating at Ku band

There is a demand from the operators of long-distance freight transport fleets for means for passing messages both ways between the headquarters dispatching office and trucks anywhere within a continent-wide area. Surveillance, enabling the location of a truck to be determined without the cooperation of the truck's crew, is sometimes a valued additional service.

Omnitracs is an LMSS system, operating in Ku band and designed to serve this market. The forward link from a hub earth station to the mobile earth stations is a high-power carrier, QPSK-modulated by a low-speed TDM system carrying data packets addressed to the mobiles. This emission is given a secondary spread spectrum modulation in order that the spectral power flux density of the downlink shall not exceed the mandatory PFD limit in this shared FSS allocation. Spectrum spreading also facilitates frequency coordination with other radio systems. Mobile earth stations transmit data packets in FDMA for the return link. The hub station may command the mobile station to report automatically its position, as indicated by a separate automatic position fixing system. Alternatively, the system design permits the incorporation of a satellite radio-determination system in Omnitracs, for which a second satellite is required.

Qualcomm opened service using Omnitracs in the USA in 1989. Positional information is provided there by the Loran C system. Service was opened in Europe in 1991 with an associated system named Euteltracs, using Eutelsat transponders and incorporating the Omnitracs radio-determination system; see [25].

Figure 14.14 *A car-mounted Inmarsat Standard M LMES. (Reproduced by permission of GEC-Marconi Communications.)*

Inmarsat operations at L band

Inmarsat Standard M antennas are available in a form that can be built on to a motor car roof. A relatively simple satellite tracking system enables the terminal to be operated on the move; see Figure 14.14. This provides what was probably the first practical satellite carphone. The Inmarsat Standard C terminal, with its omnidirectional antenna, needs no tracking system; it may be mounted on the roof of a vehicle, typically a truck, and used en route for store-and-forward telex and data. If coupled with a radio navigation system, such as a GPS receiver, a Standard C terminal provides a form of surveillance, since it can be remotely polled, causing its position to be reported automatically. Inmarsat vehicle-mounted paging receivers are in service. A new Inmarsat service now under development is a personal paging receiver, fed by a powerful satellite signal which is capable of penetrating into buildings.

14.5.3 Lower orbits for low-power earth stations

Geostationary satellites are used for LMSS applications, as is shown above, and they exhibit the usual advantages of the GSO. A single satellite can provide continuous two-way service to a vast coverage area, space segment costs are relatively low and there are no significant hand-over problems. However, the GSO has disadvantages; above all the basic free-space transmission loss between satellite and earth station is very high. High transmission loss combined with low or zero earth station antenna gain demands relatively high earth station transmitter power; this is objectionable for radiotelephones mounted in motor cars, impracticable for hand portable ones and a potential health hazard for users and bystanders in both applications.

Various lower orbits have been considered for satellites serving radiotelephones and light duty data systems. Low circular Earth orbits (LEO) at heights between about 800 and 1400 km, polar or inclined at perhaps 70°, provide global coverage, although about 50 satellites is the minimum necessary for continuous coverage. Medium altitude circular Earth orbits (MEO) at heights around 8000 km, inclined at about 50°, provide continuous global coverage with perhaps 12 satellites. Elliptical orbits, inclined at 63.4° and with a height at apogee of around 8000 km, would serve an area of continental extent with fewer satellites. There are various options for using orbits such as these for telephone and data services.

A basic Little LEO store-and-forward system

The narrow frequency allocations below 1 GHz are suitable for data store-and-forward systems of relatively low information capacity. Simple store-and-forward data systems have been developed and demonstrated over a number of years using, for example, the amateur radio Amsat satellites and the University of Surrey's series of Uosat satellites [26, 27]. Such systems are likely to be found useful, not by the general public, but by members of closed groups with special requirements, working in locations remote from conventional telecommunication facilities.

In such a system, data packets protected by FEC, each with a header containing an address and a packet identification number, are prepared and stored for transmission at a participating earth station. When a satellite in an LEO passes in sight of the earth station the packets are uplinked, to be stored in an on-board memory along with all the other packets that the satellite has received but not yet disposed of. The satellite retransmits the contents of its memory in endless repetition. When another participating earth station, receiving the satellite transmission, notes that it contains one or more packets addressed to itself, it records those packets and then commands the satellite to delete the message from its memory.

In this simple form a store-and-forward system connects a participating LMES directly to another participating LMES. It makes no use of feeder links and requires no support from terrestrial infrastructure. The earth stations would have omnidirectional antennas and are cheap, portable and need little power. As can be seen from Figure 2.18, a single satellite in a polar LEO has brief sightings of every place on the Earth's surface every day. However, using only one satellite, there may be a delay of several or many hours before the originating earth station has an opportunity to uplink the message to the satellite and there may be a similar delay before the message is downlinked to the destination earth station. Thus, the transmission time, end-to-end, is long. The average delay could be cut and the potential information throughput could be increased by using additional satellites in suitably phased orbits. Volunteers in Technical Assistance (VITA) plans to set up a system of this kind to provide access to medical information in developing countries.

Commercial Little LEO store-and-forward systems

In his presentation of proposals for the Orbcomm system Hardman [28] considers 18 satellites in circular orbits at 970 km altitude, six in each of three orbits inclined at 40°. With this constellation, virtually all the Earth between 50° north and 50° south latitude would be in sight of a satellite for most of the time, and interruptions of visibility are brief. The geographical coverage could be extended by adding polar-orbiting satellites to the constellation. See also [29].

With a constellation of satellites giving good coverage of the region to be served, a participating LMES would be able to uplink packets at any time with little or no delay. These packets would be downlinked, immediately or after a short delay, to an LES. The LES would read the packet headers and route the packets either via the terrestrial network to a fixed point address, or back into the satellite constellation for downlinking to an addressed LMES. In a similar way, a packet originating from a fixed address could be passed, via the terrestrial network to the LES, to be passed by satellite to the addressed LMES.

A substantial number of proposals have been made for systems of this kind. Authority has been given for the Orbcomm system to be set up, its constellation expanded to 48 satellites. No doubt there will be others.

Radiotelephone systems

There are a number of problems that must be solved for systems which serve radiotelephones which do not arise for store-and-forward data systems. In particular, to satisfy its customers a telephone system must be available, anywhere in the service area, all of the time. By comparison, interruptions of the availability of a data system, whether due to the absence of precise station-keeping in orbit, obstruction of the LMES/satellite path by hills or buildings or the temporary lack of a free channel, are likely to be unimportant and may not even be perceived by users.

Satellites for radiotelephones are likely to be complex, multi-beam devices, using the wider spectrum of the 'big LEO' frequency bands. Systems can obtain some protection from various kinds of service interruption by raising the number of satellites above the minimum required for continuous coverage, so that there will usually be, not one, but two or three satellites in sight of any locality. By this means, it can be expected that at least one satellite will have an unobstructed view of a user's earth station at any time, provided that the user takes care to avoid obstructions in choosing a point from which to access the system.

No big LEO system is likely to be in service before about 1998 but a substantial number of proposals have been made. Like most MSS systems, these would use satellites to connect LMESs through feeder links to LESs which provide gateways with the PSTN. Most proposed systems would use transparent transponders and CDMA, although one proposed system opts for TDMA and compatibility with terrestrial GSM systems and another would make extensive use of on-board switching. The most visible of these proposals at present are as follows:

(1) The Inmarsat affiliate company ICO Global Communications proposes a system using five satellites (plus one spare) in 6-hour MEOs in each of two planes inclined at 45°.
(2) The proposed Odyssey system would also use 12 satellites in 6-hour MEOs; an inclination of 55° is preferred in this case.
(3) and (4) The proposed Iridium and Globalstar systems would use 66 and 48 satellites respectively in LEOs.

These four proposed systems have an estimated aggregate space segment cost exceeding US$10 billion. No doubt other proposals will emerge from elsewhere. However, it seems unlikely that all of these proposals would secure frequency assignments. Furthermore, it is not at present clear whether the market, in a world already well supplied with terrestrial cellular radiotelephone facilities, will be sufficient to support such a heavy space segment investment.

References

1. Paragraph 66 of the *Radio Regulations* (1994). Geneva: ITU.
2. *Propagation data required for the design of Earth–space maritime mobile telecommunication systems* (ITU-R Recommendation PN.680-1, 1994), ITU-R Recommendations 1994 PN Series. Geneva: ITU.
3. *Propagation data required for the design of Earth–space aeronautical mobile telecommunication systems* (ITU-R Recommendation PN.682-1, 1994), ITU-R Recommendations 1994 PN Series. Geneva: ITU.
4. *Propagation data required for the design of Earth–space land mobile telecommunication systems* (ITU-R Recommendation PN.681-1, 1994), ITU-R Recommendations 1994 PN Series. Geneva: ITU.
5. Lipke, D. W., Swearingen, D. W., Parker, J. F. *et al.* (1977). Marisat–A maritime satellite communications system. *Comsat Technical Review*, **7**(2).
6. Berlin, P. (1986). Inmarsat's second-generation satellites. *Proc. IEE*, **133**, Part F.
7. Reference radiation pattern for ship earth station antennas (CCIR Report 922-1, 1990), Reports of the CCIR, 1990, Volume VIII, Annex 3. Geneva: ITU.
8. *Reference radiation pattern for ship earth station antennas* (ITU-R Recommendation M.694, 1994), ITU-R Recommendations 1994 M Series Part 5. Geneva: ITU.
9. *Noise objectives in the hypothetical reference circuit for systems in the maritime mobile–satellite service* (ITU-R Recommendation M.547, 1994), ITU-R Recommendations 1994 M Series Part 5. Geneva: ITU.
10. Warry, T. and Hawkes, J. (1992). Inmarsat-C satellite data communications for land and maritime mobiles. *British Telecommunications Engineering*, **11**.
11. *Transmission characteristics of a satellite emergency position-indicating radio beacon (satellite EPIRB) system operating through a low polar-orbiting satellite system in the 406 MHz band* (ITU-R Recommendation M.633-1, 1994), ITU-R Recommendations 1994 M Series Part 5. Geneva: ITU.

12. *Transmission characteristics of a satellite emergency position-indicating radio beacon (satellite EPIRB) system operating through geostationary satellites in the 1.6 GHz band* (ITU-R Recommendation M.632-2, 1994), ITU-R Recommendations 1994 M Series Part 5. Geneva: ITU.
13. Inmarsat annual review and financial statement 1993. Inmarsat, 1994.
14. *International Convention for the Safety of Life at Sea, 1974, consolidated with the 1978 SOLAS Protocol and the 1981 and 1983 SOLAS Amendments* (1986). London: The International Maritime Organization.
15. *Global Maritime Distress and Safety System (GMDSS)* (1986). London: The International Maritime Organization.
16. *Amendments to the 1974 SOLAS Convention concerning Radiocommunications for the Global Maritime Distress and Safety System* (1989). London: International Maritime Organization. 1989.
17. Dement, D. K. (1988). AVSAT, the first dedicated aeronautical satellite communication system. *Proc. Fourth International Conference on Satellite Systems for Mobile Communication and Navigation*. London: IEE.
18. *Aeronautical Satellite News*, No. 53, October–November 1996, London: Inmarsat.
19. Cox, B. J. (1992). World-wide aeronautical communications. *Avionics*, August.
20. *An automatic transmitter identification system (ATIS) for analogue modulation transmissions for satellite news gathering and outside broadcasts* (ITU-R Recommendation SNG.1070, 1994), ITU-R Recommendations 1994 SNG Series. Geneva: ITU.
21. *Uniform technical standards (analogue) for Satellite News Gathering (SNG)* (ITU-R Recommendation SNG.722-1, 1994), ITU-R Recommendations 1994 SNG Series. Geneva: ITU.
22. *Uniform technical standards (digital) for Satellite News Gathering (SNG)* (ITU-R Recommendation SNG.1007, 1994), ITU-R Recommendations 1994 SNG Series. Geneva: ITU.
23. *Auxiliary coordination satellite circuits for SNG terminals* (ITU-R Recommendation SNG.771-1, 1994), ITU-R Recommendations 1994 SNG Series. Geneva: ITU.
24. *Uniform operational procedures for Satellite News Gathering (SNG)* (ITU-R Recommendation SNG.770-1, 1994), ITU-R Recommendations 1994 SNG Series. Geneva: ITU.
25. Colcy, J.-N., Hall, G. and Steinhäuser, R. (1995). Euteltracs, the European mobile satellite service. *Electronics and Communications Engineering Journal*, **7**(2).
26. Ward, J. W. (1991). Microsatellites for global electronic mail networks. *Electronics and Communications Engineering Journal*, December.
27. Allery, M. N. and Ward, J. W. (1993). The potential for 'Store-and-Forward' communications using small satellites in low Earth orbits. *Proc. IEE third European Satellite Communication Conference*. London: IEE.
28. Hardman, G. E. (1991). Engineering Orbcomm: a digital satellite communications system exploiting a range of modern technologies. *Proc. Third IEE Conference on Telecommunications*, Conference Publication 331. London: IEE.
29. Lansard, E. and Corain, A. (1991). A store-and-forward micro-satellite system for autonomous packet terminals. *Proc. the Second European Conference on Satellite Communications*, Liège.

15 Satellite broadcasting

15.1 Direct broadcasting by satellite for television

15.1.1 Introduction

Statsionar-T satellites transmitted powerful television signals at about 700 MHz on a regular basis for reception by the public, in the USSR and elsewhere, starting in 1970. Two transponders on board the NASA ATS6 experimental satellite, launched in 1974, relayed wideband emissions at high power at about 650 MHz and 2.5 GHz respectively. Both satellites were used for long series of television transmissions to groups of TVRO earth stations with small antennas, demonstrating the feasibility of satellite broadcasting to isolated communities.

The use of space for broadcasting was the subject of anxious debate in the United Nations and the ITU in the 1960s and 1970s. Many countries, lacking established nation-wide TV broadcasting coverage, perceived that satellites might provide a domestic television service at less cost than terrestrial broadcasting. Other countries, having exhausted the VHF and UHF bands allocated for terrestrial broadcasting, saw in satellite broadcasting an opportunity to provide additional domestic channels using SHF bands that were not then heavily loaded.

However, it was also recognized that international TV broadcasting, which is difficult to achieve terrestrially, would be feasible by satellite and could be commercially rewarding for the providers. On the other hand it might be considered politically subversive and might offend against religious, moral and cultural sensitivities. It was also perceived that satellite broadcasting, as a medium, was not inexhaustible. In the absence of a spectrum utilization agreement, there were fears that the countries that used satellite broadcasting early might establish claims to all of the available orbit and the allocated spectrum, denying its use to countries coming later.

The outcome of this debate can be summarized from three documents:

(1) WARC-71 resolved, and subsequent ITU conferences have reaffirmed [1], that frequencies to be assigned for broadcasting from satellites should be planned at international conferences in which the governments of all countries that might be affected could participate.
(2) An Article of the ITU Convention of 1982 called on governments to bear in mind that radio frequencies and the GSO are limited natural resources which must be used efficiently and economically, taking into account the special needs of the developing countries. This article was agreed at the Plenipotentiary Conference in 1973 and was incorporated without significant change in the ITU Constitution in 1992 [2].
(3) A resolution adopted by the General Assembly of the UN in 1982 [3] requires states to bear responsibility for international direct television broadcasting carried out within their jurisdiction. Any intention to set up international broadcasting is to be notified in advance to the other states concerned, and implementation is to be within the terms of agreements reached as a result of the notification. It was recognized that unintended overspill of domestic satellite broadcasting on to the territory of another country may be technically

unavoidable but such overspill was not to exceed criteria to be determined by the ITU.

15.1.2 Frequency assignment plans at 12 GHz

The frequency assignment planning process having been set in motion by WARC-71, another WARC met in 1977 to make plans for the BSS frequency allocations listed in Table 15.1. An agreement including frequency assignment plans for BSS downlinks was drawn up for countries in ITU Regions 1 and 3. Plans could not be made for feeder uplinks at that time because suitable frequency allocations were not available. WARC-77 decided to postpone planning for Region 2. However, a RARC met in 1983 and drew up an agreement and plan for Region 2. New FSS uplink allocations had been made by that time and the Region 2 plans covered both uplinks and downlinks. Finally, one of the tasks of WARC-88 was to draw up a feeder uplink plan for Regions 1 and 3. The full text of these agreements and plans is to be found in the ITU *Radio Regulations* [4, 5]. The principal technical provisions of these plans are reviewed below and the administrative provisions of the agreements are outlined in Section 6.3.1.

Table 15.1 *Frequency bands planned for DBS*

Region	*BSS down link bands*	*Feeder link bands (Note 1)*	*Approximate satellite longitude range*
1	11.7–12.5 GHz	14.5–14.8 and 17.3–18.1 GHz	37° W to 20° E
2	12.2–12.7 GHz	17.3–17.8 GHz	166° W to 34° W
3	11.7–12.2 GHz	14.5–14.8 and 17.3–17.8 GHz	34° E to 170° E

Note 1. The use of 14.5–14.8 GHz for feeder links is reserved for countries outside Europe.

The BSS had not been allocated exclusive use of these frequency bands: most of this spectrum is also allocated with primary status to the terrestrial fixed and mobile services. It was therefore a matter for consideration, in planning for the BSS, whether the needs of the terrestrial services should be taken into account. However, at WARC-77, a majority of the Region 1 administrations insisted that the plans should be designed to provide for as large a number of television programmes as was feasible, to be distributed in equal numbers to every one of the 98 sovereign states in the region.

By adopting comprehensive, uniform and stringent system parameters it was found possible to provide for five channels per country in Region 1, and five channels per time zone within large countries. For Regions 2 and 3, containing fewer sovereign states, a less rigid scheme of channel distribution was adopted. A few groups of countries with close cultural affinities chose to have one or more of their assignments designed to serve the whole group, but with these few exceptions the coverage of the planned assignments was limited, as closely as was considered technically feasible, to national boundaries.

Very similar mandatory constraints and assumptions were applied world-wide. It was decided that all broadcasting satellites operating in the bands to be planned should be geostationary and should use circular polarization. Such differences as there are in the parameters adopted for Regions 1 and 3 on the one hand, and Region 2 on the other, arose mainly from changes in technical expectations emerging between 1977 and 1983. The main changes had been a higher expectation for the performance of domestic receiver LNAs and a more pessimistic view of the polarization discrimination achievable off-axis in the main lobe of satellite antennas.

It was assumed that broadcasting carriers would be frequency modulated by a baseband consisting of a standard definition (625-lines or 525-lines) colour video signal plus one (Regions 1 and 3) or two (Region 2) subcarriers frequency modulated by analogue sound channels. Pre-emphasis was to be used, as shown in Figure 8.15.

The plans for Regions 1 and 3 were designed to serve domestic receiving systems with a figure of merit assumed to be 6 dB K^{-1}, having antennas with an assumed half-power beamwidth θ_0 of 2°. This combination of G/T and θ_0 would be achievable, for example, by an antenna 80 cm in diameter and a receiver with a noise temperature of 1400 K. The corresponding figures assumed for Region 2 were 10 dB K^{-1} and 1.7°. Given these parameters for the domestic receiver, the first objective of the plan for Regions 1 and 3 was to enable satellites to deliver a PFD of -103 dB(W m^{-2}) per emission at the edge of the service area for 99 per cent of the rainiest month of the year in a bandwidth of 27 MHz. For Region 2 the corresponding figures were -107 dB(W m^{-2}), 99 per cent and 24 MHz. This would deliver a received C/N ratio of at least 14 dB for all but 1 per cent of the worst month. The second objective was to limit interference from all planned emissions to any one of them to ensure that the combined effect of noise and interference on reception quality would be acceptably small.

The other main requirements and assumptions used in planning can be summarized as follows.

(1) *Satellite station-keeping and attitude.* Satellites are required to remain within ±0.1° of their planned longitude in all regions. In Regions 1 and 3 the same standard of station-keeping is also to be maintained in the north–south plane; in Region 2 this is recommended but not required. The deviation of a satellite transmitting antenna beam from its nominal direction is not to exceed 0.1° in any direction. In Regions 1 and 3 the angular rotation of a beam about its nominal orientation is to be limited to ±2°; in Region 2 the mandatory figure is ±1°.

(2) *Satellite transmitting beams.* The satellite transmitting antenna beams assumed in planning are, in principle, the smallest beams of elliptical or circular cross-section having a -3 dB footprint which encloses the service area. The latter consisted of whatever part of the national territory (almost invariably the whole of it or a time zone of it) had been stated to be required to be served. For small countries, the rigid application of this rule would have led to a requirement for large, and in some cases impracticable, satellite antennas. Accordingly the smallest beams actually assumed for Regions 1 and 3 had a half-power beamwidth of 0.6°; the corresponding figure for Region 2 was 0.8°.

The downlink e.i.r.p. required to produce the target PFD was calculated and listed in the plan. It is of interest to note, for example, that the power required to be delivered to the satellite transmitting antenna to serve a small country like Luxembourg from a satellite at 19° W is a readily achievable 50 W. However, for a large country such as Chad, to be served by a single beam from 13° W, the transmitter power required is a highly problematic 1.6 kW. It is not mandatory that these power levels should be implemented but sufficient protection against interference might not be obtainable with less.

(3) *Satellite transmitting antenna off-beam characteristics.* Outside the −3 dB gain contour, the gain of the main lobe of the satellite transmitting antenna and the side-lobes is required to lie below the reference patterns shown in Figure 15.1.

(4) *Orbital plan.* For Regions 1 and 3, satellites are to be stationed in clusters 6° apart in the GSO. There was concern at RARC-83 that this pattern of satellite locations would not allow effective reuse of the spectrum by means of dual polarizations. Accordingly, a less rigid cluster location plan was used for Region 2, and each cluster is divided into two subclusters, spaced 0.2° to either side of the nominal location of the cluster; the easterly subcluster is to be occupied by satellites using LHCP and the westerly subcluster is for RHCP.

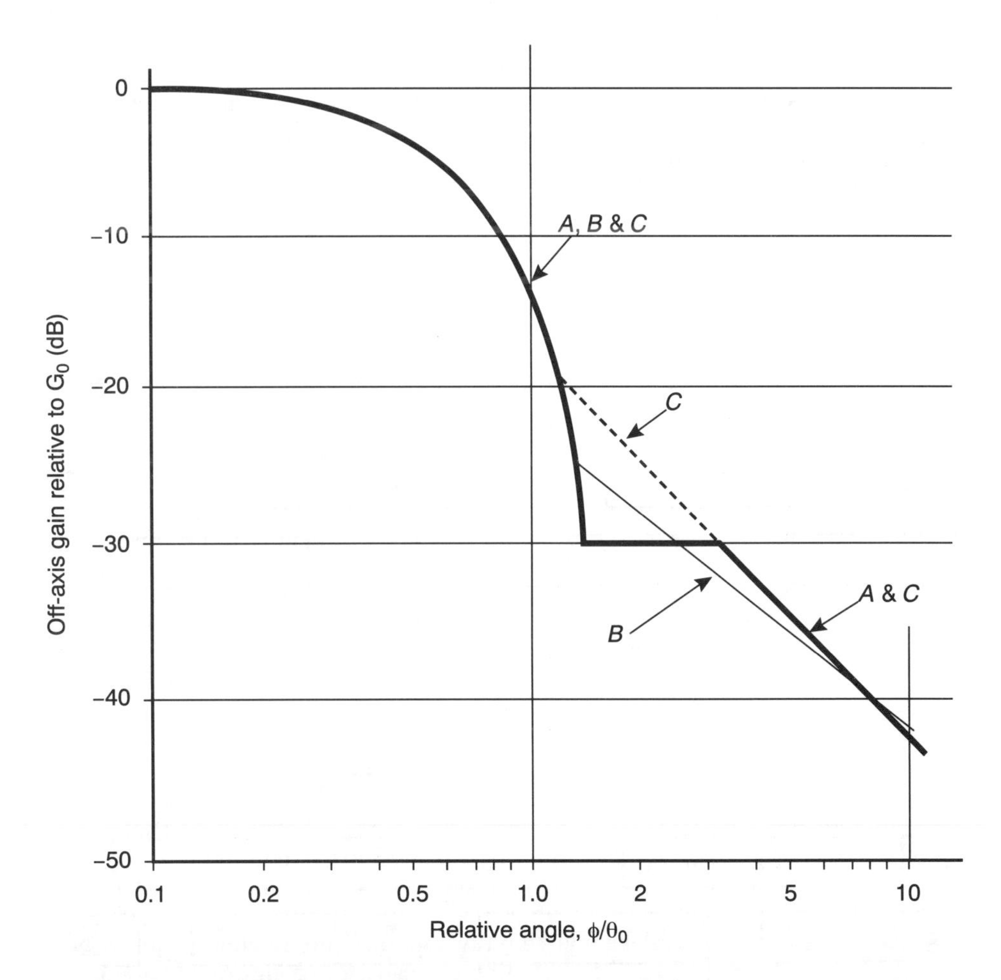

Figure 15.1 *Reference co-polar gain patterns, relative to the on-axis gain, assumed for satellite antennas in the 12 GHz DBS plans, as a function of ϕ/θ_0. Curve A: Regions 1 and 3, transmitting antenna. Curve B: Region 2, transmitting and receiving antennas. Curve C: Regions 1 and 3, receiving antenna. In each case the minimum reference gain assumed is 0 dBi. (After the* Radio Regulations, *Appendices 30 and 30A [4, 5].)*

(5) *Frequency plan*. Regular but different frequency plans were used at the two planning conferences. For Regions 1 and 3, where almost all requirements were for 625-line picture standards, emissions were assumed to have a necessary bandwidth of 27 MHz and the planned carrier frequencies are spaced by 19.18 MHz. Thus there would be considerable overlap of the outer parts of the spectra of adjacent emissions; see, for example, channels n and $n+1$ in Figure 15.2. For Region 2, with a necessary bandwidth of 24 MHz for 525-line picture standards, the carrier frequencies are spaced by 14.58 MHz, leading to greater spectrum overlap, for example, of channels m and $m+1$ in Figure 15.2.

(6) *Feeder links*. The feeder link earth stations were assumed, for planning purposes, to have antennas 5 metres in diameter at 17 GHz and 6 metres in diameter at 14 GHz. Smaller antennas, down to a minimum of 2.5 metres, or antennas larger than 5 metres may be used. The planning conferences devised different means of ensuring that this departure from uniform conditions would not cause adjacent channel interference between feeder links to exceed the levels assumed in planning. See Figure 15.3. There is provision in the plans for adaptive control of uplink power to compensate for uplink propagation degradation. Here again there were differences in the means adopted by the planning conferences to ensure that this would not increase interference.

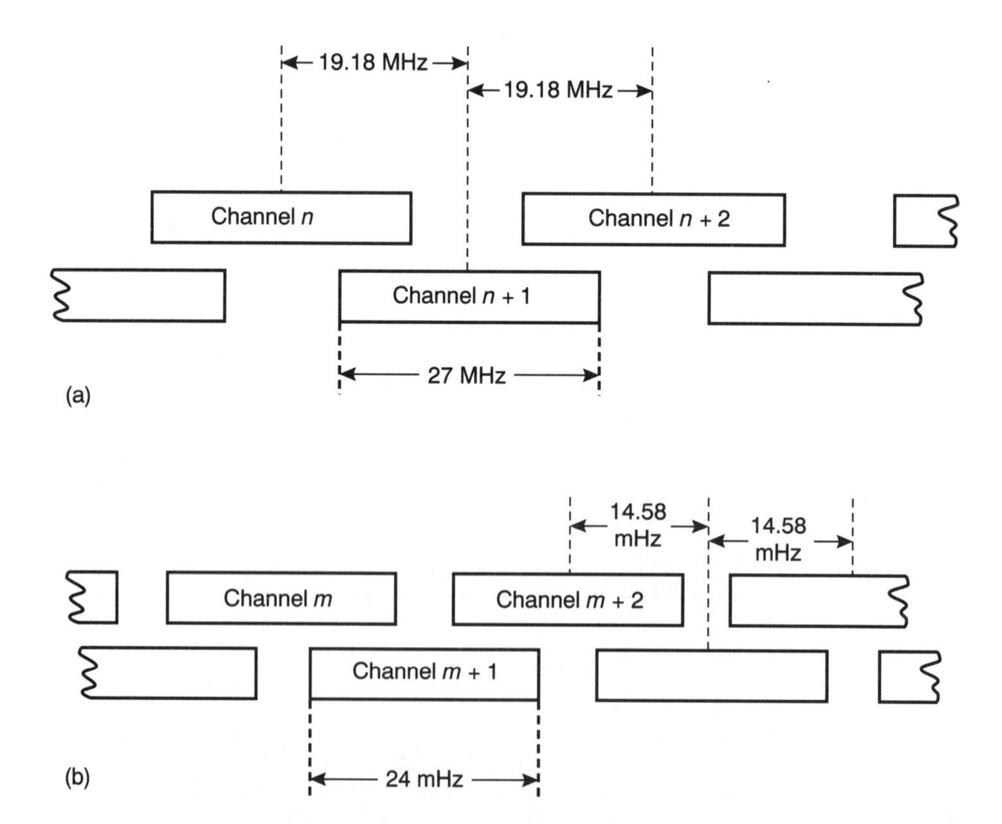

Figure 15.2 *Frequency plan for satellite broadcasting at 12 GHz: (a) Regions 1 and 3; (b) Region 2*

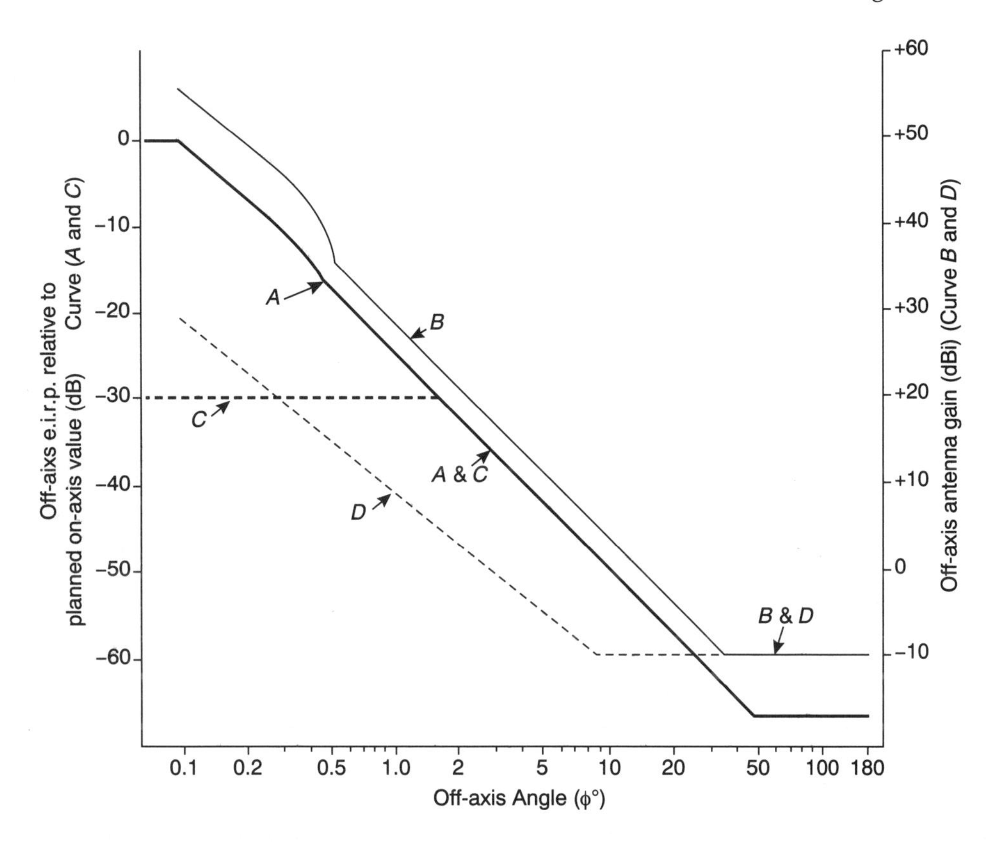

Figure 15.3 *Constraints on off-axis radiation from feeder link earth station antennas. (After the* Radio Regulations, *Appendix 30A [5].) Curve A (left-hand scale) shows the limits for co-polar off-axis e.i.r.p. in Regions 1 and 3, relative to the on-axis value defined by the plan; Curve C shows the corresponding cross-polar limit. Curve B (right-hand scale) is a mandatory co-polar reference gain pattern for Region 2; Curve D shows the corresponding cross-polar pattern*

The beam size of satellite feeder link receiving antennas was assumed to be the same as for the satellite transmitting antenna. Thus, the option was left open for the feeder link earth station to be located anywhere within the national territory. The satellite receiver system noise temperatures assumed for Regions 1 and 3 were 1500 K at 14 GHz and 1800 K at 17 GHz. The corresponding figure at 17 GHz for Region 2 was 1500 K. The off-axis gain of the satellite antenna is required to lie below the curves shown in Figure 15.1. To enable polarization discrimination to make a substantial contribution to the efficiency of the plan, cross-polar response characteristics were laid down for each of the antennas involved, and these are shown in Figures 15.3, 15.4 and 15.5.

Finally it was assumed that the off-axis gain of domestic receiving antennas would be no worse than the reference patterns shown in Figure 15.5. With the foregoing requirements and assumptions, the production of the plan involved computing a pattern of orbital position, channel and polarization assignments such that interference would be acceptably small. The computer techniques used at WARC-77, and the considerably more sophisticated techniques used at RARC-83, are discussed, for example, by Sauvet-Goichon [6] and Fortes [7], and in ITU-R Report BO.812 [8].

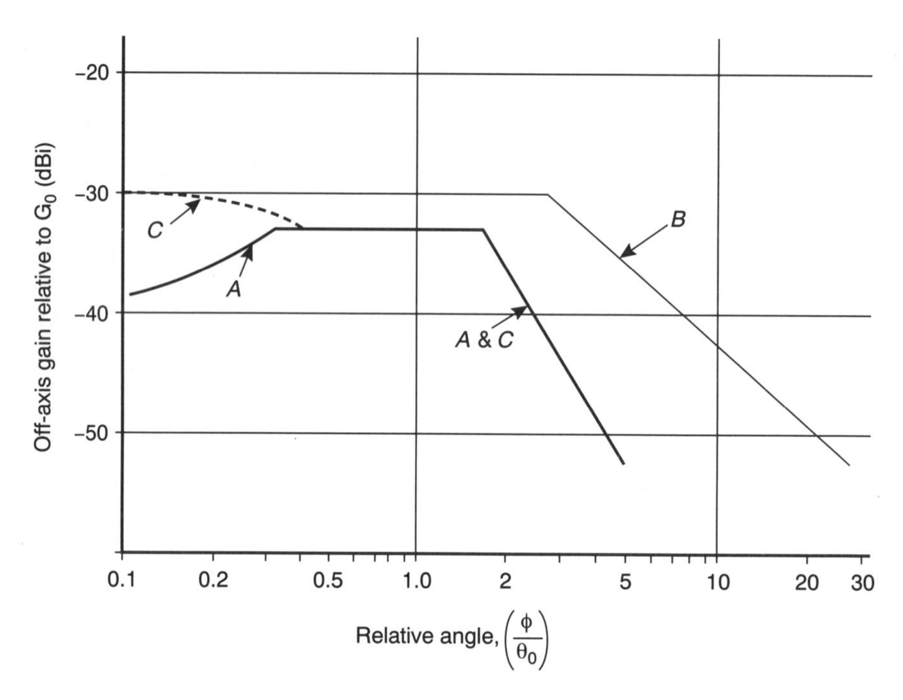

Figure 15.4 *Reference cross-polar gain patterns, relative to the on-axis co-polar gain, assumed for satellite antennas as a function of ϕ/θ_0. Curve A: Regions 1 and 3, transmitting antenna. Curve B: Region 2, transmitting and receiving antennas. Curve C: Regions 1 and 3, receiving antenna. In each case the minimum reference gain assumed is 0 dBi. (After the* Radio Regulations, *Appendices 30 and 30A [4, 5].)*

Implementation of the DBS agreements

Some countries have implemented the assignments planned for them under these agreements. However, the take-up of planned assignments has been quite limited and some at least of the reasons for the slow take-up can be identified:

(1) Global political orientations that were strong in 1971 have become more relaxed. Also while the leading broadcasting institutions in 1971 were mostly public service broadcasters providing national coverage under close government oversight, there are now several broadcasting organizations which are global in range and commercial in outlook. Consequently, limitation of satellite coverage to national territories, no longer politically imperative, has become commercially unattractive.
(2) The per-programme cost of building and launching satellites conforming to the technical parameters required by the plans has proved to be high. A main cause of this high cost has been the high signal power level that the down-link must set up on the ground to avoid the risk of interference. Technical advances in equipment design, above all in the production of low-cost LNAs for domestic receivers, have made it feasible to design satellites providing an acceptable standard of performance, covering wider areas with less satellite transmitter power and consequently at much lower capital cost.
(3) At a time when the picture standards, modulation techniques and transmission media used in television broadcasting are in a state of flux, there is seen to be high commercial risk in investing in satellites designed for these plans.

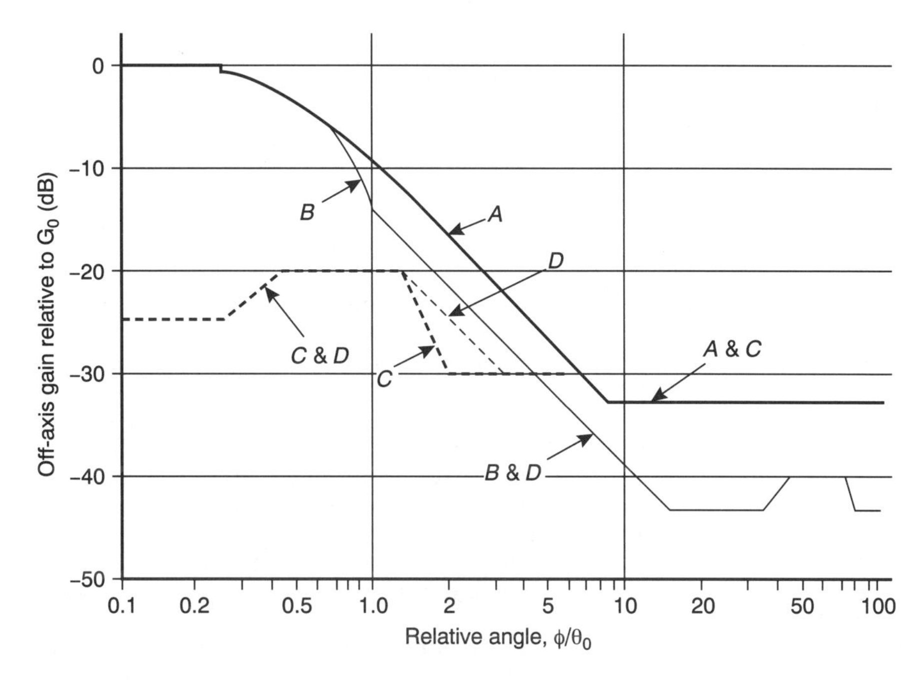

Figure 15.5 *Reference gain patterns, relative to the on-axis gain, assumed for domestic receiving antennas in the 12 GHz DBS plans, as a function of ϕ/θ_0. Curve A: Regions 1 and 3, co-polar. Curve B: Region 2, co-polar. Curve C: Regions 1 and 3, cross-polar. Curve D: Region 2, cross-polar. (After the* Radio Regulations, *Appendix 30 [4].)*

If plans for using a satellite should be abandoned, there may be little prospect of alternative use for a costly asset.

Recent WARCs and WRCs have expressed an interest in revising these plan agreements although without fundamental change to the principles involved. Work is already in progress in preparing the technical bases for new plans; see in particular a resolution of WARC-92 [9], ITU-R Report BO.810 [10], and ITU-R Recommendations BO.792 [11], BO.793 [12], BO.794 [13] and BO.795 [14].

15.1.3 Other frequency allocations for the BSS

The frequency bands 21.4–22 GHz in Regions 1 and 3 and 17.3–17.8 GHz in Region 2 were allocated for the BSS in 1992. The allocations come into full effect on 1 April 2007, by which time other arrangements will have been made, where necessary, for the systems which have assignments in these bands at present. The allocation in Regions 1 and 3 were intended specifically to provide for high-definition television (HDTV) broadcasting. There is provision for the introduction of HDTV in these regions in advance of 2007, subject to safeguards for existing services. No intention favouring HDTV was included in the Region 2 allocation.

15.2 Satellite television direct to home

Despite the limited implementation of the agreements and plans reviewed in Section 15.1.2, the reception of television programmes directly by the public from satellites operating in other frequency bands has grown rapidly.

FSS regional and domestic systems are used extensively for the distribution of television programme signals to terrestrial broadcasting stations, the head-ends of cable broadcasting networks and multi-user reception installations such as those serving isolated communities and hotels; see Chapter 12. However, in addition, members of the public wanting to receive these programmes at home have set up their own antennas, a few metres in diameter in the first instance, obtaining a quality of reception 'direct to home' (DTH) that was often acceptable.

DTH reception started in USA. The practice has now spread to many other countries, above all in Europe and South-east Asia, mostly in Ku band but with C band preferred in high-rainfall areas. DTH having emerged as an important market, satellite operators have specified higher power transponders for new and replacement satellites and have optimized downlink beam footprints to meet the new application. This has permitted the use for DTH of receiving antennas no bigger than those used for DBS. While the downlink transmissions are still, in regulatory principle, addressed to nominated fixed earth stations and therefore proper to the FSS, they are treated in practice as an alternative form of satellite broadcasting. National regulatory authorities have tolerated this departure from the intention of the *Radio Regulations*, although usually without accepting responsibility for preventing interference from terrestrial radio services sharing the FSS frequency bands. In practice this interference does not usually present a major problem.

There have been disputes regarding unauthorized interception of copyright material, but solutions, typically involving signal coding systems with access conditional upon the payment of a fee, have been found. Broadcasters are glad to have the use of a transmission medium that is transnational in coverage and cheaper that DBS. In 1996 several hundreds of television channels are being distributed in this way to tens of millions of domestic receivers, most notably via the Intelsat, Eutelsat and Astra systems but also by other regional and national systems.

The Astra satellites are well suited for traffic of this kind and provide a good example of current practice. Three satellites have been launched so far. All are located at 19.2° E. Each has 16 transponders with a nominal bandwidth of about 26 MHz, with frequency reuse by orthogonal linear polarizations and a single carrier output power rating of about 50 W per transponder. Downlinks are in the 10.7–11.7 GHz band with uplinks at 14.0–14.5 GHz. The transponder downlink frequency plan is shown in Figure 15.6. The transmitting antenna footprints vary

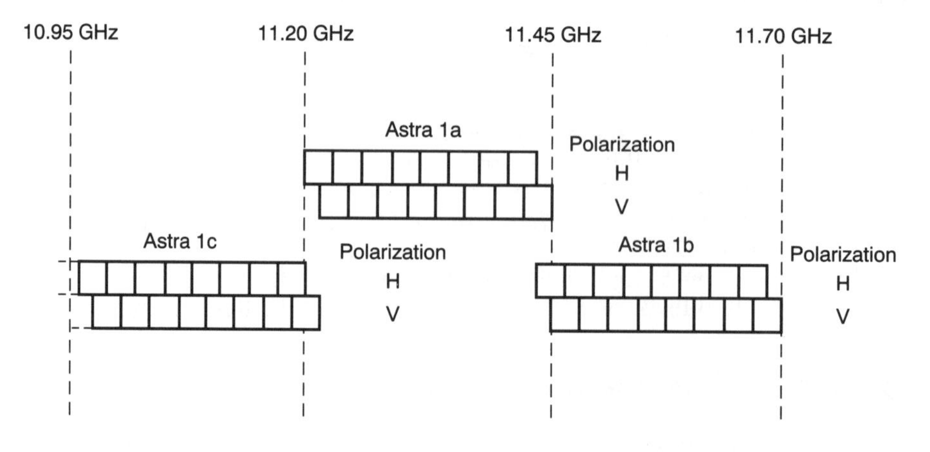

Figure 15.6 *The transponder downlink frequency plan for Astra satellites 1a, 1b and 1c*

in detail but, to a first approximation, coverage at 52 dBW or more includes the Benelux countries, France, Germany, Switzerland, and much of England, Wales, Poland, Austria and Italy. This enables good reception to be obtained using a domestic antenna 60 cm in diameter. Beyond the 52 dB contour a somewhat larger receiving antenna is required, but the 45 dBW contour includes most of western Europe. A large measure of carrier energy dispersal is needed to reduce the spectral PFD density on the ground to the value required by the ITU Radio Regulations. The Eutelsat Hotbird series of satellites include similar facilities.

15.3 Television transmission techniques

The conventional PAL, Secam or NTSC colour television basebands for standard definition picture signals have a wideband analogue luminance signal. Two or more subcarriers above the luminance signal in frequency carry chrominance information and the sound channel(s). This format is well suited to the vestigial sideband amplitude modulation transmission technique for which it was developed and it is used for terrestrial broadcasting at VHF/UHF; see Section 8.2.3. Satellite broadcasting demands the signal-to-noise ratio advantage which wide deviation frequency modulation provides, but with rare exceptions the same baseband structure has been used. One key advantage of this arrangement is that the signal can readily be converted by a small set-top unit into the form that a conventional domestic TV receiver is designed to receive.

Transmission with multiplexed analogue components (MAC)

The use of subcarriers high in the baseband is not ideal with FM, in particular because it limits the benefit that can be obtained from pre-emphasis. The multiplexed analogue components (MAC) family of systems was developed to provide better transmission with FM emissions. MAC transmits the various elements of the television signal, line by line, in sequence, using no subcarriers.

The MAC family includes various options, such as B-MAC, C-MAC, D-MAC and D2-MAC, each optimized for one of the various potential fields of application, but they differ only in detail. As an example, the 64 μs duration of a line of a 625-line D-MAC signal is time-divided to carry four components:

(1) The line starts with a period of about 11 μs during which a series of digital packets is sent, phase shift keying the carrier. These packets may contain picture synchronization code words, one or more digitally encoded sound channels, teletext, data broadcast channels and whatever code words are needed to implement a conditional access system.
(2) A brief interruption of the signal follows, for clamping.
(3) Then chrominance information for the whole line, time-compressed in the ratio 3 : 1, is transmitted in analogue form in the next 17.6 μs, frequency modulating the carrier. Alternate lines carry the two colour-difference signals.
(4) Finally, the luminance information for the whole line, time-compressed in the ratio 2 : 1, is transmitted, also in FM/analogue form, in the 34.8 μs that remains.

See Figure 15.7. See CCIR Report 1074 [15]; there is also a detailed description of various MAC system configurations in an ITU special publication [16].

In addition to the elimination of subcarriers, the MAC technique provides various other benefits relative, for example, to a standard PAL transmission. MAC avoids the picture distortions that arise from the overlapping of the spectra of

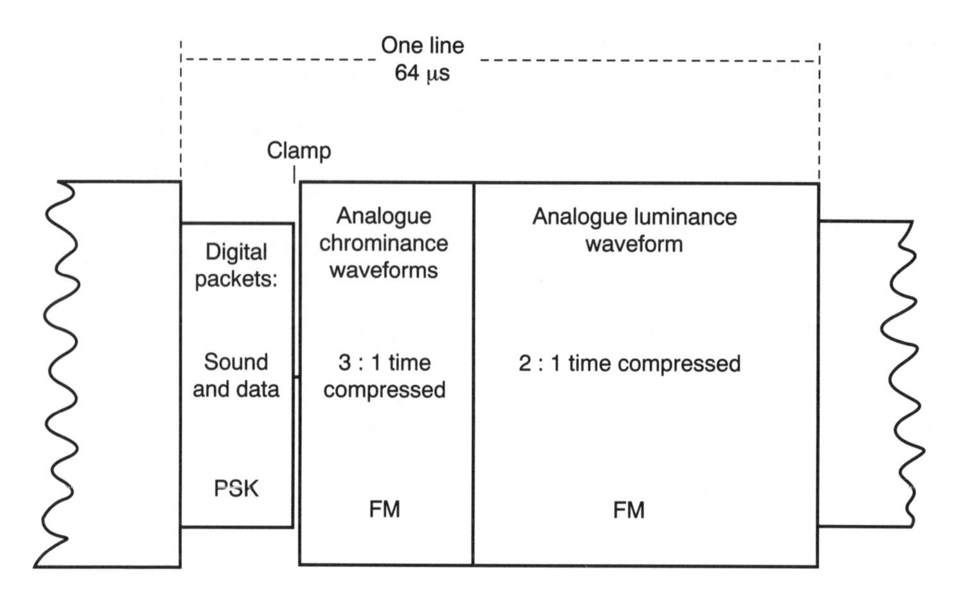

Figure 15.7 *The structure of an active line of a D-MAC signal*

the luminance and chrominance signals in the conventional baseband. It offers improved security of encryption, multiple sound channels of high quality and greater capacity for teletext and data broadcast facilities. The picture definition also is better.

The use of MAC systems is recommended for satellite broadcasting in ITU-R Recommendation BO.650 [17]. Strong support has also been given to their use, and in particular D2-MAC, in satellite broadcasting within the European Union; see in particular EEC Directive 86/529/EEC. However, relatively few channels use MAC so far.

Future options for satellite television

In the short term the basic objectives of satellite broadcasters are many more programmes and less space segment cost per programme, to enable the medium to compete in particular with cable networks and video-on-demand. In the longer term the objectives are the wide-screen format (aspect ratio = 16 × 9) and better picture definition (HDTV). And all major developments are confronted by the need for innovations to be introduced at a time, and in such a way, that the public, the end-user, will be prepared to buy new receivers or signal converters at a rate that will enable manufacturers to sell the new products at an acceptable price. Separately, some of these longer-term innovations are already reaching the market. Others are under development. The main currently visible developments are summarized below.

Compressed digital standard-definition TV systems

The obvious way to reduce space segment costs and at the same time to increase the number of channels available is to increase the number of channels that a transponder can relay. It is feasible to do this through digitization and bit-rate compression. Section 9.3.4 refers to compression of digitally encoded 625-line and 525-line video signals for transmission over

FSS links. By various means outlined there, standard definition signals can be compressed to 34 Mbit/s without significant loss of picture quality and to about 8 Mbit/s without major picture degradation. Such techniques are used already in, for example, SNG.

International coordination of further development of systems providing better picture quality at 8 Mbit/s, and greater compression without unacceptable loss of quality, is being carried out in the Motion Picture Experts Group (MPEG) under the aegis of the International Standards Organization (ISO) and the International Electrotechnical Commission (IEC). Two sets of standards have been published, MPEG1 and the more sophisticated MPEG2. Each provides a fully acceptable picture quality when operating at 8.448 Mbit/s. Each can function at bit rates down to 1.5–2.0 Mbit/s, with a picture quality that remains acceptable for many applications. MPEG2 performs better than MPEG1 at low bit rates. These bit-reduced signals must be protected by highly effective FEC. For transmission by satellite a number of channel signals are combined in TDM to modulate a carrier by BPSK or QPSK to load fully the bandwidth or the downlink e.i.r.p. of the transponder.

There is an extensive literature on this topic; see, for example, [18-21].

High definition and the wide screen

There is widespread agreement that the future of television broadcasting to the home rests with larger screens (a minimum of 1 metre diagonal), the wide screen format (16 × 9 aspect ratio) and higher definition. Some of these innovations are emerging already, although it may be doubted whether large screens will come into general use before wall-hung flat screens become available with high performance and at prices acceptable to the mass market. However, the HDTV systems so far conceived have relatively wide baseband widths. For this reason it is not assumed at present that HDTV will be distributed extensively by terrestrial radio broadcasting in the clearly foreseeable future. It is therefore important that systems for HDTV, broadcast by satellite, should be made technically compatible by means of set-top adaptors with receivers designed for the standard definition systems that will continue to be broadcast by terrestrial radio.

Pioneering development work done by NHK in Japan has produced the Multiple sub-Nyquist Sampling Encoding (MUSE) HDTV system, which is already in service by satellite. Development work on HD-MAC, a high definition version of MAC, coordinated through the Eureka 95 project, has reached the full demonstration stage. Both of these systems were developed for satellite broadcasting, using the bandwidths made available by the frequency assignment plans reviewed in section 15.1.2. Both use the 16 × 9 aspect ratio. See below for a brief note on these systems. For an account of the development of policy in this area in Europe up to 1991, see [22]. A number of projects in North America which are also developing narrowband high definition systems are reviewed in Part 1 of ITU-R Report BO.1075 [23]. Part 2 of Report BO.1075 reviews longer-term and more general studies preparing for systems that would provide virtually studio quality transmission to the domestic receiver, although requiring considerable wider bandwidth.

The MUSE HDTV system

MUSE has a scanning rate of 1125 lines per frame and 30 frames per second, presented as 60 interlaced fields per second. It uses motion-compensation and 4:1 dot-interlaced subsampling. Two or four digital sound channels and data signals are inserted in TDM into the field blanking interval. Time-compressed analogue colour difference and luminance signals are transmitted in sequence, line by line. The baseband signal has a bandwidth of 8.1 MHz. For transmission by satellite, FM is used with a peak deviation of 10.2 MHz, occupying an RF bandwidth of 27 MHz or less. The system is described in detail, for example, in Part 1 of ITU-R Report BO.1075 [23] and in ITU-R Recommendation BO-786 [24].

The HD-MAC system

HD-MAC has a scanning rate of 1250 lines per frame and 25 frames per second, presented as 50 interlaced fields per second. The structure of the baseband signal is similar to that of standard definition MAC systems. The data burst has capacity for up to eight high-quality or 16 medium-quality sound channels, or the equivalent in data packets for other applications. The picture signals are compressed by motion-adaptive subsampling with motion compensation and are transmitted in analogue form. The transmitted bandwidth of the baseband is 10.125 MHz. When used in satellite broadcasting the baseband frequency modulates a carrier with a peak deviation of 9.35 MHz, occupying an RF bandwidth of 27 MHz; see [25]. The system is described in detail in Part 1 of ITU-R Report BO.1075 [23] and in ITU-R Recommendation BO-787 [26].

15.4 Satellite sound broadcasting

Systems for reception at fixed domestic locations

A national frequency assignment planned for DBS TV in the 12 GHz bands (see Section 15.1.2) may be used instead for sound broadcasting subject to two conditions. The sound broadcasting emissions may not cause more interference to television carriers operating in accordance with the plan than would have arisen from the television emission for which the assignment was planned and the level of interference from television carriers in conformity with the plan falling on the sound broadcasting emissions must be accepted. A number of countries have set up such systems. No doubt transponders like those used for DTH TV could also be used in this way.

ITU-R Recommendation BO.712 [27] gives details of three multi-channel standard systems, all digital, for use in this way.

(1) The digital satellite radio (DSR) system was developed in Germany. It provides 32 monophonic or 16 stereophonic channels (or any equivalent combination of monophonic and stereophonic channels) in one multiplex aggregate. Analogue channels with a top frequency of 15 kHz are sampled at 32 kHz with the equivalent of 16 bits per sample. The digital channels are combined by TDM and protected by FEC, producing an aggregate bit rate of 20.38 Mbit/s. The aggregate modulates a carrier with QPSK.
(2) Systems based on the MAC signal format have been designed to carry groups of sound channels. A variety of sound channel options are available, the number of equivalent monophonic channels made available ranging from 14 up to 53.
(3) The multi-channel digital sound/data (MDS) system was developed in Japan. There are two levels of TDM. At the lower level, the bit rate is 2048 kbit/s and this aggregate may be made up of a mixture of high-quality sound and

data channels. To produce the higher multiplex level, 6 or 12 of these lower level aggregates are multiplexed together. The carrier is modulated by the higher multiplex level signal using MSK.

Services for portable receivers and car radios

The VHF/FM terrestrial sound broadcasting band (87.5–108 MHz) provides generally acceptable audio quality and stereophonic reception is satisfactory at fixed domestic antennas provided that the signal strength is high enough. However, reception in moving vehicles is degraded by multi-path transmission. Furthermore, the number of programmes that can be accommodated within the 20.5 MHz bandwidth allocated is no longer enough to satisfy public demand in some countries. Moreover, the spread of compact disc (CD) audio recordings for domestic and mobile use has led to discontent with the sound quality that VHF/FM terrestrial radio provides.

These factors have led to plans for replacing FM with a new technology for terrestrial VHF sound broadcasting. There is also interest in a satellite sound broadcasting system capable of providing an audio quality equivalent to that provided by CD when received in vehicles or with hand-portable receivers, even in remote land areas where the population density is insufficient to justify coverage by terrestrial broadcasting stations.

Portable receivers and simple car radio systems have antennas of negligible gain and, for reasons analogous to those given in the mobile radio context in Section 14.1, the frequencies used for satellite broadcasting to fixed domestic locations are too high to serve such receivers economically. It is recognized that multi-path propagation will also be a problem when receiving in moving vehicles from a satellite.

The frequency band 1452–1492 MHz was allocated for the BSS in 1992, world-wide except for the USA. For USA a corresponding allocation was made at 2310–2360 MHz. These allocations, which are limited to digital audio broad-casting (DAB), do not enter fully into effect until 1 April 2007. It is proposed that a conference should be convened, not later than 1998, to plan the use of this band.

ITU-R Report BO.955 [28] summarizes the results to date of studies of many aspects of the provision of very high quality sound broadcasting from satellites, in particular to moving vehicles. Among much else it describes the two main digital systems currently under consideration for the purpose.

(1) One system has been developed by the EUREKA 147 (DAB) consortium. This system overcomes the effects of multi-path distortion by using the coded orthogonal frequency division multiplex (COFDM) technique to distribute the bit streams from several (typically six) high quality stereophonic chan-nels between a very large number of closely spaced subcarriers. The symbol period on each of the subcarriers is increased in this way to a value (about 0.3 ms) at which multi-path propagation is not likely to degrade perfor-mance. The RF bandwidth required for the aggregate of six channels is about 1.5 MHz. See Shelswell [29].
(2) The other system is more conventional, using powerful source encoding and FEC techniques to overcome multi-path distortion.

Both systems seem likely to be used for both terrestrial broadcasting and satellite broadcasting.

References

1. Relating to the establishment of agreements and associated plans for the broadcasting-satellite service. Resolution No. 507 of *Radio Regulations* (1994). Geneva: ITU.
2. Article 44, paragraph 196 of the Constitution of the International Telecommunication Union. Final Acts of the Additional Plenipotentiary Conference (1992). Geneva: ITU.
3. *Principles governing the use by states of artificial earth satellites for international direct television broadcasting.* Adopted by the General Assembly of the United Nations, 10 December 1982.
4. Provisions for all services and associated plans for the broadcasting-satellite service in the frequency bands 11.7–12.2 GHz (in Region 3), 11.7–12.5 GHz (in Region 1) and 12.2–12.7 GHz (in Region 2). Appendix 30 of *Radio Regulations* (1994). Geneva: ITU.
5. Provisions and associated plans for feeder links for the broadcasting-satellite service (11.7–12.5 GHz in Region 1, 12.2–12.7 GHz in Region 2 and 11.7–12.2 GHz in Region 3) in the frequency bands 14.5–14.8 GHz and 17.3–18.1 GHz in Regions 1 and 3 and 17.3–17.8 GHz in Region 2. Appendix 30A of *Radio Regulations* (1990). Geneva: ITU.
6. Sauvet-Goichon, D. (1976). A method of planning a satellite broadcasting system. *EBU Technical Review,* **155**.
7. Fortes, J. M. P. (1986). Assignment of channels and polarizations in a broadcasting satellite service environment. *Proc. IEE,* **133**, Part F, 133, No. 4.
8. *Computer programs for planning broadcasting-satellite services in the 12 GHz band* (ITU-R Report BO.812-4, 1995), ITU-R Reports 1995 BO Series. Geneva: ITU.
9. Future consideration of the plans for the broadcasting-satellite service in the band 11.7–12.5 GHz (Region 1) and the band 11.7–12.2 GHz (Region 3) in Appendix 30 and the associated feeder-link plans in Appendix 30A. Resolution 524 of the Final Acts of WARC-92 (1992). Geneva: ITU.
10. *Transmitting and receiving antenna technology and reference patterns for the BSS* (ITU-R Report BO.810-4, 1995), ITU-R Reports 1995 BO Series. Geneva: ITU.
11. *Interference protection ratios for the broadcasting-satellite service (television) in the 12 GHz band.* (ITU-R Recommendation BO.792, 1994), ITU-R Recommendations 1994 BO Series. Geneva: ITU.
12. *Partitioning of noise between feeder links for the broadcasting-satellite service (BSS) and BSS downlinks* (ITU-R Recommendation BO.793, 1994), ITU-R Recommendations 1994 BO Series. Geneva: ITU.
13. *Techniques for minimizing the impact on the overall BSS system performance due to rain along the feeder-link path* (ITU-R Recommendation BO.794, 1994), ITU-R Recommendations 1994 BO Series. Geneva: ITU.
14. *Techniques for alleviating mutual interference between feeder links to the BSS* (ITU-R Recommendation BO.795, 1994), ITU-R Recommendations 1994 BO Series. Geneva: ITU.
15. Satellite transmission of multiplexed analogue component (MAC) vision signals (CCIR Report 1074-1, 1990), Reports of the CCIR 1990, Annex to Volume X/XI-2. Geneva: ITU.
16. *Specifications of transmission systems for the broadcasting-satellite service* (CCIR special publication, 1988). Geneva: ITU.
17. *Standards for conventional television systems for satellite broadcasting in the channels defined by Appendix 30 of the Radio Regulations* (ITU-R Recommendation BO.650-2, 1994), ITU-R Recommendations 1994 BO Series. Geneva: ITU.
18. Cominetti, M. and Morello, A. (1994). Direct-to-home digital multi-programme television by satellite. *Proc. International Broadcasting Conference* Amsterdam.
19. Griffiths, J and Spencer, R. (1995). International compressed video by satellite. *Proc. Tenth International Conference on Digital Satellite Communications,* Brighton, UK. London: IEE.
20. Abdel-Nabi, T., Meulman, C., Kennedy, D. J. and Kanunartne, W. (1995). Design considerations for compressed digital TV satellite systems.
21. Keesman, G., Cotton, A., Kessler, D. *et al.* (1995). Study of the subjective performance of a range of MPEG-2 encoders. *Proc. International Broadcasting Conference,* Amsterdam.
22. Bridit, G. (ed.) (1991). *European Broadcasting Standards in the* 1990s. Blackwell.
23. *High definition television by satellite* (ITU-R Report BO.1075-2, 1995), ITU-R Reports 1995 BO Series. Geneva: ITU.
24. *MUSE system for HDTV broadcasting-satellite services* (ITU-R Recommendation BO.786, 1994), ITU-R Recommendations 1994 BO Series. Geneva: ITU.

25. Storey, R. (1985). HDTV motion adaptive bandwidth reduction using DATV. Report 1986/5, BBC Research Department.
26. *MAC/Packet based system for HDTV broadcasting-satellite services* (ITU-R Recommendation BO.787, 1994). ITU-R Recommendations 1994 BO Series. Geneva: ITU.
27. *High quality sound/data standards for the broadcasting-satellite service in the 12 GHz band* (ITU-R Recommendation BO.712-1, 1994). ITU-R Recommendations 1994 BO Series. Geneva, ITU.
28. *Satellite sound broadcasting to vehicular, portable and fixed receivers in the range 500–3000 MHz* (ITU-R Report BO.955, 1995), ITU-R Reports 1995 BO Series. Geneva: ITU.
29. Shelswell, P. (1995). The COFDM modulation system: the heart of digital audio broadcasting. *Electronics and Communication Engineering Journal*, **7**(3).

Index